LIFE HISTORIES
OF NORTH AMERICAN
JAYS, CROWS, AND TITMICE

by

Arthur Cleveland Bent

TWO VOLUMES BOUND AS ONE

Dover Publications, Inc.

New York

This Dover edition, first published in 1988, is an unabridged republication in one volume of the work first reprinted by Dover Publications, Inc. in 1964 in two volumes. The work was first published in 1946 by the United States Government Printing Office, as Smithsonian Institution United States National Museum *Bulletin 191*.

International Standard Book Number:
0-486-25723-1

Manufactured in the United States of America

Dover Publications, Inc.
31 East 2nd Street
Mineola, New York 11501

PART ONE

CONTENTS

INTRODUCTION

This is the fifteenth in a series of bulletins of the United States National Museum on the life histories of North American birds. Previous numbers have been issued as follows:

107. Life Histories of North American Diving Birds, August 1, 1919.
113. Life Histories of North American Gulls and Terns, August 27, 1921.
121. Life Histories of North American Petrels and Pelicans and Their Allies, October 19, 1922.
126. Life Histories of North American Wild Fowl (part), May 25, 1923.
130. Life Histories of North American Wild Fowl (part), June 27, 1925.
135. Life Histories of North American Marsh Birds, March 11, 1927.
142. Life Histories of North American Shore Birds (pt. 1), December 31, 1927.
146. Life Histories of North American Shore Birds (pt. 2), March 24, 1929.
162. Life Histories of North American Gallinaceous Birds, May 25, 1932.
167. Life Histories of North American Birds of Prey (pt. 1), May 3, 1937.
170. Life Histories of North American Birds of Prey (pt. 2), August 8, 1938.
174. Life Histories of North American Woodpeckers, May 23, 1939.
176. Life Histories of North American Cuckoos, Goatsuckers, Hummingbirds, and Their Allies, July 20, 1940.
179. Life Histories of North American Flycatchers, Larks, Swallows and Their Allies, May 8, 1942.

The same general plan has been followed, as explained in previous bulletins, and the same sources of information have been utilized. The nomenclature of the 1931 check-list of the American Ornithologists' Union and supplements has been followed.

An attempt has been made to give as full a life history as possible of the best-known subspecies of each species and to avoid duplication by writing briefly of the others and giving only the characters of the subspecies, its range, and any habits peculiar to it. In many cases certain habits, probably common to the species as a whole, have been recorded for only one subspecies. Such habits are mentioned under the subspecies on which the observations were made. The distribution gives the range of the species as a whole, with only rough outlines of the ranges of the subspecies, which in many cases cannot be accurately defined.

The egg dates are the condensed results of a mass of records taken from the data in a large number of the best egg collections in the country, as well as from contributed field notes and from a few published sources. They indicate the dates on which eggs have been actually found in various parts of the country, showing the earliest and latest dates and the limits between which half the dates fall, indicating the height of the season.

The plumages are described in only enough detail to enable the reader to trace the sequence of molts and plumages from birth to maturity and to recognize the birds in the different stages and at the different seasons.

No attempt has been made to describe fully the adult plumages; this has been well done already in the many manuals and State bird books.   Partial or complete albinism is liable to occur in almost any species; for this reason, and because it is practically impossible to locate all such cases, it has seemed best not to attempt to treat this subject at all.   The names of colors, when in quotation marks, are taken from Ridgway's Color Standards and Color Nomenclature (1912).   In the measurements of eggs, the four extremes are printed in boldface type.

Many who have contributed material for previous volumes have continued to cooperate.   Receipt of material from nearly 500 contributors has been acknowledged previously.   In addition to these, our thanks are due to the following new contributors: J. R. Arnold, B. W. Baker, E. R. Blake, R. M. Bond, W. P. Bonney, F. W. Braund, N. R. Casillo, A. D. Cruickshank, D. E. Davis, O. E. Devitt, P. F. Eckstorm, J. H. Ennis, H. S. Gilbert, W. E. Griffee, J. G. Griggs, B. E. Harrell, C. F. Holmes, J. S. Y. Hoyt, W. A. Kent, A. J. Kirn, J. M. Linsdale, W. H. Longley, D. D. MacDavid, E. A. Mason, M. L. Miles, A. B. Miller, D. L. Newman, E. P. Odum, F. M. Packard, E. J. Reimann, A. C. Reneau, Jr., H. G. Rodeck, W. F. Smith, M. Sullivan, W. P. Taylor, J. K. Terres, Jr., H. O. Todd, Jr., B. W. Tucker, H. E. Tuttle, H. D. and Ruth Wheeler, and J. B. Young.   If any contributor fails to find his or her name in this or in some previous Bulletin, the author would be glad to be advised.   As the demand for these volumes is much greater than the supply, the names of those who have not contributed to the work during the previous ten years will be dropped from the author's mailing list.

Dr. Winsor M. Tyler rendered valuable assistance by reading and indexing, for these groups, a large part of the literature on North American birds, and contributed two complete life histories.   Dr. Jean M. Linsdale and B. W. Tucker, of the Department of Zoology, Oxford University, contributed two each; and Edward von S. Dingle, Dr. Alfred O. Gross, and Alexander Sprunt, Jr., contributed one each.

Egg measurements were furnished, especially for this volume, by the American Museum of Natural History, Griffing Bancroft, Herbert W. Brandt, Frank W. Braund, California Academy of Sciences, Colorado Museum of Natural History, Charles E. Doe, Field Museum of Natural History, James R. Gillin, Wilson C. Hanna, Ed. N. Harrison, Turner E. McMullen, Museum of Comparative Zoology, Museum of Vertebrate Zoology, Laurence Stevens, George H. Stuart, 3d, and United States National Museum.

Our thanks are also due to William George F. Harris for many hours of careful work in collecting and figuring the egg measurements and for handling and arranging the vast amount of data used in making up the paragraphs on egg dates. Dr. Eugene E. Murphey did considerable work on one species, and Stephen Waldron helped with typewriting. Through the courtesy of the Fish and Wildlife Service the services of Frederick C. Lincoln were again obtained to compile the distribution and migration paragraphs. The author claims no credit and assumes no responsibility for this part of the work.

The manuscript for this Bulletin was completed in August 1941. Contributions received since then will be acknowledged later. Because of the war, publication was delayed, and in September 1945 the author recalled the manuscript to bring it up to date although only information of great importance could be added. The reader is reminded again that this is a cooperative work; if he fails to find in these volumes anything that he knows about the birds, he can blame himself for not having sent the information to—

THE AUTHOR.

ADVERTISEMENT

The scientific publications of the National Museum include two series, known, respectively, as *Proceedings* and *Bulletin*.

The *Proceedings* series, begun in 1878, is intended primarily as a medium for the publication of original papers, based on the collections of the National Museum, that set forth newly acquired facts in biology, anthropology, and geology, with descriptions of new forms and revisions of limited groups. Copies of each paper, in pamphlet form, are distributed as published to libraries and scientific organizations and to specialists and others interested in the different subjects. The dates at which these separate papers are published are recorded in the table of contents of each of the volumes.

The series of *Bulletins,* the first of which was issued in 1875, contains separate publications comprising monographs of large zoological groups and other general systematic treatises (occasionally in several volumes), faunal works, reports of expeditions, catalogs of type specimens, special collections, and other material of similar nature. The majority of the volumes are octavo in size, but a quarto size has been adopted in a few instances in which large plates were regarded as indispensable. In the *Bulletin* series appear volumes under the heading *Contributions from the United States National Herbarium,* in octavo form, published by the National Museum since 1902, which contain papers relating to the botanical collections of the Museum.

The present work forms No. 191 of the *Bulletin* series.

ALEXANDER WETMORE,
Secretary, Smithsonian Institution.

LIFE HISTORIES OF NORTH AMERICAN JAYS, CROWS, AND TITMICE

ORDER PASSERIFORMES (FAMILIES CORVIDAE AND PARIDAE)

By Arthur Cleveland Bent

Taunton, Mass.

Order PASSERIFORMES

Family CORVIDAE: Jays, Magpies, and Crows

PERISOREUS CANADENSIS CANADENSIS (Linnaeus)

CANADA JAY

PLATES 1-3

HABITS

The name Canada jay, accepted by ornithologists, is seldom used by the backwoodsman, the hunter, the trapper, and the wanderer in the north woods, who know this familiar bird by a variety of other common names. The name most commonly applied to the bird is "whisky jack," with no reference, however, to any fondness for hard liquor; the old Indian name, "wiss-ka-chon," or "wis-ka-tjon," has been corrupted to "whisky john," and then to "whisky jack." It is also often called "camp robber," "meat bird," "grease bird," "meat hawk," "moose bird," "lumber jack," "venison hawk," and "Hudson Bay bird," all of which are quite appropriate and expressive of the bird's character and behavior.

Although cordially disliked by the trapper and the hunter, because it interferes with their interests, this much-maligned bird has its redeeming traits; it greets the camper, when he first pitches camp, with demonstrations of welcome, and shares his meals with him; it follows the trapper on his long trails through the dark and lonesome woods, where any companionship must be welcome; it may be a thief, and at times a nuisance, but its jovial company is worth more than the price of its board.

Throughout the breeding season at least the home of the Canada jay is in the coniferous forests, among the firs and spruces, or not far from them. Dr. Thomas S. Roberts (1936) says of its haunts in northern Minnesota:

During the late winter and the early spring, which is the nesting-season, it is confined closely to dense spruce, arbor vitae, and tamarack swamps and is rarely

1

seen unless such places are explored. After about the first of July, family parties, consisting of the two parents and four or five sooty-headed young, may be encountered roving through the open uplands and forests, keeping near together in their search for food. With the approach of winter, when the young resemble the adults, it seeks the vicinity of lumber camps, hunters' and squatters' cabins, and settlements, where it becomes very tame and fearless.

The above is mainly true of its haunts elsewhere, though it is not always closely confined to coniferous swamps, even in the nesting season. In the more northern portions of its range it is often found in the opener upland forests, nesting sometimes in solitary trees or in clumps of willows. In Labrador and in Newfoundland I found it common wherever there was any kind of coniferous growth, even where it was scattered or stunted.

Dr. Samuel S. Dickey tells me that in northern Alberta, where this species is common, it is often found in the higher, drier stands of aspen, balsam poplar, canoe birch, mountain-ash, spruce and fir trees, and in pure stands of jack pine *(Pinus banksiana).*

Nesting.—The Canada jay nests so early in the season, while the snow is still deep in the northern woods, that few of us have been able to observe its nesting habits, in spite of the fact that it is an abundant bird over a wide range. Its nesting site is usually remote from civilization, the nest is usually well hidden in dense coniferous forests, and extensive traveling on snowshoes is very difficult at that season. Moreover, the birds, though exceedingly tame and sociable at other seasons, are quiet, retiring, and secretive during the nesting season.

One of the earliest and most interesting accounts of the home life of the Canada jay is that given by Oscar Bird Warren (1899), who, on February 22, 1898, found a pair of these birds building their nest near Mahoning, Minn. (see Barrows, 1912, p. 416). The birds were discovered while Warren was walking down a railroad track through a spruce swamp:

Looking up, what should I see but a pair of Canada Jays pulling beard moss and spider nests from some dead trees and making short trips to neighboring live spruce about 150 feet from the railroad track, where they were evidently building a nest.

Taking a short circuit I reached a position where I could watch their movements better without attracting attention. They brought small sticks, beard moss, spider nests and strips of bark from the trees and sphagnum moss from about the base of the trees where not covered with snow, and deposited all of this in a bunch of branches at the end of a limb,—a peculiar reversed umbrella-shaped formation commonly seen in the small spruce trees, probably caused by some diseased condition of growth. The female arranged the material, pressing it into the proper shape and weaving it about the small twigs to form a safe support. Though the birds obtained the material so near, where it was abundant, yet they carefully picked up any which accidentally fell from the nest, and there were no signs of sticks or any fragments of nesting material at any time during the construction of the nest. * * *

By the 3rd of March the nest was well formed and smoothly lined with fine grass and thin strips of bark. On the 12th it was completed, being beautifully and warmly lined with feathers picked up in the forest and representing several species of birds. Those of the Ruffed and Canada Grouse were in greatest evidence, a feather of the latter being stuck in the edge of the nest where it showed quite conspicuously. These birds had spent nearly a month building their nest, and as a result the finished abode was perfectly constructed. It was large and substantial and yet not bulky, being a model of neatness and symmetry. The bulk of the nest was composed of strips of bark, small sticks, an abundance of dry sphagnum moss, some beard moss and grass, the whole being fastened securely together by small bunches of spider nests and cocoons. The first lining was made of thin strips of bark and fine grass, and this received a heavy coating of feathers, making a nest so warm that a temperature far below the zero mark would have no effect on the eggs it was to receive, as long as the mother brooded over them. The small twigs growing from the cluster of branches in which the nest was built gave it a rough appearance from below, but they served the purpose of secure supports and as a screen for concealment. As there were dozens of similar masses of limbs in the trees all about, a good observer might pass underneath this tree a score of times, and never see the nest, though but a few feet above his head.

The nest described above is unusual in its location, out at the end of a branch; most nests have been found on horizontal branches against the trunk, or in an upright crotch; but otherwise the nest construction is fairly typical of the species. Bendire (1895) says of a nest taken by MacFarlane at Pelican Narrows:

It was placed in a small spruce tree, near the trunk, about 9 feet from the ground. It is composed of small twigs, plant fibers, willow bark, and quite a mass of the down and catkins of the cottonwood or aspen, this material constituting fully one-half of the nest. The inner cup is lined with finer material of the same kind and Jays' feathers, which are easily recognized by their fluffy appearance. * * * A nest taken near Ashland, Aroostook County, Maine, is composed externally of bits of rotten wood, mixed with tree moss, plant fibers, and catkins, and is lined with similar but finer materials.

Oliver L. Austin, Jr. (1932), records several Labrador nests; one was lined with "down, feathers, hair, fur and strips of the inner bark of willow felted together." Of another, he says: "Nest of juniper twigs, wood moss, rotten wood, grass, and lined with partridge feathers [doubtless spruce grouse]; 4 feet from the ground in a white spruce, no other tree within ten yards."

The above descriptions would apply very well to half a dozen or so nests that I have examined in museums and in my own collection. There is a nest in the Thayer collection in Cambridge, taken near Innisfail, Alberta, on March 1, 1903, when the thermometer was 32° below zero; it was 6 feet from the ground in a willow and was made largely of *Usnea barbata,* reinforced at the base and on the sides with twigs; it was profusely lined with feathers, mostly those of the sharp-tailed grouse, with a few of the pinnated and ruffed grouse. Macoun (1909) mentions three other Alberta nests, all of which were in willows; perhaps it is customary for the jays of this region, where there is comparatively little coniferous forest, to nest in willows. But I have also four records of Alberta nests in spruces. Elsewhere, nearly all the nests reported have been in spruces, with an occasional nest in a larch, firm, or hemlock.

Nests have been reported at various heights above the ground, from 4 feet to 30 feet, but the majority of the nests are placed 6 to 8 feet above ground, and very few have been found above 12 feet up. All the nests that I have seen have been well made, the materials being compactly felted; they are neatly finished around the rim and more or less decorated on the exterior with plant down and with cocoons and nests of spiders, wasps, and other insects; the walls are thick, and the inner cavity is warmly lined with feathers, fur, and plant down, furnishing a warm and cozy cradle for the young, to protect them from the low temperatures of late winter in northern latitudes. I have seen one nest, taken in Nova Scotia late in April, that was profusely lined with pine needles; perhaps the warmer lining was not needed at that season. The outer diameter of the nest varies from 6 to 10 inches, but most nests measure 7 or 8 inches; the outer height varies from 3 to 5 inches; the inner cavity measures 3 to 3½ inches in diameter and is 2 to 2½ inches deep.

There are four sets of eggs of this jay in my collection, now in the United States National Museum, two from Labrador and two from Newfoundland. The latter two were collected by J. R. Whitaker, about one of which he wrote to me as follows:

The nest was firmly built on some small twigs of a spruce and placed close to the trunk of the tree at about 18 feet from the snow level. There was no noticeable litter on the snow under the nest. The nest was partly constructed on February 26, 1920, it held one egg on April 10, and was collected, with its complement of three eggs, on April 15. The nest is a very compact structure composed largely of larch twigs, for which the bird would have to go some distance, as the clump of trees in which the nest was placed is composed of nothing but fir and spruce. Mixed with the larch twigs is a good deal of Spanish moss and a large number of spider nests; there are also quite a few feathers in the structure; the lining is composed of moss, rabbit fur, caribou hair, etc., and next to the eggs quite a few jay feathers.

Robie W. Tufts has sent me the following notes on a nest of this jay that he discovered on April 4, 1919, in Annapolis County, Nova Scotia:

The bird was seen to fly to the nest and settle down as if incubating. At my close approach it left the nest, hopping about the twigs at close range and showing no sign of fear or excitement. Its mate was with it. On examination, the nest was found to be empty. Immediately upon my leaving, the bird was observed to fly back and nestle down again. The nest was not visited again until April 20, when the bird was sitting (on two half-incubated eggs) and the mate perched nearby. The sitting bird was loath to leave the nest, and not until the slender spruce was shaken did it hop off, sailing on outspread wings to a dead stub a few inches from the ground. During the two hours spent about the nest, one of the pair never left us, while the other had an uncanny way of vanishing and reappearing unannounced at intervals of about 20 minutes. The behavior of the birds was characterized by a furtive silence. The nest was placed about 12 feet up in a slender spruce in woods of open growth in a wilderness district some miles from human habitation. Little, if any, attempt was made at concealment.

Eggs.—The Canada jay lays ordinarily three or four eggs, but five have been reported, as well as full sets of two. They are normally ovate in shape, rarely short-ovate, and they are usually somewhat glossy, occasionally quite so. The ground color is grayish or greenish white, sometimes very pale gray or pearl gray, and rarely nearly pure white. They are usually quite evenly covered with small spots or fine dots of "deep olive-buff," "dark olive-buff," "olive-buff," or "buffy olive"; Bendire (1895) calls the colors "different shades of brown, slate gray, and lavender." The largest spots that I have seen on any of the eggs that I have examined are not over one-sixteenth of an inch in diameter, and these were grouped chiefly about the small end of the egg. Some eggs are very finely peppered. The measurements of 40 eggs average 29.4 by 21.3 millimeters; the eggs showing the four extremes measure **33.0** by 20.3, 29.0 by **22.8, 26.4** by 20.4, and 28.2 by **20.1** millimeters.

Young.—Mr. Warren (1899) found the period of incubation to be between 16 and 18 days; it was performed by the female alone. Both parents assisted in the care and feeding of the young, which remained in the nest for about 15 days. He writes:

The food given to the young was always in a soft, partially digested state, and was placed deep in the mouths of the young by the old birds. I often watched them feeding the young when my eyes were not three feet from the birds, thus giving a chance for the closest possible observation. I have held my hand on the side of the nest while the mother unconcernedly fed her babies, but I was never able to take as great liberties with the male.

During the first few days after the nestlings were born, the male brought most of the food, the female remaining at the nest and, when the male returned, assisting in giving the food to the young by putting her bill into their mouths and forcing down any troublesome morsels. As the birds grew older the female took a more active part in carrying the food. I have timed them during the feeding hours and found that they came and went about every fifteen minutes with great regularity until the young were satisfied. When the male had discharged his burden he left immediately without waiting for the return of the female, but the mother always stayed until the male had returned or was in sight. The male was never seen on the nest during the period of incubation, nor afterwards, and as his color is much darker than the female's there was never any trouble in distinguishing between them, even at a distance.

The female cleaned the nest often and very carefully, keeping it perfectly free from any filth. It seems this was done both for cleanliness and for the purpose of keeping the nest dry and warm. * * * The male always picked up any droppings which were cast over the nest and had clung to the branches, carrying all away almost every time he left the nest. By this constant care no trace of the presence of the nest was allowed at any time. It should also be added here that the young never made any noise excepting a weak chirp while with open mouths they waited their turn to be fed.

Ben East sent me an article he wrote for the Grand Rapids Press telling of his experience with brood of young Canada jays, near Isle Royale, Mich., on April 30, 1935. The nest was about 10 feet from the ground in a small balsam. He climbed a nearby birch to examine the nest, and the disturbance caused one of the young birds to flutter out and down to the ground. "I gave up my climbing attempt," he says, "and slid back to the ground. Instantly I was the center of a spirited attack by two distraught, angry gray jays. They did not actually strike me, but they flew back and forth over me, darting at me from behind with angry excited cries, fluttering less than a foot above my head and doing all they could to drive me away."

The youngster fluttered and ran along the ground, but it was captured and finally became quite tame and contented, perching on the fingers and heads of Mr. East and his two companions. They placed the young jay on a low branch of the balsam and took several photographs (pl. 3) of it while it was being fed by its parents. It was finally returned to the nest, where it seemed glad to nestle down among its nest mates.

In Newfoundland, in June, and in Labrador from Hopedale to Okak, in July, we found jays of this species common wherever there was coniferous timber. They were traveling about in family parties, and, although the young were fully grown and fully feathered in their dark juvenal plumage, they were still being guarded and probably partially fed by their parents. Both old and young birds were stupidly tame, often coming too close to shoot, but after one of the family had been shot the others immediately vanished. Young birds collected around the first of August were beginning to molt into their first winter plumage.

Mr. Tufts tells me that Ronald W. Smith records having seen a flock numbering from 25 to 30 birds in Kings County, Nova Scotia, on June 19, 1932, and another flock of about 25 birds on July 20, 1937. "This latter flock was seen several times during the same afternoon and evening."

Plumages.—I have seen no very young Canada jays; all that I have seen in life, or in collections, have been fully grown and in full juvenal plumage. This has been very well described by Dr. Dwight (1900) as follows:

Everywhere brownish slate-gray, darker on the crown, paler on the abdomen and crissum. The feathers are lighter basally and faintly tipped with brown producing an obscurely mottled effect. Lores, region of eye and forehead dull black. Malar region whitish with a dull white spot anteriorly. Wings dull clove-brown with plumbeous edgings on secondaries and inner primaries, all the remiges tipped with grayish white, the greater coverts with smoke-gray. Tail slate-gray tipped with brownish white.

Young birds in this plumage are so unlike adults, that Swainson and Richardson (1831) considered them to be another species. As the Canada jay breeds very early in the season, it also begins to molt early in the summer. Young birds begin their postjuvenal molt in July, and some have nearly finished molting their contour plumage before the end of that month, though this molt often continues up to the middle or end of August, or even later. I have collected young birds in Labrador in full juvenal plumage as late as August 9. This molt includes all the body plumage, but not the wings and tail, which are retained until the next postnuptial molt. At this molt old and young become practically indistinguishable in first winter plumage, though the forehead in the young bird is usually somewhat tinged with brownish and the back is darker and more brownish than in the adult. Adults have a complete postnuptial molt beginning early in July, which is generally completed in August.

Food.—The Canada jay is almost omnivorous; it has been said that the "camp robber" will eat anything from soap to plug tobacco, for it will, at least, steal and carry off such unsavory morsels; some Indians have said: "Him eat moccasins, fur cap, matches, anytink" (Bendire, 1895). About camps the "whisky jack" is an errant thief; it will eat any kind of meat, fish, or food left unprotected, will carry off what it cannot eat, and will damage or utterly ruin what is left. It will even enter the tent or cabin in search of food, prying into every open utensil, box, or can. It comes to the camper's table at mealtime and will grab what food it can with the utmost boldness, even seizing morsels from the plates or the frying pan. It shares the hunter's or the fisherman's lunch at noontime, confidently alighting on his knee or hand. It steals the bait from the trapper's traps, sometimes before his back is turned; and it often damages the trapped animal.

William Brewster (1937) wrote in his journal:

After the leaves fell, they were met with chiefly about openings, pastures, etc., hunting apparently for grasshoppers, often going out into the fields several hundred yards. * * * For about two weeks we fed them generously with all sorts of refuse from our table, placing this in one spot. After they had become accustomed to our presence, they spent the greater part of each day in carrying food back into the woods, coming sometimes together, but usually alternately every two or three minutes, filling their throats and bills to the utmost capacity, then by short flights, passing out of sight. They seemed to prefer baked beans to any other food which we had to offer them, and next to beans, oatmeal. They would take bread or cracker when nothing else offered, carrying pieces of large size in their bills, after having stuffed their throats with smaller fragments. They did not seem to care for meat when the things just mentioned could be had. Of baked beans they regularly took four at one load, three in the throat and one held in the bill. * * *

We spent the greater part of one day in following them in order to ascertain what they did with the great quantity of food which they carried off. * * * They took it various distances and to various places, rarely or never, so far as we could ascertain, depositing two loads in the same place. They would place a mouthful of oatmeal perhaps on the horizontal branch of a large hemlock, three or four crumbs of bread on the crotch of a dead stub, a large piece of bread on the imbricated twigs of a living fir. On one occasion we saw one deposit four beans carefully on the top of an old squirrel's nest.

On another occasion they found two of their storehouses: "One in the top of a pine stub where a piece of wood was started off at angle contained about a pint of bisquit and brownbread. The other in a larch stub in three peck holes of either *Colaptes* or *Hylotomus,* the three holes all crammed full of bread packed tightly, in all nearly a quart." As soon as these latter birds learned that their storehouse had been discovered, they immediately removed every vestige of the food.

During spring, summer, and fall, this jay is largely insectivorous, feeding on grasshoppers, wasps, bees, and various other insects and their larvae. Mr. Warren (1899) saw them gathering "grubs from floating logs" and says he has "often seen them chasing a Woodpecker away from the trees just when he had uncovered the worm he had worked so hard to dig out."

W. H. Moore (1904) dissected a Canada jay "and was much surprised to find that nearly one thousand eggs of the Lorset tent-caterpillar had been taken for breakfast. The chrysalids of this caterpillar are also fed upon, and in the autumn while the birds are migrating south they feed largely upon locusts, beetles, etc. The young taken in June feed upon beetles and caterpillars."

Nuttall (1832) says that it "lays up stores of berries in hollow trees for winter; and at times, with the Rein-deer, is driven to the necessity of feeding on Lichens." Audubon (1842) reports that "the contents of the stomach of both young and old birds were insects, *leaves of fir trees,* and eggs of ants."

Behavior.—The most striking and characteristic traits of the Canada jay are its tameness or boldness, one could almost call it stupidity, and its thieving propensities. Its tameness often makes it an interesting and a welcome companion in the lonesome woods, but its boldness, coupled with its thieving habits, has caused many travelers to regard it as a nuisance. Manly Hardy expressed it very well when he wrote to Major Bendire (1895):

They are the boldest of all our birds, except the Chickadee *(Parus atricapillus)*, and in cool impudence far surpass all others. They will enter tents, and often alight on the bow of a canoe where the paddle at every stroke comes within 18 inches of them. I know of nothing which can be eaten that they will not take, and I had one steal all my candles, pulling them out endwise one by one from a piece of birch bark they were rolled in, and another pecked a large hole in a cake of castile soap. A duck which I had picked and laid down for a few minutes had the entire breast eaten out by one or more of these birds. I have seen one alight in the middle of my canoe and peck away at the carcass of a beaver I had skinned. They often spoil deer saddles by pecking into them near the kidneys. They do great damage to the trappers by stealing the bait from traps set for martens and minks and by eating trapped game; they will spoil a marten in a short time. They will sit quietly and see you build a log trap and bait it, and then, almost before your back is turned, you hear their hateful "ca-ca-ca" as they glide down and peer into it.

Curiosity is another characteristic trait of this jay. One can hardly ever enter the woods where these birds are living without seeing one or more of them; the slightest noise arouses their curiosity, and they fly up to scrutinize the stranger at short range, often within a few feet, and they will then follow him to see what he will do. The sound of an ax always attracts them, for it suggests making camp, which means food for them; and the smoke of a campfire is sure to bring them.

William Palmer (1890) relates the following case of unusual curiosity, or stupidity: "After spending the day on one of the Mingan Islands, which is very densely wooded, we started to drag our dory down to the water, necessarily making considerable noise. While doing so, and glancing towards the woods, I observed a jay perched upon the top of the nearest tree, evidently interested in our proceedings. I immediately shot him, and the report had hardly died away when another jay took his place. He, too, followed the first, when instantly another flew to the very same tree, only, however, to meet the same fate." This is in marked contrast to the behavior of these birds when they have young with them; for whenever I shot one of a family party the others immediately vanished.

The flight of the Canada jay is easy and graceful but not vigorous or prolonged. It seldom indulges in long flights in the open. It floats lightly from tree to tree on its broad wings, making very little noise and seldom flapping its wings except when rising from the ground into a tree. Its ordinary method of traveling through the woods is to sail down from the top of one tree to the lower part of another, and then to hop upward from one branch to another, often in a spiral fashion, until it attains sufficient height to make another scaling flight. Its broad wings and fluffy plumage seem to make it very buoyant and enable it to float upward at the end of a sailing flight.

Dr. Dickey says (MS.) that Canada jays seem to like to associate with such small birds as myrtle warblers, winter wrens, chickadees, purple finches, and some of the northern flycatchers. Lucien M. Turner (MS.) tells of feeding one on meat until it became so tame as to perch on one hand and eat out of the other.

Voice.—William Brewster (1937) writes:

> It has a variety of notes, most of them shrill and penetrating, the commonest a loud, hawk-like whistle, very like that of the Red-shouldered Hawk, but clearly not, as in the case of one of the Blue Jay's calls, an imitation of it. Another common cry is a succession of short, rather mellow whistles, eight or ten in number all given in the same key. It frequently utters a loud *"Cla, cla cla, cla, cla, cla, cla,"* not unlike the cry of the Sparrow Hawk. It also scolds very much like a Baltimore Oriole. Twice I heard one scream so nearly like a Blue Jay that I should probably have been deceived had not the bird been very near and in full sight of me. In addition to these notes, it also has a low, tender, cooing noise which I have never heard except when two birds are near together, evidently talking to one another.

The Canada jay is credited with being something of a mimic, imitating more or less successfully the notes of the red-tailed, red-shouldered, and broad-winged hawks, as well as the songs of the small birds that it hears. Several writers have referred to its rather pleasing, twittering song, of which Mr. Warren (1899) writes: "On pleasant days the male trilled from a spruce top a song of sweetly modulated notes wholly new to my ears. He always sang in *sotto voce,* and it required an acquaintance with the songster to realize that he, though so near, was the origin of those notes which seemed to come from somewhere up in the towering pines which surrounded this strip of swamp, so lost was the melody in the whispering, murmuring voices of the pines."

Ernest Thompson Seton (1890) has heard it give a *chuck, chuck* note, like that of a robin; Knight (1908) says that "their cry is a querulous 'quee-ah' 'kuoo' or 'wah,' uttered as they perch on top of some tree or take flight." Langille (1884) adds to the list a note "sounding like *choo-choo-choo-choo.*"

Field marks.—The Canada jay is not likely to be mistaken for anything else in the region where it lives. It is a little larger than a robin and much plumper. Its general color is gray, with a blackish hood and a white forehead. It looks much larger than it really is on account of its fluffy plumage; in cold weather, especially, its soft plumage is so much expanded as to exaggerate its size. Its small bill, fluffy plumage, and confiding manners suggest an overgrown chickadee. The only gray bird of similar size in the north woods is the northern shrike, whose black wings and tail and larger bill are distinctive.

Fall.—A. Dawes DuBois writes to me: "During my 11 years of residence in the Lake Minnetonka region, in Hennepin County, Minn., I have seen Canada jays in the fall of one year only. They visited us in October and November 1929. On November 24 two of them were attracted to a chunk of suet fastened to the trunk of a tree close to our house. Their method was to cling either on top of the suet or to the bark of the tree, at one side of it. They took turns at this repast. One waited in the tree while the other was eating; then it flew down to take its share. In this manner they alternated, with some regularity; but one of them seemed dominant over the other."

The Canada jay is supposed to be permanently resident in the north woods, where it breeds; and it probably does usually remain there during ordinary winters, provided there is no failure in its food supply. It undoubtedly wanders about more or less in search of food and at times has made quite extensive migrations to points south of its breeding range. The two following quotations illustrate this point. On September 5, 1884, Napoleon A. Comeau wrote from Godbout, Quebec, to Dr. C. Hart Merriam (1885) as follows: "We have lately had a most extraordinary migration of the Canada jay *(Perisoreus)*. One afternoon I counted over a hundred in the open space near the old Hudson's Bay Company's house here; and almost every day since the first of this month it has been the same. I believe this unprecedented flight must be owing to scarcity of berries in the interior, and, since they happen to be plentiful along the coast this fall, the birds follow the shore to feed on them."

M. Abbott Frazar (1887) writes from Quebec Labrador:

On my return to Esquimaux Point, the first week in September, * * * I was soon made aware of an immense migration of these jays which was taking place. Right directly back from the house the low hills terminated in a straight line at right angles with the coast, and in a path which ran along the foot of these hills I took my stand and waited for the jays as they came straggling down the hillside. The flocks varied in size from a dozen to fifty or so individuals and kept following each other so closely that an interval of ten minutes was a rarity and they never varied their line of migration but kept right on, taking short listless flights from tree to tree. I devoted but two forenoons to them and although I had nothing but squib charges of dust to kill them with, being out of medium sized shot, I killed ninety and could easily have trebled that number had I wished. How long the force of the migration kept up I cannot say but I know there were still a few passing by when I left the country ten days later.

Winter.—There are numerous winter records for various points in New England and New York, but Pennsylvania seems to be about the southern limit of its wanderings in the eastern part of the country. Todd's "Birds of Western Pennsylvania" gives but one record for that region, in February 1923. But N. R. Casillo writes to me that the Canada jay comes down into that part of the State "more or less regularly," as borne out by his observation of two individuals in Lawrence County over a period of 4 years. The locality where these birds were seen, New Castle, Pa., is about 70 miles southwest of Forest County, where the previous record was made. It is flat or rolling country and sparsely wooded, with conifers conspicuously absent. The first bird was seen from a distance of 12 feet, on November 26, 1936, while it was feeding on the berries of a Virginia creeper that grew over a porch trellis near Mr. Casillo's kitchen window. He observed the second bird, apparently a younger bird, in the same vine on December 8. One or both of these birds were seen there on January 12 and February 4, 1937, three times in November and on December 14, 1939, and on January 1 and 13, 1940.

DISTRIBUTION

Range.—Northern North America south to New York, Minnesota, New Mexico, Arizona, and Oregon; not regularly migratory.

The range of the Canada jay extends **north** to northern Alaska (Kobuk River, Chandlar River, and Demarcation Point); Mackenzie (Horton River, Dease River, Fort Enterprise, and Fort Reliance); northern Manitoba (Du Brochet Lake, Fort Churchill, and York Factory); northern Quebec (Great Whale River and Chimo); and Labrador (Nain). **East** to Labrador (Nain and Rigolet); Newfoundland (Fogo Island and Salmonier); and Nova Scotia (Baddeck, Kentville, and Barrington). **South** to southern Nova Scotia (Barrington and Grand Manan); Maine (Milltown, Foxcroft, and Norway); northern New Hampshire (Mount Washington); northern Vermont (St. Johnsbury and Mount Mansfield); northern New York (Long Lake, Fulton Chain, and Watertown); southern Ontario (Latchford, North End, and Sudbury); northern Michigan (Pickford and McMillan); northern Wisconsin (Spring Creek); Minnesota (Mille Lacs Lake and White Earth); southwestern South Dakota (Elk Mountain); northern New Mexico (Cowles and Baldy Peak); and east-central Arizona (White Mountains). **West** to eastern Arizona (White Mountains); northeastern Oregon (Blue Mountains); southeastern Washington (Blue Springs); northern

Idaho (Coeur d'Alene and Clark Fork); British Columbia (Midway, Horse Lake, Hazelton, Flood Glacier, and Atlin); and Alaska (Iliamna Pass, Nushagak, St. Michael, Nulato, and Kobuk River).

The range as outlined is for the entire species. Three races are currently recognized and others have been proposed. The typical subspecies, known as the Canada jay *(Perisoreus c. canadensis),* occupies the major portion of the range from northern Mackenzie and central British Columbia east to Labrador, Newfoundland, and Nova Scotia; the Rocky Mountain jay *(P. c. capitalis)* is found in the Rocky Mountain region from southern British Columbia and Alberta south to Arizona and New Mexico; while the Alaska jay *(P. c. fumifrons)* occurs throughout Alaska except the coastal region east and south of the Alaska Peninsula.

Casual records.—In fall or winter the Canada jay will sometimes The ranges of other subspecies are given under their respective names. wander south of its usual range. There are several records for Massachusetts (Amesbury, Quincy, Mount Greylock, Arlington Heights, Bernardston, Cambridge, and Newton). It was recorded from Utica, N. Y., in the winter of 1868-69 and was seen repeatedly at Cortland, in that State, during January 1928. Audubon reported one from Philadelphia, Pa., in October 1836. In Minnesota it has been noted in winter irregularly in the southern part of the State (Hennepin, Ramsey, Washington, and Anoka Counties); and there are several records for Nebraska (West Point in the late winter of 1886, near Fort Robinson in April 1891, head of Monroe Canyon in February 1896, near Belmont in the spring of 1889, and Antioch on February 2 and 26, 1930).

Egg dates.—Alaska: 2 records, April 10 and May 13.

Alberta: 56 records, March 11 to April 21; 28 records, March 18 to April 8, indicating the height of the season.

Colorado: 6 records, March 17 to April 30.

Labrador: 5 records, March 20 to April 23.

Newfoundland: 17 records, April 4 to 30; 9 records, April 10 to 16.

Nova Scotia: 9 records, April 2 to May 7.

PERISOREUS CANADENSIS ALBESCENS Peters

ALBERTA JAY

Based on the study of five specimens from Red Deer, Alberta, James Lee Peters (1920) describes this race as "similar to *P. c. capitalis,* but smaller; paler above, much paler below; lower breast, flanks, and abdomen, pale smoky gray, with only a faint drab wash. Similar also to *P. c. canadensis* in size and in the extent of white on the crown, but much paler throughout. * * * This form is strikingly paler than any of the known races of *Perisoreus canadensis*. The contrast between the white throat and the fore neck and the drab lower parts, so noticeable in the other subspecies, is quite lacking."

The above common name appears in the nineteenth supplement to our Check-list (1944), where the present known range of the race is said to be "central and southern Alberta." Its habits, so far as known, are included in those of the type race.

PERISOREUS CANADENSIS NIGRICAPILLUS Ridgway

LABRADOR JAY

At long last the A. O. U. (1944) committee on nomenclature has decided to admit to the new Check-list this fairly well marked subspecies, which Robert Ridgway (1882) described many years ago as "similar to *P. canadensis fumifrons* in darkness of coloration, but forehead, lores, chin, throat, and sides of neck distinctly white, in marked and abrupt contrast with the dark color of adjacent parts; crown, occiput, and upper part of auricular region decidedly black, with little or no admixture of slaty anteriorly. Differing from true *canadensis* in much darker coloration throughout, much blacker crown, black auriculars, less extensive white area on forehead, and more marked contrast of the white portions of head and neck, with adjacent darker colors."

Dr. H. C. Oberholser (1914) proposed the name *P. c. sanfordi* for the birds of this species found in Newfoundland, but this name has never been recognized by the A. O. U. committee. I have collected birds of this species in both Newfoundland and Labrador and have examined large series of both in the museum at Cambridge, where we all agreed that the Newfoundland bird is not sufficiently different from that of the Labrador Peninsula to warrant its recognition in nomenclature.

What we know about the habits of this race in Newfoundland and Labrador is included in our life history of typical *canadensis*, which was written before *nigricapillus* was formally recognized.

PERISOREUS CANADENSIS BARBOURI Brooks

ANTICOSTI JAY

The Anticosti jay was described by Winthrop Sprague Brooks (1920) as a distinct species, but is now to be admitted to our Check-list as a subspecies. Mr. Brooks gives its characters as follows: "Size about as in *P. canadensis nigricapillus* Ridg. of Labrador. In color this jay differs at a glance from *P. canadensis nigricapillus* in that the upper parts, including lesser wing-coverts and upper tail-coverts, are plain slate-color (instead of mouse gray), the black of crown and occiput slate-black (instead of brownish black), and the under parts deep gray, less brownish or smoky."

Its habits probably do not differ materially from those of the species in Newfoundland and Labrador, as described under the type race.

Braund and McCullagh (1940) make the following interesting comment on the Anticosti jay: "Contrary to the usual antagonistic feeling of Canadian trappers and woodsmen in general, the native Anticosti Islanders have a friendly attitude toward the bird. It seems that during the cold winter months when supplies are low on the mainland the 'poachers' visit the island to obtain deer and trap mammals. The Canada Jay's characteristic habit of being a camp follower has often led the island game protectors to these 'poacher' camps."

PERISOREUS CANADENSIS PACIFICUS A. H. Miller
PACIFIC CANADA JAY

This dark race of the Canada jay has been named and described by Dr. Alden H. Miller (1943a), based on a series of 11 birds collected in "the Rainbow Mountain area at the headwaters of the Dean and Bella Coola rivers, in the central coast range of British Columbia," which constitutes its present known range. He describes it as "similar to *Perisoreus canadensis canadensis,* but dorsal coloration darker and sootier (near Dark Mouse Gray of Ridgway, Color Standards and Color Nomenclature, 1912), hence less brown; in fresh plumage, dorsal gray collar of neck inconspicuous and in some individuals obsolete; white of forehead of same extent and comparably suffused with gray in fresh plumage, but not noticeably buffy as in *P. c. fumifrons;* size as in *P. c. canadensis.* * * * The race *P. c. pacificus* shows no approach in characters to *Perisoreus obscurus* of southwestern British Columbia. The coloration dorsally is blue or neutral gray, rather than brown as in *obscurus,* the shaft streaks of the back feathers are no more apparent than in any race of *P. canadensis,* and the underparts are deep gray posterior to the throat, not whitish and uniform as in *obscurus.* * * * Compared with *P. c. fumifrons, pacificus* is not only distinctly darker but less brown."

Nothing seems to have been published about its habits.

PERISOREUS CANADENSIS CAPITALIS Ridgway
ROCKY MOUNTAIN JAY

PLATE 4

HABITS

This western race of our familiar "whisky jack" is described by Ridgway (1904) as similar to our eastern bird, but it is larger and lighter colored; the whole head is white, except immediately around and behind the eyes, which, together with the hind neck, are slate-gray; the upper and under parts are paler gray. Young birds, in juvenal plumage, are paler than the young of *canadensis* or *fumifrons,* the pileum being much paler gray or grayish white and the feathers of the under parts more or less tipped with white or pale grayish. Ridgway gave it the appropriate name of "white-headed jay," and, on account of this prominent character, it is locally called "baldhead," "tallowhead," or "whitehead"; it is also commonly known as the "camp robber," and many of the popular names applied to our eastern bird are also used to designate it. The name "jay" is usually used by the westerner for one of the races of the Steller's jay.

The Rocky Mountain jay is appropriately named, for it is confined mainly to the boreal zones in the Rocky Mountain region from southern Canada to Arizona and New Mexico. Its breeding range seems to be limited to the heavily forested regions in the mountains, from the lower limit of coniferous forests up to timberline, the altitude varying with the latitude. Referring to its haunts in the Yellowstone National Park, Wyo., M. P. Skinner (1921) says: "While I have frequently found Rocky Mountain Jays in the smaller meadows and openings, still it is apparent they like the forests best. Forests of lodgepole pine, limber pine, fir, spruce, cedar, and even aspen groves and willow thickets constitute their chosen haunts. Their nests are in the lodgepole pine belt between the 7500 and 8000 foot levels."

Aretas A. Saunders writes to me: "This bird is a common species all through the mountainous parts of Montana. Though a resident, and present throughout the year, it is much more in evidence from early in August to late in February than from March to July. From about August 5 on, these birds are likely to be encountered daily until late in February. But through the spring and early summer a sight of one of these birds is a rare thing. I find that I have records of their occurrence in every month of the year, but the records are very few, as compared to late summer and fall."

Fred Mallery Packard writes to me of the status of this jay in Estes Park, Colo.: "One of the commonest birds of the Canadian and Hudsonian forests in summer; usually found between 8,500 feet and timberline at 11,000 feet, but occasionally as high as 13,000 feet. In winter most of these jays descend to the lower edge of the Canadian and Upper Transition Zones (8,000 to 9,000 feet), some to Estes Park village at 7,500 feet, while a few winter as high as timberline."

Referring to Colorado, Dr. Coues (1874) quotes Mr. Trippe as saying: "I have never seen the Canada Jay below 9,000 feet, even in midwinter; and but rarely below 9,500 or 10,000. During the warmer months it keeps within a few hundred feet of timber-line, frequenting the darkest forests of spruce, and occasionally flying a little way above the trees."

In New Mexico its range seems to be mostly above 11,000 feet, where Mrs. Bailey (1928) says that "it belongs among the hemlocks and spruces of the Hudsonian Zone." She gives a number of records for north-central New Mexico, ranging from 7,800 up to 12,000 feet, but says that 9,500 feet is about the usual lower limit of its range in fall and winter.

Nesting.—W. C. Bradbury (1918), after several unsuccessful attempts and much heavy traveling in deep snow, finally succeeded in 1918 in securing three nests of the Rocky Mountain jay in Colorado. One nest was taken in Grand County on May 2 at an altitude of 8,600 feet; the nest, which contained only two heavily incubated eggs, was in a lodgepole pine "about twenty-five feet from the ground, in a rather bushy top, located close to the trunk on a small limb. Some of the strings used in the nest were neatly bound around the limb upon which it rested. The outside framework is "composed chiefly of pine and other twigs." The "nest proper" is "composed of fine grasses, cotton strings of several sizes, and large amounts of unravelled rags and white rabbit hairs; lined with same material and feathers. There are several pieces of cotton cloth spread between the twig foundation and the nest proper."

Another nest was at "about 8,700 feet altitude in Saguache County, Colo., in open stand of lodge-pole pine. The nest was on the south side of a tree fifteen feet high, located on two limbs two and one-half feet from the trunk and five feet from the ground. * * * Nest proper composed of fine grasses and bark fiber neatly and closely woven together, and warmly lined with chicken and occasional grouse and jay feathers." This nest contained three slightly incubated eggs on April 26.

The third nest was taken in Gunnison County, at an elevation of 10,600 feet, on April 21, containing two eggs with well-developed embryos. "The nest was in the top of a white spruce, fifty-five feet above the surface of the snow, which was fourteen feet deep on the level. * * * The entire structure is composed of spruce twigs and tree moss, with a small amount of coarse wood fiber and an occasional feather, all very closely and firmly intermixed and woven together. The cup is lined with tree moss, grouse and a few other feathers."

These nests are evidently quite similar to those of the Canada jay, and the size is about the same, though the inner cup seems to be shallower, 1¾ inches. The over-all outside diameter varies from 6½ by 7 to 7 by 9 inches; the outside height is 3 to 4 inches; and the inner diameter of the cup varies from 3 to 3½ inches.

Mr. Skinner (1921) says that in Yellowstone National Park "nests are built in tall lodgepole pines during early April at from 7500 to 8000 feet elevation. They are about thirty feet up, or two-thirds of the distance from ground to tree top, and made of straw placed in the angles between the trunk and a limb about two inches in diameter. The inner nest is mostly of pine needles."

Alfred M. Bailey tells me that he and R. J. Niedrach found two nests in the mountains of Colorado; one was 20 feet from the ground in a small Douglas fir, at 9,000 feet; and the other was 25 feet above ground in an Engelmann spruce, at an elevation of 11,000 feet (pl. 4).

Eggs.—The Rocky Mountain jay seems to lay usually two or three eggs, perhaps sometimes four. These are practically indistinguishable from those of the Canada jay, though some are more heavily marked. The measurements of 20 eggs average 29.9 by 21.7 millimeters; the eggs showing the four extremes measure **33.0** by 24.0, 32.0 by **24.5,** and **26.6** by **20.0** millimeters.

Food.—This "camp robber" has practically the same feeding habits as others of the species, frequenting the camps to steal, eat, or carry off almost anything edible. It often does considerable damage to food left in camp or to baited traps. Wilbur C. Knight (1902) writes: "Some years ago while deer hunting we had several carcasses hanging in the trees near by and some quarters that had been skinned. I noticed the birds flying away from the meat whenever I came into camp and upon examining the quarters that were skinned, I found that they had made several holes through the dried surface, large enough to admit their heads, and that they had eaten from each opening from one to two pounds of meat and had entirely destroyed the quarters."

Mr. Skinner (1921) says: "Truly omnivorous eaters, the Rocky Mountain Jays pick up oats dropped about stables or along the roads; catch caterpillars, black worms, and grasshoppers; and once I saw a Jay try for a locust, although he missed and did not try again that I could see." Mrs. Bailey (1928) adds "wild fruits, including elderberry, bearberry, sumac, and viburnum; also scattered grain in corrals; insects, especially grasshoppers and caterpillars; small mammals, meat, and camp food." On the Upper Pecos River she (1904) saw them eating toadstool.

Mr. Munro has sent me some notes on the stomach contents of Rocky Mountain jays taken in British Columbia. In four stomachs collected on September 20, 1939, one contained seeds of Rosaceae to the extent of 70 percent; two others contained 70 percent insects, including a large dipterous pupa, parts of two large Diptera, and other insect remains;

fragments of a beetle and seeds of the serviceberry were found in some of the stomachs. In the three stomachs taken December 3, 1926, seeds of Rosaceae figured largely, from 95 to 98 percent; mixed with them were a few seeds of serviceberry and a few insect fragments. Mr. Packard writes to me:

"Every camp and cabin in the higher parts of the park has its coterie of jays that depends to some extent upon food discarded by campers to supplement their own forage. A site may be used but once or twice a year, yet within 5 minutes of a person's arrival there the camp robbers are hopping on nearby trees in anticipation of a handout.

"On July 5, 1939, shrill cries coming from the top of a small Engelmann spruce near timberline at Milner Pass disclosed two ruby-crowned kinglets darting at a pair of camp robbers, each of which had a nearly grown kinglet in its claws. The jays paid little attention to the agitated parents, but calmly devoured the nestlings while we watched. In each case, the jay opened the stomach of its prey, ate the viscera, and then pecked at the head. I have also observed camp robbers carrying nestling Audubon's warblers in midsummer."

Behavior.—In general habits and behavior the Rocky Mountain jay is much like its better-known northern relative; it has the same thieving habits, is equally bold and inquisitive, and is quite as sociable and friendly, the camper's companion and a nuisance to the trapper or the hunter. Mr. Skinner (1921) calls attention to two points, not mentioned under the preceding race; he says:

The flight of a Rocky Mountain Jay seems weak. A few wing strokes carries the bird along slowly and upward slightly, then a sail carries him down at about the same angle, and this sequence is repeated over and over again, resulting in a slow flight of long, shallow undulations. * * * Birds of the air and of the tree tops as they are, when they are on the ground they move somewhat awkwardly in a series of long hops, a little sideways perhaps, a good deal like crows and ravens.

Its migrations, if they may be called such, are more altitudinal than latitudinal. It wanders to lower elevations in winter and often seeks the vicinity of permanent camps and settlements in search of food, retiring to the higher altitudes at the approach of the breeding season. During the nesting season it is very retiring and secretive but is much more in evidence during fall and early in winter.

Mr. Saunders writes to me: "They stay around lumber camps or other places, feeding on garbage, particularly scraps of meat or fat, but also bread. At such seasons, if one stops to eat lunch anywhere in the evergreen forests, the birds will appear shortly, and are very tame, and ready to share all the lunch one is willing to give them."

Mr. Munro says in his notes: "In the heavy still forest on a snowy day, they came fluttering silently from the heavy timber in response to an imitation of the pigmy owl call—soft, fluffy birds like overgrown chickadees."

Voice.—Mr. Saunders (MS.) says that this bird is an exceedingly quiet one, in contrast to other jays; only once or twice has he heard one make a sound.

Mr. Munro writes in his notes sent to me: "These birds were heard imitating the call of pine grosbeaks, which were nesting in the vicinity. They also imitated the calls of pigmy owl and red-tailed hawk. I was impressed by the exact imitation of the pigmy owl made by two pairs which were called up at different places by an imitation of the owl call. I was sure that a pigmy owl was answering me until the jays appeared. Both the single *hook* note and the quavering tremolo were given. In one instance, both were given after I had whistled only the single note."

PERISOREUS CANADENSIS BICOLOR A. H. Miller

IDAHO JAY

In naming this subspecies, Dr. Miller (1933) gives the following comparison with the type race: "Size, and tone and hue of coloration as in *P. c. canadensis* of central British Columbia, but dark color of occiput not surrounding or even extending to orbit; entire pileum, anterior breast and throat white, the white of head above and below standing in sharper contrast to dark grays and black of occiput and body; collar purer white and broader."

He designates its principal range as "in the relatively humid forest regions of northern Idaho, the principal trees of this forest being Engelmann spruce, western white pine, western larch and Douglas fir."

The A. O. U. committee (1944) applied the above common name to this race and gave its range as "southern British Columbia to central Oregon and central Idaho." A little farther south this race evidently intergrades with *capitalis,* and its habits are probably similar to those of this Rocky Mountain form.

PERISOREUS CANADENSIS FUMIFRONS Ridgway

ALASKA JAY

HABITS

Ridgway (1904) characterizes this northwestern race as "similar to *P. c. canadensis,* but dusky hood extending over the crown, leaving only the forehead white; the latter often more or less tinged with smoky gray; the general color of upper and lower parts browner, and size less. Agreeing with *P. c. nigricapillus* in greater extent of the dusky hood, but this browner, with the anterior portion more distinctly ashy, the forehead less purely white, and the general color, both above and below, decidedly paler, the under tail-coverts dirty whitish or very pale brownish gray."

The Alaska jay is known to inhabit the wooded portions of Alaska, except the coast region east and south of the Alaska Peninsula, but just where it intergrades with typical *canadensis* in eastern Alaska or western Mackenzie does not seem to be definitely known.

In the interior of Alaska, Lee Raymond Dice (1920) found this jay "common in white spruce-paper birch forest, in black spruce forest, in burned timber, and in lowland willows along the streams. In the winter they also frequent the neighborhood of cabins and camps." Dr. Joseph Grinnell (1900a), referring to the Kotzebue Sound region, writes: "During September and October, in my tramps across the tundras lying along the base of the Jade Mountains, I frequently met with two or three jays far out on the plains a mile or more from timber, feeding on blueberries. * * * Later, in the coldest days of mid-winter, I found them in the dense willow thickets."

Herbert Brandt (MS.) writes: "Throughout the great wooded interior of Alaska, where for eight months the snow and cold reign, the only conspicuous living thing that gladdens the camp and trail of the dog-musher is the Alaska jay. Wheresoever he may go and make his camp in the snow, it is sure to find him; and by his friendly manner this jocund jay gives to the cheerless by-places a touch of life that the naturalist always remembers. At every habitation that we visited and at every camp we made from Nenana to the tundra rim, where we left the trees behind, the Alaska jay was always present. Those hardy pioneers that live in this vast wooded area are outdoor people, with all the keenness and skill in woodcraft that such a life produces, yet in spite of the fact that this neighborly bird is very plentiful, and that the timbered cover in which it lives is mostly open, we did not meet a single person who had seen its egg. Often along our trail the actions of this species made obvious the fact that it was nesting, but I could find no clue that would direct me to its abode. It is evident that during nesting time the bird forsakes the immediate vicinity of habitations, where it is wont to congregate, and retires to a secluded area, which it enters and whence it departs with great caution."

Nesting.—The natives in Alaska, and in other parts of the range of this species, are unwilling to collect the nests and eggs of this jay, as they are suspicious that some evil will befall them if the nest is disturbed or even if the eggs are counted. François Mercier (Nelson, 1887) offered a tempting reward which resulted in persuading a native to bring him two nests. The older natives in the vicinity "prophesied that the weather would turn cold, and that a very late spring would ensue as a result of this robbery. As chance would have it the prophesies of the old sooth-sayers came true in a remarkable degree, and the spring was the coldest and most backward by nearly a month of any year since the Americans have had possession of the country." After that, he was never able to persuade the natives to hunt for nests. This may be one reason why so few nests of this species have found their way into collections.

Dr. Grinnell (1900a) found a pair of Alaska jays building a nest on March 20.

It was ten feet above the snow in a dense young spruce growing among a clump of taller ones on a knoll. * * * Although I did not disturb the nest in the least, a visit two weeks later found it covered with snow and apparently deserted. * * * Not until May 13, however, did I finally find an occupied jay's nest, and its discovery then was by mere accident. It was twelve feet up in a small spruce amongst a clump of larger ones on a low ridge. There were no "tell-tale sticks and twigs on the snow beneath"; as Nelson notes, and in fact nothing to indicate its location. The nest rested on several horizontal or slightly drooping branches against the south side of the main trunk. The foliage around it was moderately dense, so that it could be seen from the ground, though only as an indistinct dark spot. The bird was sitting on the nest when I discovered it. Her head and tail appeared conspicu-ously over the edge of the nest, and she remained on until I had climbed up within an arm's length of her. She then left the nest and silently flew to a nearby tree where she was joined by her mate. They both remained in the vicinity, but ostensibly paid little attention to me. * * * The nest proper was built on a loose foundation of slender spruce twigs. The walls and bottom consist of a closely felted mass of a black hair-like lichen, many short bits of spruce twigs, feathers of ptarmigan and hawk owls, strips of fibrous bark and a few grasses. The interior is lined with the softest and finest-grained material. The whole fabric is of such a quality as to accomplish the greatest conservation of warmth. Which certainly must be necessary where incubation is carried on in below-zero weather!

Mr. Brandt found his first nest near Flat, Alaska, on April 9, 1924; it contained four eggs ranging evenly in incubation, showing that incubation had begun after the laying of the first egg, which is probably necessary to prevent freezing. For the first 6 days of April the temperature had ranged from 16° below to 35° below zero, though on the 8th it had risen to 30° above. He says in his notes: "The nest of the Alaska jay is placed usually in a spruce tree in a river or creek bottom, and, in the two in-stances of which I have information, they were poorly concealed; yet the forest at that time was so snow-laden that an object as small as a jay's nest is not at all conspicuous near a tree trunk. The incubating bird sits very close, is quiet about the nest, and its mate stays away from the vicinity during the entire time that an intruder is about. The incu-bating bird did not leave its charge until the climber was but a foot distant.

"The nest found at Flat was 9 feet above the ground and was placed against the trunk of a small scrubby tree 3 inches in diameter, which it partly encircled. It was built on a whorl of very thin branches but was supported largely by a pendant spruce limb, the branchlets of which were woven into the outer rim of the nest and which served also as a snow shelter for the brooding bird. The structure of this abode is just what one would expect of a bird that chooses the severest time of the year to breed. The nest is very bulky, of a silver-gray color, and like most nests of jays is of two distinct parts, a loosely made platform, in this case of tamarack twigs, and a very well-made, compact inner nest. It has very closely built walls varying in thickness from 1 inch, where it rested against the tree trunk, to 3 inches on the opposite side. It is composed largely of cotton from an old quilt and is lined with feathers of the Alaska spruce grouse and willow ptarmigan, some thread, string and fine strips of bark, with an inner lining of dog hair and feathers in liberal quantities. All this is matted and felted together in such a manner as to make the interior well insulated against the cold, and when the incubating bird is snuggled down into the close-fitting rim, but little warmth is radiated, even in the most rigorous weather. The dimensions of the nest are as follows: Height, 7 inches; total outside diameter, 12; outside diameter of primary nest, 8; inside diameter, 3.5; depth of cup, 3; thickness of wall, 1 to 3 inches."

Eggs.—The Alaska jay apparently lays three or four eggs, probably most commonly three. Dr. Grinnell's set consisted of three eggs, and Mr. Brandt collected one set of three and one of four. The latter describes his seven eggs as follows (MS): "The egg in outline is ovate to short-ovate, and exhibits considerable luster. The egg of this species is very distinctive, and resembles that of the shrikes, but has the typical shape and texture of the egg of a jay. The ground color is conspicuous because half of it is exposed, and the markings are of a neutral color. These markings appear like freckles on the egg and are most heavily concentrated about the larger end, sometimes taking the form of a wreath. The spots are small, angular in shape and irregular in size. Those of the underlying series are of a lavender hue, ranging from grayish lavender to pale violet gray and to pale purplish gray; while the overlying spots are reddish and richer, ranging from Saccardo's olive to burnt umber and Rood's brown."

The measurements of 20 eggs average 29.5 by 20.8 millimeters; the eggs showing the four extremes measure **31.2** by 21.0, **28.4** by **21.5**, and 30.8 by **20.1** millimeters.

Behavior.—In a general way the behavior of Alaska jays does not differ materially from that of the species elsewhere. But the following incidents are worth mentioning. Mr. Dice (1920) states that "in several instances these jays were seen to attack hawks and hawk owls." Joseph S. Dixon (1938) relates the following:

On June 1, at Savage River, Wright and I watched a pair of Alaska jays being chased away from camp by a red squirrel. Every time a jay would alight in the top of a spruce tree near camp the squirrel would look up at the bird, select the proper tree, and would run up the tree and jump at the jay, driving him away. This was repeated many times. If the spruce trees were close together the squirrel would jump from one tree to the next. If this was not possible he would go down and run across on the ground climbing the tree the jay was in. After the squirrel had driven the jays away, we saw the former take a bit of food—old discarded cheese—that he had kept hidden in the crotch of a tree. Then he carried it down the tree and hid it under an old rotten log.

There is considerable competition about the camps among the Alaska jays and red squirrels to see which will get the choicest bits of discarded table scraps.

On May 26, 1926, a robin was found trying to drive a jay away from its nest. Investigation showed that the jay was doing his best to steal the robin's eggs.

Early in May 1924 Otto W. Geist (1936) witnessed a fight between a pair of Alaska jays and a full-grown weasel. Thawing during the day and freezing at night had formed a crust on the snow, but there were some bare spaces and small holes in the crust. He says, in part:

The snow under the crust was to a great extent "honey-combed," leaving spaces through which small animals such as mice and weasels could find easy passage.

From not far away I heard the shrieks of birds which seemed to be coming closer. I decided to wait. Soon I saw two Alaska Jays flying from tree to tree, diving frequently at something on the ground. I kept still in order to see what was the matter. Soon a weasel, evidently full grown and still in his white winter coat which, however, was soiled with blood, ran toward a patch of snow directly in front of me and disappeared under it. Both birds were close behind and they rested on a limb of a small tree under which I was standing. * * *

Both flew excitedly over the patch of snow. They soon returned to the tree and to my astonishment ceased shrieking. All was quiet for possibly a minute or more when through one of the smaller holes in the snow there appeared the head and forepart of the weasel. The two birds became highly excited and again flew out over the patch. They would sweep down over the weasel, first one, then the other, striking with its beak. The weasel seemed cowed and ducked low after each strike. There were blood spots on the snow and it seemed to me the birds were doing very effective work.

I now moved a little closer, but neither the birds nor the weasel seemed to notice me. I talked aloud and whistled but they paid no attention. At times one of the birds would fly out, almost stop over the weasel, using the wings to brake with, and try to see how close he could get to the weasel with his feet. Each time this was done the weasel would stretch out, sticking his head and front of the body into the air. However, he did not seem to snap at the birds. On the contrary, frequently the weasel's mouth was open and it seemed to be panting and fairly well worn out. * * *

It seemed that not a single movement of the weasel was missed by the birds. At one place where the crust on the snow was thin, the weasel managed to work under and in doing so broke some of the crust. Both birds saw this and flew down to the place where the snow moved and crumbled. The fight was on again. The weasel rushed out and made a few jumps, one a very long one of about four feet, with the birds after him at once. However, by now the weasel had reached a pile of brush. The last glimpse I had of his coat it seemed bloodier than ever.

The weasel was now safe from further attack, and the jays had put up a brave fight to protect their young from one of the fiercest fighters in the woods.

PERISOREUS CANADENSIS OBSCURUS Ridgway

OREGON JAY

HABITS

The Oregon jay, according to Ridgway (1904), inhabits the "Pacific coast district, from Humboldt County, California, to southern British Columbia (Vancouver Island and coast of opposite mainland)." The 1931 Check-list gives its range as "Pacific coast from Western Washington to Mendocino County, California." Perhaps there is something yet to be learned as to where it intergrades with the other race, *griseus.*

It is a true "whisky jack," or "camp robber," replacing the Canada jay and its subspecies to the westward of the Rocky Mountains. It closely resembles the Canada jay in general appearance and habits, though it is smaller, browner on the back, where the feathers have distinct whitish shaft-streaks, and whiter on the underparts.

It seems to be confined mainly to the heavy coniferous forests at the higher altitudes in the mountains, at least during the breeding season.

G. Buchanan Simpson (1925), who gives an account of 8 years' friendship with Oregon jays, says: "In this district (Lake Cowichan, B. C.), these birds are usually to be found in the wilder mountainous regions. In winter, however, they often come down to within a few hundred feet of Lake level in the dense forest."

Courtship.—Mr. Simpson (1925) writes: "In February the male makes very pretty love to his spouse. The latter sits on a nearby perch, ignoring any food that is thrown out. She flaps her wings in a coy way, after the manner of a nestling being fed by its mother, making plaintive little cries. The male bird scrambles for the most pleasing bit of food which is to be found, and gallantly carries it to his wife, who receives it in her beak and eats it with a great show of satisfaction."

Nesting.—A. W. Anthony was evidently the first to record the finding of the nest of this jay, near Beaverton, Oreg. He wrote to Major Bendire (1895) about it as follows:

The birds were discovered building on March 4, 1885; one of them was seen clinging to the side of a dead stub, about 75 feet from the ground. He was tearing out bits of moss, which did not seem to suit, for they were dropped again as fast as gathered; but at last, finding some to his fancy, he flew off and I saw him go into a thick fir and disappear. I could as yet see nothing of a nest, but as both birds were flying in with sticks, moss, etc., I was sure one was being built there. Both birds worked hard, were very silent, and did not come very near the ground, getting nearly all of their building material from the tree tops, I think. On the 16th I again visited the place, and with the aid of a field glass discovered the nest, which was to all appearances complete, but the birds were not seen. On the 21st I took a boy with me to climb the tree, and found the nest finished, but no eggs. On March 31 we visited it again and found the set complete and the female at home. She stayed on the eggs until the climber put his hand out for her, when she darted off with a low cry and was shot by me. The eggs, five in number, were but slightly incubated; the nest was placed about 85 feet from the ground and 10 feet from the top of the tree; it was built close to the trunk, and was very well hidden.

Mr. Anthony generously presented this set of eggs, with the nest, to the United States National Museum. Major Bendire (1895) describes the nest as follows: "This nest, now before me, is compactly built and rather symmetrical, measuring 7½ inches in outer diameter by 4½ inches deep; the inner diameter is 3 inches by 2½ inches deep. Externally it is composed of fine twigs, dry grass, tree moss, and plant fibers, all well interlaced, and the inner cup is composed exclusively of fine, darklooking tree moss."

The major goes on to say:

Mr. C. W. Swallow writes me that he took a set of four eggs of this species in Clatsop County, Oregon, on May 8. This nest was placed in a small hemlock, about 10 feet from the ground. I believe as a rule they nest in high, bushy firs. I saw a pair of these birds evidently feeding young, in a very large fir tree, near the summit of the Cascade Mountains, on June 9, 1883, while en route from Linkville to Jacksonville, Oregon, but could not see the nest, which must have been fully 60 feet from the ground. But one brood is reared in a season.

S. F. Rathbun (1911) writes:

On April 18, 1909, the writer while looking through a dense strip of second growth of young red firs (*Pseudotsuga mucronata*) in a heavy wooded tract a few miles east of the city Seattle, found a nest of this species. The young fir in which it was built was alongside an old and seldom used path through the second growth, on the edge of a small open space about ten feet in diameter, having a further undergrowth of salal (*Gaultheria shallon*) and red huckleberry (*Vaccinium parvifolium*) shrubs. The tree was five inches in diameter tapering to a height of thirty-five feet, and the nest was placed close against its trunk on four small branches, at a height of twelve feet. It was outwardly constructed of dead dry twigs, next a thick felting of green moss into which was interwoven some white cotton string, and was lined with dry moss, a little dead grass and a few feathers, among the latter some of the Steller's Jay, and is a handsome compact affair. Dimensions: average outside diameter 6½ inches, inside diameter 3½ inches; depth outside, 5 inches; inside 2 inches.

There is a nest of the Oregon jay, with a set of four eggs, in the

Charles E. Doe collection, taken by J. C. Braly near Sandy, Oreg., April 20, 1932. It was about 30 feet from the ground close to the trunk of a small fir in a fir grove.

Eggs.—Major Bendire (1895) describes the eggs as "pearl gray or light greenish gray in ground color, spotted and flecked with smoke and lavender gray, and these markings are pretty evenly distributed over the entire egg. In shape they are ovate; the shell is smooth, close grained, and only moderately glossy."

Mr. Doe (MS.) describes his eggs as "ground color dark gray, boldly marked with almost black and a few lavender marks—very striking eggs."

The measurements of 21 eggs average 27.0 by 20.6 millimeters; the eggs showing the four extremes measure **29.7** by 21.6, 27.6 by **21.8**, and **25.4** by **19.3** millimeters.

Plumages.—Ridgway (1904) describes the young bird in juvenal plumage as follows: "Entire pileum and hindneck dull sooty brown or grayish sepia, the feathers narrowly and indistinctly margined with paler; no whitish collar across lower hindneck; sides of head similar in color to pileum, the auricular region with indistinct dull whitish shaft-streaks; nasal tufts sepia brown; chin and anterior portion of malar region dirty brownish white; throat dull grayish brown, intermixed with dull grayish white; rest of under parts pale broccoli brown, some of the feathers with indistinct paler shaft-streaks; wings, tail, back, etc., essentially as in adults; bill partly light-colored."

I have noticed in young birds I have examined that the wing coverts are not tipped with white, as they are in adults. A partial postjuvenal molt, including the contour feathers and wing coverts, but not the rest of the wings and tail, begins early in July and may continue well into September in some individuals; I have seen one bird that had not quite completed the molt on September 10. The complete postnuptial molt of adults begins early in July and is probably completed in August in most cases.

Food.—Very little has been published on the food of the Oregon jay, but it is apparently just as omnivorous as other members of the genus *Perisoreus.* It is a frequent visitor at the camps and feeds freely on anything edible, scraps from the table or any other food that it can beg or steal. Mr. Simpson (1925) writes about the birds around his camp:

We have tried the birds with all kinds of food and their undoubted favourite is cheese, of which they are passionately fond. * * *

The birds each had their morning morsel of cheese today. They hold it in their mouths for a long time, turning it over and over with their tongues, as if the taste were most pleasing to them. The cheese is often shifted to the 'pouch' under the chin and held there for some time. Then it may be deposited carefully on some safe perch, licked and mouthed again with intense satisfaction, then finally eaten. They do this with no other food. * * *

The Whiskey Jacks eat bread, porridge, uncooked rolled oats, cake, farinaceous food in general, and, of course, meat, raw or cooked. They have taken an occasional bite of apple or pear. Sometimes one will catch and eat an insect, but they will not look at an earth-worm. Fish, either raw or cooked, they dislike.

Mr. Rathbun has sent me the following note, made at Lake Crescent on April 21, 1916: "Today, while slowly rowing along the shore of the lake, I came across eight Oregon jays feeding among the growth of willows and the debris strewn at the water's edge. The birds were engaged in capturing the newly hatched stoneflies, very many of which were fluttering about in the air or crawling on the rocks and broken branches on the shore. The jays took the flies in the air as easily as flycatchers capture insects. In early April, at the time the stoneflies appear, these birds resort each morning to the shore of the lake to take the insects named, and evidently capture a great number of them. We have watched them do this time and time again."

Behavior.—The Oregon jay seems to be fully as fearless, sociable, and mischievous as our more familiar "whisky jack." Major Bendire (1895) quotes from Mr. Anthony's notes as follows: "'Fearless' is an appropriate term to use in relation to this bird; it seems utterly devoid of fear. While dressing deer in the thick timber I have been almost covered with Jays flying down from the neighboring trees. They would settle on my back, head, or shoulders, tugging and pulling at each loose shred of my coat until one would think that their only object was to help me in all ways possible. At such times their only note is a low, plaintive cry." Mr. Simpson (1925) writes:

On the approach of a Hawk, whose presence is usually detected by these birds from afar, they at once become perfectly motionless in the thickest part of a bush, uttering a low, plaintive, warning cry. When hard pressed, they will successfully fight off a Cooper's or a Sharp-shinned Hawk. After all these years on the Lake shore, they continue to take a Gull or a Heron for a Hawk, and display the same symptoms of fear.

Unlike some of our race, they have a passion for soap! We cannot leave a piece of soap outside the house for a short time without it being carried off by the Jays. The camp soap suffers the same fate at 5000 ft. level in summer. When carrying anything beyond the capacity of their beaks, they use both feet with which to hold the object, the legs hanging straight down beneath the body as they fly.

As far as we have observed, these birds are a gentle, most lovable company, minding their own business in bird-land, and never robbing a small bird's nest (of which there are always several nearby of Song Sparrow, Yellowthroat, etc.).

The Jays make free with our small garden, in which we find them admirable companions, respecting all our cherished alpine plants as well as salads, tomatoes, berries and the like.

Ralph Hoffmann (1927) thus describes the behavior of these jays about a camp: "A bird * * * flits noiselessly out of the forest and starts to investigate the camp. With a soft *whee-oo* another follows, flying to the ground, hopping about or carrying back a scrap of refuse to a limb. A flock keeps constantly drifting on through the trees, flying now to the ground, then to a branch or even clinging to the side of a tree trunk. The soft, fluffy plumage gives the bird a gentle look in keeping with its fearlessness and soft voice. Let a hawk appear, however, and the Oregon Jays will mob him with loud screaming cries, *ke-wéep, ke-wéep*."

Field marks.—The Oregon jay is not likely to be confused with any other bird within its habitat, as its range does not overlap that of the Rocky Mountain jay. It is about the size of a robin; it has a white forehead, a white collar around the hind neck, a whitish breast, and a brownish back; the crown and back of the head are blackish. Its soft, fluffy plumage and its confiding habits are also distinctive. At close range the whitish shaft streaks on the back may be seen.

Fall.—Mr. Rathbun tells me that "this jay is resident throughout the entire region but is found more commonly in the higher altitudes from early in spring until late in fall, after which period many individuals come to the lowlands, and here the species will be often met with during the winter months. This movement from the mountain regions begins about the middle of October, and from that time on Oregon jays will be met with from time to time in nearly any part of the region."

DISTRIBUTION

Range.—The Northwest, from southern British Columbia south to central California; nonmigratory.

The range of the Oregon jay extends **north** to southern British Columbia (Della Lake, Malaspina Inlet, Alta Lake, and Lillooet). **East** to south-central British Columbia (Lillooet, Hope, and Chilliwack); Washington (Mount Baker, Kacheos Lake, and Bumping Lake); Oregon (Mount Hood, Crater Lake, and Lakeview); eastern California (Fort Bidwell, Warner Mountains, and Summit); and west-central Nevada (Glenbrook). **South** to central Nevada (Glenbrook); and northern California (Summit and Mendocino). **West** to western California (Mendocino, Cape Mendocino, and Orick); western Oregon (Applegate River, Sweet Home, and Beaverton); western Washington (Camas, Grays Harbor, Quinault Lake, and Lake Crescent); and southwestern British Columbia (Victoria, Mount Douglas, and Della Lake).

The range as outlined is for the entire species, which has been separated into two subspecies. The typical Oregon jay (*Perisoreus obscurus*

obscurus) is found only in the coastal mountainous regions from probably northern Vancouver Island, British Columbia, south to northwestern California. The remainder of the range is occupied by the gray jay *(P. o. griseus)*.

Since the above was written, the A. O. U. (1944) committee has ruled that the two recognized races of *obscurus* are considered as subspecies of *P. canadensis.* See Hellmayer, Cat. Birds Amer., vol. 7, p. 69.

Egg dates.—Oregon: 4 records, March 31 to April 20.

PERISOREUS CANADENSIS GRISEUS Ridgway

GRAY JAY

PLATE 5

HABITS

Ridgway (1904) describes this jay as "similar to *P. o. obscurus,* but decidedly larger (except feet), and coloration much grayer; back, etc., deep mouse gray, instead of brown, remiges and tail between gray (no. 6) and smoke gray, instead of drab-gray, and under parts grayish white instead of brownish white." He gives as its range "interior districts of northern California (northern Sierra Nevada, upper Sacramento Valley, Mount Shasta, etc.), north through central Oregon and Washington to interior of British Columbia."

The above range evidently includes the entire length of the Cascade Mountains in Oregon, Washington, and British Columbia and probably both sides of these mountains.

The gray jay had not been separated from the Oregon jay when Major Bendire (1895) wrote his account of the latter. Apparently all his personal observations refer to *griseus.* He first met with it on the summit of the Blue Mountains, between Canyon City and Camp Harney, Oreg., at an altitude of about 6,500 feet. This is well within the range of *griseus,* being far to the eastward of the Cascades. Here, he says, "they are found only on the highest portions of the mountains, which attain an altitude of about 7,000 feet. I did not see any in the neighborhood of Camp Harney."

Nesting.—I can find no information on the nesting habits of the gray jay except the following from Dawson and Bowles (1909):

The eggs of the Gray Jay have not yet been reported from this State [Washington], but it is known that the bird builds a very substantial nest of twigs, grasses, plant fibre, and mosses without mud, and that it provides a heavy lining of soft gray mosses for the eggs. The nest is usually well concealed in a fir tree, and may be placed at any height from ten or fifteen feet upward, altho usually at sixty or eighty feet. Only one brood is reared in a season, and family groups hunt together until late in the summer.

There are a nest and three eggs of the gray jay, formerly in the collection of J. H. Bowles, now in the Ferry Museum of the Washington
State Historical Society in Tacoma. W. P. Bonney writes to me that it
was collected in Deschutes County, Oreg., but no date is given. He says
that the nest "is about 6 inches in diameter, well built from few small
sticks, some fiber, grasses, soft moss, some feathers and scraps of wool.
The eggs are dingy gray, with *small* brown spots scattered all over."

Food.—Dawson and Bowles (1909) say: "Hunger is the chief characteristic of these docile birds, and no potential food is refused, nuts,
acorns, insects, berries, or even, as a last resort, the buds of trees. Meat
of any sort has an especial attraction to them; and they are the despair
of the trapper because of their propensity for stealing bait."

Food taken from campers is mentioned by Taylor and Shaw (1927)
under behavior, but they also add:

When food is scarce the camp robbers sometimes visit the garbage pile. They
are fond of dead mice, often stripping off the skin before eating them. Fresh
meat of any kind is relished also. Ben Longmire found a nest of young juncos
on a small tree that had been cut down to make way for a new trail. He
removed the nest with the young birds to another tree in plain sight so that
the parent birds could find the young. The camp robbers carried off the young
birds and devoured them. In some localities they are said to be called butcher
birds.

Behavior.—Taylor and Shaw (1927) write of the habits of this jay in
Mount Rainier National Park:

The vocal versatility and freedom from shyness of the gray jay, together with
his occurrence in the deep woods where other birds are scarce, help to sustain
his reputation as one of the park's most interesting bird citizens. The bird is
likely to be heard, first, in the upper branches of the firs or hemlocks at some
little distance. In a moment, perchance, a wisp of gray smoke seems to float
into camp and there is the saucy whiskey jack, very quiet now, perched on a
branch of the tree to which the camp table is nailed, and not 6 feet from where
you are sitting. Cocking his bright eye at you in a knowing manner, he scans
you with much circumspection. Then down he drops, as likely as not, right onto
the table, and before you know it has seized a piece of butter from a plate at
arm's length and made off with it! * * * By this time three or four more of the
birds are waiting for a turn. Scraps of meat, bacon rind, bread, potatoes, butter,
oatmeal, or almost any other foods are prized. When one breaks camp a
company of four or six gray jays is usually on hand, patiently waiting to pick up
any scraps which may be left over. Sometimes, but not often, the birds are
shy. * * *

They are very jealous and have many a severe family fight. As a rule no
two camp robbers will eat out of the same dish, though at other times friendly
enough. They do not like to have the chipmunks too close to their food supply,
either, and often combine forces to drive them away. The Steller jay and varied
thrush are admitted to their company on equal footing, but woe be to the owl
or the hawk that invades their preserves.

J. A. Munro writes to me: "Family groups, consisting of adults and three or four dark young, visited our camp and became tame enough to take food from the hand."

Voice.—Bendire (1895) says that "while some of their notes are not as melodious as they might be, the majority are certainly quite pleasing to the ear, and I consider this species a very fair songster. I have listened to them frequently, and have been surprised to find so much musical ability."

Mrs. Bailey (1902) writes: "The voices of the jays were heard around the log house on Mount Hood from morning till night. Their notes were pleasantly varied. One call was remarkably like the chirp of a robin. Another of the commonest was a weak and rather complaining cry repeated several times. A sharply contrasting one was a pure, clear whistle of one note followed by a three-syllabled call something like *ka-wé-ah.* The regular rallying cry was still different, a loud and striking two-syllabled *ka-whee.*"

Taylor and Shaw (1927) give the gray jay credit for "a truly impressive variety of calls and whistles." They refer to the robinlike call, and add: "A cackling note *whut whut kadakut* is sometimes given. Very unusual ejaculations are their *retezzt, ritizzt* or *reckekekekz.* Their whistled calls may be rendered *wheet wheet, tseeuk* or *wheeup,* and very commonly *wheeoo wheeoo.* The notes are clear and can be heard for some distance through the forest."

CYANOCITTA CRISTATA BROMIA Oberholser

NORTHERN BLUE JAY

CONTRIBUTED BY WINSOR MARRETT TYLER

PLATES 6-10

HABITS

The blue jay is a strong, healthy-looking bird, noisy and boisterous. He gives us the impression of being independent, lawless, haughty, even impudent, with a disregard for his neighbors' rights and wishes—like Hotspur, as we meet him in Henry IV, part 1.

To be sure, the jay has his quiet moments, as we shall see, but his mercurial temper, always just below the boiling point, is ever ready to flare up into rage and screaming attack, or, like many another diplomat, beat a crafty retreat. He is a strikingly beautiful bird—blue, black, and white, big and strong, his head carrying a high, pointed crest which in anger shoots upward like a flame. Walter Faxon long ago told me of a distinguished visiting English ornithologist who was eager to see a live blue jay because he considered it the finest bird in the world. He was surprised to find that this beauty, as he called it, is one of our common birds.

Originally a wild bird of the woods, the jay was canny enough to adapt itself to civilization, and nowadays it often builds its nest close to man, even in our gardens.

Spring.—Although the blue jay is considered a permanent resident over a large portion of its breeding range, and instances are known of a banded bird visiting a feeding station throughout the year, there is plenty of evidence that as a species the jay is highly migratory. In New England we detect little actual migration in spring, as a rule. Although jays become more numerous and noisier as summer approaches, they steal in without attracting much attention. E. A. Doolittle (1919) cites an observation in Ohio that may account for the inconspicuousness of the jay in its northward migration. He says: "By chance I looked up and saw five Blue Jays flying about fifty feet above the tree tops, and before my glance had ended others came into view and still others behind them. They were flying northeast and keeping very quiet. I began to count them, and in about fifteen minutes' time had seen ninety-five Jays. And this does not begin to number those that passed, for, on account of the trees, my view to each side was much restricted, and there is no telling how many had gone on before I casually looked up. They were in a long stream, with now and then a bunch of five to fifteen."

W. Bryant Tyrrell (1934) describes a striking assemblage of jays at Whitefish Point, Mich., which were preparing "to cross the eighteen miles of Lake Superior to the Canadian shore"—a favorable migration route. He says:

Extending south, back of the dunes—along the Lake Superior shore, is a wooded region composed mostly of Jack pine, broken by small swampy areas. In this wooded region the birds [of various species] congregate by the thousands before migrating north across Lake Superior. It was in these Jack pines that I saw hundreds—if not thousands—of Blue Jays *(Cyanocitta c. cristata)* on the morning of June 5, 1930. It was a dull cloudy morning with a chilly northwest wind blowing off Lake Superior. When we arrived at the Point, soon after daylight, the birds, mostly Blue Jays * * * were exceedingly restless, apparently waiting to go north but not caring to venture across in a northwest wind. The Blue Jays made very little noise but were constantly milling around, usually in flocks of varying size. A flock would form and fly off towards the lighthouse, circling and rising all the time until they were over the lighthouse several hundred feet high. They would continue to circle and then would come quietly but quickly back to the pines, only to repeat the same procedure in a short while. By the middle of the morning they had broken up into small flocks and gone off into the woods for the day to feed, congregating again in the evening. Each morning the same maneuvers took place until the morning of June 11 when the wind changed to the northeast and the weather became much warmer. On this date the birds were again circling though flying so high that at times they were almost out of sight. I did not see a single flock actually start and fly off across the lake, but on the morning of the 12th there was hardly a bird to be found in the Jack pines.

Courtship.—A survey of the literature brings little to light in regard to the courtship of the blue jay. We may infer therefore that courtship is not a conspicuous feature of the bird's behavior. Mr. Bent describes in his notes some actions having the appearance of mild courtship. He says under date of April 30, 1940:

This morning about 7:30 I saw a flock of 7 or 8 blue jays having a merry time in the top of a large oak in my yard. They were apparently courting. I could not distinguish the sexes, of course. Perhaps there was only one female, and the males were all following her, just as male dogs follow a female in heat. Several of them, presumably males, were bobbing up and down as they do when they make that musical note often heard at other times, but I heard no notes. They were constantly changing places in the tree and chasing each other about. At least one was evidently trying to escape, or perhaps starting a game of 'follow the leader.' Finally, one did fly away and all the others trooped after it. Perhaps they were only playing a game; if so, it was a lively one.

I saw (Tyler, 1920) a similar gathering of jays at about the same time of year (April 6, 1913) acting in much the same way. "Ten of the birds were sitting in a bare tree. A few were mounting toward the top of the tree by stiff upward leaps; the others, well scattered high in the tree, sat quiet; most of the company were screaming. Every few seconds came the growling note, a sound which suggested a 'snoring' frog, the quick tapping of a Woodpecker, or the exhaust from a distant motorcycle —*g-r-r-r*. During the growl, and immediately after it, one or two birds, and perhaps more, moved up and down as if the branch on which they sat were swaying. There was none of the teetering motion of a Spotted Sandpiper; the whole bird rose and sank as a man would move up and down on his tiptoes. Soon the birds flew off [as did Mr. Bent's] in a screaming company and were joined by other Jays."

Hervey Brackbill sent the following account of "Courtship Feeding" to Mr. Bent: "About sunset, 7.06 p. m., May 9, 1939, I noticed three jays in the top of a tall oak but paid no attention to them until I saw one feed another. Then I began to watch and shortly saw another feeding. For a long time at least one of the birds frequently uttered the little note that sounds like *quick,* and for a while one sang much like a catbird. This went on for some minutes, but as the birds kept moving about in the treetops and were often hidden in thick foliage, I could not tell how many feedings there were or whether there was copulation."

Nesting.—Bendire (1895), in his excellent account of the blue jay, says: "It prefers mixed woods to live in, especially oak and beech woods, but for nesting sites dense coniferous thickets are generally preferred; oaks, elms, hickories, and various fruit trees, thorn bushes, and shrubbery overrun with vines are also used, the nests being placed in various situations, sometimes in a crotch or close to the main trunk, or on the extremity of a horizontal limb, among the outer branches. They are placed at distances from the ground varying from 5 to 50 feet, but usually below 20 feet. * * * I believe but one brood is usually reared in a season, but in the South they may occasionally raise two."

Describing typical nests, he says: "The nests are generally well hidden, and are rather bulky but compactly built structures, averaging from 7 to 8 inches in outer diameter by 4 to 4½ inches in depth; the inner cup measures about 3½ to 4 inches in diameter by 2½ inches in depth. Outwardly they are composed of small twigs (thorny ones being preferred), bark, moss, lichens, paper, rags, strings, wool, leaves, and dry grasses, the various materials being well incorporated and sometimes cemented together with mud, but not always; the lining is usually composed exclusively of fine rootlets. Occasionally the Blue Jay will take the nest of another species by force."

John R. Cruttenden writes to Mr. Bent from Illinois: "A peculiar habit of this bird is to line its nest with a piece of cloth or waste paper. This is true in the majority of nests placed near dwellings or in the city, undoubtedly because of the more abundant supply of materials in the city, although the habit is not unusual in nests situated away from man." Henry Mousley (1916) reports: "Evidently the Blue Jay betakes itself to very secluded spots during the breeding season, as I have only succeeded so far in finding one nest, in May of the present year (1915), and had never seen the bird before during the months of June, July and August." Mr. Mousley is speaking here of his experience in Hatley, Quebec. Farther to the south, in New England and the Middle Atlantic States, however, the jay commonly breeds in thickly settled regions, often near houses, as the following observations show.

Frederic H. Kennard (1898) writes: "We have a pair of Blue Jays *(Cyanocitta cristata)* in Brookline, Mass., that have this year built their nest in a most conspicuous place, between the stems of a Wistaria vine and the capitol of a pillar, supporting a piazza roof. This piazza is in almost daily use, and the path leading immediately beside it is also used constantly." Charles R. Stockard (1905), writing of Mississippi, says: "With the exception of the English Sparrow the Blue Jay is probably the most abundant bird in the State. The shade trees bordering the streets of towns, the groves near dwelling houses, trees along road sides, orchards, pastures, and pine woods as well as thick woods, are nesting localities of this bird. One nest was placed in a tree crotch not more than six feet from a bed-room window, thus one might look out on the bird as she sat calmly upon her eggs, and later she was not noticeably nervous while feeding her nestlings before an audience of several persons who observed the performance from the window."

I remember some years ago seeing a nest containing eggs in a situation with no concealment whatever—on the cross-beam of an electric-light pole. The pole stood near a flight of steps used continually by pedestrians in crossing over the tracks at the main railroad station in Lexington, Mass. From the steps I might have touched the sitting bird with an umbrella. Needless to say, the nest was soon knocked down, presumably by boys.

On June 12, 1942, in Tiverton, R. I., Roland C. Clement showed us a most unusual blue jay's nest under the overhang of a cutbank beside a woodland road, which held at that time a brood of nearly fledged young. As he did not get a chance to photograph it, he has sent us the following description of it: "The recessed face of the cutbank in which the nest is placed lies only 10 feet from the farm road, the cut itself being about 6 feet high and its concavity amounting to about 10 inches two feet below the overhanging brink. In this sheltered recess two stout oak roots of 1 inch diameter reach out horizontally into space, intersecting past their exerted centers, and in this crotch our adaptable jays have firmly anchored an otherwise typical nest. The nest is thus about 4 feet from the ground below and, though not absolutely secure from molestation by terrestrial predators which could probably clamber up to it without undue difficulty because of the moderate incline of the bank, it is indeed inconspicuous among the pendant roots and rootlets of the vegetation above, which presently consists merely of shrubs such as *Corylus* and *Myrica*.

"The nest itself is well and firmly woven of long, pliant dead twigs of various species, including some spiny stems of *Smilax* and a few culms of coarse grass, as well as a long strip of paper; and it is lined with fine rootlets, probably those of the brake fern *(Pteris)*, which abounds nearby. The nest cavity is 4½ inches long, parallel to the bank, and 4 inches wide."

Mrs. Harriet Carpenter Thayer (1901) watched the family life of a pair of blue jays at a nest at close range and states that the male aided in making the nest and that both birds incubated, "each relieving the other at more or less regular intervals. And the bird at play did not forget its imprisoned mate, but returned now and then with a choice bit of food, which was delivered with various little demonstrations of sympathy and affection."

Jays are very quiet about their nest. I knew of a nest near the center of the city of Cambridge, Mass., and if I had not happened to see the nest, I should not have suspected that jays were breeding near.

Bendire (1895) quotes W. E. Loucks, of Peoria, Ill., as saying: "A nest of a pair of Robins, built in an elm tree, was stolen and appropriated by a pair of these birds. It was fitted up to suit their needs, and eggs were deposited in it before the eyes of the angry Robins."

A. D. Dubois sent the following note to Mr. Bent: "While listening to the Memorial Day exercises in the auditorium at Chautauqua Grounds (a large pavilion with open sides) I noticed a jay which flew in from the side and up to a nest in one of the roof trusses, where it fed its young and flew out again. This is the first jay's nest I have ever found in a building of any kind."

Dr. Samuel S. Dickey (MS.) reports that nests found by him have been in the following trees: 20 in white pines, 18 in hemlocks, 2 in red spruces, 2 in intermediate firs, 12 in white oaks, 5 in alders *(Alnus incana* and *rugosa),* and one each in a pitch pine, sour gum, Cassin's viburnum (only 3½ feet from the ground), and flowering dogwood.

Eggs.—[AUTHOR'S NOTE: The northern blue jay ordinarily lays four or five eggs, sometimes as few as three, frequently six, and very rarely as many as seven. These are quite uniformly ovate in shape, with occasionally a tendency toward short or elliptical ovate; they have very little or no gloss. The ground color is very variable, and shows two very distinct types, an olive type and a buff type, with a much rarer bluish type; the olive type is by far the commonest. In eggs that I have examined, I have noted the following colors: "Olive-buff," "deep olive-buff," "dark olive-buff," pale "ecru-olive," "pale fluorite green," pale "lichen green," "pale glaucous green," "sea-foam yellow," "light buff," "light ochraceous-buff," "pinkish buff," "pale pinkish buff," and pale "vinaceous-buff." There are also many intermediate shades of pale olives, buffs, greens, and very pale "wood brown," down to pale dull blue, bluish white, or greenish white.

The eggs with the pinkish-buff ground color are often very pretty, being sparingly marked with small spots of bright or purplish browns, and with underlying spots of pale quaker drab or lavender. The pale greenish and bluish types are also sparingly marked with pale, dull browns or olives and a few underlying spots. The olive types are usually, but not always, more heavily marked with spots and small blotches of darker browns and olives of various shades. Some eggs are evenly marked over the entire surface with spots or fine dots, and in others the markings are concentrated at one end; an occasional egg has a few black dots.

The measurements of 135 eggs in the United States National Museum average 28.02 by 20.44 millimeters; the eggs showing the four extremes measure 32.8 by 19.6, 25.9 by **22.4, 25.2** by 20.1, and 25.9 by **18.8** millimeters.]

Young.—From a comprehensive, carefully prepared study of the blue jay, a thesis for the degree of doctor of philosophy, sent to Mr. Bent in manuscript by John Ronald Arnold, the following observations are abstracted: The period of incubation was found to be 17 or 18 days in the vicinity of Ithaca, N. Y., and 17 days in New Jersey. The young at the time of hatching were limp, blind, and entirely naked. When 3 hours old they were able to raise their heads to the rim of the nest. By the fifth day the eyes were just beginning to open, and the birds grasped the lining of the nest with their claws. "During the eighth and ninth days the feathers in all the body tracts except the head and neck regions begin to break from their sheaths." By the seventeenth day the nestlings begin to resemble a blue jay and are almost ready to leave the nest. They leave 17 to 21 days after hatching.

In close agreement with these dates, Isabella McC. Lemmon (1904) gives the incubation period between May 2 and 19 and reports that the young flew on June 6.

Donald J. Nicholson (1936) writes, referring to the young Florida blue jay:

They leave the nest in from fifteen to eighteen days, at which time the tails are quite short, and the feathers not fully developed on any part of the body or wings. Their power of flight is not by any means strong when they first leave the nest, and only short spaces can be covered. Many a young bird at this time of the year falls an easy prey to cats and various snakes. * * *

In three weeks to a month, it is difficult to distinguish the young from the adults, but the face and throat is a smoky, dark color, instead of the rich black of the adult, and the bill is horn-colored, instead of black as in the parents; otherwise the plumage is apparently the same to all outward appearances. By the following spring no difference is seen. Even by fall I can not discern a particle of difference. A fledgling when caught, if caught by anything, emits terrified screeches as if in mortal agony, bringing the parents to its defense at once.

Apparently the voice develops early; I have heard a young bird on leaving the nest shout almost as loudly as its parents.

Francis Zirrer has sent us the following note: "Although considered more or less a raptor, the young blue jay must learn about the various small game before it will touch it. At our woodland cabin we were greatly annoyed by various species of wild mice, especially *Peromyscus.* Throughout the winter many were caught and deposited on the feeding table in the morning. We noticed, however, that the majority of the blue jays, apparently the young birds of the previous summer, were plainly afraid of the mice. Coming to the table they would, with all signs of fright, jerk back, flutter with the wings and fly away. It was up to the old birds to take the mouse, fly with it to a nearby branch, and begin to tear it to pieces. And then the young birds would come near and were fed by the adults."

Plumages.—[AUTHOR'S NOTE: The following abstracts are taken from the manuscript thesis of John R. Arnold, referred to above. He has made a thorough study of the plumages of the blue jay, and says that the young are hatched naked and have no natal down at all. On the eighth and ninth days the body plumage begins to break the feather sheaths, and when the young bird leaves the nest, at an age of about 20 days, the juvenal plumage is largely grown and the bird is able to fly. He describes this plumage as follows:

"Pileum between cadet gray and columbia blue. Feathers of forehead black at base with bluish-white tips. Superciliary line grayish white. Throat bluish white to white. Nuchal band black. Black of lores less pronounced than in adult. Back and lesser wing coverts light to deep mouse gray, tinged with blue. Wing and tail feathers as in first winter and similar to adult. Breast and flanks smoke gray, belly and under tail coverts white."

A partial postjuvenal molt takes place when the bird is between 50 and 90 days out of the nest; this produces a first winter plumage that is hardly distinguishable from that of the adult, though somewhat paler and less violet on the head and neck, and with the bars on the tail less pronounced. This molt involves the contour plumage and the lesser wing coverts only.

Adults have a complete postnuptial molt between June and September.]

Food.—The blue jay eats almost every kind of digestible food; like its relative, the crow, it may be considered omnivorous. F. E. L. Beal (1897), in an exhaustive study to determine the exact economic status of the jay, published the results of an examination "of 292 stomachs collected in every month of the year from 22 states, the District of Columbia, and Canada." He says:

One of the first points to attract attention in examining these stomachs was the large quantity of mineral matter, averaging over 14 per cent of the total contents. The real food is composed of 24.3 per cent of animal matter and 75.7 per cent of vegetable matter, or a trifle more than three times as much vegetable as animal. The animal food is chiefly made up of insects, with a few spiders, myriapods, snails, and small vertebrates, such as fish, salamanders, tree frogs, mice and birds. Everything was carefully examined which might by any possibility indicate that birds or eggs had been eaten, but remains of birds were found in only 2, and the shells of small birds' eggs in 3 of the 292 stomachs. * * *

Insects are eaten by blue jays in every month in the year, but naturally only in small quantities during the winter. The great bulk of the insect food consists of beetles, grasshoppers, and caterpillars. * * * The average for the whole year is nearly 23 per cent.

Under vegetable food Professor Beal lists corn, wheat, oats, buckwheat, acorns, chestnuts, beechnuts, hazelnuts, sumac, knotweed, sorrel, apples, strawberries, currants, blackberries, mulberries, blueberries, huckleberries, wild cherries, chokecherries, wild grapes, serviceberries, elderberries, sour-gum berries, hawthorn, and pokeberries. He continues: "Grain is naturally one of the most important groups, and may be considered first. Wheat, oats, and buckwheat occur so seldom and in such small quantities (1.3 per cent of the whole food) that they may be dismissed with slight comment. Wheat was found in only eight stomachs, oats in two, and buckwheat in one. The wheat was eaten in July, August, and September; oats in March and July, and buckwheat in October. Corn was found in seventy-one stomachs, and aggregates 17.9 per cent of the food of the year. This is less than that eaten by the crow (21 per cent) or by the crow blackbird (35 per cent)." Professor Beal summarizes his findings thus:

The most striking point in the study of the food of the blue jay is the discrepancy between the testimony of field observers concerning the bird's nest-robbing proclivities and the results of stomach examinations. The accusations of eating eggs and young birds are certainly not sustained, and it is futile to attempt to reconcile the conflicting statements on this point, which must be left until more accurate observations have been made. In destroying insects the jay undoubtedly does much good. Most of the predaceous beetles which it eats do not feed on other insects to any great extent. On the other hand, it destroys some grasshoppers and caterpillars and many noxious beetles, such as Scarabaeids, click beetles (Elaterids), weevils (Curculionids), Buprestids, Chrysomelids, and Tenebrionids. The blue jay gathers its fruit from nature's orchard and vineyard, not from man's; corn is the only vegetable food for which the farmer suffers any loss, and here the damage is small. In fact, the examination of nearly 300 stomachs shows that the blue jay certainly does far more good than harm.

William Brewster (1937) describes jays collecting acorns thus: "1898, September 30.—Several Jays spent the entire day harvesting acorns in a Red Oak that shades a village street of Bethel, Maine, taking them thence across open fields to rather distant woods. They invariably plucked them from the twigs while hovering on fluttering wings and not when perched. The acorns were still green where the cups covered them. Each Jay apparently always carried two at once, one in the mouth or throat, the other held in the tip of the bill."

Mr. Brewster (1937) also speaks of the jay as a flycatcher: "1888, September 9.—At sunset this evening when the air was warm, damp and calm, I saw about a dozen Blue Jays scattered about in the tops of small aspens growing by the Lake-shore where they were catching flying insects. In pursuit of these they would mount straight upward from ten to twenty feet and then return to their perches by swooping downward on set wings. Their flights were altogether so very like those of King-birds similarly engaged that I mistook them for birds of the latter species at first glance."

G. Gill (1920) tells of a blue jay trying to catch a mouse. "On Feb. 2, 1918," he says, "the scream of a Blue Jay rang out through the air, and, looking toward the barn, I saw the bird swooping down to the ground after something. I was interested at once, and at first I could not see what was running across the snow; when it reached the barn, where it was clear, I saw that it was a mouse.

"The Blue Jay boldly followed it right into the barn, dodging in and out of the wagons and pecking at the mouse at every chance it got. About this time the Blue Jay's mate joined the chase, but she was just a little too late. The mouse, nearly beaten, hopped into a friendly hole and escaped. For a little while the pair watched the hole, and then gave it up."

This would appear strange prey for a jay, but F. E. L. Beal (1897) states that "the jay kept in captivity by Mr. Judd showed a marked fondness for mice, and would devour them apparently with great relish."

W. L. McAtee (1914) calls attention to a bizarre feeding habit of the jay apparently seldom resorted to. He quotes Grace Ellicott from the *Guide to Nature,* 1908, p. 168, as follows:

The occupants of a recently disturbed ant hill were excitedly crawling about the hill and the adjacent cement walk. They were large, and to a blue jay in a neighboring tree they must have looked luscious, for flying down, the jay began to pick them up with an eagerness that seemed to say that this was an opportunity that might come his way but once. As rapidly as he could do it he seized the ants, with each capture lifting a wing, sometimes one, sometimes the other, and seemed to deposit his prey amongst the feathers back of and under-neath it. So quickly he worked and with such evident eagerness to make the most of this rare occasion that, as he lifted the wing, putting his bill amongst the feathers, it often seemed that he must lose his balance and topple over backwards. But he kept his poise, worked on with all speed and had laid in quite a store when a passerby frightened him from his task. Whether this jay had only just discovered the most convenient of all storehouses for his use or whether this food was to be carried to the nest for the young, for it was nesting time, he was most interesting.

McAtee comments on the observation as follows:

This Blue Jay was therefore taking advantage of the instinct of ants when disturbed to fasten their jaws onto any object that presents itself. * * * These three most interesting observations suggest that numerous birds may have the same or other wonderful habits about which we are ignorant. They should stimulate minute and careful research and comfort those who fear that all the interesting things have already been discovered. [See Auk, vol. 57, pp. 520-522, 1940, and vol. 58, p. 102, 1941, for other similar performances.]

Behavior.—The jay commonly progresses through the air steadily and rather slowly, although with full and regular quick flips of the wings. He keeps on an even keel and maintains a characteristically level flight. The long axis of the body is parallel to the ground, although his beak appears to point slightly downward, perhaps only because his crest gives the impression of a downward-sloping profile.

A company of jays, like their small relatives the chickadees, almost always fly across a wide, open space one at a time, at some distance from each other. They generally fly directly to the place where they are about to alight, rarely deviating from their course by swerving from side to side, and, on arriving at their perch, often come to a stop deftly upon it in perfect balance, although they may sometimes alight, with head held proudly high, after a short upward-slanting sail. I have seen a jay come to rest on a slender vertical rod (a radio aerial) as neatly as any kingbird.

Sometimes, in making short flights, jays will undulate a little, sailing with wings held open longer than in the steady, level flight. Now, as they fly overhead, slowly and silently, they flap the wings back and without an instant's pause fan them out full again. Here there is a short pause with the wings expanded, during which the bird sinks a little in the air before the next stroke carries him on and upward again—very different from the undulating flight of a woodpecker, which closes its wings on the downward plunge.

William Brewster (1937) describes an unusual method of flight which he observed at Lake Umbagog. He writes: "1895, September 20.— About eight o'clock this morning I was standing on a wooded knoll near our camp at Pine Point, watching some small birds, when a sound resembling that of strong wind blowing through pine-tops came from directly overhead. It could not be ascribed to such an origin, however, for the air was then perfectly calm. The mystery remained unsolved until an hour or so later when I saw a dozen Blue Jays mount in a compact flock, by a spiral course, to a height of several hundred feet above the tallest trees and then dash almost straight down together, with half-closed wings, like so many stooping falcons, thereby producing a loud rushing sound exactly like that heard earlier in the morning."

The motions of a company of jays as they flit about among the branches of a tree are surprisingly easy, light, and graceful. The wings move slowly, like great moth's wings, yet the birds alight accurately on the branches, or float to the ground from which they often almost bounce up to a high perch again. With all their energy, alertness, and spirited behavior, jays seldom seem to be in a hurry; we never see them move with that intensely rapid, flashlike speed which is characteristic of many birds.

Nowadays we regard the blue jay as rather a tame bird—almost as tame as the robin—but Witmer Stone (1926) states that in Germantown, Pa., the bird's habits have changed in the last 3 or 4 decades. He says: "When studying birds in Wister's woods and vicinity from 1880 to 1897, the Blue Jay was a very wild species occurring only during autumn flights, but upon returning to reside in the old neighborhood after some thirty-five years absence I found the bird's habits totally changed. I was surprised to find a pair of Jays present about the end of May, 1922, acting as if they were located for the summer. Later, I detected them constructing a nest in a beech tree close to the railroad station about ten feet above a path along which hundreds of persons passed to and from the trains, and not over fifty feet from the tracks."

On the other hand, Nathan Clifford Brown (1879), writing of Coosada, Ala., says that the blue jay is a "very common resident, and, to one who has known the species only at the North, remarkably tame. I observed them feeding in the streets of Montgomery, and unsuspiciously flying about much after the manner of the domestic pigeons of Northern cities."

Individual jays react differently in the presence of man. Wilbur F. Smith (1905) gives an instance of remarkable tameness in a sitting bird. He says:

To those knowing the Blue Jay only as a wild, shy bird of the tree-tops, so hard to approach, or, by reputation, as a thief or a robber of other birds' nests, there remains a pleasure like unto finding some new and rare bird, to watch a pair of Jays through the nesting season and to find them so devoted to their nest and young that they lose much of their shyness and allow a familiarity which very few other birds will tolerate. One pair of Jays built for several years in a tangle of briers near my home, and the female became so tame, through constant visiting, that I could at last spread her wings and tail-feathers without her leaving the nest, and even stroke her back with no further sign of disapproval than a settling lower in the nest and a parting of the bill.

Mrs. Harriet Carpenter Thayer (1901) says of a pair which nested in her garden: "The Jays were not at all shy, but on the contrary were very valiant and determined in standing by their home. Soon after the eggs were laid, the house-painters began work opposite the nest, and many sharp pecks they received on their ears and backs."

In its relation to small birds, consensus classes the blue jay as an outlaw and robber. Bendire (1895) says:

Few of our native birds compare in beauty of plumage and general bearing with the Blue Jay, and while one can not help admiring him on account of his amusing and interesting traits, still even his best friends can not say much in his favor, and though I have never caught one actually in mischief, so many close observers have done so that one can not very well, even if so inclined, disprove the principal charge brought against this handsome freebooter. He is accused of destroying many of the eggs and young of our smaller birds, and this is so universally admitted that there can be no doubt of its truth. * * *

Mr. Manly Hardy, of Brewer, Maine, fully corroborates these statements, writing me as follows: "It is a great robber of birds' nests, taking both eggs and young. I also feel quite sure that in some cases it kills adult birds. * * * There is little doubt that they destroy many nests of eggs and young; all of the *small birds say so.*"

Mrs. Marie Dales (1925) thus adds her testimony against the jay: "I saw a Blue Jay harassing a Mourning Dove, eighteen or twenty feet up in a tree. He would pluck out a mouthful of feathers and then retreat for a moment. When the dove had settled down, back would come the jay to torment her again. On closer observation I discovered the nest, wonderfully well hidden for a Mourning Dove's nest. The jay kept up his attacks for several minutes and finally the dove left the nest and went to her mate sitting on a limb farther out. This was just the opportunity the jay was waiting for. He hopped to the nest, pecked a hole in the egg and carried it off." William Brewster (1937) states that he "saw a Blue Jay take an egg from a Robin's nest and fly off with it, hotly pursued by the outraged Robin."

The following story, sent to Mr. Bent by Dr. Daniel S. Gage, gives an interesting sidelight on the blue jay's character: "I once saw a demonstration that animals note the warning cries of the blue jay. I was walking on a trail in the Flat Top Mountains of Colorado. A porcupine was waddling along ahead of me. The trail ran through an open space several hundred yards across, dense woods bordering it on all sides. The porcupine was going away from me and did not notice me, as he could not see behind him as he waddled along. He stopped repeatedly to nibble at some plants at the side of the trail. I halted each time he stopped to bite at a plant, and he did not note me at all, although I was only a few feet behind him. Suddenly, from the woods some hundred of yards away, a blue jay shrieked his *jay, jay, jay.* He had seen me. Instantly, the porcupine raised his quills, rose to his hind feet and sniffed in each direction. Then he noticed me, although I was standing perfectly still, eyed me carefully, his quills erect. Then finally, with angry mien and raised quills, he dropped down and ran as fast as he could into the forest."

Aretas A. Saunders (MS.) reports: "A family of jays came to my bird bath fairly regularly late in summer. Six birds would come together and stand about the edge of the bath while each one in turn bathed," and Mr. Bent (MS.) calls attention to the jay's habit of sunning itself. He says: "I have been amused lately in watching them sunning themselves on my lawn, even on the hottest days. Usually the bird turns over on one side, with its breast toward the sun and the upper wing partially raised, so as to let the sun in on its under plumage, remaining in this position for several minutes. At other times it lies prone on the ground, breast down, with both wings widely spread, so as to sun the wings."

Henry C. Denslow sent to Mr. Bent a record of a banded blue jay that lived for 15 years. Other banded individuals have been recorded as 9, 11, and 13 years old.

The jay's tendency to pester owls and hawks is one of its best-known habits. If a jay comes upon an owl hidden in the daytime, he sets up an outcry to which all the jays within hearing respond, and, collecting in a screaming mob, they drive the owl from tree to tree. It is sometimes to our advantage to follow up such a gathering when their voices rise to the high pitch of anger, for the jays may have found a rare bird.

In regard to the jay's habit of storing food for future use, Bendire (1895) says: "Where they are resident they lay up quite a store of acorns, corn, and nuts in various places for winter use, but where they are only summer visitors they do not resort to this practice."

Dwight W. Huntington (MS.) reports the following observation: "I had many small pheasants running at large in my gardens, and one day a blue jay lit on a small tree just above a bantam with a brood of golden pheasants. He evidently had his eye on the little birds, and the bantam led them away. The jay followed, lit in another tree, and this was repeated several times until, much to my surprise, he struck at the little birds just as a hawk does. The bantam flew up at him as he came down. The birds came together, and a fight was on. Blue feathers and black from the bantam soon covered the ground. The bantam won, and, seeing that the jay was dead, she proudly led her little brood away. I was dumbfounded and amazed at what I had seen and called a gamekeeper to come and see the dead jay and the feathers scattered about."

Voice.—It is the blue jay's voice, more than his gay color, that makes him conspicuous. We cannot be long in the open air before we hear him—in woodland, in open country, in the suburbs of our large cities. At the least alarm he begins to shout, and often, with no apparent cause, even a lone bird will break out, like a schoolboy, it seems, out of pure joy in making a noise. Especially in autumn the jays shout so loudly that they fill all outdoors with sound.

The note we hear oftenest is a loud, clear cry often written *jay* or *jeer,* well within the range of human whistling and readily imitated by the human voice. *Peer* or *beer,* with no *r* sound, is perhaps a closer rendering, because the note lacks the hard *j* sound at the beginning. It is long drawn out, falling in pitch at the end and is generally repeated a number of times. This is the note we hear all through the autumn from screaming companies of jays traveling through the woods. It suggests to us various emotions or states of mind—remonstrance, taunting defiance, whining complaint, anger, but never, I think, fear. The tone of voice varies too. It may be harsh, hard and flat, or musical and delicate; sometimes it has a tin-whistle quality; and rarely it is pitched so high that it resembles a killdeer's piercing whistle.

The jay uses a great number of calls—too many for us to describe them all in detail—and the fact that they tend to run into each other makes enumeration difficult. Even dissimilar notes, by a slight alteration in inflection or tone, will often merge into one another. For example, when the *jay* call is produced in its purest musical form, and uttered as two notes, it becomes the well-known, bell-like *tull-ull* or *twirl-erl.* When a bird is near us we can sometimes detect the transition as one note is gradually converted into another.

Francis H. Allen (MS.) terms the *tull-ull* the anvil call, an apt comparison, and says that in making the note "the jay raises and lowers its head twice, once for each part of this dissyllabic note. This bobbing of the head is up and down, not down and up."

During the warmer months the jay often utters a pleasing whistled note that sounds like *teekle,* pronounced like our word tea-cup. Over and over he sings it as he flies about, sometimes giving it in pairs or series. It seems to reflect a quiet, happy mood in which the bird is free for the moment from antagonism. This note is allied to the creaking, wheel-barrow call, commonly written *whee-oodle.*

Frequently heard in the autumn gatherings is a chuckling, conversational *kuk.* This note differs widely in its mode of delivery. It may be extended into a bubbling chatter—a sort of tittering laugh—or, ranging up and down in pitch, it may run off into pretty, rambling phrases. The voice is not loud, and we have to be near the bird to appreciate the charm of the phrasing. Jays give a modification of the *kuk* when they are feeding in trees or when they visit feeding stations.

Quite different from the shouted or whistled notes is a dry, wooden rattle, almost a growl. A lone jay may give it, or one or more in a large company. The notes are often accompanied by an odd rising up and down on the perch. Francis H. Allen (MS.) speaks of it as "a grating, pebbly *r-r-rt,* generally given twice, but sometimes three times. The repetition is in the manner of most of the calls of the species. The grating quality I express by the *r,* but of course the *it* sound ran all through the note. 'Pebbly' seems to express it rather well."

Comparatively few observers are familiar with the song of the blue jay. When he sings, the jay throws off his boisterous demeanor. He retires to the recesses of a wood or seeks seclusion in a thick evergreen tree and there, all alone, sings his quiet solo. I have sometimes heard a song from a bird hidden in a tangle of second-growth, and have not at first recognized the author as a jay at all. The song is a potpourri of faint whistles and various low, sweet notes, some in phrasing and pitch, suggesting a robin's song—a mockingbird might be singing, *sotto voce*. But as the song goes on one realizes that most of the notes are clearly in the blue jay's repertoire but are disguised by being jumbled together and delivered gently and peacefully.

Francis H. Allen has noted the song several times in his journal. He heard it first on February 28, 1909, the notes "coming from a row of large hemlock trees. The bird was keeping in the very heart of the tree, near the trunk. The notes sounded not unlike the goldfinch's song, but very subdued in tone. The song consisted of sweet lisping notes and chippering, and was continuous and long." Again he says: "Sweet and rather loud song notes from a jay in one of our Norway spruces this morning (September 4, 1933). One was a sort of short descending trill, rather high pitched, that suggested a mockingbird." And on March 22, 1935: "Long subdued song from a jay in a hemlock. It lasted two or three minutes, I should say, and was absolutely continuous, with no pauses between phrases. Some notes were very suggestive of *Spinus tristis*, both the long upward-slurred note and a succession of short notes resembling *per-chic-o-pee*. The whole remarkably soft and sweet. The bird remained hidden among the foliage, as is the jay's custom in this sort of singing."

Isabel Goodhue (1919) speaks of the song as "sweet, tender and quite lovely; delivered * * * with a retiring modesty not perceptible in the Blue Jay's deportment on other occasions."

The jay's loud cry often sounds exactly like the *teearr* of the red-shouldered hawk. I have sometimes been misled and have mistaken one note for the other. On more than one occasion I have supposed I was listening to a hawk screaming in the distance but found that a jay near at hand was the author of the notes.

This similarity to the scream of the red-shouldered hawk and the resemblance of some of the jay's notes to those of other birds have given him a reputation as an imitator. It is difficult, perhaps impossible, to be sure that such cases are not coincidence, especially when we recall the multiplicity of the jay's vocabulary.

Enemies.—Jays are subject to attack from the smaller, quick-moving hawks but appear in the main to be able to protect themselves.

Taverner and Swales (1907), in their studies at Point Pelee, say: "During the hawk flights of 1905 and 1906 they were much harassed by the Sharp-shins but, as they are perfectly able to take care of themselves and kept pretty close in the grape vine tangles, it is not probable that they suffered much. * * * In fact once within the shrubbery, they seemed to rather enjoy the situation, and from their safe retreats hurled joyous epithets at their baffled enemies. * * * We have only once found the remains of a hawk-devoured bird of this species."

Frank Bolles (1896) speaks thus "of an encounter between a sharp-shinned hawk and a flock of blue jays":

The hawk arrived when several flickers were in the tree and hurled himself upon them. They fled, calling wildly, and brought to their aid, first a kingbird, which promptly attacked the hawk from above, and then a flock of blue jays, which abused him from cover below. When the kingbird flew away, as he did after driving the hawk into the bushes for a few moments, the jays grew more and more daring in approaching the hawk. In fact they set themselves to the task of tiring him out and making him ridiculous. They ran great risks in doing it, frequently flying almost into the hawk's face; but they persevered, in spite of his furious attempts to strike them. After nearly an hour the hawk grew weary and edged off to the woods. Then the jays went up the tree as though it were a circular staircase, and yelled the news of the victory to the swamp.

Henry C. Denslow sent the following note to Mr. Bent: "It is said that shrikes sometimes attack blue jays, but in one case the tables were turned. A shrike came to a feeding-table where eight blue jays were feeding and met a warm reception. The shrike alighted on a branch a little above the jays. They looked at him for an instant and then all started for him. He flew into a hedge for protection, but was driven out, then started for an evergreen tree, but the jays were so hot on his trail that he took flight—all the jays trailing after, each one screeching his loudest, until the sound of battle faded away in the distance."

Feathers and other remains of blue jays are often found in and about the nests of the duck hawk.

Dr. Herbert Friedmann (1929) mentions two records of the blue jay being imposed upon by the cowbird but suggests that "the eggs of the Blue Jay are so much larger than those of the Cowbird that there is little probability of the latter ever hatching if present."

Harold S. Peters (1936) lists as external parasites on the blue jay four lice (*Degeeriella eustigma, Menacanthus persignatus, Myrsidea funerea,* and *Philopterus cristata*), one fly (*Ornithoica confluenta*), one mite (*Liponyssus sylviarum*), and one tick (*Haemaphysalis leporis-palustris*).

Fall.—The migration of blue jays in autumn is much more conspicuous than the northward movement in spring. P. A. Taverner and B. H. Swales (1907) describe a flock leaving Point Pelee on their southward journey on October 14, 1906. They say:

We noticed a very interesting migration across the lake. All morning long we saw large flocks passing out the Point. In the afternoon we followed them to the end and, though most then had passed, we witnessed one small bunch of perhaps fifty birds essay the passage. The day was fine and clear and but very little wind blowing, but when they came out to the end of the trees they turned back and sought a large tree-top, where they settled to talk the matter over at the top of their voices. Then, reassured, they started out, rising above gun shot from the ground and making for the Ohio shore, not for Pelee Island as we supposed they would. When they got far enough out to see the blue water under them they slowed up, and when we waved our hats and shouted at them a few wavered, paused and then fled back to the shore to their tree again, followed a moment later by the whole flock. Another pow-pow was held and again they started, with great determination and seemingly filled with the motto, "Ohio or bust." This time they had hardly got well out over the lake when a Sharp-shin was discerned far in the distance, but it was enough to again send them shrieking back to their oak tree. This time the consultation lasted a little longer than before, but at last the coast seemed clear and they started once more. Again, as they drew over the water, they slightly paused as though doubtful, but no one shouted, there was not a hawk in sight and, as there was no possible excuse for backing out this time, they kept slowly and gingerly on until well started and away from land, when they settled into their pace and, when lost sight of in our glasses, were continuing on their way in a straight line that would carry them several miles to the east of Pelee Island.

William Brewster (1937), under date of September 21, 1895, gives this account of migrating jays at Lake Umbagog, Maine:

As I was bathing in the Lake at seven o'clock this morning a flock of seventeen Blue Jays started from the woods on Pine Point and rose above them to a height of fully *two thousand* feet, by a spiral course not less than a half mile across, making only one complete and another half, lateral turn during the entire ascent. They then started off towards the southwest and kept straight on, with ceaseless flapping; until lost to sight in the distance, thereby accomplishing what was obviously the initial stage of a diurnal migratory flight. * * * An hour later the members of another flock, seventeen in number, appeared over the Point at a height of about two hundred feet, probably arriving from somewhere further north. Setting their wings they came hurtling down altogether, precisely like those seen yesterday and making the same sound as of rushing wind. [Quoted under Behavior.] It was loud enough to bring Jim Bernier, my guide, running forth from his tent with the expectation, as he afterwards admitted, of seeing a big flock of Scoters pitching down into the Lake. That the first flock of Jays should have apparently started on a migratory journey, and the second have completed one at so nearly the same time of day seems very interesting, and also suggestive of the inference that these flights may often be of no great duration. While engaged in them the birds remain severely silent, in this respect differing from migrating Crows. Such, at least, has been the case with all that I have observed for not one has ever uttered a vocal cry of any kind within my hearing, when on wing.

I remember seeing, several years ago in mid-September, a migration of jays that covered a wide area. During a drive of 50 miles northwest out of Boston, Mass., jays continually crossed the road in front of my car. I soon noticed that all of them crossed from the right to the left side of the road and were therefore flying south. Most of them were single birds, but occasionally two or three flew near together. I noticed them for 20 miles or so. Again, also in September, I saw a flock of 15 or 20 jays fly southward across the parade ground on Boston Common, which is surrounded on all sides by miles of closely built-up city. These birds were so closely packed that I mistook them at first for a flock of grackles.

William Brewster (1937) speaks of a similar observation thus: "1888, September 13.—During the last three days I have seen many flocks of Blue Jays, containing anywhere from a dozen to twenty birds each, flying southward in the daytime over open country, not in scattered order, but as compactly 'bunched' as so many Blackbirds correspondingly employed. Without doubt they were migrating."

Rev. J. J. Murray writes to us: "In the Valley of Virginia, they are certainly migratory. Here they are much commoner in summer than in winter, being very scarce indeed during some winters. Migration is more noticeable in fall than in spring. Through October, and sometimes up to the middle of November, migrating flocks are seen moving south. I have seen as many as 25 or 30 blue jays pass a favorable location in an hour, usually in strung-out flight."

Maurice Broun (1941) reports heavy migrations of blue jays at Hawk Mountain, Pa., "from the third week in September until mid-October." He says:

The jays may be seen in loose flocks, or in orderly processions, on either side of the ridge, and at any elevation, in numbers varying from twelve to three hundred or more birds. I have noticed each season that jays are on the move by 7 a. m., but by mid-afternoon their flights terminate. As a rule, the birds keep just above the tree tops, and seldom is there much fuss or noise; indeed, observers at the lookout must be keenly alert to detect each passing group of jays. * * *

During a sixteen day period beginning September 24, 1939, I made an approximate count of 7,350 Blue Jays. Doubtless *many* jays slipped by uncounted. The majority of the birds passed through in a constant stream regardless of weather conditions, from September 30 to October 6. The peak of the migration came on October 1, a day of alternating rain and mist, with raw northerly winds; at least 1,535 birds passed the lookout, even during the rain, in groups of from 100 to 350. Again on October 3, despite obliterating mists during the forenoon, and fresh easterly winds all day, I counted several large flocks at various parts of the Sanctuary, and the far from complete count for the day was 1,250 birds.

Winter.—The blue jay is an attractive winter bird. He fits well into the wintry scenery—bright, clear sky, and the blue shadows on the snow. After his burst of noise in the autumn, he becomes comparatively quiet, and during the colder months uses mainly his *jeer* call, and this not overmuch. But on soft mornings in January and February, when the temperature is rising, we may hear his sweetly whistled *teekle* note. *Tea-cup, tea-cup,* he sings—a sure sign of a mild day.

DISTRIBUTION

Range.—The United States and southern Canada, chiefly east of the Great Plains; partially migratory.

The range of the blue jay extends **north** to central Alberta (Stony Plain, Lac la Biche, probably Poplar Point, and Battle River); southern Saskatchewan (probably Prince Albert, Regina, and McLean); southern Manitoba (Fort Ellice, probably Chemawawin, Gypsumville, and West Selkirk); Ontario (Indian Bay, Lac Seul, and Cobalt); New Brunswick (Restigouch Valley and Bathurst); and southeastern Quebec (Magdalen Islands). The **eastern** limit of the range extends southward along the Atlantic coast from southeastern Quebec (Magdalen Islands) to southern Florida (Miami). **South** to southern Florida (Miami and Fort Myers) and west along the Gulf coast to southern Texas (Houston and Atascosa County). **West** to central Texas (Atascosa County, Waco, and Decatur); Oklahoma (Norman); eastern Colorado (Lamar and Wray); eastern Wyoming (Torrington); western North Dakota (Killdeer Mountains and Charlson); and Alberta (Red Deer and Stony Plain).

While the blue jay is generally resident, it partly withdraws during some winters from the extreme northern parts of the summer range. It has been recorded in winter north to southern Alberta (Red Deer); southern Manitoba (Lake San Martin); northern Michigan (McMillan); southern Ontario (Plover Mills, Toronto, and Ottawa); southern Quebec (Montreal and Bary); and Maine (Foxcroft and Machias).

The range as outlined is for the entire species and is occupied largely by the northern blue jay *(C. c. bromia)*. The southern blue jay *(C. c. cristata)* is found in the Southeastern United States (except the southern half of the Florida Peninsula) north to North Carolina and west to Louisiana, while the lower part of Florida is occupied by Semple's blue jay *(C. c. semplei)* *C. c. cyanotephra* is found from eastern Colorado and Nebraska to northern Oklahoma and the panhandle of Texas.

Migration.—Because of the fact that as a species the blue jay is found in winter throughout most of its breeding range, dates of arrival and departure of migrating individuals are difficult to obtain. A migratory movement is, however, evidenced in autumn, when troops of jays may be seen working southward through the trees, while a corresponding northward movement may be detected in spring. More positive evidence of the partially migratory habits of this species is found among the recovery records of banded individuals. In the files of the Fish and Wildlife Service there are many cases that show definite fall travel in the year of banding from Massachusetts to North Carolina; from New York to Virginia; from New Jersey to Virginia, North Carolina, and South Carolina; from Ohio to Alabama; from Wisconsin to Arkansas; from Minnesota to Missouri, Arkansas, Oklahoma, and Texas; and from South Dakota to Oklahoma, and Texas. Data illustrative of the spring movement are not so numerous, but records are available showing spring flights from North Carolina to New York; from the District of Columbia to Rhode Island; from New York to New Brunswick; from Massachusetts to Prince Edward Island; and from Iowa to northern Wisconsin.

Casual records.—The blue jay appears to be extending its range westward, as there are several records for the vicinity of Denver, Colo., that have accumulated during recent years. There are a few records for the southern part of Newfoundland made during the period from the last of June to the last of September. Apparently a specimen was taken at Moose Factory in northern Ontario, in 1862, while at Fruitland, N. Mex., one was seen on October 17, 1908, and three were noted the day following.

Egg dates.—Florida: 69 records, March 17 to August 29; 35 records, April 11 to May 11, indicating the height of the season.

Illinois: 62 records, April 18 to July 12; 32 records, May 5 to 26.

Kansas: 29 records, April 28 to July 29; 15 records, May 23 to June 12.

Massachusetts: 54 records, April 30 to June 17; 28 records, May 14 to 28.

Minnesota: 20 records, April 27 to June 11; 10 records, May 12 to 22.

Nova Scotia: 9 records, May 6 to June 16.

South Carolina: 14 records, April 5 to June 1; 8 records, April 20 to May 18.

Texas: 6 records, April 5 to June 18.

CYANOCITTA CRISTATA CRISTATA (Linnaeus)

SOUTHERN BLUE JAY

PLATE 11

HABITS

Dr. Oberholser (1921) has shown that the range of this race has been found to extend much farther north, including South Carolina, the type locality of *Cyanocitta cristata cristata* (Linnaeus), which, of course, necessitates the relegation of the subspecific name *florincola* and the common name Florida blue jay to synonymy. This would leave the northern blue jay without a name, for which Dr. Oberholser proposed the subspecific name *bromia*. The 1931 Check-list admits this extension of range as far north as North Carolina but entirely ignores the fact that the name *C. c. cristata* (Linnaeus) was based on Catesby's bird, which undoubtedly represents this southern race.

I prefer to use Arthur H. Howell's (1932) names, *Cyanocitta cristata cristata* (Linnaeus) and "southern blue jay," for this race. He says that its range covers approximately the northern half of the State, at least as far south as Volusia and Lake Counties, where it probably begins to intergrade with the extreme southern form, *semplei*. W. E. Clyde Todd (1928), who described *semplei,* seems to think that specimens taken north of the Everglades are not typical of either race, thus allowing a large area of intergradation.

This race is smaller than the northern race, with coloration paler and duller and with the white tips of the greater wing coverts, secondaries, and tail feathers smaller.

Nesting.—The nesting habits of the southern blue jay are not very different from those of the northern blue jay, with due allowance made for the difference in environment. Major Bendire (1895) writes:

Two nests found by Dr. Ralph were placed in low, flat pine woods, 25 and 30 feet, respectively, from the ground; these were composed of twigs, Spanish moss, pine needles, and pieces of cloth, and lined with fine roots. In some of the nests the material were cemented with mud. A third nest was placed in an orange tree standing within a few feet of a house, near the banks of the St. John's River, about 20 feet from the ground; it was composed of twigs, catkins, plant fibers, weeds, grasses, pieces of string, and a little Spanish moss, and these materials were cemented together with mud; the lining consisted entirely of wire grass *(Aristida).* Another nest was placed among some small branches at the end of a limb of an orange tree, about 11 feet from the ground, and was composed of similar materials outwardly, but no mud was used in its construction, and it was thickly lined with fine rootlets of the orange tree.

The average measurement of two nests is about 8 inches in outside diameter by 4 inches in depth, the inner cup measuring about 4 inches in diameter by 2¼ inches in depth.

Mr. Howell (1932) says that the "nests are placed in trees—commonly oak, orange, or pine—at a height of 8 to 35 feet above the ground"; and he quotes D. J. Nicholson as saying that two or three broods are raised in a season, beginning late in March and ending in August. There is a set of eggs in my collection, taken in Leon County, Fla., from a nest in a magnolia tree; it was made of materials similar to those mentioned above, including the mud.

In Texas, according to George F. Simmons (1925), the nests are placed in various oaks, hackberry, pecan, cedar elm, and cedar trees. Wayne (1910) says that in South Carolina it seems partial to live oaks.

Eggs.—The eggs of the southern blue jay are indistinguishable from those of its northern relative and probably would show in a large series all the wide range of variation in shape, color, and pattern exhibited in eggs of the northern blue jay. According to Mr. Nicholson (Howell, 1932), "the first sets nearly always comprised 4 eggs, the second sets either 3 or 4, the third sets nearly always 3." The measurements of 40 eggs average 27.1 by 20.4 millimeters; the eggs showing the four extremes measure 29.6 by 20.7, 27.4 by 21.3, and 23.2 by 18.4 millimeters.

Food.—The feeding habits of the southern blue jay are similar to those of the northern bird; it lives on such varieties of nuts, wild and cultivated fruits, grains, insects and their larvae, and other small forms of animal life as it can find within its range. It is said to do some damage to small cultivated fruits and to rob birds' nests of their eggs and young.

Dr. Walter P. Taylor tells me that in Walker County, Tex., these jays are useful in providing food for bobwhite quail, by dropping pieces of acorns that they have broken up.

Behavior.—Wherever I have been in Florida I have been impressed with the fact that the local blue jays are among the commonest and most familiar birds. They are not at all shy and seem to enjoy living in the towns and villages, in the gardens and trees close to houses, in the shade trees along the streets, and in the citrus groves, as well as in the open country. Mr. Howell (1932) says that they "are found less commonly in pine woods, hammocks of oak or mixed timber, turkey-oak scrub, and the borders of small cypress swamps. The birds are noisy and restless during the greater part of the year, moving about in small companies or loose flocks, calling vigorously as they go. While for the most part indifferent to the presence of man, they nevertheless retain a degree of caution and can scarcely be tamed enough to eat from one's hand, as can the Florida Jays. They take great delight in worrying owls whose retreats they may discover, and their reputation for robbing the nests of smaller birds is rather bad."

M. G. Vaiden writes to me from Rosedale, Miss.: "The blue jay is a very destructive bird to other small birds' nests in this area; many times have I seen the blue jay in the act of destroying cardinal, mourning dove, mockingbird, and Maryland yellowthroat nests. My notes give an instance of the determined and persevering way these birds have when bent upon the destruction of another species' nest. On May 5, 1928, a blue jay was caught in the act of robbing a cardinal's nest located in a crape-myrtle bush in my front yard. He was sitting on the nest with one egg in his bill when first noticed. I secured a 22 rifle and shot very close to the jay; he flew away with the egg, yet returned in a short while for another; and three shots were made as close as possible to the bird, with no intention of hitting it, before it would leave the nest locality. After a short interval it returned again and five shots were made, each shot hitting very close to the bird, yet it hopped to the cardinal's nest and secured another egg. The next shot made it leave, but it carried the second egg along, as with the first. I awaited his return, fired one shot which failed to make him fly away, so I proceeded to kill the bird with the next shot. The cardinals, both, produced only mild scolds toward the jay and failed at any time to put up a fight. The nest was left with the two eggs remaining, but the cardinals quit the nest after the following day."

CYANOCITTA CRISTATA SEMPLEI Todd

SEMPLE'S BLUE JAY

My friend John B. Semple collected some blue jays in southern Florida and sent them to the Carnegie Museum. Based on this material from extreme southern Florida, some 11 specimens, W. E. Clyde Todd (1928) described this bird as a new subspecies and named it in honor of Mr. Semple, who collected the type near Coconut Grove. He characterizes it as "similar to *Cyanocitta cristata cristata* (Linnaeus) of the South Atlantic and Gulf States, but general coloration paler, the under parts white, with less grayish suffusion, the lower throat with less bluish wash, and the upper parts paler and duller blue, with less purplish tone." He continues:

This new form is as much different from *C. c. cristata* as the latter is from the northern race *C. c. bromia.* Its pale coloration stands out well as the two series lie side by side. While occasional specimens from peninsular Florida (north of the Everglades) approximate in their pallor the characters above specified, it is only in the extreme southern part of the State that these characters become sufficiently constant and pronounced to justify giving a name to individuals showing them. * * *

In the average example of *cristata* the upper parts are "deep dull bluish violet No. 2" of Ridgway (as seen with the eye between the bird and the light), while the pileum is brighter, between "grayish blue violet No. 2" and "dull bluish violet No. 2." In the new race these parts are respectively "deep madder blue" and "deep plumbago blue."

Arthur H. Howell (1932) gives the range of this race as "southern and central Florida, from Osceola and Hillsborough Counties south to Key West."

Its habits seem to be similar in every way to those of the other Florida race.

CYANOCITTA CRISTATA CYANOTEPHRA Sutton

WESTERN BLUE JAY

Based on a study of some 49 specimens of blue jays from Colorado, extreme western Oklahoma, and Kansas, Dr. George M. Sutton (1935) named this pale western race and described it as "similar to all races of *Cyanocitta cristata* found to the eastward of the Mississippi, but coloration paler, especially on the crest and back; paler even than *C. c. semplei* Todd, from which it differs also in being decidedly larger and relatively smaller-billed; and much paler than birds from Michigan; Minnesota; Ontario and southeastern Canada; and the northeastern United States. White markings of wings and tail noticeably more extensive than in *semplei,* and somewhat more extensive than in breeding birds from Georgia, Louisiana, and northern Florida." He says further:

All available Minnesota specimens are far too dark for the present race; Manitoba specimens apparently tend to be a trifle paler than eastern Canadian birds; and a single mile from Alberta (Lac la Nonne, June 28, Canadian National Museum No. 21512) is decidedly paler than any other Canadian specimen at hand, especially on the crest.

It is my present belief that the most typical examples of *cyanotephra* are to be found in extreme western Oklahoma, where the Blue Jay is decidedly rare as a breeding species, in eastern Colorado; in western Kansas; and in the northwestern corner of the northern Panhandle of Texas; but that the race ranges throughout Kansas and northern Oklahoma (save in treeless regions); throughout Nebraska (save presumably in the northeastern part where the race found in Minnesota should occur); and along the eastern foothills of the Rocky Mountains to the northwestward of Nebraska.

CYANOCITTA STELLERI STELLERI (Gmelin)

STELLER'S JAY

HABITS

Jays of the *stelleri* group are widely distributed in western North America and Central America, from the Rocky Mountains to the Pacific coast and from the Alaska Peninsula southward to Nicaragua. They are the crested blue jays of this vast region, where they replace our familiar blue jay of the East and share many of its interesting habits and some of its bad manners. The subject of this sketch is the northern race, extending its range in the Pacific coast region only as far south as Washington. There are five other races that are found north of the Mexican boundary.

Steller's jay is the oldest known race of this species, named by Gmelin in 1788, yet after more than 150 years it is far from being the best-known subspecies. It was known by description to all the early writers on American ornithology and was figured by Swainson and Richardson (1831), Wilson and Bonaparte (1832), and Audubon (1842). Bonaparte says, in his continuation of Wilson's "American Ornithology," that "it is mentioned by Pallas as having been shot by Steller, when Behring's crew landed upon the coast of America. It was first described by Latham from a specimen in Sir Joseph Bank's collection, from Nootka Sound."

The haunts of Steller's jay are chiefly in the coniferous forests of southern Alaska, British Columbia, and Washington as far south as the Columbia River, where it begins to intergrade with the subspecies *carbonacea*. But it is not wholly confined to the forests, as it often ventures out into the clearings, orchards, and farms on its mischievous raids for food. Bendire (1895) says: "It is usually a constant resident and breeds wherever found. It is an inhabitant of the canyons and pine-clad slopes of the higher mountains, and is not as often seen in the deep forests as on their outskirts near water courses."

Nesting.—We waited nearly 3 weeks for our ship to sail from Seattle to the Aleutian Islands, but we made our headquarters in the meantime at the little town of Kirkland, across Lake Washington from the city, and spent our time profitably by collecting in the vicinity. At that time much of this region was heavily wooded with a primeval forest of lofty firs, but the greater part of it had been lumbered and had grown up to small or medium-sized second-growth firs. Much of it had been cleared and cultivated, with houses and little farms scattered through it. There were two or three species of firs forming the principal forest growth, with a considerable mixture of hemlock and a very handsome species of cedar; the deciduous growth consisted of large alders and some maples and flowering dogwoods. Here we found Steller's jays quite common and discovered several of their nests between April 30 and May 20. The first nest we found, on April 30, was new but still empty; it was placed about 10 feet from the ground against the trunk of a small fir in the coniferous woods. It had a bulky foundation of large sticks, on which was a layer of dead leaves and mud and then a firmly woven, deeply hollowed nest of coarse rootlets. Another nest was 14 feet up in a thick fir in an open situation in the coniferous woods; it held four young only a few days old; the old bird remained on the nest until I almost touched her, when she flew off and scolded me with a mewing squawk. We found one other new nest and several old ones, all similarly located and constructed.

There are three sets in my collection from the same general region. In one case the nest was 10 feet from the ground and 15 feet out on a limb of a small lone fir on the edge of a prairie pond, near some mixed fir and oak growth, and 100 yards from a house. One of the others was 12 feet up in a green spruce; and the nest from which the other was taken was said to have been on a shelf in a woodshed!

Published accounts of the nesting of Steller's jay are not numerous, but the following from D. E. Brown (1930) is worth quoting: "This species usually nests at a moderate height. The majority of nests will be found from eight to fifteen feet from the ground, but the writer has found them only two feet up, and has seen them well over one hundred feet from the ground on the horizontal branches of giant firs.

"In the early part of the season coniferous trees are used almost exclusively. * * * Later when the deciduous trees are in full leaf they are quite often used. This fact is brought out by the number of old nests that are found in the fall when the leaves have shed."

The location of one nest, well within the city limits of Seattle, puzzled him until it was found "only two feet from the ground in the center of a mass of salal bushes and blackberry vines. He continues:

The birds nest regularly in Seattle city parks often on trees or branches that lean over trails that are used by hundreds of people daily. I have seen at least three such nests that were so low that they could be touched with the hand from the trail. * * *

The nest is usually very large and sometimes composed of twigs so large it hardly seems possible that the birds could handle them. A very thick layer of mud weighs down and cements the nest together, and it is lined with rootlets that are worked in while wet. The very start of the nest is always some light colored material such as cedar bark, leaves of the maple tree, shreds of decayed wood or pieces of newspaper. Samuel F. Rathbun of Seattle once found a nest in one of the parks that had a handkerchief worked into its foundation, a variation somewhat unusual in nest material.

Mr. Rathbun has sent me his notes on several nests of Steller's jay, and says: "With one exception all the nests of this jay I have found were placed in coniferous trees, usually firs of not large size, and oftener the location of the tree would be in a rather dense part of the wood. Often the place selected for a nest is the fork formed by several small branches jutting from the trunk wherein are lodged a number of dry, dead leaves, and on these is placed a little platform of twigs that forms the base of the nest, as if the jay attempted to convey the idea that the structure is only some rubbish caught by the branches. In fact, more than once our attention has first been caught by noticing dead leaves in what might be considered an out-of-the-way place, and on a nearer approach the material was seen to be the commencement of a jay's nest.

"As to the height of the nests, the lowest was only 8 feet from the ground, the highest 40, and the average of all 20 to 25 feet. I have always found Steller's jay to be quiet and secretive in the general locality where it was breeding—one would not know that there was a jay anywhere around; but when its nest is disturbed the jay makes a great outcry, and then silently leaves the place."

He gives the dimensions of one nest as follows: Extreme outside diameter, 14 inches; outside height, 6 inches; inside diameter, 5 inches; and inside depth, 2¼ inches.

Eggs.—Steller's jay lays three to five eggs, usually four. These are ovate and only slightly glossy. The ground color is pale greenish blue, or pale bluish green, "pale turquoise green" to "pale Nile blue" or paler, or "pale sulphate green" to "microcline green" or paler, sometimes almost greenish white. Some eggs are more or less sparingly marked with fine dots; others are more or less irregularly spotted with small spots, fine dots and markings of indefinite shape. The markings are in different shades of dark browns or purplish brown, or shades of olive, more or less evenly distributed.

The measurements of 40 eggs average 31.4 by 22.5 millimeters; the eggs showing the four extremes measure **34.5** by 22.8, 33.5 by **24.0**, **27.8** by 22.3, and 30.4 by **20.6** millimeters.

Plumages.—Ridgway (1904) describes the juvenal plumage as follows: "Wings and tail as in adults, but the blue usually more greenish (china blue to cerulean blue) and usually (?) without distinct black bars on secondaries or rectrices; under parts, rump, and upper tail-coverts dull slate-grayish, the former becoming darker and more sooty anteriorly; head and neck plain sooty or dark sooty slate, the forehead without any blue streaks."

Young birds begin the postjuvenal molt late in July or early in August; this molt, which involves everything but the wings and tail, is usually completed during August, but it sometimes continues until after the middle of September. In this first winter plumage young birds are practically indistinguishable from the adult female, having the black barring on the secondaries and rectrices less distinct than in the adult male, or sometimes entirely wanting.

The complete postnuptial molt of adults begins in July and is often completed before the middle of August; I have records of adults in fresh winter plumage as early as August 10. The molts average somewhat later in the more southern subspecies. The sexes are alike in all plumages, except that the females are somewhat smaller and have less distinct bars on the secondaries and tail feathers.

Food.—As the food of the California subspecies has been much more thoroughly studied than that of this northern race, and as the feeding habits of the species probably vary but little in the different portions of its range, this subject has been more thoroughly discussed under the blue-fronted jay. Two reports, however, are worth quoting here. Referring to Vancouver Island, Harry S. Swarth (1912) writes: "At Errington, in September, the jays were exceedingly abundant, particularly about the edges of the pastures and grain fields. Harvesting operations were in progress at this time, and a wheat field near our camp had just been cut and the grain piled in shocks. On those nearest the edges of the field, close to the shelter of the woods, the jays were feeding by scores; when startled most of the birds departed, carrying one or more long straws with them, to be thrashed out at their leisure in the nearby woods. Certain favorite stumps and logs were well covered with straws from which the grain had been eaten."

Ford Dicks (1938) reports considerable damage done by these jays in filbert orchards near Puyallup, Wash., and says: "As a matter of fact, the writer has known of instances where the entire nut crop was lost due to the depredations of Steller's Jays in late summer and early fall, at which time the fruit is approaching maturity."

Behavior.—Although bold in the defense of their nests and rather tame about camps and houses, where their intelligence tells them that they are not in danger, they are very shy in the open woods, much shier than our eastern blue jay, and difficult to approach or shoot when pursued. They often escape by "climbing" some tall spruce or fir, starting on one of the lower branches and hopping or flitting upward from branch to branch around the trunk, as if climbing a spiral staircase, until the summit is reached, when off they go with a derisive scream. At such times their movements are so lively that it is not easy to shoot one. They sometimes travel through the forest in this way, descending from the top of one tree to the lower part of another and so on from tree to tree, until out of sight. The best way to outwit them is to remain well concealed and imitate their notes, to which their curiosity will generally lead one or more of them to respond. They are notorious as nest robbers and seem to be cordially hated and dreaded by the smaller birds, but they are not always guilty of this practice. William L. Finley (1907) says of a pair that he watched: "If this pair of jays carried on their nest robbing, they did it on the quiet away from home, for in the thicket and only a few yards away I found a robin's nest with eggs, and the nest of a thrush with young birds. Perhaps the jays wanted to stand well with their neighbors and live in peace. I am sure if the robins had thought the jays were up to mischief, they would have hustled them out of the thicket. I think we give both the crow and the jay more blame for nest robbing than they deserve."

Alfred M. Bailey (1927) writes:

They are robbers of the first order, and steal anything edible about camp. I do not know whether we are able to give birds credit for a sense of humor, but if we do, then the Jays surely must come in for first place. I have watched a pair of these fellows tease a spaniel. They would alight in a path, only to be chased away by the dog, and they kept returning so often as to completely exhaust him; then, when the dog refused to chase them longer, they would alight over his head and talk to him,—undoubtedly they were cursing him, until he finally got up and walked away. The same performance was carried on daily. This species is not particularly in favor among hunters, for when one is quietly crossing a muskeg in the hope of jumping a deer, it is the usual thing to have a couple of Jays open a serenade, and then keep just ahead of the hunter, talking all the time.

Voice.—Dawson and Bowles (1909) give the best description of the varied notes of this jay, as follows: "The notes of the Steller Jay are harsh and expletive to a degree. *Shaack, shaack, shaack* is a common (and most exasperating) form; or, by a little stretch of the imagination one may hear *jay, jay, jay.* A mellow *klook, klook, klook* sometimes varies the rasping imprecations and serves to remind one that the Jay is cousin to the Crow. Other and minor notes there are for the lesser and rarer emotions, and some of these are not unmusical."

Leslie L. Haskin writes to me that, like so many other jays, it has a scream like that of the red-tailed hawk, which may be a true jay note rather than an imitation of the hawk. He says further: "Steller's jay also has a true song of his own. I have heard it only a few times, but it is very sweet in tone. In many ways it resembles the 'whisper songs' that many birds indulge in in winter. Because of the extreme shyness of these birds, and the softness of the song, it is very hard to hear. Only when the bird is entirely unaware of observation will it be given. I would compare it with the 'whisper song' of the American robin, as I have heard that bird on cold winter days singing in red cedar tangles in the East. In it are also some tones that suggest the song of the ruby-crowned kinglet, but not so loud. Heard without seeing the performer, it could easily be mistaken for the kinglet. Altogether it is a very interesting and surprising performance."

Theed Pearse writes to me that he has heard it mimicking the crow's spring falsetto song, as well as the cry of the red-tailed hawk; and has heard one "really singing a song of its own, and a very delightful one; I could not recognize any other bird's notes, except perhaps the trill of the junco. When first heard it was a song that could not be identified as that of any local species, a strong warble, consisting of various notes with some trills; one feature was the number of different notes that the bird could go through without repetition."

Mr. Rathbun tells me that one of this jay's notes "is a gritting, rasping one, as rough as the edge of a saw," and unlike any other of the bird's notes. Dr. Samuel S. Dickey (MS.) adds the following to the bird's vocabulary: "Ordinarily the birds vented raucous, blue jay-like *cahs,* but they would vary such outbursts with *kirk-kirk, kirk-perk, perk-er, perk-er,* or wheezy magpie-like notes, such as *ca-phee, ca-phee, pheeze-ca.*"

Fall.—These jays are supposed to be resident all the year round in the region where they breed. They probably do not make any regular migration, although Mr. Swarth (1922) says that, late in August, "at Sergief Island many were seen, under circumstances suggesting migration. They were frequently in small gatherings, seven or eight together, and often on tidal marshes, far from timber, apparently traveling in a definite direction. When thus seen they were flying by easy stages from one drift log to another, in a southerly direction."

During fall and winter they are given to erratic wanderings, probably in search of food, throughout the open country and about the farms and villages. They may be very common during some seasons at certain places and scarce or entirely absent there at other seasons. Though quiet, retiring, and secretive during the nesting season, they are much more noisy, bolder or tamer, and more aggressive during the fall and winter, traveling about in family parties or small groups.

J. A. Munro tells me that Steller's jays were very abundant on Vancouver Island during the fall of 1913; on September 30 they were industriously carrying acorns from the Garry oaks; one collected on this date had two in its gullet and one in its bill. During the winter of 1921-22 they were also unusually numerous; several hundred were caught in quail traps. On February 1, 1923, there was an invasion of these jays at Victoria; 50 were strung on a wire at the game farm, and the operator mentioned catching seven at one time in a quail trap.

DISTRIBUTION

Range.—Western North America and Central America south to Nicaragua; nonmigratory.

The range of Steller's jay extends **north** to Alaska (Lake Aleknagik and Northeast Bay); British Columbia (Flood Glacier, Poison Mountain, Parsnip River, Moose River, and Yoho Park); and Montana (McDonald Lake and Big Snowy Mountains). **East** to central Montana (Big Snowy Mountains); Wyoming (Yellowstone Park and Torrington); central Colorado (Fort Collins and Colorado Springs); New Mexico (Halls Peak, Capitan Mountains, and Guadalupe Mountains); western Texas (Guadalupe Mountains and Davis Mountains); Chihuahua (Tomochic and Pinos Altos); Durango (La Ciénaga de las Vacas and Arroyo del Buey); Veracruz (Mirador and Orizaba); Honduras (Seguatepeque and San Juancito); and southeastern Nicaragua (Greytown). **South** to Nicaragua (Greytown); El Salvador (Chalatenango); and Guatemala (Tecpam and Volcán de Fuego). **West** to western Guatemala (Volcán de Fuego and Quelzatenango); western Oaxaca (Cieneguilla); western Sonora (Sonoyta); casually northern Baja California (San Pedro Mártir Mountains); western California (Palomar Mountains, Santa Barbara, Santa Lucia Mountains, San Geronimo, and Turner); western Oregon (Pinehurst, Prospect, and Dayton); western Washington (Camas, Grays Harbor, and Seattle); western British Columbia (Nootka Sound and Massett); and Alaska (Baranof Island, Sitka, and Lake Aleknagik).

The range as outlined is for the entire species, which has been separated into eight currently recognized geographic races. The northern part of the range along the Pacific coast (except the Queen Charlotte Islands) is occupied by the typical race *(Cyanocitta s. stelleri)* from the Alaskan Peninsula south, probably to northwestern Oregon; the Queen Charlotte jay *(C. s. carlottae)* is found only on the Queen Charlotte Islands, British Columbia; the coast jay *(C. s. carbonacea)* occupies the coastal zone from northern Oregon south to the Santa Lucia Mountains and Napa Valley in California; the blue-fronted jay *(C. s. frontalis)* is found from the Mount Shasta region of northeastern California, south through the Sierra Nevada and San Bernardino Mountains to northern Baja California; the black-headed jay *(C. s. annectens)* occurs in eastern British Columbia and south through the Rocky Mountains to Wyoming, Idaho, and casually northern Utah; the long-crested jay *(C. s. diademata)* is the form found in the Rocky Mountain system from northern Utah and southern Wyoming south to central Mexico; the Aztec jay *(C. s. azteca)* is found in south-central Mexico; and the blue-crested jay *(C. s. coronata)* occurs from southern Mexico south to Nicaragua.

The Nevada crested jay *(percontatrix)* is found in the Sheep and Charleston Mountains, Clark County, Nevada.

Casual records.—Among the cases where Steller's jays have been recorded outside the normal range are the following: There are a few records for western Nebraska, some of which have been recorded as the race *annectens,* but two specimens, one taken at Mitchell in October 1916 and the other at Oshkosh on March 5, 1920, proved to be *diademata.* A specimen *(annectens)* was taken at Indian Head, Saskatchewan, on May 24, 1923; one *(diademata)* was shot in Lincoln Park, Chicago, Ill., on June 12, 1911; and a specimen *(diademata)* was taken at Cap Rouge, Quebec, on November 8, 1926.

Egg dates.—Alaska: 3 records, May 12 (2) and July 7.

California: 103 records, April 12 to June 24; 51 records, April 30 to May 15, indicating the height of the season.

Colorado: 29 records, April 23 to June 3; 15 records, May 8 to 24.

New Mexico: 4 records, April 27 to June 6.

Oregon: 16 records, April 4 to June 5; 8 records, April 11 to May 4.

Washington: 20 records, April 1 to June 20; 10 records, April 19 to May 3.

CYANOCITTA STELLERI CARLOTTAE Osgood
QUEEN CHARLOTTE JAY

This island form was named by Dr. Wilfred H. Osgood (1901) and described as "similar to *C. stelleri,* but larger and darker colored; abdomen and flanks deep Berlin blue instead of Antwerp or China blue as in *C. stelleri;* frontal spots much reduced; black of head extending on breast and merging into blue of abdomen without sharp demarcation. * * *

"The large size and dark color of this jay were noticed in the field, and subsequent comparison of specimens in the museum showed these characters to be amply sufficient to distinguish it from the mainland form *C. stelleri.* * * * Jays are not very common on the islands. They were seen only occasionally and were generally in family parties of four to six adults and young" [June 13 to 25].

Almost nothing seems to have been published on the habits of this jay, as very little ornithological work has been done on the Queen Charlotte Islands, to which the subspecies seems to be confined. Though I have no notes on the subject I have no reason to think that its habits differ in any respect from those of the mainland from, which lives in a similar habitat. So far as I know, there are no authentic eggs of this race in collections.

Clyde A. Patch (1922) found it fairly common on Graham Island, "usually moving about in family parties. Frequently seen feeding on green fruit of the Skunk Cabbage which they manage to remove from its stem and carry to a comfortable spot on a trail, roadway or log. On one occasion a Jay was observed to capture a young wood mouse."

CYANOCITTA STELLERI CARBONACEA Grinnell

COAST JAY

The coast jay, or Grinnell's jay, as Ridgway (1904) calls it, is very evidently quite intermediate in characters between *stelleri* on the north and *frontalis* on the east and south, for Dr. Grinnell (1900b) in naming it described it as "intermediate in size * * * between *C. stelleri* and *C. stelleri frontalis*. Dorsal surface sooty-black as in *stelleri*, but with blue on forehead nearly as extended as in *frontalis*. Tint of blue of posterior lower parts paler than in *stelleri*, and extending further forward into pectoral region, as in *frontalis*." Furthermore, its range, the "coast region of Oregon and California, from the Columbia River south to Monterey County," is just where one would expect to find the two previously named forms to intergrade. This is just another case, like *annectens*, where the naming of an intermediate immediately produces two more sets of intergrades. Dr. Grinnell (1900b) says further: "*C. stelleri annectens* from Idaho resembles *carbonacea* somewhat closely, but the white spot over the eye distinguishes both *C. s. annectens* and *C. s. macrolopha* [= *diademata*] of the Northern and Southern Rocky Mountain regions, respectively, from the parallel Pacific Coast races, *carbonacea* and *frontalis*, neither of which has any trace of such a marking."

What has been written about the nesting habits, eggs, plumages, food and general habits of the other Pacific coast races of the species would apply very well to this subspecies. The measurements of 40 eggs average 30.6 by 22.5 millimeters; the eggs showing the four extremes measure 33.1 by 22.4, 31.1 by 24.0, and 26.6 by 21.2 millimeters.

CYANOCITTA STELLERI FRONTALIS (Ridgway)

BLUE-FRONTED JAY

PLATE 12

HABITS

This is the "crested blue jay" of the Sierra Nevadas and the inner coast mountain ranges of northern California. Ridgway (1904) gave it the common name of Sierra Nevada jay, which seems a more appropriate designation than blue-fronted, as the blue stripes on the forehead are not conspicuously more prominent than in some of the other races of the species. He describes it as "much lighter colored, and average size decidedly less" than in *C. s. carbonacea*, which, in turn, he calls "paler throughout and averaging slightly smaller" than *C. s. stelleri*.

The chosen summer haunts of the blue-fronted jay are in the coniferous forests of the Transition and Canadian Zones of mountain ranges, mainly in California. In the Lassen Peak region, according to Grinnell, Dixon, and Linsdale (1930), these jays "in summer were found in and about clumps of closely growing small coniferous trees, often as forming dense thickets of undergrowth in old forest that is thinning out, or at edges of forests. Kinds of trees that formed such suitable clumps and which were frequented by the jays were white fir, red fir, yellow pine, and hemlock. At the western frontier, the occasional pairs seen were usually in tracts of small yellow pines. The birds were seen most often at heights of close to four meters above the ground."

The range of this jay in the mountains extends upward to about 8,000 or 9,000 feet, or to the lower limit of the Hudsonian Zone, where Clark's nutcracker is found. It finds the lower limit of its range where the coniferous mountain forest gives way to the foothill oaks and chaparral; here it mingles to some extent with the California jay, but sticks mainly to the pines.

Professor Beal (1910) says: "It sometimes ventures to the edges of the valleys and occasionally visits orchards for a taste of fruit, of which it is very fond, but in general it keeps to the hills and wilder parts of the canyons. It is fond of coniferous trees and is likely to be found wherever these abound. Where ranches have been established far up the canyons among the hills, this jay visits the ranch buildings."

Nesting.—J. Stuart Rowley writes to me: "I have located many nests of this bird in the Sierra Nevadas, from Tulare County in central California to San Bernardino County in southern California. The most frequently used nest site seems to be a young conifer, with the nest placed about 10 feet up near the main trunk and supported by horizontal branches. The incubating females are rather close sitters and make quite a fuss when flushed from the nests."

W. E. Griffee sends me the following note: "While cruising timber in the lower Sierras, about 20 to 30 miles east of Placerville, Eldorado County, Calif., I found several nests of this subspecies. All were high on dry hillsides in rather dense reproduction of ponderosa and sugar pines and incense cedar, at elevations of 8 to 12 feet from the ground. Nests were, of course, easy to see and readily accessible, but to find them, had I not been climbing over the timbered hills as a part of my work, would have required a tremendous amount of walking."

J. G. Suthard tells me that he found a nest containing four fresh eggs on April 21, 1940, in the San Bernardino Mountains. "The nest was situated 8 feet up in a willow along a mountain stream at 9,000 feet elevation. It was shielded from view by a cluster of branches growing up from the slanting trunk of the willow. There were plenty of pine and fir trees in the vicinity, but the jays seemed to prefer the willows, which are possibly less frequented by squirrels. At the time, there were numerous patches of snow along the stream, and a few hundred yards higher the whole range was completely blanketed in white. The willows were still dormant. As will be noted from the photograph (pl. 12), the nest is lined with pine needles, which are those of the Jeffrey pine (*Pinus jeffreyi*)."

Rollo H. Beck wrote to Major Bendire (1895): "I have found about a dozen of their nests, placed in oaks, buckeye, laurel, and holly bushes, at various distances ranging from 7 to 40 feet from the ground." Some unusual, or unexpected, nesting sites have been recorded. Col. N. S. Goss (1885) found quite a number of the nests near Julian, Calif., "and in all cases but one in holes and trough-like cavities in trees and stubs, ranging from four to fifty feet from the ground, generally ten to twenty feet up. The nest found outside was built upon a large horizontal limb of an oak close beside a gnarl, the sprout-like limbs of which thickly covered the nest overhead, and almost hid it from view below. * * * The nests are quite bulky, made loosely of sticks, stems of weeds, and lined with fibrous rootlets and grasses, and as they are all built at or near the opening, the tell-tale sticks project and make the findings of their nests an easy matter."

Walter E. Bryant (1888), on information received from A. M. Ingersoll, writes:

A strange departure from the usual habits of jays was noticed in Placer County, Cal., where they had persisted in building within the snow-sheds in spite of the noise and smoke of passing trains. The destruction of their nests by the men employed on the water train, which makes two trips a week through the sheds during the summer, sprinkling the woodwork and tearing down the nests of jays and robins with a hook attached to a pole, seemed not to discourage them. So accustomed do the jays become to the passing of trains, that they will often remain on their nests undisturbed. In one season more than two hundred nests of jays and robins were destroyed, so the train men say, between Cisco and Summit, a distance of thirteen miles.

These, like all jays, are very secretive in their nesting activities and use the greatest stealth in approaching the nest while building it or when it contains eggs or young. But Grinnell and Storer (1924) were able, under favorable circumstances, to observe a pair building their nest. They say:

One of the jays was seen to fly into a black oak, obtain a twig, and carry it off, upward, through the adjacent trees to the nest site, at the top of a yellow pine, fully 40 feet above the ground. Then the other member of the pair came, broke off a twig, dropped it, evidently by accident, and sought another. * * * Pieces dry enough to break off readily, and a little longer than the jay's body, were chosen, and twisted off by a wrench with the bill. The twig would be worked along between the mandibles until held across the middle and then the jay would ascend by the usual vertical hopping and short flights to the nest. Following the taking of black oak twigs the two jays, together, flew across the river which flowed close by the nest tree, and there, descending quickly to the ground, sought material in an azalea thicket at the edge of the water. Each took a quantity of twigs and grass and apprently also some mud, and flew again to the nest tree. Again they took twigs from the black oak.

They say of a nest examined in the Yosemite region:

It was solid in construction, with a large external basal framework of dead and more or less weathered twigs of irregular shape and small diameter (2 millimeters or less). Many of these were black oak twigs while others were of a very furry herbaceous plant. All of the material of this outer framework, as was attested by the clean, fresh-appearing ends of the pieces, had been freshly broken off by the jays. This suggests that, save for the small amount of herbaceous material, all the outer constituents were gathered above the ground. The outside framework measured about 300 millimeters (12 inches) in one direction and 400 millimeters (16 inches) in the other.

The inner cup of this nest was composed of dry needles of the yellow pine, held together by enough mud to give the structure a firm resistant feel. The mud, however, did not extend to the inner surface. The interior of the cup consisted solely of pine needles, which crossed and recrossed so as to make a porous interior lining. This cup was 100 millimeters (4 inches) in diameter at the rim and 68 millimeters (2⅝ inches) deep at the center.

Eggs.—The blue-fronted jay lays three to five eggs, usually three or four; Mr. Rowley, who has examined a number of nests, tells me that sets of four are found as frequently as three but that he has found only one set of five. The eggs are usually indistinguishable from eggs of Steller's jay, but Rollo H. Beck (1895) describes some variations, as follows: "In a series of these eggs now before me there is considerable variation in shape and markings. One set closely resembles those of the California Thrasher, another is marked exactly like the eggs of the Yellow-billed Magpie, and others the eggs of the California Jay. Some have but few spots, principally about the larger ends, while others have the ground color nearly obscured, so thickly are they spotted. The usual ground color is light-blue, which is spotted with various shades of brown and not infrequently with lavender and purple."

The measurements of 52 eggs in the United States National Museum average 30.22 by 22.61 millimeters; the eggs showing the four extremes measure 34.0 by 24.0, 27.6 by 23.2, and 29.5 by 21.2 millimeters.

Young.—Bendire (1895) says that "an egg is deposited daily, and incubation lasts about sixteen days. The male assists in these duties, and usually but one brood is raised in a season." Grinnell and Storer (1924) say: "During the nesting season the jays are to be seen in devoted pairs, and after the broods leave the nest the full-grown young and their parents remain for a time in family parties. With the coming of fall, the parental and filial instincts wane, these family parties break up, and the individuals scatter out rather uniformly through the forest."

Food.—Prof. F. E. L. Beal (1910), in his study of the food of this jay, examined 93 stomachs and found that the animal food amounted to 28 percent and vegetable matter to 72 percent. The animal food consists largely of insects; beetles, a little more than 8 percent; Hymenoptera, about 11 percent, the largest item of animal food; grasshoppers and crickets, about 3.5 percent; caterpillars and moths, a little more than 2 percent; other insects were found only in insignificant amounts. Of the Hymenoptera, he says: "They were found in 30 stomachs altogether, and 2 were entirely filled with them. Ants were found in only 2 stomachs. Three honey bees were identified, one in each of three stomachs. One was a worker, another a drone, and the third indeterminate. None of the smaller parasitic Hymenoptera were identified. The greater part of this item of food consisted of wasps and wild bees, which would indicate that this bird is an energetic and expert insect catcher." Miscellaneous creatures identified were spiders, sowbugs, raphidians, hair and skin of a mammal, "two bits of bone, probably of a frog," and eggshells were found in 13 stomachs. "Only 6 of these egg-eating records occurred in June, the nesting month. All the rest were in September or later and were probably old shells picked up in abandoned nests or about ranch buildings or camp grounds."

Of the vegetable food, "fruit amounts to 22 percent and was found in 55 stomachs. Prunes were identified in 2 stomachs, cherries in 2, grapes in 2, Rubus fruits in 15, strawberries in 1, elderberries in 15, bay laurel fruit in 1, unknown wild fruit in 2, and fruit pulp, not fully identified but thought to be of cultivated varieties, in 16 stomachs. Thus 38 stomachs held fruit supposed to be cultivated. This number contains all containing Rubus fruits, which probably were not all cultivated— perhaps none of them were. * * *

"Grain amounts to 5 percent, and was found in 15 stomachs, distributed as follows: Wheat in 7, oats in 9, and barley in 1. * * * The chief food of this jay, however, is acorns, though occasionally it eats other nuts or large seeds. Mast amounts to 42.5 percent of the yearly diet, and was found in 38 stomachs. * * * In October and November it amounted to 76 percent, in December to 90, and in January to 99 percent."

He considers the economic status of this jay as of minor importance:

In destroying beetles and Hymenoptera it performs some service, but it destroys only a few. Of the order of Hemiptera, which contains most of the worst pests of the orchardist and farmer, it eats scarcely any. The Orthoptera, which are almost all harmful insects, are eaten only sparingly, and the same applies to the rest of the insect food. The destruction of birds' eggs is the worst count against the jay. But none were found, except in June, until September, when it was too late in the season for fresh eggs to be obtainable. In June 17 birds were taken, and 6 of them, or 35 per cent of the whole, apparently had robbed birds' nests. Now, it is evident that if 35 per cent of all the Steller jays in California each rob one bird's nest every day during the month of June the aggregate loss is very great.

So far as its vegetable food is concerned, this bird does little damage. It is too shy to visit the more cultivated districts, and probably will never take enough fruit or grain to become of economic importance.

In his paper on Modoc County birds, Joseph Mailliard (1927) writes:

In September, 1924, this jay was so numerous in Eagleville as to be a pest in the many small apple orchards of the settlement. These orchards are small, for home supply only, and the inroads made by the jays upon the apple crop assumed serious proportions. With the crop limited as it was by the drought of that year, the owners of such orchards as were bearing fruit waged incessant warfare upon the jays, both of this species and of the following one. Hundreds were shot, but those that were left soon became expert in dodging their pursuers and the slaughter lessened.

In fall and winter, while wandering about in the foothills and valleys, these jays become quite omnivorous, picking up any scraps of food, bread, crackers, meat, or anything edible, that they can find around the camps or ranches; what they cannot eat on the spot they carry off and hide; they have even been known to steal a piece of soap. They probably store some acorns and other nuts for future use and are suspected of robbing the stores of the California woodpecker.

Behavior.—There seems to be nothing in the behavior of the blue-fronted jay that differs materially from that of other races of the species, to which the reader is referred.

Voice.—Its vocal performances are apparently similar also to those of other races, though some different descriptions of its various calls have appeared in print. Ralph Hoffman (1927) says that "besides the ringing *tchek*, a little lower in pitch than the cry of the California Jay and generally given in flight, the Crested Jay utters from its perch a loud *kweesch, kweesch, kweesch*. It has besides a deeper *chu-chu-chu* and a note resembling a squeaking wheelbarrow, *kée-lu, kée-lu*. * * * Occasionally from the cover of dense foliage, it utters a formless succession of liquid, pleasing notes quite unlike its usual discordant notes, or a purring or rolling note."

Grinnell and Storer (1924) give slightly different renderings of what are apparently the same as the above notes, and add that "when two jays of a pair are hunting close together a low crackling or growling *ker'r'r'r'* is uttered."

Field Marks.—Any of the jays of the *stelleri* group may be easily recognized by the long, brownish-black crest, so conspicuous at all times and giving the bird an entirely different outline from that of the flat-headed jays of the genus *Aphelocoma*. The dark brownish-black head, neck, upper breast, and upper back, contrasting with the blue of the lower back and abdomen, are also distinctive; and the blue of the wings and tail is conspicuous in flight.

Winter.—In winter these jays desert to a large extent their summer haunts in the mountain forests and wander about in the foothills and valleys, visiting camps and ranches in search of food. John G. Tyler (1913) says that "during the winter of 1900-01 large numbers of these jays invaded the valley, being found literally by hundreds everywhere eastward from Fresno, where they frequented the trees bordering the vineyards, roadsides and ditches. Their large size and gay plumages rendered them very noticeable, and no doubt not a few of their number were missing when the blue-coated host returned to its Sierran home. The species has not been observed in the valley since that time."

CYANOCITTA STELLERI ANNECTENS (Baird)

BLACK-HEADED JAY

The crested jay of the northern Rocky Mountain region is apparently a connecting link between the long-crested jay *(diademata)* to the southward and Steller's jay *(stelleri)* to the westward, as it combines some of the characters of both races, in about equal proportions in the center of its range. Baird recognized this fact when he suggested the appropriate name of *annectens* (Baird, Brewer, and Ridgway, 1874). However, as it is an abundant form, covering a considerable range, it may be well to give it subspecific status, rather than to consider it as merely an intermediate, which, in fact, it really is.

Ridgway (1904) describes it as "similar to *C. s. stelleri*, but with a distinct (though sometimes small) elongated spot of grayish white immediately above the eye; streaks on forehead (if present) paler blue or bluish white; chin and upper throat more conspicuously streaked (the streaks grayish white rather than gray); back and scapulars rather paler and grayer, and the blue of rump, upper tail-coverts, and under parts of body paler and greener (nearly verditer or china blue)."

The 1931 Check-list gives the range of this race as the "Boreal and Transition zones of the Rocky Mountains from British Columbia south to eastern Oregon, Idaho, and Wyoming." This is probably the main breeding range of typical *annectens*, but it evidently intergrades gradually into typical *stelleri* from eastern British Columbia westward, and into *diademata* from southern Wyoming southward. Major Bendire (1895) sent 11 skins, taken near Walla Walla, Wash., to William Brewster, "who pronounced five of them typical *Cyanocitta stelleri annectens,* and two nearly typical *Cyanocitta stelleri,* and four intermediate between these two forms."

I cannot find that the black-headed jay differs materially in its haunts or in any of its habits from other races of the species. J. A. Munro, who has sent me some notes on it, says that in British Columbia it breeds in the Canadian Zone above 3,500 feet and comes down to the lake region in October, remaining until May. Seldom more than three or four are seen in a day's walk. There is apparently a limited migration from the northern part of its range, both southward and eastward, perhaps nearly or quite to the coast. He has heard it give the tremolo call of the loon and a perfect imitation of the redtail's scream, as well as the call of the raven. He says that black-headed jays were common all through the winter of 1921-22 about Okanagan Landing, wintering in the shore brush and coming to the kitchen door for scraps that were thrown out. He says that they are very curious and come readily to the pygmy-owl call.

The eggs are indistinguishable from those of other races of the species. The measurements of six eggs in the United States National Museum average 30.7 by 22.0 millimeters; the eggs showing the four extremes measure **31.8** by 21.6, 30.3 by **23.1, 29.5** by 21.6, and 31.3 by **21.1** millimeters.

CYANOCITTA STELLERI DIADEMATA (Bonaparte)

LONG-CRESTED JAY

PLATE 13

HABITS

The long-crested jay is the representative of the species that is found in the southern Rocky Mountain region, from southern Wyoming and Utah southward throughout a large part of Mexico. It is described by Ridgway (1904) as "similar to *C. s. annectens,* but lighter colored, with white superciliary patch much larger (or else purer white), forehead more conspicuously streaked with bluish white, greater wing-coverts distinctly barred with black, and the deep black crest very strongly contrasted with the clear brownish gray (nearly mouse gray) of the back and scapulars; rump, upper tail-coverts, and under parts of body light glaucous-blue."

Coues (1871) pays the following tribute to the long crest of this jay, from which it derives its name:

The imposing crest of this jay merits more than a passing allusion. * * * It grows to be two inches and a half long, and is composed of many slender feathers with loosened barbs. The longest ones grow from the crown, while shorter ones fill in from behind and before, to make an elegant pyramid when standing close together, or a bundle of plumes when shaken apart. * * * The crest can be raised or lowered, and opened or shut at pleasure; and its rapid movements, when the bird is excited, are highly expressive. The jay seems to be proud of his top-knot, and generally holds it pretty high, unless he happens to be on a birds'-nesting expedition, which I am sorry to say is not seldom, when he lowers his standard, and makes himself as small as possible, as he skulks silently about, looking, and no doubt feeling, like the thief that he is.

The haunts of the long-crested jay during the breeding season, at least, are in the coniferous forests of the mountains, ranging up to 10,000 or 11,000 feet among the pines. In the Huachuca Mountains, Ariz., in May, we found the long-crested jays very common from 6,000 feet upward. We frequently saw them about our cabin in Ramsey Canyon, evidently foraging for scraps in the little group of summer camps; this was far below the pine belt where the tree growth consisted mainly of sycamores, maples, walnuts, and other deciduous trees. But their main summer haunts were on the steep hillsides that rose abruptly from the sides of the canyon, where there was an open growth of large and small pines, and from there up to the pine-clad summit at 9,000 feet. H. S. Swarth (1904), referring to the same locality, says that "up to the middle of April they were most abundant in the oak regions and along the canyons from 5,000 to 7,000 feet, usually in flocks of a dozen or more; but after that time they gradually withdrew to the higher parts of the mountains to attend to their domestic duties."

Fred M. Packard tells me that in Estes Park, Colo., this jay is a permanent resident, "most common in the upper Transition zone, not uncommon in the lower Canadian, and occasionally seen in the Hudsonian in late summer."

Nesting.—The nesting habits of the long-crested jay are practically the same as those of the blue-fronted jay of the mountains of California. The only nest I have seen was found in the Huachuca Mountains, Ariz., on May 30, 1922; it was well up toward the summit, above Ramsey Canyon, and was placed near the end of a branch of a "bull pine," 8 feet from the ground (pl. 13); it was the usual nest of sticks, reinforced with mud, and was lined with rootlets. My companion, Frank C. Willard, records in his notes three other nests found in the same region; one was 15 feet up in a small oak, another on a horizontal branch of a large fir tree, about 25 feet from the ground, and the third was between 50 and 60 feet above ground in the top of a pine tree. All these nests were at altitudes above 7,000 feet.

Bendire (1895) says that "their nests are usually placed in small bushy pines or other conifers, at no great distance from the ground, varying mostly from 8 to 15 feet." But he mentions a nest, taken by

Denis Gale in Boulder County, Colorado, that was "in a black willow, 9 feet from the ground, at an altitude of 5,500 feet." In his description of a nest he says that "the inner lining consists mostly of small rootlets, in one instance considerable horsehair being intermixed, while in another the lining consists principally of grass and pine needles."

Aiken and Warren (1914) tell of a Colorado nest that "was 6 feet from the ground in a Douglas's fir sapling, only 2 inches in diameter at the base, and on a branch close to the stem of the tree. The outside diameter of the nest was about 10 inches, and it was 5 deep, the nest cavity being 4½ inches in diameter inside, by 3 deep."

Eggs.—The long-crested jay lays three to six eggs, usually three or four, though five is not a rare number. These are practically indistinguishable from those of other races of the species. The measurements of 40 eggs average 31.1 by 22.5 millimeters; the eggs showing the four extremes measure **34.5** by 22.9, 34.0 by **24.0**, **27.9** by 21.6, and 28.8 by **21.2** millimeters.

Plumages.—The sequence of plumages and molts is the same as in other races of the species, but Mr. Swarth (1904) calls attention to some points in which the plumages seem to differ from those of *stelleri.* He says that in the young male in juvenal plumage, "there is some whitish on the chin, an indistinct whitish line over the eye, and the faintest suggestion of bluish white markings on the forehead. A juvenile female is essentially the same in coloration but lacks the whitish markings about the head." Of the adults he says: "Specimens in fresh, unworn plumage have the upper parts of a decidedly bluish tinge, in marked contrast to the brown dorsum of late spring and summer birds."

Food.—What has been written about the food of the blue-fronted jay will apply equally well to the long-crested. Clinton G. Abbott (1929) writes entertainingly of watching these jays at a feeding shelf: "Soft food would be gobbled on the shelf, but the roughly broken pieces of toast were invariably carried in the bill to a distance. Here, either on a branch or on the ground, the jay would place the morsel under one foot (the other foot sometimes also adding its grasp) and then with strong pecks would break off fragments. It is evident that this bird cannot swallow without raising its bill, and, also, its gullet must be surprisingly narrow. I have seen the upward jerk of the bill several times repeated, and each time the piece of toast was returned, to be whittled a little smaller, before finally disappearing out of sight."

Of its feeding on the ground he says: "Hopping, hopping methodically the bird would seem to examine every square inch over which it passed. Sometimes the head would be held high and the gaze directed downward, the long crest almost bobbing forward; at other times the attitude would be more one of sneaking and peering, with head near the ground and crest drawn back. With incomprehensible intuition, a certain spot would be selected, and a hole dug with powerful strokes of the bill, each stroke accompanied by a side motion of the head. In this way the miniature mattock would make quite a little excavation (sometimes as deep as the bird's bill was long) and something edible would be found, as the up-jerk of the bill would plainly show. * * * The bird's bill is its constantly used tool. It turns over small stones with its bill and, especially, it scratches among dead leaves with its bill."

Dr. Coues (1871) says that "in the mountains where the Long-crested lives, pine-seeds contribute in large part to his nourishment. I have often watched the bird hammering away at a cone, which sometimes he would wedge in a crotch, and sometimes hold with his feet, like a hawk with a mouse. Though most at home in the depths of the pines where the supply is pretty sure, he often strays into the adjoining patches of scrubby oak and juniper after the acorns and berries, or to pick a quarrel with Woodhouse's jay, and frighten the sparrows."

Dr. Walter P. Taylor tells me that in Texas these jays do considerable damage to strawberries.

Behavior.—In a general way the habits of the long-crested jay are similar to those of other members of the species, or of most other jays as well. Dr. Coues (1871) gives us a good sketch of jay character, as follows:

All the jays make their share of noise in the world; they fret and scold about trifles, quarrel over nothing, and keep everything in a ferment when they are about. The particular kind we are talking about is nowise behind his fellows in these respects; a stranger to modesty and forebearance, and the many gentle qualities that charm us in some little birds and endear them to us, he is a regular fillibuster, ready for any sort of adventure, that promises sport or spoil, even if spiced with danger. Sometimes he prowls about alone, but oftener has a band of choice spirits with him, who keep each other in countenance—for our jay is a coward at heart like other bullies—and share the plunder on the usual principle in such cases, of each taking all he can get. * * * But withal our jay has his good points, and I confess to a sneaking sort of regard for him. An elegant dashing fellow, of good presence if not good manners; a tough, wiry, independent creature, with sense enough to take precious good care of himself, as you would discover if you tried to get his skin.

Mr. Abbott (1929) was evidently impressed with his vigorous character, for he says: "In fact the word 'vigorous' aptly fits most of the activities of the Long-crested Jay. He will alight in a tree and hop up, up, up as though ascending the rungs of a ladder, from sheer energy. He wipes his bill on the branch with the utmost vigor. He loves to 'flick' his wings and tail. When he launches himself into flight from a small tree, he leaves it trembling with the force of his push-off. Even during the noonday siesta, when I have seen the jays resting like balls of blue in the branches on all sides, the head is never still; there is no hint of sleepiness."

Winter.—Mr. Packard tells me that, in Estes Park, Colo., these jays spend the winter from the upper Transition Zone (9,000 feet) to as low as 5,000 feet. "During the winter these birds frequent the feeding stations and cabins of Estes Park village, where they obtain food to supplement their forage. They do not associate as closely with man as do the camp robbers, but can be induced to feed from a person's hand."

At Cragmore, near Colorado Springs, at an elevation of 6,300 feet, in January, Mr. Abbott (1929) found the long-crested jay "to be the tamest and most abundant bird inhabitant of the open, landscaped grounds of this institution. I have learned that these beautiful jays may commonly be seen in the parks of Colorado's high-lying cities. * * * At Cragmore, they make themselves so thoroughly at home that they pay practically no attention to the passing motor-car or pedestrian, and settle as readily on buildings or electric wires as on the branches of trees. * * *

"Even when water is available, the Long-crested Jays seem to prefer to drink snow. I have seen one perch on a branch covered with soft snow and literally 'guzzle' the snow beside him, billful after billful. On the ground, too, I have watched them gobble far more fresh-fallen snow that [sic] seemed to be necessary. After thaws, when the snow remains only in frozen patches in sheltered spots, it is a different story. I have observed a jay at the edge of such a patch hammer away with all the energy of a woodpecker, raising his whole body with each stroke, in order to add strength to his efforts, and thus break off icy fragments, which he eagerly swallowed."

CYANOCITTA STELLERI PERCONTATRIX van Rossem

NEVADA CRESTED JAY

A. J. van Rossem (1931) obtained four specimens of crested jays in southern Nevada, three from the Charlestons and one from Sheep Mountain, to which he gave the above scientific name and which he describes as "similar in head markings and in general body coloration to Arizona, New Mexico, and Colorado specimens of *Cyanocitta stelleri diademata* (Bonaparte), that is with the supra-orbital region extensively white, the lower eyelid narrowly white and frontal streaks white or bluish white, but differing from that form in having the back and sides of neck 'deep neutral gray' (color terms in quotations from Ridgway, Color Standards and Color Nomenclature, 1912) instead of 'mouse gray.' Differs from

Cyanocitta stelleri annectens (Baird) of the northern Great Basin in decidedly paler coloration throughout, more extensively white eyelids and longer crest."

He gives the known range as "Transition Zone in the Sheep and Charleston Mountains, Clark County, Nevada." This race seems to be closely related to the long-crested jay, which it probably resembles in habits.

APHELOCOMA COERULESCENS COERULESCENS (Bosc)
FLORIDA JAY

PLATES 14-16

CONTRIBUTED BY ALEXANDER SPRUNT, JR.

HABITS

Some birds are so thoroughly typical of certain habitats that one looks for them almost automatically when passing through such places. Perhaps of no species is this more true than the Florida jay. Indeed, so true is it that the local term for the *habitat* is applied to the bird itself, and thus we have the "scrub jay," the universal name of the species in Florida.

No visitor to that fascinating State can have failed to notice the topographical divisions that distinguish it, and the "scrub" is essentially Floridian. The scrub consists, according to Arthur H. Howell (1932), of a type of vegetation peculiar to Florida that occupies scattered areas of whitish sand in the lake region, a narrow strip along the east coast, and smaller tracts on the west coast from Manatee County south to Collier County. The characteristic plants of the scrub are the sand pine *(Pinus clausa)* and shrubby oaks of several species *(Quercus myrtifolia, Q. geminata, Q. catesbaei)*. These oaks, with saw palmetto and rosemary *(Ceratiola ericoides)*, form dense and almost impenetrable thickets.

Proceeding south from Jacksonville one encounters the scrub just south of St. Augustine on the seacoast. Along the Ocean Shore Boulevard the great stretches of saw palmetto behind the dunes of the sea beach reach away illimitably in front of the car.

Here and there roadside signs, advocating the advantages of hotels, camps, and fishing guides, rear themselves above the gray-green fronds, and on these structures, as well as on the lines of telephone wires, one is almost certain to see that characteristic blue and gray dweller of the low growth perched in plain view of pedestrian or motorist, its crestless head and long tail in sharp silhouette against the sky. As many as two dozen "scrub jays" may be seen between St. Augustine and Daytona Beach any day from a speeding car, as an introduction to Florida's thrilling bird life.

Yet, a person could very well spend a lifetime in Florida and never see a single specimen of this bird. It is so partial to the type of vegetation it inhabits that it is utterly useless to look for it anywhere else. In former years it was possible to meet with it almost from the State line at the St. Marys River, southward along the entire east coast, but this is the case no longer. There has been a gradual recession of the range to the north and south of Jacksonville probably because of the elimination of the typical habitat—as S. A. Grimes (MS.) says "to make room for beach houses." This recession is to be noted in even short periods of time, for, as he adds, "the northern limit of the range has receded 20 odd miles within the past year (1939-40). I am no longer able to find a single jay in Duval or northern St. Johns Counties. Ten years ago there were five or six pairs in Duval County."

The present northern limit of the bird's distribution, therefore, is St. Augustine on the east coast. From that point southward along the coastal scrub it is quite common. There is considerable scrub on Merritts Island south of New Smyrna, and Hugo H. Schroder (MS.) states that he has "found more of these birds on Merritts Island than anywhere else in the State." The narrow scrub area between the Indian River and the tracks of the Florida East Coast Railway is doubtless the best part of the State (including Merritts Island) for the visitor to study this interesting species. Quoting Schroder again, "Florida jays are quite numerous south of Indian River City between the highway (U. S. No. 1) and the railway tracks." In my monthly trips to Florida throughout the year, with the exception of midsummer, I have found this to be invariably true. However, in 6 years of intensive field work on the Kissimmee Prairie I have yet to see a single specimen. This is strange, as much of that country seems well suited to their needs and inclinations. Nevertheless, they do not occur there. Records exist only in one area about Lake Okeechobee, that of the Fish-eating Creek section in Glades County on the west side of the Lake. On the east coast, this jay is found as far south as Miami, Dade County, but stops at about that point. Many observers have not noted them that far. The southernmost record comes from what was once known as Rockdale, a station on the Florida East Coast Railway, 10 or 12 miles south of Miami (Howell, 1932).

On the west coast, doubtless because of the abundant mangroves and scanty scrub, it occurs only as far south as Naples, Collier County, according to all records but one. This one, representing the southernmost point of the west coast, is an observation by Edward J. Reimann, a former Audubon warden of the Southwest Coastal Patrol. He writes me that on nearly 2 years' duty in the field from Fort Myers to Cape Sable he saw the Florida jay but once, and that was on Marco Island (Collier County) on October 27, 1936. Marco is about 15 miles south of Naples. Concerning this occurrence he states: "I saw this individual near the cemetery on the north end of the island and am inclined to believe it was a straggler. I searched the same locale numbers of times and also the piny woods a great deal, with the sole purpose of digging up resident birds. Near Caxambas (southern end of Marco) are wonderful live-oak thickets where I hunted them to no avail."

Here and there throughout Florida in suitable areas, inland as well as coastal (some in the very middle of the State), one can find this species up to Gainesville (interior) and Pine Point (west coast) just to the north of the mouth of the Suwannee River. It does not occur at all in the western "handle" of Florida. It is also absent from the open Everglades as well as the Kissimmee Prairie—Lake Okeechobee region. It has been noted sparingly in the Big Cypress Everglades about the village of Immokalee.

There are no records of this jay outside of Florida. I can find but one instance of a sight record beyond the confines of that State and that is considered unreliable by contemporary and present ornithologists. Not only is this jay confined to Florida exclusively, but very definitely to certain portions of that State.

Courtship.—Nothing in the literature I have seen throws any light on the courtship of *coerulescens*. Even those who live in its range and know the bird intimately say nothing about it. Personally, I have seen no evidence of it, and cannot speak from experience. S. A. Grimes, of Jacksonville, who knows the bird as well as any ornithologist living, states that it is his belief that pairs remain mated throughout the year. This is very probably the case and would account to a large degree for the lack of any literature on this phase of the bird's habits.

Nesting.—The Florida jay is gregarious in its nesting habits to the extent of gathering in small, scattered colonies. Perhaps half a dozen pairs will sometimes occupy a tract of scrub of limited extent, but again a nest may be found at some distance from any other pair.

Material is usually the same in all cases, viz, oak twigs of varying shapes and thickness, formed into a substantial, thick-walled cup lined with fine rootlets. It is much like the nest of the blue jay *(Cyanocitta cristata)* in appearance and structure but, unlike that species, does not occupy such elevation, for it is usually constructed at 4 to 12 feet above ground. Probably a high Florida jay nest would be about at the elevation of a low blue jay's. Necessarily, it is rather limited in the choice of a site because of the sameness of the scrub, but the myrtle (*Myrica cerifera*), sand pine (*Pinus clausa*), and various oaks (*Quercus*) are the shrubs and trees most used. S. A. Grimes (MS.) states that the wild olive *(Osmanthus)* "seems to be the favorite site, for it affords the best cover. It is a thick-branched and densely foliaged plant when the dune vegetation is in the 'scrub-jay stage'."

The seasonal range of nesting is extensive, a characteristic of many of the Florida forms, and this jay may be found any time between late March and late May, with eggs. Strange discrepancies in dates may be noted in the same locality, fresh eggs being possible in a nearly 2 months' range of time.

Both parents are assiduous in all domestic duties. Grimes (MS.), who has paid much attention to the scrub jay, writes that "both gather nest material and work it into the nest; both incubate; both brood; both feed and attend the young in and out of the nest. I have seen the adults swallow the cloacal sacs of the nestlings and at other times carry them away and drop them. The female probably does the greater part of the incubating, but the male sees to it that she does not want for food while she is so engaged."

Incubation occupies a little more than 2 weeks, 15 to 17 days. Again quoting from Grimes' notes: "One nest that I kept under observation was in use 45 days, including the ten it was under construction. The last egg in this set of five was laid on April 1, and the three eggs that hatched did so in the night of April 16-17. The young left the nest on May 5."

One interesting fact noted by Grimes in the northern limit of the range (Duval County) is that there is always a percentage of unhatched eggs. "In fact," he writes, "I have never known all the eggs to hatch in a set of scrub jay. If that condition is general, it must indicate some form of decadence in the species. Perhaps it is a normal condition at a border extremity of range, due to inbreeding."

In his comments on the recession of this jay's range from its former northern limit about Jacksonville in the past few years, Grimes has noted another unusual condition. "When the Florida jays," he says, "were down to the last five or six individuals here, for two successive seasons I found three birds attending one nest. Two were males. Even so, in the nest that I followed up, only two eggs out of four hatched."

Though evincing tameness to a degree at times, under almost any conditions, the Florida jay is particularly indifferent to human beings about its nest. Its behavior under these circumstances is often remarkable. When investigating a nesting *coerulescens,* one is reminded strongly of the primitive unconcern displayed by the noddy (*Anous stolidus*) on the Dry Tortugas. It is quite possible to handle the bird freely, and in certain cases there is not even an attempt made to peck at the intruder. Neighboring birds occasionally show more concern than the sitting individual!

Well illustrative of this trait are some interesting notes sent to me by Hugo H. Schroder (MS.), who says that on "April 25, 1932, I found a female on nest in scrub oak and vines about 5 feet up, in Orange County, northwest of Orlando. She remained while vines were opened so nest and occupants could be photographed, and she allowed herself to be picked up and placed in a different position whenever we desired her to be in a better pose; she uttered no protest and made no attempt to bite when picked up. Even when she hopped from the nest she allowed us to replace her, when she remained. The male came to scold while we were handling his mate, and once he came within a foot of my head. A number of neighboring jays added their voices of protest and one of these allowed me to reach within a foot of his body without moving away.

"May 3. Same nest, two young nearly ready to leave. Female allowed herself to be handled but did make a weak attempt to peck at my finger. One of the youngsters wanted to leave the nest, and I held him down while a photo was made; the female was perched on the other side of the nest at the time, her feathers puffed out a little but otherwise giving no sign of resenting the effort to restrain her youngster. Several times when I picked her up she uttered a very soft, low-pitched sort of song."

The Florida jay sometimes shows a decided preference for the nest even after the young have left. Both adults and young return to it for varying periods, and observations on this trait should be more extensive. An interesting instance is furnished by Wilbur F. Smith (MS.), who has had 9 years of experience with this jay near Englewood, Fla., on the lower west coast. He says: "My most thrilling experience with it was about four years ago when a pair built their nest in a hedge of Cherokee roses in a friend's yard. The nest was placed well in the middle of the hedge where light conditions prevented a picture. Three young birds hatched, and when they were about grown I took a friend to see them. We found the nest empty and no birds in sight, though the young had been in the nest the day before. The old birds had been fed all winter and were very tame. The owner of the place had left for his Kansas home, and no doubt the jays missed the daily supply of food, so it was not so surprising that one of them appeared on a wire above us, looked down expectantly, and dropped to proffered food in an outstretched hand. Then the other bird (adult) appeared, and on looking again at the nest we found that two of the young had climbed through the vines and were sitting on the edge of it, while the third was nearby.

"One of the old birds went to the nest with the young and resented my trying to so part the vines as to let light in for a picture. So I braved its displeasure by bringing the nest forward about 2 feet to the outside of the hedge on the chance of the birds following. The inclosed photo shows both birds sitting on the nest in the changed position and one of them taking food from the hand, when we stood, without any effort at concealment, about 3 feet away. We ran out of film, when the nest was returned to its old site, and before we left two young had climbed back into it."

Eggs.—According to Bendire (1895), "the eggs of the Florida Jay range from three to five in number, and their ground color varies from pea green to pale glaucous green. They are blotched and spotted with irregularly shaped markings of cinnamon rufous and vinaceous cinnamon, these being generally heaviest about the larger end of the egg. They are usually ovate in shape, though an occasional set may be called elongate ovate; the shell is smooth and compact, and shows but little gloss."

The measurements of 46 eggs average 27.5 by 20.3 millimeters; the eggs showing the four extremes measure **30.8** by 20.6, 26.9 by **21.3**, **24.6** by 20.1, and 27.0 by **18.6** millimeters.

Plumages.—Immature Florida jays are much like the adults in appearance, but the colors are duller, with less blue on the breast, and the top of the head is lighter. The sexes are alike in all plumages.

Food.—The Florida jay maintains the family tradition for a rather wide choice of food, deserving the term omnivorous, but leaning toward selections of animal matter to an extent of somewhat more than 60 percent. The tendency of this bird to become familiar with humanity and accept its offerings leads to the inclusion of many items that would not otherwise appear, notably such food as bread, cake, and peanuts, which are invariably accepted with apparent avidity. Any such food, however, is highly artificial in nature and should not enter strictly into any summary of normal consumption. So strongly has the bird become entrenched in many parts of its range as a semidomestic species that these items are mentioned because of their frequent offering and equally accepted status.

Dr. Clarence Cottam, of the U. S. Fish and Wildlife Service, has kindly furnished me with a detailed account of the stomach findings of 16 specimens of *coerulescens* taken in January, March, April, May, and September. The conclusions from this study reveal that the food is: "Animal matter 60.63 percent, plant matter 39.37 percent, gravel 6.38

percent, trace of feathers." The breakdown of the above is worthy of note. Though the exact percentages are not given, the findings include the remains of grasshoppers, locusts, crickets, termites, burrower-bugs, squash bugs, leafhoppers, earwigs, beetles, weevils, butterflies, moths, caterpillars, cutworms, bees, wasps, ants, anglewings, flies, millipeds, and centipedes. Also included were spiders, scorpions, ticks, mites, mollusks, snails, turtles, frogs, and lizards. Vegetable matter was represented by wheat *(Triticum)*, crowfoot grass *(Dactyloctenium aegyptium)*, acorns *(Quercus)*, purslane *(Portulaca)*, milkwort *(Polygala)*, huckleberry *(Gaylussacia)*, blueberry, cranberry *(Vaccinium)*, and fogfruit *(Lippia)*. Portions of vegetable debris and indeterminate matter (mast?) and wood pulp were also present.

Audubon (1842) states that the seeds of the saw palmetto are a favorite food, so much so, indeed, that "no sooner have the seeds of that plant become black, or fully ripe, than the Florida jay makes them almost its sole food for a time." He adds that the method of feeding is like that of the blue jay, for *coerulescens* "secures its food between its feet, and breaks it into pieces before swallowing it, particularly the acorns of the *live oak,* and the snails which it picks up among the *sword palmetto.*" Nuttall (1832) also gives the seeds of the saw palmetto as being eaten "largely."

Bendire (1895) adds another item in his summary of the food as "offal." He also mentions wood ticks specifically, as does Maynard (1896), the latter stating that "upon examining the contents of its stomach, found that it was filled with ticks or jiggers which infest the skin of all quadrupeds in this section of Florida." These references to ticks substantiate, without saying so, of course, the observations of N. B. Moore on the habit of this jay of alighting on the backs of cattle and securing ticks in that manner. "Jigger" is the universal name of the redbug in the southeast, an even worst pest than ticks in many ways.

Another food habit of this jay, not hitherto mentioned and something of an indictment against the bird, is its fondness for the eggs and young of other birds, and even of poultry. Just how much this is indulged in does not seem clear, but there is certainly abundant evidence that predation of the sort occurs. Bendire (1895) states that this jay is "charged with being very destructive" in this way. A writer whose name I am unable to determine, but whose initials are C. S. C., writing in the Chicago Field, says that they "eat and drink with poultry, having an eye on eggs and young chickens." M. M. Green (1889) states: "Stomachs of two shot contained insect food. The birds' bills were smeared with yolk of eggs. Several people told me that the jays were nest robbers." Nuttall (1832) notes that it "destroys the eggs and young of small birds, despatching the latter by repeated blows on the head."

Grimes (MS.) says: "I know they like crickets for I saw a male pass up four, one after the other, to his sitting mate. * * * In the fall and winter they feed to a large extent on the little acorns of Chapman's oak."

Behavior.—The Florida jay is a true representative of its family in traits and character. Individual variations occur, but essentially it resembles its better-known relative *Cyanocitta cristata,* in actions and habits. As its local name implies, it is not a high-ranging bird in any sense. One of its outstanding characteristics is its love of the ground and low elevations, which must impress anyone observing the bird for any length of time. Along roadsides it is frequently seen on the "shoulders" of the highway, particularly in sandy stretches, where it feeds commonly. Passing cars often flush it from such situations, whence it dashes off amid the scrub palmetto or ascends to a convenient telegraph wire. The flight is strong and without particular character unless the frequent sweeps with wide-open wings could be called such. The long tail is often fully expanded. On the ground it hops with strong, sure jumps, planting the feet firmly. In searching for food under such conditions it is given, according to Howell (1932), to probing the loose sand with the bill.

One often sees the moods of the bird expressed in the action of the tail. Usually, in repose, it hangs fairly straight down, offering a good field mark, but under stress of excitement this member is jerked and twitched in a highly expressive manner. Approach is not difficult most of the time, though easier during the nesting period. Sometimes an almost utter disregard of the human intruder is evidenced. In former years these birds were caged, and they proved easy to keep. Audubon (1842) gives an account of a pair he saw in captivity at New Orleans (!), which he states "had been raised from a family of five, taken from the nest, and when I saw them had been two years in confinement. They were in full plumage, and extremely beautiful. The male was often observed to pay very particular attentions to the female, at the approach of spring. They were fed upon rice, and all kinds of dried fruit. Their cage was usually opened after dinner, when both immediately flew upon the table, fed on the almonds which were given them, and drank claret diluted with water. Both affected to imitate particular sounds, but in a very imperfect manner. These attempts at mimicry probably resulted from their having been in company with parrots and other birds. They suffered greatly when moulting, becoming almost entirely bare, and required to be kept near the fire. The female dropped two eggs in the cage, but never attempted to make a nest, although the requisite materials were placed at her disposal."

A reference to the adaptability of this species to captivity is referred to by Nuttall (1832) when he states that it is "easily reconciled to the cage." Since caging of wild birds is now a thing of the past, the above may probably be all we will know about this species in private captivity, but successful attempts to tame it at large have been often accomplished. A striking example is noted by Howell (1932) just 100 years after Nuttall's observation, as follows:

Miss Edith Werner, who in the spring of 1923 was operating a tea house on the shore of Lake Jackson, near Sebring [Fla.], has been remarkably successful in taming the Florida Jays, which are abundant in the scrub close to her house. She whistles a bright little tune and in a few minutes the Jays appear from all directions and without hesitation alight on her arm or shoulder to take the pieces of bread she offers them. She told us she had been a year or more taming the birds, and that it was a month or more before she could get them near her. At the time of our visit however, they had become so used to strangers that they allowed us to feed them and even alighted on our heads and shoulders. On hearing a note of alarm from one of the Jays in the brush, they all deserted us and flew into the scrub. Miss Werner says the birds always have a lookout posted on a high bush, which sentinel remains there while the rest are feeding and gives warning of danger. She added that they often frolicked together in the morning, at which times they snap the bill continuously as they shake their bodies. Occasionally they sing very softly, under their breath, "like a canary."

The indifference of the Florida jay to human presence is alluded to by Hugo H. Schroder (MS.) in the following note: "While I was eating lunch beside the road south of Indian River City, Fla., a male jay landed on ground near my car. When I threw down some bread he picked it up and flew off with it. As soon as he returned, I threw more pieces of bread; each time the bird would fly off with it. More than a half dozen pieces were taken away; whether each one was eaten I could not see."

Wilbur F. Smith (MS.) states that "Florida jays become tame about the houses of winter visitors, taking peanuts and bread from the hand or on the head, or even from between the lips of some."

This bird appears to run true to corvine traits in its predilection for making away with odds and ends of property. This is a characteristic overlooked by many, or at least not referred to. Such articles are, as usual with avian thievery, bright and shiny as a rule, easily seen and attractive. Buttons, tops of small tins, spoons, bits of glass, china, and the like are among the hoards. A reference to this habit, the only one in fact that has come to my attention from the literature, appears in the Chicago Field of May 1880 and states that these birds "bury such food as they cannot immediately consume, and also spoons, thimbles, or any shining object that attracts their attention."

Another interesting habit is also apparently not well known and may have been more frequently indulged in during past years than now, though I know of no reason why it should have now ceased. This concerns the picking of ticks from cattle, a habit shared by some of the Florida herons. I have never observed it, nor can I find anything in the literature about it, but N. B. Moore (MS.) in writing of this jay says: "A common habit of this species during the time when cattle have many ticks upon them, and this happens through the greater part of the year, is to perch upon their backs, move or hop upon their rumps and hip bones, and pick them off and eat them, or, if they have young, carry them to the nest or to a tree or fencepost, where the young are perhaps waiting for food. It reminds one of the habits of the *Buphaga* of Africa to see this jay riding about on the backs of cattle and feasting on these disgusting parasites. The jay often obtains the ticks by hopping on the ground about the legs of the cows jumping by the help of its wings up to the buttocks, flanks, or brisket and seizing the most palatable ones. The cattle seem not in the least annoyed by those on their backs, and yet the pretty constant switching of their tail and throwing back their horns keep the jays constantly on the alert, and they often quit their place to avoid a blow, perching either on another cow or on a tree or a fence."

With the even greater prevalence of cattle in Florida today than when Moore wrote (about 1870), it seems strange that this habit has not been commented on more by recent observers. To many persons' surprise Florida is one of the greatest cattle-raising States in the Union, but in recent years there has been a definite effort, attended by marked success, to eliminate ticks, and this may have resulted in such a sharp decrease in the parasites that the jays have largely abandoned this source of food and the method of obtaining it.

On monthly investigations on the Kissimmee Prairie, I see literally thousands of cattle, but as mentioned previously the jay does not occur on the open prairie and therefore could not be expected in the largest cattle concentrations. It is known that the Florida crow (*Corvus brachyrhynchos pascuus*) procures ticks and other insects from the backs of cattle, and occasionally some of the smaller herons do the same thing, which reminds one of the African cow heron.

Somewhat contrary to the accepted opinion that jays are domineering and quarrelsome, there is evidence that *coerulescens* is an exception to the rule. Though at times seen to drive off such species as blackbirds and mockingbirds, it appears to live in considerable harmony with its avian neighbors, with little bickering and interference. Wilbur F. Smith (MS.) in noting this trait says: "The Florida jay has far better manners than other members of the family. I have photographed it with quail (*Colinus virginianus floridanus*), ground doves (*Columbigallina pas-*

serina), meadowlarks *(Sturnella magna argutula),* red-winged blackbirds *(Agelaius phoeniceus mearnsi),* and grackles and have never seen it bully or disturb these birds, a fact worth noting in view of the family reputation. In the section where I have known it for nine years [Englewood, Fla.] it is a general favorite, giving ground only to the quail that feed about the homes." It is supposed that in his expression "giving ground" Mr. Smith refers to the popularity of the quail over the jay, not that the jay retreats before the other.

Voice.—Notoriously noisy as are most of the jays, this species is not unusually so. Compared at least to *Cyanocitta* and its forms, it is decidedly less vocal. The notes are essentially jaylike, which is not too general a term to employ for quick recognition, but are given at greater intervals and not so constantly. Certain calls are loud and have a harsh, rasping quality, and it is probably some of these that Howell (1932) likens to notes of the boat-tailed grackle *(Megaquiscalus mexicanus major)* and that he terms "churr." I cannot say that they ever impressed me in such a way, but bird calls sound different to different hearers.

The song, if one can designate the attempt as such, is widely at variance with the call or alarm notes. It is a rather surprising performance really, and would puzzle many not seeing the singer. Wetmore (MS.) describes it as "a mixture of low, sweet-toned calls, high in pitch, mingled with others that were variously slurred or trilled in utterance." It is next to impossible to describe most bird notes in words. However, the above seems to me to be as good an interpretation as can be given. Not in character or similiarity, but in that one would not expect such a song from such a bird, it recalls some of the performances of the loggerhead shrike *(Lanius ludovicianus)*! On some occasions, I have heard low, subdued notes that cannot be described otherwise than as a chuckle, delivered rather rapidly and having an abrupt quality. It is an agreeable delivery and imparts the distinct impression that the bird is in a thoroughly contented mood at the time.

Field marks.—The outstanding field marks of this species, aside from the characteristic color pattern, are the crestless head and the long tail. The name *"Aphelocoma"* is from the Latin meaning "smooth hair," referring, of course, to the lack of a crest in a crested family; *"coerulescens"* refers to the prevailing color, blue.

DISTRIBUTION

Range.—The peninsula of Florida; nonmigratory.

The range of the Florida jay extends **north** in that State to **Port**

Richey, Fruitland Park, and Ormond. **East** along the Atlantic coast from Ormond south to Lemon City. **South** to Lemon City, Immokalee, and Fort Myers. **West** along the Gulf coast from Fort Myers north to Port Richey.

Egg dates.—Florida: 49 records, March 21 to June 14; 25 records, April 10 to 30.

APHELOCOMA COERULESCENS SUPERCILIOSA Strickland

LONG-TAILED JAY

PLATES 17, 18

HABITS

The California jay of the interior is now known by the above name. Under its former name, *Aphelocoma californica immanis,* the 1931 Checklist gives its range as "extreme southern Washington, valleys of Oregon between the Cascades and the Coast ranges, and the Sacramento and San Joaquin valleys of California and adjacent mountain slopes."

Dr. Grinnell (1901) in describing it, from specimens taken near Scio, Oreg., gave as its characters, "in coloration similar to *Aphelocoma californica,* but size greater and tail proportionately much longer." This description was apparently based on only four birds, at least the measurements of only four are given, all from the Willamette Valley, Oreg. Mr. Swarth (1918), with a much larger series from a much larger area does not agree exactly with Grinnell's description; he says that *immanis* is "distinguished from *A. c. californica* both by large size and *pale* [italics mine] coloration; from *oocleptica* by pale coloration, size being about the same." At the time that Dr. Grinnell described *immanis* the characters and the distribution of the California races of *Aphelocoma* were not so well understood as they are today, and the fact had not been recognized that the two coastal races are dark colored and the interior race is paler. Ridgway (1904) does not recognize *immanis* but lists it as a synonym of *californica.*

Grinnell, Dixon, and Linsdale (1930) call this bird the interior California jay, an appropriate name. They say of its haunts in the Lassen Peak region: "This species, belonging to the brush-covered portions of the section, found suitable surroundings on the western slope of the section where the following kinds of plants grew: buck-brush, scrub oak, elderberry, hazel brush, manzanita, red-bud, grapevine. Individuals were also often seen in trees, but, as a rule, in their lower portions. The kinds of trees thus frequented were: blue oak, willow, living or fire-killed digger pine, knobcone pine, cottonwood, valley oak, sycamore, box elder, and orchard trees. In the eastern part of the section the jays frequented the slopes that were juniper covered. In addition to the junipers they were seen in mountain mahogany, sage-brush, and willows (in the canons)."

Nesting.—The nesting habits of the long-tailed jay are apparently similar to those of the other California races, and the eggs are practically indistinguishable. The measurements of 40 eggs average 28.4 by 21.0 millimeters; the eggs showing the four extremes measure **31.2** by 21.3, 27.4 by **22.0**, **25.0** by 20.0, and 26.5 by **19.5** millimeters.

Young.—Grinnell and Storer (1924) have written quite fully on the habits of this subspecies, some of which may well be quoted here. While watching the two parent birds feeding their young, they noted that—

the parent birds had a particular route in approaching and leaving the nest, and this route was adhered to strictly. They would always approach through the trees of a wooded slope to the east, and then, having reached the nest tree, hop by easy stages to a position on the west side of the nest. From there the nestlings would be fed, and then the nest cleaned. After that the bird would work out of the south side of the willow, fly to a digger pine across the creek immediately above our tent, hop upward until near the top of the pine, and from there would take off in a direct course to its next forage ground. Even when the jays had been hunting insects in the open area immediately west of our camp, they would circle about when ready to return to the nest and approach it from the east. Only one adult visited the nest at a time although they often followed one another in quick succession. Save for the low crooning given when standing over the young, no calls were uttered while the parents were in the vicinity of the nest. There was a "zone of quiet" about their home, within which the owners would not call or raise any alarm.

Behavior.—Of its behavior they say: "The Interior California Jay is notoriously bold and forward in its behavior; although it is counted as a nonflocking species, individuals and pairs will gather quickly in response to the excited calls of one of their kin. The birds seem never to be so busy with their own affairs that they cannot stop and investigate any object of an unusual nature. Ordinarily this jay is the picture of animation. Perched, it stands in an attitude of alertness, its head up, tail straight back or tilted slightly upward, and feet slightly spread. Just after alighting a jay will often execute a deep bow involving the entire body, and this may be repeated a number of times and in different directions. The purpose of this bowing is not clear to us."

Mrs. Ruth Wheeler writes to me of her experience with this jay: "I had a very interesting experience last year photographing a family of the California jay. We found the parents to be extremely wary. I have never worked with birds that appeared to show as much intelligence. We set up our bird blind near their nest, which was in a young oak, and only about 4 feet from the ground. Although the birds had become used to the blind and were nowhere to be seen when we entered, still they appeared to know that we were there. They came back very quietly, slipping through the trees and alighting near the blind. Then one of them leaned over and peered through the small opening through which the camera was focused. After looking very carefully, he saw us and set up a great outcry. We were able to get only one picture of the nest, which we took with a flash. After that the birds would not come near while we were in the blind."

Voice.—Grinnell and Storer (1924) describe the notes of this jay as only slightly different from those of the other subspecies, but they add to our knowledge of the bird's varied vocabulary, perhaps a limited language. Grinnell gives *"cheek, cheek, cheek,* etc., staccato, 3 to 10 times in rapid succession; *chú-ick, chú-ick, chú-ick,* etc., usually in 3's slowly; *schwee-ick,* higher-pitched, 2 to 6 times, uttered still more slowly." Storer adds: "A series of mildly harsh notes, *kwish, kwish, kwish,* uttered usually 3 to 5 times in quick succession; a more protracted softer note, *kschu-ee,* or *jai-e,* usually given singly. Birds of a pair when foraging together, and young and adults when in family parties, utter a subdued guttural *krr'r'r'r'r.* When attending young still in the nest, the parent birds utter a low crooning, impossible of representation in syllables; and the young birds, after leaving the nest and before gaining their living independently, have a 'teasing scold' which they utter almost incessantly, in keeping their parents apprised of their need for food."

APHELOCOMA COERULESCENS OOCLEPTICA Swarth

NICASIO JAY

HABITS

Harry S. Swarth (1918) gave the above name to the flat-headed jays of "the coast region of northern California, west from Mount Diablo and the coast ranges. North to Humboldt Bay, south to the Golden Gate and the east side of San Francisco Bay." Of its distinguishing characters, he says: "Of large size and dark coloration. In color closely similar to *A. c. californica,* but size measureably greater throughout. In measurements *oocleptica* is equal to the maximum of *immanis,* from which subspecies it is distinguished by its dark coloration. Differs from *hypoleuca* both in greater size and much darker color."

Nesting.—There is not much to be said about the nesting habits of the Nicasio jay, which are practically the same as those of the other California subspecies. John W. Mailliard (1912) says that this jay is an abundant resident in Marin County and that their "nesting notes upon this species established the following sites for the eighty-three nests observed: oaks 69; bay 3; wild coffee 4; elder 2; madrona 1; gooseberry 1; toxon 1; poison oak 2. And yet in Belvedere, Marin County, where live-oaks are most plentiful, a nest has been built almost yearly, for seven or eight years, in a clematis which climbs up the side of our summer home. The nest has usually been placed within reach of, as well as observation from, the window of a constantly occupied bed room, a window opening out and frequently opened and closed daily."

Eggs.—The eggs of the Nicasio jay are practically indistinguishable from those of the California jay and show the same interesting variations. The measurements of 40 eggs average 27.8 by 20.5 millimeters; the eggs showing the four extremes measure **30.8** by 21.8, 29.0 by **22.4, 24.4** by 19.3, and 25.4 by **18.8** millimeters.

Food.—The notes that Charles A. Allen, of Nicasio, Calif., sent to Major Bendire (1895) evidently refer to this race. He writes:

No bird is more destructive of the smaller species building open, unconcealed nests than this Jay. I have seen one alight on a limb near a nest, eat the eggs that it contained, and, not satisfied with this, give the nest a down and inward stroke with its bill, ripping it open. They are especially destructive to the nests of the Black-chinned and Anna's Hummingbirds and the Ground Tit. They also become altogether too familiar about the poultry houses, and will eat the eggs as fast as the hens lay them. As soon as they hear a hen cackle after laying, three or four of these birds go to the spot at once. Even the chicken house affords no protection against these robbers, if they can find a way of entering it; shooting is equally ineffectual, for they are too numerous. They destroy vast quantities of fruit in apple, peach, pear, and plum orchards, as well as many smaller fruits. Shooting them by hundreds and hanging their carcasses in the fruit trees as scarecrows is of no avail; they do not know enough to be frightened at anything. I have tried to poison them, but never saw a dead one except when shot. They also destroy a great deal of young wheat when first sown, until it is 2 or 3 inches high. They pull it out of the ground and eat the soft, swelled grains; after the stalks begin to grow they will not molest it.

One cannot help feeling that the above bitter invective is somewhat overdrawn and perhaps a bit prejudiced. His statement that they "do not know enough to be frightened" is offset by the fact that he was never able to poison one; the truth of the matter probably is that they *know too much* to be frightened unnecessarily, and that they are crafty enough to avoid real danger. His statement that they pull up sprouting grain does not agree with Professor Beal's (1910) statement that "the jay is not known to pull up grain after it has sprouted."

But Professor Beal evidently overlooked Mr. Allen's statement, as well as the following from Joseph Mailliard (1900): "I have had acres of peas * * * practically destroyed by these birds. * * * I remember one spring when a patch of about an acre and a half was sown with a mixture of peas and oats, and the peas were pulled up as fast as sprouted, by the jays, so that the crop consisted of oats alone. * * * Some years they destroy a lot of corn and other years almost none. * * * This year the Jays, in conjunction with Towhees, Juncos and a few Flickers, badly damaged some late sown oats beside the house." He watched them through glasses and saw them pull up the sprouting grain and eat the kernels.

Professor Beal (1910) adds the following observation in the jay's favor: "But the jays do not frequent orchards entirely for fruit. During May and June the writer many times visited an apple orchard, the leaves of which were badly infested with a small green caterpillar, locally known as the canker worm. When a branch is jarred, these insects let themselves down to the ground on a thread spun for the purpose. Many jays were seen to fly into the orchard, alight in a tree, and then almost immediately drop to the ground. Observation showed that the caterpillars, disturbed by the shock of the bird's alighting on a branch, dropped, and that the birds immediately followed and gathered them in. These caterpillars were found in the stomachs of several jays, in one case to the extent of 90 percent of the contents."

Voice.—Mr. Allen mentions some notes of this jay that are somewhat different from those recorded by others; he wrote to Major Bendire (1895): "One of their notes of alarm, uttered when they see something they do not like, especially an Owl asleep in a tree, sounds like 'cŭr, cŭr, cŭr'; as soon as this is heard by others in the vicinity they will commence to gather and join in the chorus. A sort of social note of recognition sounds like 'whŭze, whŭze', given while moving about among the trees and shrubbery, and one of their common call notes sounds like 'creak, creak'."

APHELOCOMA COERULESCENS CALIFORNICA (Vigors)

CALIFORNIA JAY

PLATES 19, 20

HABITS

The three subspecies of *Aphelocoma coerulescens* that are found in California are common, and in many places abundant, over almost all the State, except the desert regions and the mountains, but the subject of this sketch, *A. c. californica,* is confined to a comparatively narrow coastal strip along the southern half of the coast. The 1931 Check-list gives its range as "from the southern arm of San Francisco Bay to the Mexican line, east to the eastern base of the Coast ranges." But Swarth (1918), who does not recognize *A. c. obscura,* extends its range into northern Lower California, as far south as the San Pedro Mártir Mountains.

Roughly speaking, the characters distinguishing the three California **races are size and color.**

Swarth (1918) says that *californica,* as compared with *immanis* (now *superciliosa*), the interior form, "is of small size and dark coloration. The blue areas are of a deeper shade, the back distinctly darker brown, and the light colored under parts have a dusky suffusion. Lower tail coverts usually tinged with blue, sometimes conspicuously so. Coloration is about the same in *californica* as in *oocleptica* [the northern coast form], from which subspecies *californica* is distinguished by smaller size throughout." The other two races are larger than *californica,* and both about the same size, but *oocleptica* is dark colored and *superciliosa* is much paler.

The habitats of the three races and their general habits are all very similar. One life history might well do for all three. They all live mainly in the Upper Sonoran Zone, with some extension of range into the Lower Sonoran and Transition Zones. Their favorite haunts are the oak and brush-covered foothills of the mountains, the brush-covered sides of the canyons, the oak and digger-pine chaparral, thickets of *Ceanothus* and poison-oak bushes, and among the small trees and shrubbery along watercourses. In such places, where there is ample concealment among the thick foliage, this handsome, flat-headed, mischievous villain is quite sure to be found; if not immediately in evidence, the well-known squeaking sound, such as one uses to call small birds, will bring all the jays within hearing of it.

Nesting.—Major Bendire (1895) writes:

The nests are usually found on brush-covered hillsides or in creek bottoms, placed in low bushes and thickets, such as blackberry, poison oak, wild gooseberry, currant, hazel, hawthorn, and scrub-oak bushes, or in osage-orange hedges; occasionally in a small piñon pine or a bushy young fir, and quite frequently on a horizontal limb of an oak, varying in height from 3 to 30 feet from the ground. In the majority of cases the nests are located near water, but sometimes one may be found fully a mile distant. Externally they are composed of a platform of interlaced twigs, mixed occasionally with moss, wheat stubble, and dry grass; on this the nest proper is placed, which consists of a lining of fine roots, sometimes mixed wih horsehair. No mud enters into the composition of their nests. One now before me * * * measures 9 inches in outer diameter by 3½ inches in height; the inner cup is 4 inches in diameter by 2 inches deep. Outwardly it is composed of small twigs of sagebrush, and the lining consists entirely of fine roots; it is compactly built and well constructed. The nests are usually well concealed, and the birds are close sitters, sometimes remaining on the nests until almost touched.

In addition to the trees and shrubs mentioned above, nests of the California jay have been found in live oaks, elders, willows, apple trees, pear trees, junipers, cypresses, and honeysuckles. W. L. Dawson (1923) says:

Taking the country over, nests built in oak trees probably outnumber all others combined, yet the component members of the chaparral, ceanothus, chamissal, and the rest, must do duty in turn, and all species of the riparian sylva as well. The thick-set clumps of mistletoe are very hospitable to this bird, and since this occurs on oaks, cottonwoods, and, occasionally, digger pines, it follows that jay-heim is found there also. * * *

The lining varies delightfully, but is largely dependent, it is only fair to say, upon the breed of horses or cattle affected on the nearest ranch. So we have nests with white, black, bay, and sorrel linings, not to mention dapple gray and pinto. One fastidious bird of my acquaintance, after she had constructed a dubious lining of mottled material, discovered a coal black steed overtaken by mortality. New furnishings were ordered forthwith. The old lining was pitched out bodily, and the coal black substitute installed immediately, to the bird's vast satisfaction— and mine.

Eggs.—Four to six eggs generally constitute a complete set for the California jay; as few as two and as many as seven have been recorded. The eggs are usually ovate in shape, rarely elongate-ovate. They are very beautifully colored and show a wide range of variation, seldom, if ever, equaled and never exceeded among North American birds' eggs. In any series of these colorful eggs there are apparent two quite distinct types of coloration, the green and the red. Lawrence Stevens, of Santa Barbara, writes to me that about half of the sets that he takes are of the green type and half of the red type. He finds eggs of the green type mostly in the creek bottoms in willows, usually in sets of four, and finds eggs of the red type mostly on the hillsides, in sets of five. As several other collectors do not agree with him on these points, there is probably no correlation of color with the locality or the size of the set. He mentions one set that has a cream-colored background with red spots, that one would hardly believe to be jays' eggs.

James B. Dixon tells me that only about two sets in ten are of the red type. Dawson (1923) describes the colors very well as follows:

The red type is much the rarer. In this the ground color varies from clear grayish white to the normal green of the prevailing type; while the markings— fine dots or spots or, rarely, confluent blotches—are of a warm sepia, bister, verona brown, or Rood's brown. The ground color of the green type varies from pale sulphate green to lichen green, and the markings from deep olive to Lincoln green. In the Museum of Comparative Zoology we have a set kindly furnished by Mr. H. W. Carriger, whose markings are reduced to the palest subdued freckling of pea-green. In another set of the red type, fine Mars brown markings of absolute uniformity cover the egg; while the eggs of another set are covered as to their larger ends with an olive-green cloud cap, which leaves the remainder of the specimen almost free of markings.

Bendire (1895) describes the eggs somewhat differently as follows: "The ground color of the egg of the California Jay is very variable, ranging from deep sea green to pea and sage green, and again to dull olive and vinaceous buff. The eggs with the greenish ground color usually have markings of a dark bottle-green tint, mixed sometimes with different shades of sage green. The eggs having a buffy ground color are spotted, blotched, and speckled with different shades of ferruginous, cinnamon, rufous, and occasionally lavender. The markings are generally scattered over the entire surface of the egg, and are usually heavier about the larger end, but nowhere so profuse as to hide the ground color."

The measurements of 50 eggs average 27.6 by 20.5 millimeters; the eggs showing the four extremes measure **30.6** by 21.8, 30.2 by **22.4, 24.4** by 19.3, and 25.4 by **18.8** millimeters.

Young.—Bendire (1895) says that the male assists "to some extent in incubation, which lasts about sixteen days. The young are able to leave the nest in about eighteen days, and follow the parents for some time." Mrs. Wheelock (1904) says that "the male assists in the nest-building, but not in the incubation. The latter requires fourteen days. * * * One of the first lessons the young Jays learn is to love the water. It requires some coaxing for the first splash, but they seem to take to their bath as do little ducks, and to find it just as necessary as food."

Plumages.—The young jay is hatched naked and blind; probably some form of natal down appears in advance of the juvenal plumage, though I have not seen it. Ridgway (1904) gives the following detailed description of the juvenal plumage: "Pileum, hind neck, auricular and suborbital regions, sides of chest, rump, and upper tail-coverts uniform mouse gray, the pileum slightly more bluish gray; back, scapulars, and lesser wing-coverts deep drab-gray; lores dusky; a broad postocular and supra-auricular space, narrowly streaked with dusky gray; anterior portion of malar region, chin, throat, median portion of chest and under parts generally white, faintly tinged across upper breast and on anterior portion of sides with very pale brownish gray; wings (except smaller coverts) and tail as in adults."

The postjuvenal molt begins early in July, and I have seen a young bird beginning to molt on the throat and upper breast as early as June 29. This molt involves all the contour plumage and the lesser wing coverts, but not the rest of the wings and tail. I have seen no specimens showing the latter part of this molt, but it is apparently completed before September, when young birds become practically adult.

Adults have a complete postnuptial molt, which seems to be accomplished mainly in August; the plumage becomes much worn and faded during late spring and early summer, the blue wearing off on the head, exposing the dusky bases of the feathers, the brown of the back fading, and the under parts becoming soiled and browner; I have seen birds molting in August and others in full fresh plumage as early as September 24.

Food.—Prof. F. E. L. Beal (1910) examined 326 stomachs of the California jay, and found that 27 percent of the food consisted of animal matter and 73 percent vegetable, though the animal matter amounted to 70 percent in April. Among the insect food, he lists predaceous ground beetles, mostly beneficial species, 2.5 percent for the year and as much as 10 percent in April; other beetles, mostly harmful, 8 percent for the year and 31 percent in April; wasps, bees, and ants amounted to less than 5 percent; honey bees were found in 9 stomachs, and all, 20 in number, were workers; Lepidoptera, mainly in the caterpillar stage, amounted to 2.5 percent; this item included 12 pupae of the coddling moth, an unexpected service that would cover a multitude of sins in other directions; grasshoppers and crickets were eaten to the extent of 4.5 percent. Of other animal food, he says: "A few miscellaneous creatures, such as raphidians, spiders, snails, etc., form less than one-half of 1 percent of the food. * * * Besides the insects and other invertebrates already discussed, the jay eats some vertebrates. The remains consisted of bones or feathers of birds in 8 stomachs, egg-shells in 38, bones of small mammals (mice and shrews) in 11, and bones of reptiles and batrachians in 13 stomachs." In destroying small mammals the jay does good service, as most of them are injurious, but the same cannot be said about its appetite for the useful reptiles and batrachians. The damage to the eggs and young of small birds is a serious matter. Some 95 stomachs were collected between the middle of May and the middle of July, the height of the nesting season, "of which 17, or 18 percent, contained eggs or remains of young birds. If we may infer, as seems reasonable, that 18 percent of the California jays rob birds' nests every day during the nesting season, then we must admit that the jays are a tremendous factor in preventing the increase of our common birds. Mr. Joseph Grinnell, of Pasadena, after careful observation, estimates the number of this species in California at about 126,000. This is probably a low estimate. If 18 percent of this number, or 22,680 jays, each robs a nest of eggs or young daily for a period of sixty days from the middle of May to the middle of July, the total number of nests destroyed in California by this one species every year is 1,360,800."

Mr. Dawson (1923) draws a still blacker picture; he figures, on the basis of suitable acreage and average population per acre, that there are 499,136 pairs of California jays in the State and says: "If we allow only one set of eggs or nest of birds to each pair of jays *per diem* for a period of two months, we shall be well within the mark of actuality. Yet that will give us in a season a total destruction of 29,948,160 nests, or, say, 100,000,000 eggs." These figures look appalling, but, in considering them, we must not lose sight of three important facts: First, jays and small birds have existed together for untold ages without any serious reduction in the number of the latter; second, the increase in small birds is limited by the amount of suitable area that will support them, and such area is probably kept filled to capacity; and third, it is a well-known fact that if a pair of birds is robbed of its eggs a second or third set will be laid; this is less likely to happen, however, if the young are taken.

But wild birds are not the only sufferers from the depredations of this jay; the eggs and young of domestic poultry are preyed upon. Professor Beal (1910) writes:

He is a persistent spy upon domestic fowls and well knows the meaning of the cackle of a hen. A woman whose home is at the mouth of a small ravine told the writer that one of her hens had a nest under a bush a short distance up the ravine from the cottage. A jay had found this out, and every day when the hen went on her nest the jay would perch on a near-by tree. As soon as the cackle of the hen was heard, both woman and bird rushed to get the egg, but many times the jay reached the nest first and secured the prize. * * *

A still worse trait of the jay was described by a young man engaged in raising poultry on a ranch far up a canyon near wooded hills. When his white leghorn chicks were small, the jays would attack and kill them by a few blows of the beak, and then peck open the skull and eat the brains. In spite of all endeavors to protect the chicks and to shoot the jays, his losses were serious.

Of the vegetable food, mast, mainly acorns, is the largest item, and during the late fall and winter months made up one-half to three-quarters of the entire food; October showed the largest amount, 88.57 percent. Acorns and nuts are carried off and stored wherever they can be hidden in cracks and crevices, but since many are dropped on the way, or hidden on the ground, the jay may be considered useful as a tree planter.

Grain constitutes an important item; in March, when grain was being sown, it amounted to 45.50 percent; and again, during the harvesting season in September, it made up 24.26 percent of the food. "Grain was found in 95 stomachs, of which 56 contained oats; 34, corn; 2, wheat; 2, barley; and 1, grain not further identified. Many of the oats were of the wild variety.

"Fruit was found in 270 stomachs. Of these, cherries were identified in 37, prunes in 25, apples in 5, grapes in 2, pears in 2, peaches in 1, gooseberries in 2, figs in 1, blackberries or raspberries in 71, elder-berries in 42, manzanita in 4, cascara in 1, mistletoe in 1, and fruit pulp not further identified in 76." He remarks, further, that "it is safe to say that half of the fruit eaten was of wild varieties and of no economic value." His table shows that fruit formed 22.05 percent of the total food for the year but averaged nearly half of it during the summer months and 61.41 percent in May. In addition to what fruit the jay eats, much more is damaged and left on the trees to rot and more falls to the ground.

Robert S. Woods writes to me: "The California jay is very destructive to almonds and finds no difficulty in cracking the harder-shelled varieties. Its raids begin before the nuts are ripe enough for human consumption and continue as long as any of the crop remains. The almond is held against a branch with the foot and vigorously pounded with the bill until an opening is made large enough to permit the kernel to be extracted piecemeal. English walnuts are broken into while still on the tree. However, the Eureka variety, at least, seems to be immune after the shell has thoroughly hardened, though some of the thinner-shelled strains or varieties could doubtless be successfully attacked even after maturity.

"Jays will often eat dry bread crumbs but greatly prefer food of a more fatty nature. When coarsely chopped suet is placed on a feeding table, it is ignored by most of the local dooryard birds, but the jays will diligently carry away and hide the pieces until all are gone."

From the foregoing evidence it may be seen that the California jay has more faults than virtues. It has few redeeming traits, and economically it does more harm than good. Its beauty and its lively manners make it an attractive feature in the landscape, but it may be that there are too many jays in California.

Behavior.—There is much in the actions of the California jay that reminds one of our familiar eastern blue jay; it is a handsome villain, but one misses the jaunty crest. It is far less shy, much bolder, more impertinent, and more mischievous. Its flight is just as slow and apparently laborious, accomplished by vigorous, heavy flappings of its wings on its usually short flights; it lives mostly at the lower levels among the trees and shrubbery and may often be seen sailing down over a brush-covered hillside with its blue wings and tail widely spread; as it glides upward to its perch it greets the observer with its harsh cries. It is quick and agile in all its movements, as it darts about through the underbrush, where it searches diligently for small birds' nests, or follows the little birds about to learn their secrets. It is not above picking a quarrel with the California woodpecker, whose stores it probably wants to steal, but, like most thieves, it is a cowardly bird and often needs the support of its fellow brigands. It is cordially disliked and dreaded by all the smaller birds. It is a nuisance, too, to the sportsman or the bird student, as its curiosity leads it to follow a human being about and proclaim his presence in such a loud voice that every creature within hearing is warned to disappear.

This jay seems to have a sense of humor or a fondness for play. Joseph Mailliard (1904) gives an amusing account of the behavior of California jays with his cats, stealing their food and teasing them. While a jay is attempting to steal food from a cat, "each has the measure of the other, and while a cat is watching, it is rarely that a jay approaches within reach of its business end, though it will do all it can to make the cat jump at it, or at least turn away. Grimalkin has learned to keep her tail well curled up when feeding, as a favorite trick of the jay is to give a vigorous peck at any extended tail and, when the cat turns to retaliate, to jump for the prize and make off with shrieks of exultation. * * * To find a cat napping, with its tail partially extended is absolute joy to one of these birds, which will approach cautiously from the rear, cock its head on one side and eye that tail until it can no longer resist the temptation, and, finally after hopping about a few times most carefully and noiselessly, Mr. (or Mrs.) Jay will give the poor tail a vicious peck and then fly, screeching with joy, to the nearest bush."

One day, after one of the cats had hidden her newly born kittens in the garden, "a faint mewing outside the window attracted the attention of someone in the kitchen when lo and behold there was a jay hauling a very young kitten out from under a young artichoke plant in the garden. The jay lugged the poor kitten along for a little way, seeming to enjoy its feeble wails, and then stopped and screeched in exultation over the find, only to repeat the process again and again. Needless to say the old cat was not present at the moment or things would have been made more lively. The bird certainly did not want to eat the kitten, and the affair seems to have been nothing else than a matter of pure mischief."

Voice.—Mrs. Bailey (1902) describes this jay's voice very well as follows: "The *Aphelocoma* voice differs strikingly from that of *frontalis,* having a flat tone and being uttered with unseemly haste. Its notes vary greatly in expression and are so emphatic and often peremptory that one cannot doubt that something important is being said. A favorite cry, used apparently to rouse attention, is quick *'quay-quay-quay-quay-quay-quay-quay.'* Another still more emphatic one is *boy'ee boy'-ee* while an inquiring *quay-kee?* is often heard. Sometimes when a jay flies down to a companion it gives its *quay-quay-quay-quay-quay* and is answered by a high-keyed *queep-queep-queep-queep*—however that may be interpreted."

Ralph Hoffmann (1927) says that "the California Jay utters a succession of harsh cries like the syllable *tschek, tschek,* slightly higher pitched than those of the Steller Jay. Another note commonly uttered when the bird is perched is a very harsh *ker-wheek.*" James B. Dixon tells me that "the male has a very pleasing ventriloquial song, which he sings during the mating season and which can be heard only a very short distance."

Field marks.— A flat, crestless, blue head, a pale brown back, blue wings and tail, and a white-streaked throat will serve to distinguish this species from the much darker crested jays of the *stelleri* group. Its jay-like behavior and its loud voice make it conspicuously different from other birds.

Enemies.—Jays are probably sometimes attacked by predatory birds and animals, but they are fairly well able to take care of themselves and defend their eggs and young. Man seems to be their bitter enemy. Large numbers are shot every year by farmers and fruit growers where the jays are damaging their crops. Organized jay shoots are popular in some parts of California, under the pretext of reducing the numbers of a destructive bird, but largely, too, as a pleasant recreation and an interesting competition for the shooters; dealers in ammunition also find it profitable. Dr. Mary M. Erickson (1937) witnessed one of these shoots and has published an interesting article on it. She says:

Jay shoots have been held in Calaveras County for many years. Two persons reported that hunts have taken place about once a year during the eleven and fourteen years they had lived in the vicinity. Two old-time residents said that occasional shoots had been held thirty or forty years previously. Recently, one or two shoots a year have been held, usually in the fall, sometimes in the spring, but the time of year and the number are irregular. The last shoot had been held on October 20, 1935, when, according to a local newspaper, 1368 jays were killed. The shoots, at least in recent years, have been conducted as contests between two teams, and after the count there has been a dinner, or as this year, a barbecue in which wives and friends shared, at the expense of the losing side.

In the shoot that she witnessed, 398 California jays, 214 Steller's jays, 1 red-tailed hawk, 1 Cooper's hawk, and 3 sparrow hawks were brought in. She estimated that an area of approximately 200 square miles was covered in the hunt, but probably not with systematic thoroughness. "On the day before the shoot, fifteen hours were spent by Mr. Hooper and me in taking a census in three sample areas of typical jay habitat, and every effort was made to get an accurate count. On this meagre basis, the population is estimated as one jay, either California or Steller, for every 5½ acres of suitable habitat, or 118 jays per square mile of such habitat. * * * In comparison with these figures, an estimate for Calaveras County of one jay for every 5½ acres, in an acre of equally good or better habitat, does not seem excessive. Assuming that only half of the total area is suitable for occupancy by jays, the jay population of the 200 square miles in which the hunting was most concentrated, would be 11,636. On this basis, the shooting of 612 jays resulted in destruction of about 5 per cent of the jay population."

The shoot in which these 612 jays were killed was at the beginning of the nesting season, when it would have the maximum effect on the breeding population. She reasons that " the shoot held in the fall of 1935 when the population was near its maximum, probably did not eliminate more than 5 per cent of the next *breeding population* [italics mine], even though twice as many were killed, for part of the kill was composed of birds which in time would have been destroyed by natural forces."

Probably this 5 percent reduction in numbers, even if accomplished every year, would have no appreciable effect on the year to year population of jays. For it is a well-known fact that every suitable habitat is filled up to its capacity to support the species; and that the removal of a few individuals makes it just so much easier for others to survive, or to drift in from outside. Natural causes probably eliminate much more than 5 percent of the species each year, but any release of pressure enables the species to expand and fill in the gap. Any attempt at a wholesale and systematic elimination of the California jays, that would be effective, would prove very expensive and would probably not succeed.

DISTRIBUTION

Range.—Western United States and Mexico; not regularly migratory.

The range of the California jay extends **north** to southwestern Washington (Vancouver); southeastern Oregon (Malheur Lake); northern Utah (Ogden); and northern Colorado (Two Bar Springs and Sedalia). **East** to central Colorado (Sedalia and Fountain); northwestern Oklahoma (Kenton); eastern New Mexico (Santa Rosa and Capitan Mountains); western Texas (San Angelo and Kerrville); Coahuila (Sierra Guadalupe and Carneros); San Luis Potosí (Chorcas and Jesús María); Hidalgo (Real del Monte); Veracruz (Perote and Orizaba); and Oaxaca (Coixtlahuaca and Mount Zempoaltepec). **South** to Oaxaca (Mount Zempoaltepec and Ejutla); and southern Baja California (Cape San Lucas). **West** to Baja California (Cape San Lucas, Llano de Yrais, San Ignacio, and San Pedro Mártir Mountains); western California (San Diego, Santa Barbara, San Francisco, Red Bluff, and Mount Shasta); western Oregon (Waldo, Corvallis, and Dayton); and southwestern Washington (Vancouver).

As outlined, the range includes the entire species, of which 10 subspecies are currently recognized. The typical California jay *(Aphelocoma coerulescens californica)* occupies the coastal region of California from

San Francisco Bay south to the Mexican border; the long-tailed jay (*A. c. immanis*) is found in the interior, from the San Joaquin and Sacramento Valleys of California north to southern Washington; the Nicasio jay (*A. c. oocleptica*) is found in the coast region of northern California; Belding's jay (*A. c. obscura*) is found mainly in the Upper Austral Zone of northwestern Baja California; Xantus's jay (*A. c. hypoleuca*) occupies the Cape district of Baja California; Woodhouse's jay (*A. c. woodhousei*) is found from southeastern Oregon and the central Rocky Mountain region south to southwestern Texas and southeastern California; the Texas jay (*A. c. texana*) is found in central Texas south to the Davis Mountains, and probably to northern Coahuila; the blue-gray jay (*A. c. grisea*), occupies the Sierra Madre region of southern Chihuahua and Durango; Sumichrast's jay (*A. c. sumichrasti*), is found in the southeastern parts of the Mexican tableland, chiefly in the states of Veracruz, Puebla, Tlaxcala, and Oaxaca; while the blue-cheeked jay (*A. c. cyanotis*) occupies the Mexican Plateau from the states of Mexico and Hidalgo north into Coahuila and Durango. There is almost endless intergradation between some of these races. The desert California jay (*A. c. cactophila*) occurs in central Baja California.

Egg dates.—California: 160 records, March 10 to July 11; 80 records, April 6 to 30, indicating the height of the season.

Mexico: 22 records, March 20 to June 24; 11 records, April 20 to May 16.

Oregon: 4 records, April 20 to June 4.

Texas: 28 records, March 14 to May 18; 10 records, March 30 to May 8.

WOODHOUSE'S JAY

Arizona: 9 records, April 5 to June 6.

New Mexico: 8 records, April 19 to May 27.

Utah: 26 records, April 6 to May 20; 14 records, April 25 to May 3.

SANTA CRUZ JAY

Santa Cruz Island: 27 records, February 6 to May 16; 13 records, March 27 to April 7.

APHELOCOMA COERULESCENS OBSCURA Anthony

BELDING'S JAY

Belding's jay now seems to be recognized as a valid race of the *coerulescens* species, inhabiting northwestern Baja California as far south as latitude 30° N. It was named by A. W. Anthony (1889) and described as "differing from *A. californica* in much darker colors and weaker feet." It was accepted by the A. O. U. committee in the 1910 and the 1931 Check-lists, and by Ridgway (1904), regardless of the fact that it is practically identical in coloration and size with the California jays found in the southern half of the coast region of California, though darker and smaller than the long-tailed jay *(immanis)* found in the interior of California, as shown by Swarth (1918). Mr. Swarth's remarks on the subject are worth quoting in full, as they throw some light on the status of this race. In his study of this genus, he says:

> The present treatment of the races of the California jay differs from that in most recent literature covering the subject (e.g., A. O. U. *Check-list,* 1910, p. 225; Ridgway, 1904, pp. 327-331) in that it does not recognize the subspecies *obscura.* This race was described by Anthony (1889, p. 75) from specimens taken in the San Pedro Martir Mountains, Lower California. In a subsequent paper (1893, p. 239) the same writer asserts that birds from the San Pedro Martir Mountains and from San Diego County, California, are indistinguishable, and for some years past the name *obscura* has been generally used to cover the bird of the San Diegan region of California. * * * Comparison of series from these points, however, with specimens from various coastal localities as far north as San Francisco Bay (including the vicinity of Monterey, the type locality of *californica*), shows that all belong to the same race, that there are no characters serving to distinguish specimens from these several places. Hence the name *obscura* must be considered a synonym of *californica.*
>
> *Aphelocoma californica obscura* was described as a smaller and darker colored bird than *A. c. californica.* Perpetuation of this error may have occurred through comparison of southern California specimens with others from the Sacramento Valley or the Sierra Nevada, in the belief that the latter were representative of typical *californica.* This assumption is wrong, however, and although jays from certain sections of California may readily be distinguished as, respectively, larger and paler, or smaller and darker, true *californica* and *obscura* both fall into the latter category.

It is significant, also, that Mr. Anthony (1893), in his subsequent paper, appears in doubt about the status of this race, for he says; "It seems, however, from the series now on hand as if *obscura* would have to be reduced to a synonym of *californica.*"

The San Pedro Mártir Mountains have produced so many new subspecies, five described by Mr. Anthony (1889) and a number more by others, that it seems worth while to quote his description of them:

> About one hundred and fifty miles south of the United States boundary, and midway between the Pacific Ocean and Gulf of California, lies a high range of mountains, which is marked upon the later maps of the peninsula as 'San Pedro Martir.' The region embraces a series of small ranges which rise from an elevated *mesa,* having a mean elevation of about 8,000 feet, and an extent of sixty by twenty miles. In these mountains are born the only streams that this part of the peninsula affords, and an abundance of pine timber is found throughout the region. Many of the ranges on the eastern side of the San Pedro Martir rise to an elevation of 11,000 feet, or even, in one or two places, to 12,000 (?) feet.

Arising as the region does from the dry, barren hills of the lower country to an elevation higher than any other on the peninsula or in Southern California, and presenting in its alpine vegetation and clear mountain streams features so different from the dry manzanita and sage-covered hills of the surrounding country, it is not unnatural to suppose that its animal life would be found to differ in some respects from that of the surrounding hills.

Mr. Anthony found this jay ranging up to 10,000 feet in these mountains. He says nothing about the habits of the bird, and nothing seems to be published on the subject elsewhere.

The measurements of 12 eggs average 27.8 by 20.8 millimeters; the eggs showing the four extremes measure **29.2** by **21.1, 26.4** by 20.6. and 26.8 by **20.2** millimeters.

APHELOCOMA COERULESCENS CACTOPHILA Huey
DESERT CALIFORNIA JAY

Laurence M. Huey (1942) gave the above names to the jays of this species found in the central portion of the peninsula of Baja California. In his description of it, he says that it "is closest to *A. c. hypoleuca* of the Cape District of Lower California. From that form it differs in the general tone of the back, which is darker, more slaty; also the under-parts are not as white as are those of *hypoleuca*, having a faint wash of gray color which is more highly exemplified in *Aphelocoma californica* [=*coerulescens*] *obscura* from the northernmost part of the peninsula and in the races farther north in upper California. The bib, or throat, and sides of neck are of a darker shade on both their blue and dusky aspects than are those of *hypoleuca*, and lighter than those of *obscura*."

Of its range, he says: "From near latitude 29° 20′, south over the width of the peninsula to the vicinity of Mulejé, on the gulf coast near latitude 27°. On the Pacific slope the range extends farther south, latitude 25° 40′ being reached before intergradation takes places."

This subspecies seems to be strictly intermediate in characters between the race to the northward of it and that to the southward of it, which might be expected in the intermediate territory that it occupies.

APHELOCOMA COERULESCENS HYPOLEUCA Ridgway
XANTUS'S JAY

HABITS

In the Lower Austral and Arid Tropical Zones of the southern half of Lower California, south of latitude 28° or 29° N., we find this smaller and paler form of the California jay. Not only is it decidedly smaller than the more northern forms of the species, but also the blue portions of its plumage are a lighter and more clearly azure blue, the under parts are more purely white, and the bill and feet are relatively larger.

William Brewster (1902) writes: "This, the only Jay known to inhabit the Cape Region, is very common and generally distributed there, being found almost everywhere from the sea-coast to the tops of the highest mountains. About La Paz it nests in March, but the birds seen by Mr. Frazar on the Sierra de la Laguna in May and early June were in flocks and showed no signs of having bred that season or of being about to breed. They probably leave the mountains before the beginning of winter and seek more sheltered haunts in the valleys and foothills at lower elevations, for Mr. Frazar did not find a single individual on the Sierra de la Laguna during his second visit, in the latter part of November, 1887."

Griffing Bancroft (1930) says of its haunts: "The habitat of these jays is arboreal associations other than those of the oases. The level country adjoining San Lucas Lagoon in places is heavily overgrown with mesquite and palo verde. The small cañons in the mountains support scattered trees. The large valleys are frequently dotted with them, especially where moisture is not too far beneath the surface. The riparian associations are almost uniformly accompanied by the taller growths. Within these limitations *hypoleuca* is common, for a jay." He also writes to me that he found it "most abundant in the vast mangrove swamps of Magdalena Bay, but did not observe it elsewhere in the mangroves."

Nesting.—Although eggs were collected by Xantus as early as 1860, Walter E. Bryant (1889) was the first to describe a nest, as follows: "A single nest of this new variety was found by myself a few miles southward from San Ignacio on April 12, 1889. The nest was built about three meters high in a green acacia near the trail. The female was sitting, and did not fly until preparations for climbing the tree had commenced. The nest was in quite an exposed situation amongst scant twigs on a horizontal branch. It is composed of small loosely laid dry twigs, and a shallow receptacle lined with fibre and horsehair."

Mr. Bancroft (1930) says of the nesting "The breeding habits of the Xantus Jay, however, are unlike those of the other races of its family, partly through choice and partly from necessity. Nearly all the nests we found were in the arrow tree whose dense growth of leaves afforded a maximum of concealment. The nest is usually in the heart of the foliage, six to ten feet above the ground. It consists of a foundation of fine twigs which support a semispherical cup. The foundation may be scanty or it may be quite pretentious, according to the requirements of its location. The cup is thin and neatly woven. It is composed of fine rootlets, tree yucca fibres, or cow-hair. It may be of one material only or the three may be used together. It is stiff enough to maintain its shape; the foundation merely serves to hold it in place."

Nesting dates seem to differ considerably in different parts of the peninsula, for J. Stuart Rowley (1935) says: "My notes show that on the last of April along the shore of Concepcion Bay on the Gulf, many nests of this jay were found, and without exception all contained newly hatched young. Then, after crossing the peninsula to the Llano de Yrais on the pacific slope, no nests were found occupied, but young were flying about in nearly full plumage (specimen of such juvenal collected there). When we reached Miraflores, in the Cape district, nesting activities were just beginning and from May 10 to 19, inclusive, at this locality eight sets of two eggs each and five sets of three eggs were taken. * * * To the northward, at San Ignacio, only one nest was found to hold even eggs, three fresh being taken on April 27; the majority of birds were apparently just building here."

One of the sets, now in the Doe collection, was taken by Mr. Rowley from a nest placed in the center of cardon growth 6 feet up.

Eggs.—Two or three eggs seem to be the usual complement for Xantus's jay, and oftener two than three constitute a full set. Mr. Rowley (1935) located over 50 occupied nests and never found more than three eggs or young in a nest, and he thinks three are "rather uncommon."

Major Bendire (1895) says that the two eggs in the United States National Museum, taken by Xantus in 1860, "have a pale bluish-green ground color and are spotted over the entire surface with small markings of grayish brown, which are slightly heavier about the larger end of the egg. The eggs are ovate in shape and slightly glossy."

Mr. Bancroft (1930) says that "laying begins in April, two eggs being the usual number. Reversing the customary order, as the season progresses the size of the clutch increases until, in June, we found three more often than two. That number represents the largest set of which we have knowledge. The eggs differ from those of any other subspecies of the California Jay in averaging a very much greener background and in being marked with decidedly finer spots."

Mr. Doe tells me that the eggs referred to above are "very dark emerald green, obscurely spotted with gray brown."

The measurements of 50 eggs average 27.2 by 20.2 millimeters; the eggs showing the four extremes measure **30.1** by **22.8, 23.4** by 17.8, and 26.8 by **13.5** millimeters.

APHELOCOMA COERULESCENS WOODHOUSEI (Baird)

WOODHOUSE'S JAY

PLATE 21

HABITS

Woodhouse's jay has long stood on our list as a distinct species and was originally described as such. There are some reasons for thinking that the original designation may be more nearly correct than the present concept as a subspecies. Mr. Swarth (1918) says: "Compared with any of the subspecies of *Aphelocoma californica, A. woodhousei* differs in coloration and in proportions of bill. The blue areas are dull and pale, the back is strongly suffused with bluish gray, and the under parts and throat with gray; the under tail coverts are blue. The general effect of these modifications is to produce a much more uniformly and inconspicuously marked bird than *A. californica*. The bill of *woodhousei* averages longer than in *californica,* but is more slender." Believing this jay to be a distinct species, he says: "The range of the Woodhouse jay in California is restricted to scattered and disconnected areas of Upper Sonoran in the Inyo region, the arid desert section of the eastern part of the state. In the late summer and fall it is a visitant to the eastern slope of the Sierra Nevada, where it comes into direct contact with *A. c. immanis,* but it apparently does not breed in this section." He claims further that, although there is intergradation between the three California races where their habitats meet, "nothing of the sort can be detected along the boundary between *immanis* and *woodhousei";* and he says that "comparison of three California specimens at hand in fresh fall plumage, with individuals taken at the same season in southern Arizona, shows no difference between the two birds."

According to the 1931 Check-list, the wide range of Woodhouse's jay extends from "southeastern Oregon, southern Idaho, and southern Wyoming south to southeastern California (east of Sierra Nevada), southern Arizona, southern New Mexico, and southwestern Texas."

Woodhouse's jay is a bird of the foothills and the lower slopes of the mountains. In Arizona we found it rather local in its distribution, mainly in the oak belts about the bases of the mountains and on the steep, brush-covered hillsides; we never saw it below 3,000 or above 7,500 feet, but noted it mostly between 5,000 and 6,000 feet. We first saw it about the base of the Mule Mountains, where the blackjack oaks grew thickly in the little valleys and gorges or were scattered over the opener hillsides. About our camp in the foothills of the Dragoon Mountains, where blackjack and other oaks dotted the gently sloping hills and grew more densely in the brush-covered gulches well up into the mountains, these jays were really common. But they were very shy. We frequently saw them flying ahead of us or sailing from one oak to another, with blue wings and tail widespread, or heard their squawking cries, but to shoot one was another matter; a long shot at one on the wing obtained my only specimen.

In New Mexico, according to Mrs. Bailey (1928), this jay is widely distributed "among the nut pines, junipers, and scrub oaks of Upper Sonoran zone." Dr. Coues (1874) says that its preference "is for oak openings, rough, broken hill-sides, covered with patches of juniper, manzañita, and yuccas, brushy ravines and wooded creek-bottoms." Dr. Jean M. Linsdale (1938b) says that, in Nevada, "resident Woodhouse jay was found regularly in small numbers about every locality worked, from a little over 6,000 feet, near the base of the mountains, up to about 9,000 feet on the ridges. Individuals or family groups were found in the thickets of willow and birch along the streams and in the piñons and mountain mahoganies on the adjacent slopes and ridges."

In Colorado these jays range from 6,000 up to 9,000 feet, where Robert B. Rockwell (1907) says that "their favorite haunt is a gulch on an open hillside, which is heavily covered with scrub-oak, service-berry and pinyon, and here they are found in numbers, flitting thru the underbrush and keeping out of sight as much as possible, but continually uttering the coarse, grating cry characteristic of so many of this family."

Nesting.—We did not succeed in finding a nest of Woodhouse's jay in Arizona; they are said to be very well concealed, and probably we overlooked some, which might easily happen, as Robert B. Rockwell (1907) says that "in the location and concealment of the nests they are evidently adepts, as in five years' observations I found but two nests, one of which was unoccupied"; and this was in a locality in western Colorado, where the birds were abundant.

He describes the finding of his nest as follows:

As my pony brushed against a peculiarly thick clump of service-berry I heard a very slight flutter and not seeing a bird fly out, I dismounted and forced my way into the clump. As I did so the bird slipped quietly out on the other side and I caught a fleeting glimpse of her as she flew, barely a foot off the ground, into a nearby bush.

The nest, for such it proved to be, was built near the center of the clump and about four feet from the ground. It was held in place by a thick net-work of small angular twigs and two larger vertical branches none over ⅜ inch in diameter. The only concealment afforded the nest was the thick mat of leaves at the extremity of the branches which formed a sort of canopy about the exterior of the bush, not a leaf being near enough to the nest to afford concealment; but right here is where I discovered the secret of their concealment. The outer structure of course so nearly resembles the network of small twigs in the service-berry bush that it was difficult to tell where the nest stopped and the twigs began.

The nest itself, which at first appeared to be a rather fragile structure, upon closer examination proved to be a remarkable piece of bird architecture. It was composed of a platform of very crooked dead twigs, thickly interlaced to form a basket-like structure, in which the nest proper was firmly placed. The latter, which was entirely separate from the outer basket was a beautifully woven and interlaced cup, composed of fine weed stalks on the outside, giving place to fine, brown, fibrous rootlets toward the interior which was sparingly lined with horsehair.

In general appearance the exterior was not unlike the nest of the white-rumped shrike, while the interior or nest proper closely resembled a black-headed grosbeak's nest. The entire structure, while not particularly artistic, exhibited a high grade of bird architecture and was remarkably strong and durable.

The nest outside measured about six inches in diameter by six inches in depth, and the interior structure measured outside 4½ inches in diameter by 2¾ inches deep; inside 3¾ inches in diameter by 2¼ inches deep.

Dr. Linsdale (1938b) thus records two Nevada nests: "An occupied nest of this bird was found on June 4, 1932, on the top of a ridge, at 9,000 feet, near Wisconsin Creek (Orr). A jay was seen to slip away through mountain mahoganies and piñons. About 100 feet from there and 30 feet down a south-facing slope a nest was found in a piñon. It was 7 feet above the ground, resting on the outer part of a limb and supported by small twigs. The outer part of the nest which was 10 inches in diameter was composed of small and medium-sized twigs of sage brush. The inner part was 5 inches in diameter and 2½ inches deep. It was composed of fine grass stems and lined with porcupine hair."

The other nest, from which the young had just flown, was found "in a mountain mahogany, 5 feet above the ground, on the east side of the tree and on a southeast-facing slope. The nest was about 9 inches in diameter and was composed of small twigs. The cup was 5 inches in diameter and an inch deep. It was made of small grass stems and horsehair."

There are three sets of eggs of Woodhouse's jay in my collection, all from Utah. Two were placed in sagebushes, one 4 feet and the other 20 inches above ground. The other was placed in a young pine next to the trunk and about 6 feet from the ground. The construction of the nests was similar to that of those recorded above from Nevada. Frank W. Braund tells me that he has a nest in his collection, taken on a cactus desert in Arizona, that was 4 feet up in a white cholla cactus. There are two nests in the Thayer collection in Cambridge; one was 5 feet up in a scrub oak, lined with black horsehair; the other was 6 feet from the ground in a cedar, lined with fine grass.

Eggs.—Woodhouse's jay lays anywhere from three to six eggs to a set, but oftener four or five. They are mostly ovate in shape, with variations toward short-ovate or elliptical-ovate. They are only slightly glossy. The ground color is light bluish green, "bluish glaucous," "pea green," or pale "sage green." They are more or less evenly marked with various shades of brown, pale shades of "ferruginous" or "tawny," in small blotches, spots or fine dots, and sometimes with a few underlying spots of pale drabs. The measurements of 50 eggs average 27.8 by 20.4 millimeters; the eggs showing the four extremes measure **31.0** by **21.5** and **24.3** by **19.2** millimeters.

Young.—Bendire (1895) says that "incubation lasts about sixteen days, both parents probably assisting; and I think that but one brood is raised in a season." Dr. Linsdale (1938b) found a family of four young birds near one of the nests described above. "Both parents," he says, "were present, and they were feeding the young ones. During this process the adult made no sound, but feeding calls were uttered by the young. Once when a young one was preparing to fly to one of its parents following the other, a slight movement by the observer resulted in a sharp call from a parent jay. Following this the young bird remained perfectly quiet for about a minute. It then began to move about, but another warning call caused it again to become silent."

Mr. Rockwell (1907) says: "The young of the year are not very much in evidence until they are well matured, but during August and September by which time the young are all able to take care of themselves the birds are particularly conspicuous and noisy. * * * As soon as the young birds are able to travel there seems to be a sort of vertical migration, during which large numbers of the birds ascend a few thousand feet into the heavier timbered country." But, he continues, "this vertical movement does not affect the entire number of the species."

Plumages.—The plumage changes of Woodhouse's jay are apparently similar to those of the California jay. But the juvenal plumage seems to be somewhat different, which Ridgway (1904) describes as follows: "Pileum plain mouse gray; rest of upper parts (except wings and tail) plain brownish gray or deep drab-gray; an indistinct superciliary line, or series of streaks, of white; general color of under parts dull light brownish gray, paler on chin, throat, chest, and abdomen, deeper and more brownish on upper portion of breast, against pale grayish jugular area; wings and tail as in adults, but smaller wing-coverts gray and lesser coverts indistinctly tipped with the same."

Food.—Like its neighbors and relatives in California, Woodhouse's jay is quite omnivorous, and its food covers the same wide range. Where oaks and nut pines, or pinyons, are abundant, the fruits of these trees evidently make up the largest percentage of the food of this jay at the proper seasons. During the summer it is said to feed somewhat on grasshoppers and other insects. Mrs. Bailey (1928) says that "in some of the few stomachs examined, three-quarters of the food consists of pinyon nuts. Acorns, wheat, ground beetles, grasshoppers, caterpillars, and ants are also eaten." This jay probably has the same bad habit as other jays of robbing small birds of their eggs and young, though perhaps not to such an extent as the California jay. Mr. Rockwell (1907) says: "I have never seen any indications of this and judging from the good feeling which apparently exists between these birds and other species I am inclined to think that their depredations are not as extensive as those of others of the jay family."

Behavior.—Based on my rather limited experience with it in the field, I should say that this is the shiest, most secretive, and most elusive of all the jays that I have seen in life. Even in its favorite haunts, where these jays are common, one is seldom seen except at a distance or as a fleeting blue shadow disappearing through the underbrush. We often saw one perched in an alert attitude on the top of some blackjack oak; but, if we attempted to approach, it bobbed its head and body, gave its harsh cry of alarm, and bounded off to a more distant tree; within about 50 yards was as near as we could come to it on the more open slopes. In the ravines and gulches, where the trees and bushes grew more thickly, we could get a closer view of it, but only for an instant, as it made an abrupt dive downward from its observation perch and faded away through the brushy tangle, not to be seen again.

The facile pen of Dr. Coues (1874) describes its flight and other movements much better than I can, as follows: "The flight of the bird is firm and direct. When going far, and high over head, in flocks, the wing beats are regular and continuous; among trees and bushes, the short flights are more dashing and unsteady, performed with a vigorous flap or two and a sailing with widely-spread wings and tail. The tail is often jerked in the shorter flights, especially those of ascent or the reverse, and its frequent motion, when the bird is not flying, is like that seen in *Pipilo* or *Mimus*. Among the branches the bird moves with agile hops, like all true Jays, and its movements when on the ground have the same buoyant ease; it never walks, like Maximilian's and other Crows."

Dr. Walter P. Taylor writes to me: "Mr. Colton, proprietor of a store at Grand Canyon, has a tame jay that he has been feeding for four years. He whistles in a certain way, and the bird flies either direct to his hand or to a nearby tree, from which it then flies to his hand. The jay stands on the hand, picking up pinyon nuts until it gets five or less. Five is the maximum. It then flies off down over the edge of the Grand Canyon with its load, returning, if called, about 10 minutes later. It ate not only from Colton's hand but also from Gilchrist's and mine."

Voice.—Dr. Coues (1874) gives the following attractive account of the vocal ability of this jay:

The ordinary note is a harsh scream, indefinitely repeated with varying tone and measure; it is quite noticeably different from that of either Maximilian's or Steller's, having a sharp, wiry quality lacking in these. It is always uttered when the bird is angry or alarmed, and consequently is oftener heard by the naturalist; but there are several other notes. If the bird is disporting with his fellows, or leisurely picking acorns, he has a variety of odd chuckling or chattering syllables, corresponding to the absurd talk of our Blue Jay under the same circumstances. Sometimes again, in the spring-time, when snugly hidden in the heart of a cedar bush with his mate, whom he has coaxed to keep him company, he modulates his harsh voice with surprising softness to express his gallant intention; and if one is standing quite near, unobserved, he will hear the blandishments whispered and cooed almost as softly as a Dove's. The change, when the busy pair find they are discovered, to the ordinary scream, uttered by wooer and wooed together, is startling.

Field marks.—Superficially, Woodhouse's jay looks and acts much like the California jay, and its voice is similar; but its coloration is much more uniform, appearing largely dull bluish gray, with less contrast between the brown of the back and the blue of the wings and tail. The under parts are much grayer, less whitish, and the blue of the flight feathers is duller.

Fall.—This jay is supposed to be nonmigratory, and it probably is mainly resident throughout the year over most of its range. Extensive fall wanderings in search of a food supply might easily be mistaken for true migration. Aiken and Warren (1914) report: "When Aiken was at his ranch on Turkey Creek in October, 1873, a migratory flight of Woodhouse's Jays was seen. They were not flying high, but making short flights from point to point, always in a southerly direction. It was estimated that there were at least 500 scattered over from 50 to 100 acres of ground, as they kept lighting after their short flights. After this flight had passed the species seemed to be fully as common during the following winter as it had been during the summer. The flight had undoubtedly come from a more northern locality. Local birds appear to be non-migratory and are found in the same localities throughout the year." Mr. Rockwell (1907) writes: "With the first frosts they congregate in small scattered flocks and perform whatever migration may be credited to them, which I am inclined to think amounts to very little, usually before the first big storm; but climatic conditions seem to have very little effect upon them, food supply alone being responsible for their migratory movements."

Winter.—Referring to their winter habits in Colorado, Mr. Rockwell (1907) says: "When the winter coat of white has entirely covered their food on the bleak hillsides, they return to their winter haunts nearer the inhabited sections where the waste from barn-yard and granary affords an abundant food supply until spring comes again. * * *

"During the winter months they are found in large numbers in the brush-clad gulches and ravines in the lower part of their range and usually not far from cultivated ground, where they feed largely upon grain and seed in the barn-yards, feedlots and fields. During this period they become very tame if not molested and will even occasionally slip into an open kitchen door in quest of some tempting morsel."

APHELOCOMA COERULESCENS TEXANA Ridgway

TEXAS JAY

HABITS

This race seems to be confined to central and central-western Texas, from Kerr and Edwards Counties to the Davis Mountains. It is a connecting link between *A. c. woodhousei* and *A. c. cyanotis,* intergrading with the former to the northward and with the latter to the southward. The blue-eared jay *(cyanotis)* was formerly supposed to occur casually in Texas, but subsequent investigation by Dr. Oberholser (1917) has shown this to be an error, and this race was dropped from our list.

The Texas jay differs from Woodhouse's jay in having the chest and lower throat very indistinctly, if at all, streaked with blue, by the paler gray of the under parts, and by the pure white under tail coverts, the latter being blue in *woodhousei.*

The type of this race was collected near the head of the Nueces River, in Edwards County, Tex., presumably by Howard Lacey (1903), who says: "In December, 1894, when deer hunting on the head of the Nueces Ricer, I shot and skinned one of these birds and sent it to the professor [H. P. Attwater]. He sent it on, I believe, to the late Captain Bendire, and it is now the type of the species." Attwater was credited with collecting the specimen.

Mr. Lacey was, evidently, the first to collect the eggs of this jay; in April 1898, near the head of one of the main branches of the Guadaloupe River. He says of the locality:

Numerous little valleys run down toward the rivers, becoming steeper and steeper as they approach the larger creek, and often forming narrow canyons with high bluffs on both sides. Large trees are not numerous, but the whole face of the country is covered with clumps of shin oak and scrubby live oak. In these clumps we found the jays' nests, generally placed near the outside of a thicket, at from four to six feet from the ground, and often conspicuous from quite a distance, as the shrubs were only beginning to put out their leaves at that time. As a rule the birds were setting and one nest contained young nearly ready to leave it. The nests were composed of an outer basket of twigs not very firmly put together, and lined rather neatly with grass, hair, and small root fibres. They were rather more bulky than mockingbirds' nests and the inner nest was saucer shaped rather than cup shaped. Most of them were placed in shin oaks, but some few were in live oaks, and I have since found several in cedar bushes. The birds are not so noisy as the common blue jay and are particularly silent when near their nests.

They have a habit of hopping upwards through a thicket from twig to twig until they arrive at the top of it, when they fly off with four or five harsh squeaks to the next clump of brush, into which they dive headlong.

Austin Paul Smith (1916) has this to say about the Texas jay:

This very local form keeps well within the Upper Sonoran, except on occasions when it descends to the streams to drink, mostly after dry weather has set in; but it quickly returns to its natural haunt—hillsides covered with a mixed growth of cedar and oak. It was found to congregate in flocks, even during the breeding season which, as Lacey has correctly stated, occupies late March and early April, so perhaps only a portion of its numbers nest annually. The Texan Jay while affecting a varied diet is very fond of the acorns of the Spanish and shin oaks, searching these out and eating them after they have sprouted. Until the plumage of this Jay is much worn, it closely resembles *A. woodhousei,* for the brown on the back is much obscured by a slaty cast in the fresh plumage while many of the adults have the under tail coverts strongly tinged with blue.

The measurements of 44 eggs average 27.0 by 20.4 millimeters; the eggs showing the four extremes measure **31.0** by 21.3, 26.0 by **21.5, 23.4** by 18.7, and 23.6 by **18.6** millimeters.

APHELOCOMA COERULESCENS INSULARIS Henshaw

SANTA CRUZ JAY

HABITS

Santa Cruz Island, one of the Santa Barbara group, off the coast of southern California, has developed this large, handsome, dark-colored jay, which once stood on our list as a full species, though it is evidently very closely related to the mainland forms of the California jay. It is confined entirely, so far as we know, to the island for which it is named. It is larger than the largest race and darker than the darkest race of *californica,* and it has definitely blue under tail coverts, which the neighboring mainland jays of the California species do not have. Mr. Swarth (1918) says: "From *A. woodhousei,* which it resembles in its blue under tail coverts, *insularis* is distinguished by greater size, darker coloration, and (like *A. californica*) in more strongly contrasted markings. * * * The Santa Cruz jay is one of the most sharply differentiated of any of the island species, and it is hard to appreciate the possibility of the development of the form under the given conditions. * * * The most striking feature of the Santa Cruz jay, as compared with the mainland species is its enormous size, so in this case a marked restriction of range, with the consequent probability of inbreeding of closely related individuals has not been productive of the dwarfed stature which such conditions are supposed to engender."

Santa Cruz Island is the only one of the group on which any jays are to be found. It is one of the larger islands, over 20 miles long and up to 5 miles in width at its wider parts; it lies about 21 miles due south across the channel from Santa Barbara on the mainland. It is a rugged, steep, mountainous island, with some of its peaks said to rise from 1,800 to 2,700 feet above sea level. The coast line is irregular and largely precipitous, with few suitable landing places except on the beaches at the mouths of the streams. A. B. Howell (1917) says: "The eastern part is very irregular, barren and almost destitute of water. The western part, however, is, in certain localities, especially near Prisoners Harbor, plentifully besprinkled with forests of Santa Cruz pine, which, in the higher parts, gives a distinctly boreal impression. At the lower edge of the pines are oaks and considerable grass land. The larger canyons are well wooded with a variety of deciduous trees, some of them quite large, and there is good water in many of them."

It was in this latter locality that we landed on June 5, 1914, and spent parts of two days exploring the vicinity of our camp. Here the deep valleys, watered by rocky streams, were well wooded with ancient live oaks, willows, and other deciduous trees and shrubs. There were scattering oaks, or small groves of them on the less exposed portions of the steep hillsides; and the opener hills were covered with wild oats, scattering bushes, and some cactus. But it was in the wooded valleys that we found the main object of our search, the grand Santa Cruz jay. Most of the other birds, towhees, vireos, flycatchers, wrens, warblers, and hummers, were concentrated with the jays in these valleys or canyons. The jays were common enough here, but were rather shy and were oftener heard than seen. We did not explore the pines at the higher elevations, but Mr. Howell (1917) says that these jays "are not equally common over the entire island, but seem to prefer the neighborhood of the pines and heavy brush."

Nesting.—Mr. Howell (1917) writes: "It is truly surprising to note the number of old jays' nests upon the island. These must either last for a greater number of years than is the case elsewhere, or else the birds are in the habit of building extra or dummy nests. The favorite sites seem to be in the tops of the local 'palo fierro' (ironwood) trees, though many were noted in low oaks or large bushes, mostly on the sides of the canyons. Construction is the same as that employed by the mainland form. The latter part of April, 1911, all the females shot had already laid, and I believe that a large majority of them had small young. Two nests that I examined on the 28th were some twenty feet up in ironwoods, and held, respectively, two small young and an addled egg, and three young, half grown."

W. L. Dawson (1923) says:

The Santa Cruz Jay nests early. The last week in March is the height of the season, counting always by fresh eggs. We have found them as early as March

10th. For nesting sites the California live oaks are leading favorites, but the birds nest indifferently throughout the scrub * * * to the tops of the ranges. Manzanita, Christmas berry, hollyleaf cherry, ironwood, mountain mahogany, scrub and Wislizenus oaks, and Monterey pines, all serve as hosts, therefore, with little preference save for shade. Nests, although bulky, sometimes being as large as a crow's, are placed at moderate heights, usually from eight to twelve feet; and are, habitually, so well made that they may be lifted clean of their setting without injury. The jays evidently have assigned beats, or ranges, of mutual adjustment, and they are very loyal to a chosen locality at nesting time. Thus, the nests of succeeding years are grouped in a single tree, or scattered narrowly in a small section of the scrub.

He says that the nest is "a bulky mass of interlaced twigs of live oak tree, into which is set neatly and deeply a cup of coiled rootlets with some admixture of grasses and, rarely, horsehair."

Eggs.—Three or four eggs, rarely five, constitute the full set for this jay. They vary in shape from rounded-ovate nearly to elliptical-ovate but are mostly ovate, with only a slight gloss. They do not show the wide variation in colors exhibited in the eggs of the California jay.

Mr. Dawson (1923) remarks on this subject: "It is in the uniform coloring of the egg that the Santa Cruz Island Jay most surely reveals its isolation, and its consequent inbreeding. The ground color of fresh eggs is a beautiful light bluish-green (microcline green), and this is lightly spotted with olive (Lincoln green to deep grape green). The green element fades quickly, however, so that eggs advanced in incubation are of a pale Niagara green color. Among a dozen sets there are no color variants worth mentioning; nor have I seen a single example of the 'red' type, which is so pleasing a feature of the mainland form."

Dawson's colored plate shows some variation in the size and shape of the markings, in pale olives and light browns; some are finely sprinkled with faint dots, others more clearly marked with small spots, and one has pale olive-brown blotches of fair size. He gives the average measurements of 140 eggs in the Museum of Comparative Zology at 29.0 by 21.3 millimeters; they range in length from 25.4 to 31.7, and in breadth from 19.6 to 22.6 millimeters.

The measurements that I have collected of 50 eggs average 29.2 by 21.6 millimeters; the eggs showing the four extremes measure 32.0 by 21.5, 29.0 by 23.0, 26.2 by 20.6, and 28.5 by 20.5 millimeters.

Plumages.—The plumages and molts of the Santa Cruz jay probably follow the same sequence as in the California jay, to which it is so closely related. The juvenal plumage seems to be only slightly different; Ridgway (1904) describes it as follows: "Pileum, hindneck, auricular and suborbital regions, and sides of chest dull slate color, slightly tinged with dusky blue; back, scapulars, rump, and smaller wing-coverts dark brownish mouse gray; upper tail-coverts dull grayish blue; chin, throat, and median portion of chest white, the last somewhat streaked with gray; under parts of body pale smoke gray, separated from the white of the chest by a narrow collar of bluish slaty, connecting the two slaty areas on sides of chest; under tail-coverts and thighs smoke gray; wings (except smaller coverts) and tail as in adults."

Food.—Practically nothing has been published on the food of this jay, which probably includes the same wide range as that of other jays. Dawson (1923) implies that it is a robber of small birds' nests and that it even invades the poultry yard. As much of its habitat is thinly settled, or entirely uninhabited by human beings, its feeding habits are not of great economic importance.

Behavior.—Its habits suggest those of the mainland forms of the genus.

During the breeding season it is quiet and secretive, but at other times it is bolder and more inquisitive, coming readily in response to the squeaking call and watching the intruder at short range in silence. Pursuing it through the brushy valleys leads to only passing glimpses of it, but its curiosity often leads it to reward the patient waiter.

J. Stuart Rowley writes to me: "It was my good fortune to be able to spend the day of April 18, 1937, on the western side of Santa Cruz Island. Here, in the scrub oak on the hillsides and in the ravines, is the home of this large, dark-blue-colored jay. Since this was my first experience with these jays, I had expected them to behave much like the mainland species. However, they were extremely shy and quiet and would duck down into the bushes as I walked along, remaining hidden from view until I was well beyond them. I was able to collect a nice series of birds by concealing myself in the brush and making a squealing distress sound with my lips on the back of my hand. This procedure brought the birds up promptly, without any voiced protest, but with wide-eyed curiosity.

"I found one nest that day, it being placed about 8 feet up in a scrub oak. When located, no sign of either bird was seen or heard; and, until I had climbed the tree and just reached into the nest, no jays would seem to be in the vicinity. However, when I started taking the four eggs from the nest, both birds burst forth with the most vociferous scolding I have ever had administered. This retiring behavior by a species of jay was a new experience for me. After the nest was reached, their restraint exploded."

Voice.—The various notes of the Santa Cruz jay are much like those of its relatives but rather harsher and suggestive of the notes of Steller's jay.

Dawson (1923) mentions a note that is strikingly like "the *rickety rack rack rack* or *shack shack shack shack shack* of the Magpies." And "exquisite warblings have I heard at a rod's remove, so delicate that a Wren's outburst would have drowned them utterly, but so musical that I had hoped the bird was only tuning his strings in preparation for a rhapsody."

APHELOCOMA SORDIDA ARIZONAE (Ridgway)

ARIZONA JAY

PLATES 21, 22

HABITS

This is the northernmost race of a Mexican species that extends its range into southern Arizona and southwestern New Mexico. Major Bendire (1895) says of its haunts in these localities: "The Arizona Jay is a common resident throughout the oak belt of southern Arizona and New Mexico which generally fringes the foothills of the mountains and ranges well up among the pines. In suitable localities these Jays are very abundant, especially so along the slopes of the Santa Catalina, Huachuca, Santa Rita, and Chiricahua mountains, in southern Arizona, and the ranges adjacent to the Rio Mimbres, in southern New Mexico. They are rarely seen any distance out on the arid plains; but after the breeding season is over small flocks are sometimes met with among the shrubbery of the few water courses, several miles away from their regular habitat."

Around the base of the Huachucas, especially where the mouths of the canyons open out toward the plains, are several large groves, or open parklike forests of large blackjack and other oaks; where the oaks extend upward into the foothills, they form a thicker growth of smaller trees, mixed with scrub oaks and other thick brush. Here, especially among the larger and opener growth at the lower levels, we found Arizona jays abundant and noisy, traveling about in groups of four to eight birds all through the nesting season. I was glad to make the acquaintance of this interesting jay, the only jay I have met that shows a tendency toward communal nesting and gregarious behavior. We saw none above 7,000 feet and found them most abundant from the base of the mountains up to 6,000 feet. W. E. D. Scott (1886) found them in the Santa Catalinas between the altitudes of 3,000 and 7,000 feet.

Nesting.—In the region referred to above we found numerous nests of the Arizona jay; nests containing eggs were found during the first three days of May, and nests containing young were seen on May 1 and May 11. On May 1 we found two occupied nests in one tree, one with eggs and one with young; in both cases the parent bird was on the nest, incubating the eggs or brooding the young. I climbed the tree to examine the nests, which caused both birds to leave and begin scolding; as soon as I retired, the bird returned immediately to the nest containing eggs, though it held but two, an obviously incomplete set. Others have seen these jays sitting on the completed nests some time before the eggs were laid.

The nests that we examined were all placed in oaks at heights ranging from 10 to 25 feet above ground, though my companion, F. C. Willard, told me that he found one as low as 6 feet in a scrub oak, and another at 9 feet. The nests were all very much alike, quite bulky and conspicuous. The base of the nest usually consists of a rough, scraggly platform, or basket, of large, coarse sticks, held in place by their crooked shapes, mixed with a mass of smaller twigs. In this basket and strongly supported by it, is a well-made cup of closely woven rootlets, lined with fine dry grasses, or with horsehair or cow's hair. Although apparently loosely built, the foundation is really very firm, and the whole structure can be removed without falling apart.

The nests we found were placed at various situations in the tree, usually on a horizontal branch or an upright crotch, but occasionally out near the end of a branch. We always saw several birds in the vicinity of the nests, and Mr. Willard told me that they live in loose colonies on a communal basis, and that three or four birds often assist in the building of a nest. Most of the nests we saw were not far apart, and all the jays in the group showed their mutual interest by flocking, with loud cries of protest, about any nest that was being investigated.

Mr. Scott's (1886) experience with the nesting of this jay is interesting and is thus related by him:

About the last of February, 1885, I noticed the birds mating, and on the 16th of March found a nest, apparently completed, but containing no eggs. There were at least half-a-dozen pairs of the birds in the immediate vicinity, but a close search did not reveal any other nests. The nest was built in an oak sapling about ten feet from the ground, and is composed of dry rootlets laid very loosely in concentric rings, and with little or no attempt at weaving together. There is nothing like a lining, and the walls of the structure have an average thickness of about three-quarters of an inch. The interior diameter is five inches, and the greatest interior depth an inch and three-quarters. The whole fabric recalls to mind a rather deep saucer. The nest was not built in a crotch, but where several small branches and twigs leave the large branch (an inch and a half in diameter) which forms the main support. All the other nests I have seen resemble this one so closely that this description will answer for them.

I did not visit the nest again until the 25th of the month, and was then rather surprised to find another nest, precisely similar to the first, only about a foot away from it on the same branch, further out from the main stem of the tree. The female bird was sitting on the nest first built, and remained there until I was about to put my hand upon her; no eggs had been laid. * * *

On the 1st of April I again visited the two nests first mentioned, and though the old bird was sitting on the nest earliest completed, it still contained no eggs. A visit to the same spot on April 7 was rewarded by finding five fresh eggs in this nest. * * * The other nest did not, at this time or afterward, contain eggs; though I visited it for several weeks, at intervals of five or six days.

The striking features developed by these observations are, first, the long period after the nest was built before eggs were laid (the nest being evidently complete on March 16, and having no eggs until later than April 1), though the old birds, one or the other, were sitting on the empty structure; and, second, the building of another nest in every way identical with the first, and very close to it, which was of no obvious use, for I never noticed either of the old birds sitting on it, as was so constantly their habit in the nest close by.

Lt. H. C. Benson, according to Bendire (1895), found this jay breeding abundantly in the vicinity of Fort Huachuca, Ariz., during April and May 1887. "All of the nests taken by him, some thirty in number, were placed in oaks, from 12 to 30 feet from the ground, usually about 15 feet high, being generally only moderately concealed." The Major describes one of his nests as follows: "It is composed outwardly of small sticks and twigs; next comes a layer of fine rootlets, well woven together—this mass is alone over half an inch in thickness—and, finally, the inner nest is lined with a liberal supply of horsehair. It is well constructed, and measures about 10 inches across externally by 4 inches in depth; the inner diameter is about 4½ by 2 inches in depth."

A nest I collected in Arizona, and afterwards presented to the United States National Museum, is somewhat larger than those described above and apparently was made with larger sticks. The outside diameter is 13 inches, or 18 inches, if the longest extruding twigs are included; the outside height is 6 inches; and the largest sticks in the foundation are from ⅕ to ¼ of an inch in diameter, rather large and long for a jay to handle.

Eggs.—The number of eggs in a set varies from four to seven; perhaps three sometimes constitute a full set; four is the commonest number, and five is not a rare number. In the United States National Museum there are 34 sets, with only one set of six and one of seven among them. The eggs are unique among jays' eggs, in being entirely unspotted. They vary in shape from ovate to elongate-ovate and are somewhat glossier than the eggs of other jays. They have been said to closely resemble eggs of the robin and the crissal thrasher, but they are larger than either of these birds' eggs and of a somewhat different color, greener than either. Bendire (1895) calls the color glaucous green. I should call it "Niagara green." The measurements of 136 eggs in the United States National Museum average 30.28 by 22.26 millimeters; the eggs showing the four extremes measure **35.1** by 22.1, 28.7 by **24.6, 26.9** by 21.6, and 30.5 by **20.3** millimeters.

Young.—Bendire (1895) says that "but one brood appears to be raised in a season, and incubation lasts about sixteen days." Whether both sexes assist in incubation and in the care of the young does not seem to have been stated, but probably the care of the young, at least, devolves on both. Swarth (1904) says that "soon after the first of June young birds begin to appear, and by the middle of the month are very much in evidence everywhere in the oak region; first sitting in the trees squalling to be fed, but very soon descending to the ground and rustling for themselves."

Dr. Walter P. Taylor tells me that he once "saw an adult feed a full-grown juvenal by regurgitation."

Plumages.—I have seen only small naked young, none with natal down. The young bird in juvenal plumage is not strikingly different from the adult in general appearance, though less blue above the browner below. The top and sides of the head are dark gray, with hardly a trace of blue; the rest of the upper parts are lighter and browner gray, including the lesser wing coverts; the rest of the wings and tail are as in the adult: the under parts are dull grayish brown; the base of the upper mandible and most of the lower mandible are yellowish horn-color in the dry skin, probably flesh-color in life. I have seen specimens in this plumage from June 2 to August 2, when the postjuvenal molt begins; this molt involves the contour feathers, but not the wings and tail; it is sometimes completed in August and sometimes not until October, but I have seen young birds in fresh, first winter plumage in September. The first winter plumage is worn until the first postnuptial molt the following summer; it can always be recognized by the yellowish base of the lower mandible, which persists through the winter and spring; I have seen it as late as May. Furthermore, young birds are always duller colored than adults, less blue above and browner below, with hardly a trace of blue on the breast; the wings and tails are also much worn by spring. Young birds become fully adult, in plumage, at their first postnuptial molt in August, but I think that some are able to breed in their first winter plumage.

Adults, with their bright blue plumage and wholly black bills, have a complete postnuptial molt, between August and October. I have seen them in worn plumage from June to August, and in wholly fresh plumage in September.

Food.—Major Bendire (1895) says: "Their food consists of grass-hoppers and insects of various kinds, animal matter when obtainable, wild fruits, seeds, and especially acorns. The latter probably form the bulk of their subsistence throughout the greater portion of the year. In the Suharita Pass, between the Santa Catalina and the Rincon mountains, near Tucson, Arizona, I noticed about twenty feeding on the fig-like fruit of the suahara, of which, like many other birds, they seemed to be vary fond."

Mr. Swarth (1904) writes: "Acorns form a staple article of diet with these birds, and they can be seen everywhere under the oak trees searching for their favorite food, progressing by means of strong, easy hops; and poking under sticks and stones, eating what they can, and hiding more for future use. On finding an acorn, a retreat is made to some near-by limb or boulder, where the prize is held between the two feet, and opened by a few well directed blows."

Mrs. Bailey (1928) adds to the list of insects eaten "beetles, true bugs, gray tree moths, and alfalfa weevils." And I once saw one at a tent caterpillar's nest, picking out and eating the caterpillars. In their fall and winter wanderings, Arizona jays come readily to feeding stations and become quite tame. Mr. Scott (1886) says that "a bone or piece of meat hung in a tree that shades my house, induced daily visits as long as the severer weather of the past year lasted." Earl R. Forrest has sent me some fine photographs (pl. 21) of these jays that came regularly to his feeding station at Oracle, Ariz. Doubtless Arizona jays, like other jays, sometimes rob the nests of small birds and eat the eggs or young, but very little positive evidence of this has been published. It apparently does little, if any, harm to cultivated fruits, or other human interests; it helps to reforest barren areas by planting acorns; and it destroys some harmful insects; its economic status is neutral, or perhaps beneficial. Dr. Taylor tells me that they do some damage to deer carcasses, or other meat hung up outdoors and unprotected; he saw them also eating and carrying off sausages that he had thrown out.

Behavior.—The Arizona jay is one of the most interesting birds of the family, unique in more ways than one. It is the only one of our jays that is markedly gregarious at all seasons, traveling about in scattered flocks of 6 to 20 or more birds; even in the breeding season it lives under semicommunal conditions, with mutual interest in all the nests in the community, helping to build and defend its neighbors' nests and young, shrieking loud invectives at the intruder, with much bobbing of heads and twitching of tails. All this is in marked contrast with the solitary and secretive habits of other jays during the breeding season. Mr. Swarth (1904) writes of its behavior:

Noisy, fussy and quarrelsome as all the jays are, I know of no other species which possesses to such an eminent degree the quality of prying into all manner of things which do not concern it, and of making such a nuisance of itself in general, on the slightest provocation or on none at all, as the Arizona Jay does. * * * A Red-tail or Swainson Hawk sitting on some limb, furnishes a little excitement until he removes to some quieter locality; but the crowning joy of all is to find some wretched fox or wild cat quietly ensconced on some broad, sheltered, oak limb. In such a case the one that finds the unhappy victim takes care to let every jay within half a mile know from his outcry that there is some excitement on hand; and it is nothing unusual to see thirty or forty birds gathered about the object of their aversion, letting him know in no undecided terms just what their opinion of him is. It is a curious sight also to see a dozen or more gathered around some large snake, which they seem to fear nearly as much as they hate. On one occasion I had an excellent opportunity of watching about twenty Arizona Jays protesting at the presence of rather a large rattlesnake which was leisurely travelling down a dry watercourse which passed our camp. The jays seemed imbued with a wholesome fear of their wicked looking antagonist, and though they surrounded it, kept at respectful distance; they were not as noisy as they often are, but kept uttering low querulous cries, quite different from their usual outbursts. Some of the boldest lit a short distance from the snake and strutted before it in a most curious fashion, head and body held bolt upright, and the tail pressed down on the ground until about a third of it was dragging. * * * Besides his vocal outbursts, the Arizona Jay makes when flying a curious fluttering noise with his wings, loud and distinct enough to be heard some little distance producing a curious effect; especially when, as often happens, a troop of them comes swooping down some steep hill side to the bottom of the canyon. Though wary and cunning to a marked degree, so that it is usually impossible to get within gun shot of them, still their curiosity leads to their destruction; for it is a simple matter for the collector, by hiding behind a bush and making any squeaking or hissing noise, to get all the specimens desired.

Bendire (1895) says: "Their flight appears to me far less laborious than that of the California Jay. It reminds me of that of some of our Raptores, rising now high in the air, partly closing their wings, and then darting suddenly down, then up again, and repeating these movements for some time."

Voice.—Herbert Brandt (1940) records the call of the Arizona jay as "a rather rasping, nasal 'wait-wait-wait', given as rapidly, and in such a key, as the occasion may warrant." My own brief field notes record a loud and incessant alarm note sounding like *wack, wack, wack;* I have also heard a soft, cooing note, like *cŏot, cŏot, cŏot,* in a conversational tone. Dr. Taylor writes to me that "one conspicuous call-note is a *weent! weenk! weenk!,* with rising inflection."

Field marks.—Within its limited range in the United States, there is no other bird with which this bird is likely to be confused. It is a large jay, much larger than its neighbor, Woodhouse's jay. It has a dull-blue head, bluish-gray or dull-blue back, wings, and tail, and bluish-gray or dull-gray under parts.

The brown or gray back and the streaked throat of the other species of *Aphelocoma* are conspicuously absent; it is decidedly more uniformly colored than the other jays, with no very conspicuous markings. It can be distinguished from the long-crested jay by the entire absence of any crest.

Winter.—It is resident practically all winter long throughout its range, though it wanders about more or less, in large or small flocks, visiting the ranches, farms, and houses to pick up what scraps of food it can find, and it will become quite familiar and friendly where it is fed regularly.

DISTRIBUTION

Range.—Southern Arizona, southwestern New Mexico, and Texas south to the southern part of the Mexican tableland; nonmigratory.

The range of this species extends **north** to central Arizona (Payson, Strawberry Valley, and Fort Apache); southwestern New Mexico (Silver City and Fort Bayard); southwestern Texas (Chisos Mountains); and northern Neuvo León (Parras). **East** to Nuevo León (Parras and Monterey); Tamaulipas (Realito and Galindo); and southeastern Veracruz (Jalapa and Mount Orizaba). **South** to southern Veracruz Mount Orizaba); Puebla (Texmelucan and San Pedro); Mexico (Toluca); Michoacán (Patzcuaro); and Colima (Sierra Madre). **West** to Colima (Sierra Madre); Jalisco (Sierra Nevada); Nayarit (San Sebastián and Santa Teresa); Durango (Salto, Arroyo del Buey, and Providencia), northeastern Sinaloa (Sierra de Chaix); eastern Sonora (Oposura, La Chumata, and Saric); and Arizona (Baboquivari Mountains, Tucson, Oracle, Salt River Refuge, and Payson).

The range as outlined includes the entire species, which has been separated into several subspecies, only two of which are found in the United States. The form known as Sieber's jay *(Aphelocoma sordida sieberii)* is found on the southern parts of the Mexican Plateau; the Colima jay *(A. s. colimae)* occupies the southwestern portion of the Mexican tableland; the San Luis Potosí jay *(A. s. potosina)* is found in the northeastern portion of the Mexican Plateau; and the Zacatecas jay *(A. s. wollweberi)* occupies the central and northwestern parts of the plateau. The Arizona jay *(A. s. arizonae)* is found from Arizona and New Mexico south to the northern parts of Sonora and Chihuahua, while Couch's jay *(A. s. couchii)* occurs from the Chisos Mountains of southwestern Texas south to the southern part of Nuevo León.

Egg dates.—Arizona: 87 records, March 25 to July 4; 43 records, April 17 to May 8, indicating the height of the season.

Texas: 3 records, April 27 to June 13.

New Mexico: 4 records, April 24 to June 2.

APHELOCOMA SORDIDA COUCHI (Baird)

COUCH'S JAY

HABITS

This is another of the *sordida* group of jays that extends its range northward from Mexico into the Chisos Mountains in central-western Texas. Van Tyne and Sutton (1937) report the capture of a specimen near Alpine, which is near the northern end of Brewster County, Tex., and about 85 miles north of the Chisos Mountains, "the only record for Couch's Jay in the United States outside of the Chisos Mountains."

Referring to the subspecific characters of this race, they write: "Since the published descriptions of this subspecies provide no very satisfactory comparison between *couchii* and *arizonae* we offer the following comments: *couchii* is brighter, richer blue above than *arizonae*, especially on the head, rump, wings, and tail; in *couchii* the gray-brown of the back is darker and more contrasted with the blue of the head and neck than in *arizonae;* in *couchii* the throat is white in rather sharp contrast with the gray of the breast, while in *arizonae* the throat is gray, shading gradually into the gray of the breast; the thighs of *couchii* are gray or blue-gray but in *arizonae* they are practically concolor with the flanks; the bill of young *arizonae* is mottled with yellowish, and this often persists for at least a year, but in *couchii* the bill becomes entirely black soon after the young bird leaves the nest."

They found this jay "common everywhere above the lower limit of trees (about 5000 feet)" in the Chisos Mountains. Herbert Brandt (1940) also found Couch's jay very common in these mountains and has written considerable about it. As to its relationship to the Arizona jay he says: "The geographically connecting link between these two birds is said to be far south in Mexico, so the territories occupied by these two distantly related subspecies are the terminals of a long, horseshoe-shaped range. The habitats of the two birds, instead of being separated by but four hundred miles—or the distance from the Big Bend to Arizona—are in reality some eighteen hundred miles apart. This may be seen by following south their respective mountainous territories—one along the western highlands of Mexico, the other on the eastern side—until these mountains blend into each other on the plateau of lower mid-Mexico. The evidence, however, from the standpoints of dissimilar eggs, nest, voices, and size, perhaps explains in part the reason for the view of the older ornithologists that the Couch and Arizona Jays were two distinct species."

Nesting.—Van Tyne and Sutton (1937) report the following nests observed in the Chisos Mountains: "On April 27, 1935, Semple found a nest and four eggs in a willow oak that stood along a dry stream bed at the lower margin of the Basin, * * * at about 5000 feet. The nest, which was compactly built, was about fifteen feet from the ground on a bough that extended over the stream bed. The eggs were fresh. The parent birds were noisy in defense of their nest. * * * On May 1, 1935, Sutton found two nests, the first in the process of construction, in a pine tree not far from the trail to Laguna at an elevation of 6000 feet; the second with two eggs, an incomplete set, in an oak near Boot Spring. Both nests were well built, with an ample cup which was neatly lined with rootlets. On May 17, 1932, Peet found a nest with three well-fledged young ten feet from the ground in a small oak tree near Boot Spring."

Mr. Brandt (1940) found a number of nests of this jay in the same region and says: "The Couch Jay of the Chisos Mountains appears to have but little individuality in its nest building, for one abode seems just like the other, and evidently it is perfectly satisfied with the appointments and furnishings of its home. The noteworthy feature of each nest is the thick horsehair lining, which is woven concentrically into a smooth, springy mattress of trim saucer shape. * * * The builder is very particular in this respect, because it never seemed to use any other lining in the nests we saw than the long black or white tail or mane hairs of horses."

He remarks that horses are scarce in that region and that the jays must work hard to procure enough hair; but he suggests that, as he found no hair lining in the old nests, the birds may rob the old nests to line the new. He describes a nest that they found in a small oak, 12 feet up toward the top, in a tricleft crotch of a level limb. "It was typical in every way of a jay's home, as it was made up of two distinct parts, the outer portion having a well-laced platform of twigs neatly laid at various angles, and the inner structure composed, to a depth of half an inch, of curly white grass, this lined in turn with a mat of circularly woven, black-and-white horsehair to the thickness of about an inch, and giving to the structure a soft gray color."

He mentions another nest, at least 30 feet from the ground in a tall pine tree. One was found "in an oak sapling, not over three feet from the ground and fully exposed to view. Another nest was revealed in an upper crotch of a pinyon, in which the sitting bird was at home, its beautiful blue tail protruding colorfully over the side until I was able almost to touch it, when it left headlong, to reveal four purple-skinned jaylets."

Eggs.—Four eggs seem to constitute the normal set for Couch's jay. One would expect this jay to lay eggs like its Arizona subspecies, but this is not so. Such an occurrence is unusual among subspecies. I have not seen any of its eggs, but they have been well enough described by others to show the difference. Mr. Brandt (1940) says: "All ten sets of Texas eggs that we observed were invariably dotted with dark, greenish spots on a paler greenish ground color, closely resembling the eggs of the California Jay group."

Van Tyne and Sutton (1937) say: "The eggs are Pale Nile Blue speckled and blotched with pale brownish markings, ranging from mere specks to blotches one and two millimeters in diameter. The markings tend to be concentrated more about the large end, and on one egg there is a distinct wreath of marks about that end."

The measurements of 28 eggs average 28.6 by 21.9 millimeters; the eggs showing the four extremes measure **30.3** by 21.3, 29.0 by **24.0, 26.0** by 22.0, and 28.5 by **20.7** millimeters.

Plumages.—The plumage changes of Couch's jay are probably similar to those of the Arizona jay, except that, as mentioned by Van Tyne and Sutton (1937), the young bird acquires an entirely black bill soon after it leaves the nest, whereas in the Arizona bird the base of the bill remains light colored for about a year. Also, whereas the Arizona bird begins the postnuptial molt about the first of August, they found that Couch's jay was "molting heavily at least a month earlier."

Food.—What small scraps of information we have on the food of this jay indicate that it does not differ materially from that of the Arizona jay, or from that of the *Aphelocoma* genus in general. Its haunts are so far removed from human settlements and agricultural regions that it is probably of little economic importance. Van Tyne (1929) says: "Their food consisted mainly of coleoptera and orthoptera, together with a few nuts and seeds. They probably also raid the nests of small birds, for I saw them repeatedly pursued by Scott's Orioles and Mockingbirds whose nests they had approached."

Mr. Brandt (1940) emphasizes the latter trait, saying: "At nesting time this winged coyote is the most competitive of all egg collectors as it moves wantonly through the wild countryside, sly, observant, and alert. It is when the callow young are in the cradle that it is most insatiable, and levies a heavy toll wherever the slightest opportunity offers."

Behavior.—Mr. Brandt (1940) noted that these jays "traveled about in scattered groups numbering up to a dozen birds, which usually moved forward through the trees by short flights, one bird flying over the other and then alighting until those in the rear repeated the maneuver. Their more extended flight is direct, and performed by quick wing-beats, with brief cessations at intervals." He found it an easy matter to call up **an unseen flock of these jays**, almost anywhere within their habitat, by using the well-known squeaking call. "If the caller be concealed beneath a tree the bird flies directly to it; its decurved wings then produce an air-drumming that in its rapidity results in a peculiar, hollow, booming whir. The birds may soon discover the human source of the unusual noises, and thereupon depart, but will return again and again to the place, each time apparently as excited and bubbling over as at the first experience. Here we find a strange anomaly, for this bird has a reputation as the most cunning, wise, and wary of the mountain dwellers, yet it is the one most easily duped by confused squeaks, usually making a ridiculously entertaining show of itself."

Voice.—Comparing the voice of this jay with that of the Arizona jay, he says: "The much different, and more agreeable, call of the Couch Jay is made up of groups of from three to six notes, and to my ear may best be written as 'oint-oint-oint,' being delivered more slowly, and evenly, in a high pitch. We learned that the Texas bird has, too, an additional and entirely different note which I have never heard from the Arizona jay; it is a loud, rattling, throaty cackle, resembling somewhat that given by the Blue Jay, although only an occasional individual in a flock seems to render this call. This particular bird, when the flock is agitated, will issue its gurgling scream repeatedly, but the others make no attempt to join."

Dr. Van Tyne (1929) noted that "they were very noisy, constantly repeating a shrill, rasping *scree, scree, scree.* I also heard them give a peculiar rattling note, not unlike the call of *Dryobates pubescens.*"

Enemies.—Baird, Brewer, and Ridgway (1874) relate the following incident, apparently quoted from the notes of Lt. D. N. Couch, for whom the subspecies was named: "Near Guyapuco a large snake *(Georgia obsoleta)* was seen pursued by three or four of this species. The reptile was making every effort to escape from their combined attacks, and would, no doubt, have been killed by them, had they not been interfered with. The cause of so much animosity against the snake was explained when, on opening its stomach, three young of this species, about two thirds grown, were found."

XANTHOURA YNCAS GLAUCESCENS Ridgway

GREEN JAY

HABITS

This brilliantly colored jay brings to that favored region of the lower Rio Grande Valley in Texas a touch of tropical color that adds much to the many thrills one feels as he meets for the first time the many new forms of Mexican bird life to be found only in that unique region.

As I sat on a log near the edge of a stream in a dense forest along one of the resacas near Brownsville, I caught my first glimpse of a green jay, a flash of green, yellow, and blue, as it flitted through the thick underbrush and the trees above me. In spite of its brilliant colors it was surprisingly inconspicuous among the lights and shades of the thick foliage. I had just been admiring the dainty little Texas king-fisher that flew down the stream and perched on a fallen snag, had been lulled almost to sleep by the constant cooing of the many white-winged doves, and awakened again by the loud calls of the gaudy Derby flycatcher. The curious chachalaca and the red-billed pigeon had their nests in the vicinity, and there were a host of other interesting birds all about me, but the green jay was the gem of the forest.

I am wondering how much longer this bird paradise will last, for I have read that huge tractors have been uprooting the forest trees, clearing up the chaparral, and plowing up the rich land to make room for the rapidly growing citrus orchards and other expanding agricul-tural interests. Thus will soon disappear the only chance we have of preserving on United States soil this unique fauna and flora; and all these interesting birds will have to retreat across the Mexican border, leaving our fauna that much poorer.

According to Baird, Brewer, and Ridgway (1874), "Colonel George A. McCall, Inspector-General of the United States Army, was the first person to collect these birds within our limits. He obtained them in the forests that border the Rio Grande on the southeastern frontier of Texas. There he found them all mated in the month of May, and he felt no doubt that they had their nests in the extensive and almost im-penetrable thickets of mimosa, commonly called chaparral."

We learn more about it from the writings a number of years later of Dr. James C. Merrill (1876 and 1878) and George B. Sennett (1878 and 1879). The latter writes (1878): "It was first met with on April 2nd, in the vicinity of Brownsville; but it was not until we reached the heavier timber about Hidalgo that we saw it in full force. They were there April 17th in pairs, and busy constructing homes. They are most frequently seen during the breeding season in the densest woods and thickets, but at other times I am told they are common visi-tors of the camp, the ranche, and the huts in the outskirts of towns, to the annoyance of all on account of their thieving propensities."

The subspecies *glaucescens* is smaller and its coloration is paler and duller than in the other four races of the species found in Mexico and Central America. Its range extends from the lower Rio Grande Val-ley southward into northern Tamaulipas and Nuevo León.

Nesting.—Most of the information we have on the nesting of the green jay comes from the two observers mentioned above. Mr. Sennett's (1878) first nest was taken on April 28 "from a mezquite-tree standing in a dense thicket not far from the river-bank, and contained four fresh eggs. It was situated in a fork about fifteen feet from the ground, and was composed of sticks lined with fine stems, and a rather bulky affair." He tells of a nest, found on April 30, that "was some nine feet from the ground on the outer branches of a small tree, and composed wholly of sticks and fine twigs. The sticks were so full of thorns that when they were crossed about among the lining branches more firmness was given to the nest than usual, and by cutting off the branches I could readily take it entire. The outside diameter is nine inches one way by eight the other; its depth is four inches; inside, three and a half inches wide by two inches deep."

Dr. Merrill (1876) reports a nest, taken on May 27 near Hidalgo, Texas: "It was placed on the horizontal branch of a waican-tree, about twenty-five feet from the ground, and was built of twigs and rootlets; the cavity was slight, and the entire structure so thin that the eggs could be seen through the bottom. These were three in number, and were quite fresh. * * * A second nest, found in the same vicinity May 8, was on a sapling seven feet from the ground; it closely resembled the first one, and contained four eggs, three far advanced in incubation; the fourth * * * was quite fresh."

Major Bendire (1895) says of the nests: "The nests are generally placed in dense thickets and well hidden among the branches at heights varying usually from 5 to 10 feet from the ground, and rarely in large trees. They are frequently found in retama, anacahuita, brasil, and hackberry bushes or trees. The outer nest consists usually of a slight platform of small thorny twigs and branches, sparingly lined with fine rootlets, small pieces of a wire-like vine, bits of moss, and occasionally dry grass and leaves. The Green Jay apparently does not use mud in the construction of its nest. * * * It is probable that two broods are sometimes raised in a season."

Eggs.—Major Bendire (1895), with a large series of eggs before him, writes: "The number of eggs laid by this species is from three to five; sets of four are most often found. The prevailing ground color of these eggs is grayish white, occasionally pale greenish white or buff color. They are profusely spotted and blotched—but never heavily enough to hide the ground color—with different shades of brown, gray, and lavender; these markings are generally more abundant about the larger end of the egg. The shell is close grained, moderately strong, and shows little or no gloss. Their shape is mostly ovate, and sometimes short ovate."

The descriptions of the eggs, as given by others, are not very different. Mr. Sennett (1878) says that "the ground-color is usually light drab, tinged faintly with green, but I have one egg out of a set of four with the color dull yellowish-white. The markings are brown, sometimes distinctly spotted or speckled or streaked, and sometimes quite indistinct and clouded." Dr. Merrill (1876) mentions one that "differed in having the markings most numerous at the smaller end." The measurements of 70 eggs in the United States National Museum average 27.31 by 20.43 millimeters; the eggs showing the four extremes measure 30.8 by 21.8, 24.9 by 20.3, and 25.9 by 19.1 millimeters.

Plumages.—Dr. Herbert Friedmann (1925) found a nest containing four young birds about four days old. "The young were still blind but the primary and secondary quills were beginning to sprout and were dull bluish in color. The top of the head and the spinal tract (both the skin and the neossoptiles) were greenish gray in color."

I have seen no very young birds, but Ridgway (1904) describes the juvenal plumage as follows: "Pileum, hindneck, and malar patch greenish blue, the forehead and palpebral spots similar but paler, and the nasal tufts darker; black of chin, throat, chest, etc., much duller than in adults; under parts of body very pale yellowish green or greenish yellow anteriorly, fading on flanks, abdomen, under tail-coverts, etc., into very pale creamy yellow; otherwise like adults."

Food.—Austin Paul Smith (1910) charges the green jay with feeding on the eggs and young of various small birds, such as thrashers, orioles, sparrows, wrens, chats and mockingbirds, and that, outside of the nesting season, it feeds "mostly on seeds and insects. In winter the seeds of the Ebony *(Siderocarpus)* is the main reliance; also in less quantity the fruit of the Palmetto, to secure which they will travel far into the open."

Cottam and Knappen (1939) have given us the most detailed account of the food of this jay as follows:

Smith's indictment is not confirmed by the only previous record of stomach contents with which the writers are familiar—that from Vernon Bailey's field notes for April-May 1900, at Brownsville, Texas: 'Stomach contained one grasshopper, beetles, small insects, and part of a kernel of corn.' Except for the corn this food listed by Bailey is similar to that found by us in the stomach of a Green Jay collected by Dr. Francis Harper near Norias, Texas, on August 17, 1929. One per cent of the food consisted of fragments of two seeds of a bristlegrass *(Setaria ? grisebachii)*, seed fragments of pricky-ash *(Xanthoxylum clava-herculis)*, and undetermined plant fiber. The remainder of the food was animal in origin, including fragments of: sixteen or more stink bugs *(Brochymena* sp.), 79%; several coreid bugs *(Acanthocephala* sp.), 8%; finely ground indeterminable bugs (Heteroptera), trace; a short-horned grasshopper nymph (Acrididae), 1%; a field cricket nymph *(Gryllus assimilis)*, 1%; a hymenopteran, trace; indeterminable insect, trace; and a fragment of a spider, 1%. Gravel formed 1% of the gross contents.

Behavior.—All jays are more or less alike in their behavior, and the green jay is no exception to the rule. It is a noisy and conspicuous bird, making itself known by its harsh notes and its gaudy plumage. As a rule it is not at all shy and has the usual supply of jay curiosity, and the collector should have no difficulty in obtaining all the specimens he needs. Dr. Merrill (1878) says that "it is often very tame and bold, entering tents and taking food off plates or from the kitchen whenever a good opportunity offers. Large numbers are caught by the soldiers in traps baited with corn, but the plumage is their only attraction as a cage-bird."

Austin Paul Smith (1910) says that the green jay—

ranks above all its North American cousins in plumage, tho not in bearing. I have yet to find a species of crestless jay that is free of cowardly disposition and sneaky manner. It is born in them. The crested members of this group, as most of us well know, are no disciples of uprightness, but they can hide their faults, in a large degree, by a dignified appearance. Unluckily, for the Green Jay, his feathers seem to accentuate his sins. * * * It is another resident species, most at home in heavy growth along the river; altho from there it will often wander on foraging expeditions, even inspecting rural barnyards when hunger be pressing. The Green Jay is the worst gourmand in its family; and this failing often causes it to lose its liberty. Its plumage makes it very attractive as a cage bird, and to secure one only requires a wicker cage, set in a conspicuous place and baited with meat of some kind; fitted with a trap door worked by a string held by some hidden Homo, who possesses the instinct to pull the string at the opportune moment. Captivity does not curtail the Jay's appetite, and they have been known to accept food immediately after being trapt. Indeed, this bird will eat all the time if food be accessible; and the indulgent owner finds it a matter of difficulty to keep the bird alive more than a week, but such individuals as are fed with discretion, will live to make interesting, altho noisy pets. In a wild state, the Green Jay is suspicious as becomes the tribe, tho as a rule it falls to a ruse quitely easily. If one be shot, the balance set up a din that can ordinarily only be stopt, either by shooting them all or decamping from the neighborhood.

DISTRIBUTION

Range.—Lower Rio Grande Valley south to Guatemala; nonmigratory.

The range of the green jay extends **north** to northern Jalisco (San Sebastián); and southern Texas (Rio Grande, Lomita, and Brownsville). **East** to Texas (Brownsville); Tamaulipas (Río Cruz); Veracruz (Tampico, Jalapa, and Presidio); Yucatán (Río Lagartos and Chichen Itzá); Quintana Roo (Chunyaxche); British Honduras (Belize and Mantee); eastern Guatemala (Santa Tomás); and northeastern Honduras (Omoa). **South** to northern Honduras (Omoa, Santa Ana, and Chamelicon); southwestern Guatemala (Zapate, Naranjo, and Patio Bolas); and Oaxaca (Santa Efigenia, Guichcovi, and Pluma).

West to western Oaxaca (Pluma); Guerrero (Rincon); and northwestern Jalisco (San Sebastián).

The range as outlined is for the entire species, which has been separated into several subspecies. One race *(Xanthoura yncas luxuosa)* occupies the eastern edge of the Mexican Plateau from Veracruz and Puebla north to southern Tamaulipas and Nuevo León; the Rio Grande green jay *(X. y. glaucescens)* is found in the lower Rio Grande Valley in Texas and south into the Mexican states of Nuevo León and Tamaulipas; the Tehuantepec green jay *(X. y. vivida)* is found in southwestern Mexico and northwestern Guatemala; the Guatemalan green jay *(X. y. guatimalensis)* occupies the country north and east from northern Honduras to the Yucatán Peninsula; while the Jaliscan green jay *(X. y. speciosa)* is apparently confined to the state of Jalisco.

Egg dates.—Texas: 47 records, April 2 to May 29; 25 records, April 15 to 30, indicating the height of the season.

Mexico: 2 records, April 23 and 30.

PICA PICA HUDSONIA (Sabine)

AMERICAN MAGPIE

PLATES 23-25

CONTRIBUTED BY JEAN MYRON LINSDALE

HABITS

The magpie has been closely associated with man for many centuries in many parts of the Northern Hemisphere, and the lore concerning it has developed in great amount. Sometimes hated for its disagreeable traits, sometimes admired for its attractive ones, this bird has remained one of the best-known to people who live near it. And wherever it occurs, the magpie tends to favor lands also occupied by man.

In America only persons living in the North and the West have opportunity to become intimately acquainted with magpies in their normal haunts. Of the close to 20 distinguishable kinds of magpies in the world, two occur in northwestern North America. The black-billed one, most like its relatives in Europe and Asia, occupies much of the mountainous country west of the prairies and north of the deserts even to the Alaska Peninsula. It avoids the deep forests and the dry, open plains, but it is at home in the canyons and on streamsides where tall thickets and scattered trees provide cover from pursuit, sites for nests, and clearings for foraging on the ground.

The magpie is one of the larger birds in any locality in which it is found. Its structure and inherited habits enable it to feed upon a wide variety of foods, including both plant and animal matter. A bird of its size is able, probably, in the region it inhabits, to find food of this nature in sufficient quantity most easily by foraging on the ground. The most productive ground is in the open where there is low-growth vegetation. A magpie's wings are short and rounded and so shaped that it cannot fly rapidly or far. Therefore, if it is to escape from pursuit, it must stay in places where it can move rapidly into thick clumps of brush. These two circumstances, then, tend to restrict magpies to places where there is open forage ground and where clumps or bushy trees and bushes are scattered over the landscape. Further limitation requires trees and bushes of sufficient strength to support the bulky nest. These suitable nesting trees are oftenest found along the streams.

A rather striking relation to climate exhibited by this bird has not been clearly explained. On the map of the dry climates of the United States published by Russell (1931) the region marked as Cold Type Steppe Dry Climate is almost exactly the range of the black-billed magpie. The boundaries coincide everywhere within a few miles. In this type of climate the mean January temperatures furnish the greatest contrast with the climate of the region occupied by the yellow-billed magpie.

Magpie, the name for this bird now used almost universally among English-speaking people, is a contraction of Magot Pie, a Middle English name for the bird. According to Swann (1913) the first part of the name appears to have no reference to the bird's habit of picking maggots from the backs of sheep (as some persons have supposed), but it is "derived from the French Margot, a diminutive of *Marguerite,* but also signifying a Magpie, perhaps from its noisy chattering, in which it is popularly supposed to resemble a talkative woman." The second part of the name is supposed to come through French from the Latin *pica,* which refers to the black-and-white coloration of the bird.

Throughout the range of the group the many allusions to magpies in folklore and the superstitions concerning them demonstrate widespread familiarity with the bird in early times. The same tendency is reflected in the large number of vernacular names, more than 400, that have been applied to the magpie.

Because a long account of all the magpies, which I once assembled, was published so recently (Linsdale, 1937) and only a little additional information has been available, this entire story has been mainly extracted from the earlier one, which may be consulted for more extensive detail.

Courtship.—The study of courtship in magpies is especially difficult because their gregarious habit, their marked shyness, and aversion to entering small traps ordinarily used for birds make it practically impossible to distinguish individuals in a natural group and then to keep track of them for any considerable length of time. The screen of their habitat further complicates any attempt to study this part of magpie life. Repeated glimpses, however, result in a composite notion that must approximate the true story. The most important items have come from observers in Europe, but they seem to apply equally as well to the American form. The available information indicates that many magpies remain together in pairs through the year, and the changes consist mainly in the replacement of lost mates and the entering of young birds into the life of the flock.

Aretas A. Saunders reports (MS.) that he saw birds going through what he believed were courting actions, on March 12, 1911, in Montana. Three he took to be males were following one supposed female.

Nesting.—Nests of magpies are substantial structures that ordinarily endure several seasons of weathering. They are so characteristic of the species that they are useful as certain indicators of the nesting of the birds in any locality where they may be found. In some instances the nests remain long after the birds have given up nesting in a place or have been driven out by human interference.

Observers generally have agreed that only a single brood is reared in a season. Bendire (1895) found that whenever a set of eggs was destroyed a second and sometimes even a third set would be laid. These were frequently in the same nest or in one close by. The second set usually contained a smaller number of eggs, five or six.

Many observers have concluded that a pair of magpies builds a whole series of nests each season, of which they actually use only one. This impression possibly resulted from the tendency of the birds to desert a nest upon slight disturbance at an early stage of its construction. Then, too, these conspicuous structures are more impressive than is the case with smaller birds which may exhibit the same trait.

Actual building in the Western States usually begins in March and requires about six weeks. Before this for several weeks the birds show indications of the coming nesting. Early stages are interrupted often by late snows, but the birds ordinarily come back and resume their work with the return of mild weather. A nest found in Washington by Averill (1895) and started on March 22 had beginnings of a roof on April 10 and held the first egg on May 1. In Colorado nests half completed by the end of March contained partly incubated eggs on April 28 (Gilman, 1907).

Black-billed magpies nest in small colonies at sites separated and strung out along some stream course or over some area of woodland or thicket. Such a group of nests may occupy an area half a mile across or a line a mile or more in length. Kalmbach (1927) found 26 broods of young within a mile along a creek in Utah. Each group of nesting pairs is separated from the next group, often by an intervening area of unfavorable habitat, but sometimes the vacant places result merely from lack of birds to fill them. At any rate the favorable habitat is seldom filled and the nesting pairs tend to be clustered rather than spread evenly.

The nest of the magpie is noteworthy for the large quantity of material contained in it. Ordinarily the structure measures about 2 feet in height by 1 foot in diameter, or slightly larger. A nest found by Silloway (1903) in Montana was 4 feet high, 4 feet long, and 40 inches wide. Dawson (1909) found one in Washington 7 feet from bottom to top, the upper one-third being the dome, and another not over a foot in diameter and scarcely that in depth.

The base and outer walls of a nest are composed of coarse material, heavy sticks, often thorny ones. The materials vary greatly, depending mainly upon the nature of the available supply. The sticks may be 2 feet long, and often they are pulled from sagebushes or cottonwood trees. Sometimes they are picked up from the ground. Inside the base is a heavy cup of mud held together with some vegetable material. Fresh cow dung sometimes is substituted for the mud. Within the cup is a lining, installed after the construction of the dome and after the nest appears completed. It is composed of rootlets, fine plant stems, or horsehair.

The dome or canopy built over the whole structure gives the magpie nest its bulky appearance and provides questions of special interest in the life of this bird. This canopy is usually made of thorny twigs, pulled from nearby plants, and it has one or more openings in its side for the birds to enter and leave. Sometimes there is no dome, but this is rare. A nest found by Potter (1927) was in a railroad bridge, directly under one of the rails and between two ties. There was no room for a dome and none was needed. The track was used by at least one train each day.

Apparently this dome serves as a protection against the raids of predators, especially birds. Throughout the range of magpies their old nests are used also by raptorial birds, often night-hunting owls. Protection from owls may be a chief function of the dome on a magpie nest. An example of this use was observed in Nevada (Linsdale, 1937). On June 6, 1933, at dusk when the long-eared owls became active, magpies near camp showed much concern. Alarm notes were heard at three different magpie nest locations, and birds were seen in flight. Whenever an owl flew up it was followed closely by a magpie. In the vicinity each owl nest was situated close to an occupied magpie nest. The owls did practically all their foraging at night when the magpies probably slept, and hence they most likely needed some special protection just as might be provided by the thorny canopy over the nest. This was near the nest-leaving time for both species.

The black-billed magpie nests sometimes in tall trees, but oftener in tall bushes, especially in thorny ones. In trees the nests are placed often on low horizontal limbs sometimes close to the trunk and nearly always less than 25 feet from the ground. In bushes the nest may be in the extreme top, and nearly always it is far enough above ground to be out of reach of large foraging mammals.

Preference for certain kinds of bushes or trees to hold the nest seems to depend largely on what ones are available in each locality. In various places the following kinds of plants have been recorded as providing sites for nests of black-billed magpies: Alder, aspen, birch, buffaloberry, cottonwood, fir, hawthorn, juniper, pine, scrub oak, and willow.

Both birds of a pair take an active part in nest-building; both carry materials and place them on the nest. However, one of the pair, probably the female, gives more attention to the actual shaping of the nest.

In some localities magpies regularly use the same nests year after year. In this case the old nest is repaired and new material is added to it. Sometimes a remarkably large structure is thus built. In other localities the birds seem never to use a nest for a second brood but always build an entirely new one or, at most, build upon an old one, using it mainly as a base. Repair of nests is not limited to reoccupation. The birds see to it that the nest is kept in repair during the incubation and brooding of the young. If a part of the canopy is torn away, it is likely to be replaced quickly.

The use of magpies' nests by other birds has been discussed fully by Rockwell (1909). The abandoned nests furnish protection during severe rain or hail storms, or other severe weather, for robins, blackbirds, bluebirds, warblers, and other species that live along the timbered streams. Some birds, the horned owl, long-eared owl, and screech owl, make use of these nests almost continuously for daytime hiding retreats. These birds, especially the first two, also lay their eggs in old magpie nests. The sparrow hawk uses these nests for laying, but nearly always it chooses nests that still have their roofs intact. Rockwell noted that sparrow hawks that utilized old magpie nests always appeared timider than the ones that nested in cavities of trees. Other species reported as using these nests for their eggs or as bases for nests of their own are the sharp-shinned hawk, at Fort Lewis, Colo., the mourning dove, at Fort Harney, Oreg., and the bronzed grackle, at Littleton, Colo. At Barr, Colo., a brood of young magpies left a nest early in May, and within a week a pair of English sparrows started to build within the structure. Afterward a cowbird's egg was found in this nest. Nearly all the birds found laying eggs in these used nests chose ones between 15 and 30 feet above the ground.

A dilapidated magpie nest 8 feet above the ground in a willow near Pyramid Lake, Nev., held a nest of a gadwall, built of down and containing nine eggs (Ridgway, 1877). Near Denver a magpie's nest found by Bradbury (1917a) held seven fresh eggs, and on its flattened roof were three eggs of the black-crowned night heron. Apparently both pairs of birds had started about the same time to nest in the same structure. Another magpie nest 50 feet away contained five long-eared owl's eggs that were being incubated.

A peculiar association between nesting magpies and other kinds of nesting birds, mainly hawks, has been detected by many observers in many localities. In Alberta Taverner (1919) found magpie nests invariably in the neighborhood of, or not more than a hundred yards or so from, nests of red-tailed or Swainson's hawks. Bowles and Decker (1931) report that almost invariably one finds a nest of the magpie in the same tree only a few feet from that of the ferruginous roughleg; there is a continual supply of food to be had from the leavings of the hawks. Rhoads (1894) cited Captain Lewis as observing "that the nests of the Bald Eagles, where the Magpies abound, are always accompanied by those of two or three of the latter, who are their inseparable attendants." On the Columbia River a magpie's nest was once found by Dawson (1909) in the "basement" of an occupied osprey's nest. He intimates that the magpies derived benefit in having access to surplus food brought in by the ospreys.

Mammals, of fairly large size, have been found occupying old nests of the magpie. In New Mexico, Gilman (1908) reported four young house cats living in a nest 16 feet up from the ground. A gray fox found resting in the daytime in an old magpie nest near Colorado Springs, Colo., was reported by Warren (1912) on authority of C. E. Aiken.

Aretas A. Saunders sends the information that magpies sometimes use their old nests as shelter in winter in stormy or windy weather, and he thinks that possibly they use them regularly at night in winter.

Eggs.—Seven is the usual number of eggs in a set; sets of 8 and 9 are not uncommon, and as many as 10, 11, and even 13 have been found in one nest. Small sets in a nest are likely to be the result of some accidental destruction.

Concerning the time of egg laying, Kalmbach (1927) writes that "in Colorado, Utah, California, and southern Oregon, egg laying begins before the middle of April, in Washington and Montana about two weeks later, while in the extreme northern part of the magpie's range it does not begin before June or even July."

Coloration of the eggs varies, but the ground color is usually greenish gray, and this is profusely blotched with different shades of brown, sometimes completely hiding the ground color. Bendire (1895) describes the majority of the eggs as ovate. Others are short-ovate, rounded, elliptical, and elongate-ovate. The shells are close grained and moderately strong and show little or no gloss. Measurements given by him of 201 eggs in the United States National Museum are as follows: Average 32.54 by 22.86 millimeters; largest 37.84 by 26.42; smallest 27.94 by 21.59.

The average weight of 17 eggs was given by Bergtold (1917) as 0.34 ounces, or 9.64 grams. Weights of 13 eggs, in two sets, given by the same writer ranged from 132 to 155 grains (8.58 to 10.07 grams). The average for the whole lot of 13 eggs was 9.39 grams. An egg studied by me in Smoky Valley, Nev., which contained a well-developed embryo, measured 31 by 24 millimeters and weighed 7.2 grams. Another (incubation estimated at 14 days) measured 37 by 23 millimeters and weighed 8.2 grams.

Young.—Bendire (1895) gives the length of the incubation period as 16 to 18 days. He observed that one egg is deposited daily, and that incubation does not begin until the full set is nearly completed. Wheelock (1904), apparently on the basis of original observations, wrote that the female incubates for 18 days.

Opportunity came to me in several seasons to study young magpies in various stages of development in Smoky Valley, central Nevada. Three young, just hatched on May 16, had colorless skins that appeared pinkish because of the blood showing through them. Each was weak and barely able to raise its head and open its bill. When first taken from the nest they made a few weak squeaks. The brooding parent allowed approach to within 50 yards and then flew off with a series of about eight loud, harsh calls. It did not return, but this one or another continued the alarm from a spot 150 yards away. At another nest in similar stage the brooding bird flew off silently at approach within 75 yards.

When removed from a nest two young magpies that had grown to more than ten times their bulk at hatching at first uttered loud food calls and opened their bills. Soon an adult came and gave sharp calls, whereupon the young became quiet and the parent moved away. Later, however, the young again gave food calls. When placed in a cardboard box they worked their legs rather violently and grasped with their claws, apparently in an attempt to raise their bodies. At the same time they opened their bills and gave cries. They were unable to support their bodies above the flat surface, but they raised their heads the full length of their necks.

A bird one-third larger than the two just described (133 grams) was able to support itself on its tarsi, and, when placed on the ground, it immediately began to move off through the brush. Through most of the day this bird was kept it was silent, but it became hungry and uttered calls in the afternoon. Other young magpies of about this size kept perfectly quiet and made no sound or move during the whole time they were being examined.

When disturbed on June 6, five large young in a nest crawled upward inside the tall dome and clung to sticks in its wall. One went through to the outside, but later it returned. Several times these young uttered alarm notes nearly as loud and harsh as those of the parent birds. The next day at noon when this nest was approached the young birds crawled out of it and perched among the branches of the surrounding thicket.

Wheelock (1904) watched a brood of young magpies through the period of nest-leaving. For several days before time to come out of the nest the four young ones poked their heads out the doorways. On the twenty-second day one hopped out and perched on a branch. The parents meanwhile showed great excitement. When the young bird was approached closely it jumped off its perch and flew or was blown out of the tree. It could not control its long tail, which opened and acted as a sail in the wind. One parent followed this bird while the other one remained with the rest of the brood in the nest.

A late time for nest-leaving was observed at 6,500 feet in Yellowstone Park by Skinner (MS.), who saw a fully feathered young magpie perched outside a nest on June 22. By July 7 there was no sign of any birds about the site.

At midday on May 18, 1932, in Nevada, I watched another group of five young magpies out of the nest and able to fly a little. Four were in one thicket and the other was 50 feet away, but all had difficulty in maintaining a balance. Except for their tails being only about 4 inches long, I could scarcely distinguish them at a distance from adults on the basis of feathering. When approached closely the birds moved off through the thicket by short flights and by hopping, and they kept up a series of harsh notes of alarm. When I was still 100 yards off I saw one parent fly away quietly, and no adult was seen while I was near the young. But as soon as I left there were feeding calls; probably one of the parents had arrived.

At noon on May 22 this brood was seen at the same place. The birds were closer to the ground and a little more scattered than they had been the first time. When they moved they were a little more sure-footed. One of them allowed me to approach within 10 feet before it flew.

In the same vicinity, late in the afternoon of June 7, I watched about 25 magpies about a spring. They were in brush clumps and willows and evidently were mostly young, representing 3 or 4 families. They did not fly off so quickly or so well as adults would have. Most of them permitted approach to within a few feet, while adults at the same time remained as shy as ever and kept well out of sight. When I approached them these young birds moved off in several directions through the buffaloberry thickets. This was the earliest group noted that season, which included more than one family. Judged from the numbers of young seen, nearly all were then out of the nest and had left their home surroundings. Magpie calls were heard much more frequently than they had been early in May.

Out of close to 50 nests examined in one season in Smoky Valley, Nev., the birds at only three were so bold as to return to the nest, even on repeated visits by persons, and to give alarm notes. Usually these pairs kept close to the nest while it was being examined. At one nest especially, which contained large young, the female would come within 3 or 4 feet of an intruder and, uttering alarm notes, would stay as long as the person remained. The male came within 15 or 20 feet but did not stay so long. At the time for the young to leave, the parent often picked at branches of the buffaloberry bush on which it perched, breaking off thorns and pieces of bark with its beak.

Plumages.—Nestling magpies have skins at first nearly free of color except that which comes from the blood near the surface, and they have no down. This evidently is an adaptation to the enclosed type of nest. The skin soon assumes a yellow tinge and then becomes grayish as the birds begin to fledge. When they leave the nest their tails are not more than 5 inches long.

Juvenal birds are much like the adults except for the loose texture of the body feathers. The black parts of the body plumage are browner and the white parts creamier, while the wings and tail are less brilliant. The wing feathers and tail feathers and primary coverts of juvenal magpies are not molted in the first autumn. The body feathers and the wing coverts are molted. The young bird then has browner and less glossy wings and tail than adults, and this distinction becomes marked in the following spring and summer.

Also, adults may be distinguished from young birds up to the time of **their** second molt by differences in the form of the outer primary. The white area is less extensive on the inner web of the primaries in the young than in the adults, and the black terminal borders are larger. Often also the delimitation of the white and black is not sharp in the young where it is distinct in the adult. The outer (reduced) primary is shorter in adults than in young, and all the primaries are more pointed in young than in adults. The rectrices, especially the outer, vary according to age. The lateral one is narrow and rounded at the end in the young; it is large and square at the end in the adults.

The time and order of replacement of the feathers seem to be nearly the same in Western United States as given for western France by Mayaud (1933). There the juvenal molt of body plumage takes place from July to October. The complete annual molt commences near the first of July, reaches its height in August, and ends from mid-September to late in October. Molt of body plumage is very rapid. Molt of the tail commences early and ends sooner than that of the remiges. The two median rectrices fall first, and the replacement proceeds regularly to the lateral ones. The upper and lower caudals fall a little after the beginning of the molt of the rectrices. The molt of the primaries begins very soon and ends very late. Its direction is from the inside out, from the first to the tenth (and eleventh). The secondary remiges molt in three series; the two proximal ones with the ninth and tenth falling in the external-internal direction and the eighth and seventh in the internal-external direction, and a third series, the first to sixth, falling in the external-internal direction. The molt of the secondaries commences after that of the primaries and that of the distal series ends a little after that time.

Abnormal plumages in the magpie attract more than ordinary interest because of the possibility, which seems almost probability, that some of the existing geographic forms arose by the preservation of this kind of character. Several examples of freakish plumage in the black-billed magpie have been reported: DuBois (1918) on July 20, 1918, near Collins, Mont., saw one "entirely of a grayish-white, or very pale gray color," but exhibiting no definite markings. Rockwell (1910) found two albino magpies in a brood, the balance of which appeared to be normal. Both were pure white except for a slight creamy tint. A specimen mentioned by Svihla (1933) from near Nampa, Idaho, has normal pattern, but the feathers, bill, tarsus, and claws, which are usually black, are rusty-brown in this one.

Food.—E. R. Kalmbach has made extensive study of magpie food in North America. His published report (1927) is based upon the examination of 547 stomachs, of which 313 were adults and 234 were nestlings. This material is fairly representative of the bird's range and is well distributed throughout the year. Chief points of interest in this report are included in the following abstract.

Food of nestlings differed from that of adults at the same time of year and was decidedly different from that of adults at other times of the year. More than 94 percent of the food of these young magpies was animal matter. This contrasted with 82 percent for the parent birds. Insects made up the greater part of the animal matter fed to nestlings, and the groups best represented were caterpillars (nearly 18 percent), grasshoppers (more than 11 percent), and flies (more than 11 percent). This last group consisted chiefly of larvae and pupae of flesh flies that the parents obtained from carrion. The indications are then that the adults visited the carrion for the "purpose of procuring the insect food for their young, even in preference to the carrion itself." Kalmbach also pointed out that "the magpie's depredations on wild birds and domestic poultry may be attributed mainly to a desire to satisfy the appetites of its young."

About three-fifths of the food of adult magpies examined was of animal origin; the greatest proportion was found in May, during the breeding season. November, December, and January mark the period of smallest consumption of animal food. Insect food constitutes the predominant item for the magpie through the year. The species is more highly insectivorous than any other of the common species of the crow family in this country. The kinds of insects eaten are chiefly ones that live on or close to the surface of the ground. Grasshoppers form a conspicuous part of the diet during the late summer and fall. Insects associated with carrion are important in the magpie's food. Kalmbach thinks that in many local outbreaks "magpies doubtless have an important controlling influence" upon insect abundance. Two other items of animal matter important in the food are carrion and small mammals. Magpies congregate on highways to eat remains of animals killed by automobiles.

The stomach examinations indicated that the magpie is "by preference carnivorous and that the vegetable portion of its diet is taken more or less as a matter of necessity and not from choice. Notwithstanding the fact that wild fruit of one kind or another is as readily available in September as in August or October, the magpie's food preferences lead it to resort extensively to grasshoppers during that month and to reduce its consumption of wild fruit. There is every indication, also, that the grain eaten by magpies during the winter months is consumed largely as a matter of necessity. Grain could be secured in quantity during July and August at many points in the bird's range, but it turns naturally to an animal diet during those months. The rigorous weather of

November, December, and January forces the magpie to adopt a diet that is more than 60 per cent vegetable, while in May the abundance of animal food permits it to reduce the vegetable portion of its diet to 8 per cent of the total."

The notion is widespread that magpies provide a serious threat to all kinds of birds in their vicinity by robbing nests and eating the eggs or young, but the evidence to support this opinion is scanty. The manner in which some smaller birds drive magpies from the vicinity of their nests indicates that they recognize a potential danger to their young. On several occasions I have seen Brewer's and red-winged blackbirds, from nesting colonies, chasing magpies. In a group of blackbird nests a magpie could reap an appreciable harvest of food.

A magpie has been observed (Wheelock, 1904) taking both eggs and young from nests of tree swallows in hollow piles of a deserted pier at Lake Tahoe, Calif. This bird, a male, would search over the colony of swallows, and wherever the size of the opening to the nest cavity permitted it would reach in and take the contents, eggs, or young. These were then carried and given to the brooding female magpie. Young domesticated chickens were also taken by this bird.

Kalmbach's (1927) thorough study of the magpie led him to the conclusion that depredations against smaller birds are primarily in the breeding season and that the "serious cases of bird destruction reported against the magpie are probably localized or due to some peculiar environmental factor, as lack of cover for the birds attacked, an over-abundance of magpies, or scarcity of other food." Only 8 of 313 stomachs contained remains of wild birds. Specific identifications could not be made. Remains of eggs of native birds were found in two stomachs, "those of a robin and what appeared to be those of a shorebird being recognized." Three young from the Bear River marshes in Utah had been fed portions of coots, "probably disabled by alkali poisoning."

In this connection the significant observation was made by Saunders (1914) in an area near Choteau, Mont., that the magpies nested earlier than other kinds of birds. After nesting they left the area so that smaller birds there were not molested by them.

Behavior.—The general manner of a magpie is that of a bird well able to take care of itself. It is extremely suspicious yet is inquisitive to a high degree. It takes alarm quickly and rushes away from threatening danger, but it responds to kindness and is easily tamed.

Much of a magpie's time is spent on the ground in search of food. The walk is somewhat jerky, but it has been characterized as being graceful. The tail is slightly elevated and is constantly twitched. When the bird is in a hurry the ordinary walk is sometimes varied to a series of hops. Small droves of magpies were watched by Fisher (1902) as they caught grasshoppers every morning in a field near Mono Lake, Calif. Their agility in dodging and circling showed how mistaken persons are likely to be in forming an estimate of a bird under ordinary conditions. "Usually nonchalant and absurdly dignified in their demeanor, these birds could at times assume the utmost interest in their occupation, and dart with surprising speed here and there."

Bendire's (1895) comments on flocking in this bird were that: "Although more or less quarrelsome, it is social in disposition and likes to be in the company of its kind. I have frequently seen from twelve to thirty feeding together near a slaughterhouse or some other locality where food was abundant; but such gatherings are oftener met with in late fall and winter than during the season of reproduction."

Kelso (1926) records magpies in British Columbia as occurring in winter singly or in small flocks of up to 8 or 10 birds. When they were unusually numerous as many as 10, 20, or even 30 or 40 individuals made up the flock. Winter flocks in Washington contain any number of birds up to 50, according to Dice (1917). In western Nebraska, Zimmer (1911) observed magpies to occur abundantly, but usually singly or in pairs, never in flocks.

It was the opinion of Goss (1891) that the small flocks so often met resulted from the social natures of the birds holding the family groups together. In Nevada, Taylor (1912) observed late in June and early in July that adults and young were traveling in company. According to Dille's (1888) observations in Colorado, after the young are out of the nest and for the balance of the year the birds roam over the country in large flocks. In the same State, Rockwell and Wetmore (1914) saw usually not more than six together, although in one November evening a straggling flock of at least 50 was seen flying across a valley. In a detailed study of a small area in Montana, Saunders (1914) learned that all the magpies left after nesting was over.

One purpose in coming together in flocks is for roosting. In central Nevada, after the nesting season, groups of magpies sometimes roost together in some suitable place. I discovered one of these just after dark late in June 1930. Toward the end of a slough was a thick border of tall willow clumps through which rose vines were tangled. Ten or twenty magpies had just settled in this thicket for roosting. When found they were still making persistent cries. As the birds were disturbed they could be heard flying ahead out of the bushes and, finally, in small groups crossing openings.

Roosting habits of black-billed magpies in western Oklahoma were reported by Sutton (1934), as follows: "Once the birds had gone to roost they were loath to leave the trees, and upon being frightened flopped about clumsily, making their way to trees nearby, where they became quiet as soon as possible. If disturbed in the early evening at a favorite roosting-place they frequently flew to the mesas, then trailed back, one by one, in a series of swift, headlong plunges, just at nightfall." Skinner (MS.) reports that he has found magpies roosting in the thick foliage of Douglas firs.

It was stated by Kalmbach (1927) that "during the winter magpies sometimes roost much after the fashion of crows, and in one instance these two species were found using the same small island in the Snake River in eastern Oregon as a place of nightly resort." In the winter of 1922, a cattle shed near Treesbank, Manitoba, was regularly used as a sleeping place by a group of six or seven magpies. The birds rested on the backs of the cattle overnight (Criddle, 1923).

Bendire (1895) characterized the flight of the magpie as "slow and wavering, and in windy weather evidently laborious. The long, wedge-shaped tail seems to be decidedly in the way and a positive disadvantage, causing it no little trouble in flying from point to point, and in such weather it will only leave through necessity the sheltered bottom lands it usually frequents." Also the flight is never very protracted.

Taverner (1926) writes of the magpie that it is "more often seen retreating up the coulée, chattering as it glides from bush to bush. * * * At other times, a small flock or family party will be seen passing noisily along the tops of the hills, from brush clump to brush clump. Again, they steal silently into camp or about the farm buildings intent on any mischief that may present itself, but flee away in consternation when disturbed."

On many occasions, however, as indicated by Goss (1891), magpies when pursued do not fly away wildly but will tempt the pursuer by fussing about just out of reach. That observer also pointed out that although the birds sustain themselves by rapid strokes of the wings, the effort is too great for extended flight.

In a strong wind magpies tend to fly low, just over the tops of the bushes, and they raise and lower their flight according to variation in level of the bush tops. They probably avoid much of the force of the wind by doing this.

In his notes supplied for this account, M. P. Skinner writes that the flight of a magpie is level, with wing beats slow and measured, yet the flight is really swifter than it appears. At times they fly with a short period of about five wing strokes alternating with a short sail on spread wings. They fly low to escape a head wind. He has seen them fly down a hill slope with a peculiar dipping, wavering flight, with head held up and long tail thrust up to act as a rudder. In flight the white on the wings is conspicuous, giving rise to the term "side wheelers" as they are sometimes called. Individuals are quite apt to follow one another, separated by several hundred feet, and not in flocks. Even when flushed from food they leave one at a time. Magpies often are seen flying across a broad valley, well up above the ground.

The black-billed magpie is usually so wary as to give little opportunity to watch any of its activities closely. I have never observed one bathing in the wild, but observations on caged young birds revealed that they take to water readily and make constant use of it in caring for their plumage. When a pan of water was first placed in the cage, each of the young birds (which had been taken from a nest and, hence, could never have had experience with water) went through almost exactly the behavior described by Pycraft (1918) for the magpies in Europe. At first, just as described for those birds, there were some misguided attempts to bathe on the dry floor, but the habit soon became regulated and adjusted so as to eliminate these wasted movements. Within half an hour after water was placed in the cage for the first time, each of the birds had discovered it and had made use of it for drinking and for bathing. In succeeding months bathing was a part of the daily routine of each of the three birds. In fact, they appeared to await their daily supply of water with as much concern as they watched for a new food supply.

Man encroaches upon the territory of the magpie in nearly every part of the range of the bird. This has been true especially in the range of the black-billed kind in North America since the comparatively recent occupation of the land by the white race. Through much of that area the requirements of these two kinds of animals overlap so much that they come to occupy common ground. This is especially striking here because in most places the combined habitats of man and magpie involve only a small part of the total area of land. Concentration of these two animals on the same ground results partly from their need for water, but their competition is mainly for food materials that are produced there.

When man settles in magpie country, he immediately begins to "improve" his surroundings. Very often this also means that the environment is improved for the magpies. The birds usually need, and they are quick to take advantage of, increased food stores that human settlement brings. This is not a new trait, for magpies have been reported as normal attendants at Indian camps. Also, it was noted by Baird, Brewer, and Ridgway (1874) that "the party of Lewis and Clark, who were the first to add this bird to our fauna, also describe them as familiar and voracious, penetrating into their tents, snatching the meat even from their dishes, and frequently, when the hunters were engaged in dressing their game, seizing the meat suspended within a foot or two of their heads." They add further that "Mr. Nuttall, in his tour across the continent, found these birds so familiar and greedy as to be easily taken, as they approached the encampment for food, by the Indian boys, who kept them prisoners. They soon became reconciled to their confinement, and were continually hopping around and tugging and struggling for any offal thrown to them."

Improvement of the habitat by the magpies, if it takes place, is usually not noticed at all by people. However, if the magpies remove, or interfere with, any article claimed by people, this is likely to be noticed immediately and to be followed by some kind of retaliation. The result often is the destruction of a certain part of the magpie population. But the magpie is a hardy kind of animal, and unless the destruction is organized and well planned the birds have a good chance to survive, at least in small numbers. It is rare that human concentration on an area within magpie range reaches a point where the continued presence of the bird is hindered, unless direct killing is resorted to by the people.

The hatred that many persons hold for the magpie has found expression in the carrying on of contests in an attempt to "exterminate" the species. An item from a newspaper in British Columbia gives some results of one of these contests as it was conducted in 1931 in the Okanagan Lake region. Two teams, of six persons each, killed a total of 1,033 magpies in one season.

In western States magpies in certain localities have hindered campaigns against predatory animals, by raids on the baited stations, to such an extent that special efforts have been made to remove the birds before spreading poison. Kalmbach (1927) reports that "during campaigns against coyotes in the winter of 1921-22 along Butter Creek, in Umatilla County, Oreg., it was conservatively estimated that 5,000 magpies were killed. In Douglas County, Colo., magpies were practically exterminated in the country covered by poison lines placed for coyotes in the winter of 1922-23. In the winter of 1921-22 a coyote campaign planned on the Pyramid Lake Indian Reservation, Nev., called for preliminary measures against magpies. On the first day after placing the baits three grain sacks full of dead magpies were picked up. An inspection of this reservation during the following winter showed not a dozen magpies, where in the previous year there were probably more than a thousand. At one poison station at Summit, Utah, 143 of these birds were accounted for within a few days."

The same writer concluded, concerning the magpie, that "over much of its range, where it appears in moderate numbers, the bird is not an outstanding agricultural pest or a serious menace to other wild birds, and the present study has revealed the fact that there are times when its influence may even be decidedly beneficial."

One of the ways magpies come into conflict with man is by interference with his domesticated animals. Magpies come about the herds of stock, in the corral or on the range, mainly for the extra food available there. A large part of this is the group of carrion- and dung-inhabiting insects that they search for and eat. But they sometimes examine the large animals for parasites they may be carrying, and when opportunity offers they open sores or cuts and eat the flesh of the animal itself. This habit has resulted in considerable losses on occasion, and the bird has a reputation for damage among stockmen, probably far beyond justification. Instances have been reported many times showing the attacks may persist until the death of the horse, cow, or sheep. A full account of injury to sheep by magpies in Montana has been published by Berry (1922).

It has occurred to many observers that such a habit as the attacking of large domestic animals must have developed first with relation to the larger wild herbivores, and some evidence has been given to verify the supposition. Packard (MS.) sends the information, from Colorado, that magpies are frequently seen picking insects off the heads and backs of Rocky Mountain mule deer and American wapiti, especially in spring, when these animals are infested with ticks. They also have been observed on Rocky Mountain bighorns. These animals appear to make no effort to disturb the birds while they are engaged in this activity, although they occasionally turn their heads to look at them. Another observer (Cameron, 1907) concluded that the deer did not appreciate this attention, after he saw a doe push a magpie from her back with her nose.

A possible indication of the close dependence of magpies on these larger animals was shown in an account of the changes in status of some of the animals in central Alberta, reported from the testimony of old-time buffalo hunters by Farley (1925). He wrote concerning this bird, as follows:

The appearance of the Magpie in large flocks in this section of the province during the last ten years has been the cause of much discussion. Until 1907, they were unknown north of the Red Deer River. In October of that year the writer observed a pair about six miles north of the town of Lacombe. The following year magpies were reported from the vicinity of Bittern Lake, and from then on, they have gradually become more numerous, until at present they are our commonest winter resident bird. Magpies were very numerous during the buffalo days when flocks would follow the hunting parties and live on the refuse of the hunt.

The bird was considered a great pest in those times on account of its habit of alighting on horses, with saddle or harness galls, and persistently pecking at the sores until the death of the animal resulted. The only means of saving the horses when thus attacked was to stable or blanket them. With the extinction of the buffalo, the magpies disappeared and the present incursion is the first which has occurred since that time. [The buffalo was plentiful in that district until 1875.]

Skinner (MS.) reports from the Yellowstone region that magpies frequently linger about horses, elk, and buffalo for the sake of the manure. He has seen them feed on carcasses of those animals, which they attacked first by picking out the eyes, next the eyelids and lips, and then the neck.

Magpies are partly dependent too for food on another group of mammals, the carnivores, but in a different way. These birds seem to know that a carnivore is likely to leave scraps of food, and they are able to take advantage of the circumstance. My attention once was directed to the location of a coyote in an aspen thicket by the commotion of magpies near it. Skinner (MS.) writes that magpies find coyote kills so promptly that he sometimes thinks they follow coyotes purposely.

An account supplied by A. A. Saunders tells how he once watched a pair of magpies scold a cat. The cat was trying to hide in a thick growth of willow and cottonwood, but the magpies followed it about, alighting a few feet above it, and calling. The cat would try to move on to another spot, only to be discovered there also, and its presence announced derisively to all the bird population.

Voice.—The ordinary call note of the black-billed magpie has been written (Bendire, 1895) as a querulous *cäck, cäck,* or *chäeck, chäeck,* uttered in a high key, and disagreeable to the ear. That writer adds that: "It frequently utters also a low, garrulous gabble, intermixed with whistling notes, not at all unpleasing, as if talking to itself, and if annoyed at anything it does not hesitate to show its displeasure by scolding in the most unmistakable manner." The distinct chatter of the usual notes impressed Ridgway (1877) as being unlike the notes of any other bird of his acquaintance. The more musical note, which he heard uttered frequently, sounded like *kay e-ehk-kay.* He could detect no difference between the notes of this bird and those of the yellow-billed kind.

Brooks (1931) writes that during four years in France he "was surprised to note the great difference in voice between the Old and New World Magpies; the latter to his regret have no call that he can imitate sufficiently well to decoy the birds to him; the former on the other hand had two easily imitated calls and decoyed readily."

Saunders (MS.) notes that the common cackling call sounds to him like *ca-ca-ca-ca-ca.* It is in a harsh, cracked voice, higher pitched than a crow's caw. When one is near the nest and the birds are scolding with this call, it has a derisive sound. On March 14, 1909, Saunders heard a magpie producing a faint, twittering sound, much as the blue jay does on some occasions, a sound he now would class as a primitive song.

The general demeanor of the black-billed magpie was characterized by Grinnell and Storer (1924) as being "decidedly quieter than that of most of the other members of the jay-magpie-crow family. Its voice is far softer than that of the jays, and it does not 'bawl out' intruders as do those birds. Many of its notes are low and pleasant chuckling sounds, recalling certain notes of the California Thrasher. On one occasion one of our party was attracted by a noise arising in a mountain mahogany bush and sounding like two of the branches rubbing together. It proved to come from a black-billed magpie. Even in early fall, when bluejays and nutcrackers are at their noisiest, the magpie is noticeably quiet."

Magpies have long been favorites among captive birds as pets. Partly this results from the ease with which the black-billed ones may be obtained from their nests as young nearly ready to leave. But also it comes from their varied mannerisms and especially their trait, so well exhibited, of learning to imitate words and phrases that they hear in the households of their keepers. The young birds can learn and repeat an amazingly long and varied vocabulary, if we can believe the accounts that have been printed. Just when the limit of plausibility in the stories of their accomplishments is reached is hard to determine. Even the most commonplace of these accounts, however, is rather remarkable.

Field marks.—Any native bird seen in the United States or northward whose length is 15 to 20 inches, with more than half of this tail and with conspicuous pattern of black and white, is a magpie. The large white patches on the wings appear and disappear as the wings open or close. Close examination reveals iridescent greenish blue and bronzy green in the feathers of the wings and tail. Also the sharp line separating the white belly from the black breast, along with the white patches on the shoulders, helps to verify the identification. And the black bill distinguishes the present form from the yellow-billed one occurring only in California. Such a strikingly marked bird should require no characterization for recognition, but a surprisingly large number of people inquire as to its identity on their first encounter with it. Once learned, it is not soon forgotten.

Predators and parasites.—Watching the behavior of black-billed magpies gives the impression that they are subject to being captured by predators, but there is little direct evidence to demonstrate it. Once in central Nevada I watched a small group of magpies flying about excitedly among the desert bushes. Then a prairie falcon flew up carrying a dead magpie, apparently freshly killed. These magpies probably were young ones and might have been captured more easily than old ones. In a report on the food habits of hawks in Canada, Munro (1929) mentions the flushing of a goshawk from the still-warm body of a magpie.

A special study of the invertebrate fauna of occupied magpie nests in Montana was made by Jellison and Philip (1933). The extreme infestations of the blood-sucking dipterous larvae of *Protocalliphora avium*, which they observed, indicated that the large, twig-canopied, mud-plastered, fiber-lined nests of magpies served as favorable habitations and the nesting birds as excellent hosts. From five of the magpie nests examined, 108, 187, 190, 343, and 373 larvae were taken, respectively. The last was the greatest number reported to that time from a single nest of any bird. Two hymenopterous parasites, *Marmoniella vitripennis* (Walker) and *Morodora armata* Gahan, both chalcids, were reared from the puparia. Larvae of the beetle *Dermestes signatus* LeConte were observed to be predacious on the puparia of the flies.

Blood-gorged midges, *Culicoides crepuscularis* Malloch, were abundant in nests built close to a stream and seemed to be at home in the debris of the nests. Two kinds of Mallophaga were taken from the nestlings, *Docophorus communis* Nitzsch and *Myrsidea eurysternum* (Nitzsch). Additional kinds of beetles found in the nests were: *Dermestes talipinus* Mannerheim, *Anthrenus occidens* Casey, *Helops convexulus* LeConte, and *Cratraea* sp.

Migration.—The migratory habit is developed to different degrees in the various kinds of magpies, but nowhere is it well marked. In the southern part of the range the birds move scarcely at all. Farther north they migrate, sometimes great distances, but always probably in immediate response to severe winter conditions. In the United States fall and winter movements are noted regularly within the general range of the bird, and in some years well-defined migrations occur outside that range.

In many parts of the range of the black-billed magpie local movements of the birds may be detected in fall, usually in September. On the Taku River, Alaska, in the latter half of September, Swarth (1911) observed numerous flocks of eight or ten individuals flying from the interior out toward the coast, where they spent the winter. The same observer (1926) reported a similar migration, also in September, in northern British Columbia. Fall dispersal of magpies was detected from August 28 to late in September by Taylor and Shaw (1927) on Mount Rainier.

The return to the interior takes place early in spring by way of the major passes through the mountains. In one example, cited by Dawson (1909), D. E. Brown saw, on March 4, several bands of magpies passing eastward at a considerable height, perhaps between 3,000 and 5,000 feet. The birds were unrecognizable until glasses were trained on them, and at least 50 were counted.

At numerous places in the mountains the fall movement is upward to high levels. Saunders (1921) reports that in Montana a fall movement into the mountains takes place in October at the time of the first cold weather and snowstorms. The birds have been seen as high as 8,000- and 9,000-foot altitudes. Skinner (MS.) points out that in Yellowstone National Park magpies appear in August and September and are numerous in winter, especially in stormy weather. Packard sends the information that in the vicinity of Estes Park, Colo., magpies are occasionally found at timberline or above, in September or October, when small flocks visit the alpine meadows and feed on grasshoppers.

Still another type of magpie movement is that mentioned by Bendire (1895), who records that along the eastern border of its range the magpie occasionally wanders eastward late in fall and in winter. He thought that the birds were driven away from their usual haunts by scarcity of food or the severe storms that so frequently occur in those sections of the country. An especially well-marked movement of this sort took place in the Missouri Valley in the fall of 1921. Record of exceptional magpie movements that year were reported for Iowa, Nebraska, Kansas, Minnesota, and, farther north, in Manitoba. The birds sometimes go beyond the ordinary limits of their range on the south too. Great flocks of magpies late in October 1919 came into Death Valley, Calif., from the north. They had never been seen there before, and they gradually drifted away until few were left by the end of December.

A possible indicator of rate of travel in these fall movements was reported by Mrs. Bailey (1928), who observed a flock of 17 magpies at a 10,000-foot altitude in New Mexico. Two days later the same flock was seen 5 miles farther up the valley at 10,700 feet.

A single recovery of a banded black-billed magpie, reported by Lincoln (1927), gives information concerning the direction and extent of movement. The bird, banded at Laramie, Wyo., on May 30, 1925, was recovered on January 14, 1926, at Rosita, Colo. The latter locality is approximately 220 miles from the station of banding and is almost directly south of it.

DISTRIBUTION

Range.—Alaska and the Western United States and Canada; casual in the eastern parts of the North American Continent; nonmigratory, **but given to erratic wanderings.**

The normal range of the American magpie extends **north** to Alaska (Kings Cove, Nushagak, and Little Savage River); Yukon (Fort Selkirk and Carcross); southern Alberta (Glenevis and Mundare); and southern Saskatchewan (Wiseton and Qu'Appelle). **East** to southeastern Saskatchewan (Qu'Appelle); eastern North Dakota (Devils Lake and Dawson); South Dakota (Faulkton and Rosebud); central Nebraska (Niobrara Refuge, Kenesaw, and Red Cloud); western Kansas (Coolidge); western Oklahoma (Brookhart Ranch and Kenton); and northeastern New Mexico (near Raton and Watrous). **South** to northern New Mexico (Watrous, Santa Fe, and Shiprock); southern Utah (probably Bluff and Panquitch); central Nevada (Toquima Mountains, Peavine Creek, and Carson); and east-central California (Bridgeport). **West** to eastern California (Bridgeport, Markleeville, Beckwith, Susanville, and Tule Lake); Oregon (Klamath Lake, Drews Creek, Prineville, and The Dalles); Washington (Kalama, Yakima Valley, and Seattle); British Columbia (probably Port Moody, Ashcroft, and Atlin Lake); and Alaska (Yakutat, Kodiak, and Kings Cove).

Casual records.—Among the records that are well outside the normal range are the following: One was taken on Hotham Inlet, Alaska, sometime prior to 1900; one was taken at Forty-mile, Yukon Territory, on October 15, 1899; in northern Manitoba a specimen was obtained at Brochet, on Reindeer Lake on October 12, 1934, and one was taken at York Factory sometime prior to 1891; in Ontario, one was taken at Odessa on March 12, 1898, one at Port Sidney in the summer of 1898, and it was reported at Kingsville on March 31, 1916; in Quebec, one was recorded about 1883 from Montreal, another was seen in that vicinity on May 17, 1920, while two were seen near Hatley on October 17, 1915.

There are several records for each of the States of Minnesota, Wisconsin, Iowa, and Missouri. Among the records for Illinois is a report of seven seen at Knoxville on May 16, 1896, one seen at Philo on April 26, 1914, and a specimen taken at Lake Forest on November 10, 1918; one was seen at Bicknell, Ind., on December 24, 1907, and again on February 10, 1908; and a specimen was obtained at Toledo, Ohio, on May 9, 1937. There are a few records for Atlantic coastal localities, all of which must be open to suspicion as escapes from captivity. Among these are one seen at Point Lookout, Md., on June 28, 1931, one seen on Edisto Island, S. C., in May 1934, and one seen near Palm Beach, Fla., on January 17, 1934.

Egg dates.—Alaska: 3 records, May 28 to June 19.

California: 30 records, April 9 to May 13; 16 records, April 14 to May 6, indicating the height of the season.

Colorado: 80 records, March 26 to May 29; 40 records, April 24 to May 8.

Montana: 17 records, March 28 to May 23; 8 records, May 6 to 12.

Washington: 20 records, April 1 to May 15; 10 records, April 7 to 19.

PICA NUTTALLII (Audubon)

YELLOW-BILLED MAGPIE

PLATE 26

CONTRIBUTED BY JEAN MYRON LINSDALE

HABITS

California contains within its borders the whole range of the yellow-billed magpie. Localities occupied are known with exhaustive detail. They are restricted to that part of the State west of the Sierra Nevada from Shasta County, at the north end of the Sacramento Valley, southward to Ventura and Kern Counties, and are chiefly in the Sacramento and San Joaquin Valleys and the coastal valleys south of San Francisco. The area occupied is less than 150 miles wide and extends for about 500 miles from north to south.

The yellow-billed magpie is obviously a close relative of the black-billed magpie. Some persons like to think of this relationship as subspecific; others consider the two kinds as distinct species. Probably it makes little difference which way we think of them so long as we recognize the nature of the characters and ranges of the birds, insofar as they represent the true relationship, for it is scarcely possible to prove the correctness of either opinion. The most nearly obvious distinctions have to do with the possession of the yellow pigment, which shows in the bill, claws, and some places in the skin of the yellow-billed form, and its generally smaller size. Some differences in habits also may be seen on close study of the two birds. The ranges do not overlap; in fact, the gap separating them is about 50 miles wide at its narrowest place.

The situation is not much different on the opposite side of the range of the black-billed bird, where it approaches the nearest representative of the group in Asia. The relationship indicated there seems even more remote than that with the yellow-billed bird, but it is commonly recognized as a subspecific one. It is true that the birds in western Europe appear structurally to be much like the American black-billed ones, although these two kinds are separated by several forms that differ considerably from them. Somewhat similar problems arise in determining the phylogenetic status of the magpies in southwestern Europe and northern Africa. The anatomical, behavioral, and geographical situations there are much like the ones encountered in California. The whole question of relationships and trends of variation in the magpies is an attractive one for the person who may be able to collect the pertinent evidence.

A feature common to situations inhabited by the yellow-billed magpie is the presence of tall trees, usually in linear arrangement bordering streams or in parklike groves, either on valley floors or on hills. Another is open ground either bare, as in well-kept orchards, or comprising cultivated fields or grassy pastures and slopes. This particular kind of magpie appears not to extend its range into lands where there is frequent high wind, long, snowy, and cold winters, or especially dry and hot summers. The nature of the restriction in each case is more or less obscure—sometimes, as with the strong wind, it is evidently some direct influence of the environment upon the birds. Again, the limitation may act indirectly by so reducing the available supply of food that magpies could not exist for the whole year or for a time sufficient to rear their young. Restrictions of water supply may be important in preventing spread of these birds into desert regions. Water seems to be necessary for the birds to drink and also as an aid to nest-building.

Courtship.—Observations pertaining to courtship in the magpie are hard to separate clearly from those concerned with competition and with intolerance exhibited toward other individuals. Until the suitable distinctions between these types of behavior can be made, it may be better to consider all the observations made in the early part of the breeding season, which appear to indicate special attention of one member of a mated pair toward the other member. Also it may be permissible to extend the scope of this topic to include examples of attention directed toward other birds, even of other species, exhibited at this season.

Observations made on the Frances Simes Hastings Natural History Reservation, in Monterey County, Calif., where most of the material contained in this account was obtained by numerous watchers, show that activities connected with nesting begin early in fall. In one instance a magpie in a morning early in October (3d) flew from a sycamore into a locust, carrying a piece of sycamore bark 3 by 2 inches in dimensions. It hopped about, eyeing four other magpies already in the tree. After visiting three twig clusters such as provide nest sites, the bird dropped the bark and paid no further attention to it as it fell to the ground. The other birds ignored it also. One on October 31 was carrying a 6-inch-long stick or root in its bill. This behavior was considered to be a sign of early nest-building.

Another and more easily interpreted example was noted on the morning of November 12, 1937, when one of two magpies in a regularly used blue-oak nesting tree was carrying a dead twig about 10 inches long as it moved from branch to branch. At about the same hour the next morning the two magpies were seen again in this tree and one of them was carrying a 10-inch stick. A few seconds later there was excited calling and then a pursuit flight round and round in the top of the tree. Several magpies flew to the tree and joined in the chase.

Fifteen minutes later two magpies flew into this blue oak and wiped their bills on branches, apparently returning from a foraging expedition. After a few seconds they flew, together, downslope to where two others were foraging, but when these flew up toward them and called, the first two circled back and returned to the tree. Later they left together in another direction. This behavior surely indicates segregation by pairs.

After another return to the tree these two magpies flew down and joined the two they had approached earlier. All four strutted about with tails held high for a few minutes, and then the first two flew back to the oak.

Further evidence of segregation into pairs in fall was recorded on November 18. Two birds were foraging in a stubble field and occasionally flying up to fence posts. One took some food, probably a large insect, to a post where the morsel was picked to pieces. When a person approached the birds, one flew off to willows along the adjacent creek, while the other, engrossed in its foraging, stayed a moment longer and then flew off to an oak in the opposite direction. The two then called to each other with several rattling calls, which were answered. Other calls heard up and down the canyon indicated that magpies were scattered over an area extending for several hundred yards. Several times birds in flight were in pairs.

In the afternoon two pairs were perched in a locust tree, the birds of each pair sitting within a few inches of each other. Low, musical notes, a primitive song, were heard. Both pairs flew off at about the same time. Other magpies, also active and noisy in the vicinity, were flying from one tree top to another, usually in pairs. Even when they started for the roosting place in the evening, magpies on this day seemed still to be segregated by pairs. Two birds started off across the canyon, followed immediately by ten others from various points. These converged until all were strung out in a single file along the same line of flight.

It is desirable to keep in mind, in considering these happenings, that they took place before the start of winter, long before the start of nest-building and several months before time for egglaying. Through much of this interval most of the magpie activities, too, are centered about the flock, in the daytime as well as at night.

Fighting among magpies takes place generally through the spring months, beginning, in the yellow-billed form, usually in January. On one of these mornings, January 20, when rain was threatening, magpies seemed to be more quarrelsome than usual on the Hastings Reservation. Three were flying at one another. One would make a short dash at another, which would fly out of reach and then, perhaps, return the attack. Then the third one would attack one of the others. Later, one of five magpies in a tree seemed to be the object of occasional attacks by two of the others. Three or four times when this one hopped to a lower branch, the two jumped at it, causing it to hop out of the way. Occasionally it picked a leaf and let it fall. Once it tugged at a twig but was rather apathetic about it and was driven off by the others.

Apparently members of a mated pair do not always join together in the pursuits. Often a third bird follows the quarreling two but takes no part in the conflict. Once one of three birds was seen to fly to and enter a nest, not taking any nesting material, but it was routed out almost immediately by one of the other two, and all three flew away together. This casualness seems to be a normal feature of the fighting. Often in the nest-building season a magpie can be seen making a short dash toward another, evidently without serious intent, for no response can be detected, and there is no further indication of enmity. Sometimes excited vocal sounds among a small group are the only indication of quarreling, and there is no apparent reason for conflict.

After a brief fight between two magpies on the ground beneath a nesting-tree, one bird called loudly about five times. Several magpies within 30 feet of this spot then quickly flew there, whereupon one of the first two flew away about 20 feet. This ended the disturbance, and the magpies scattered and resumed their foraging.

Fighting in the nesting season does not always take place at the nest. One morning six magpies were foraging in a small pasture beside a creek. Two birds started fighting, and soon two more out of the group were fighting. When the fighting, which lasted only a short time, stopped, one of each pair of fighters went off with one of the other pair. Apparently the two males were fighting, and at the same time the two females were in combat.

Encounters between pairs from adjacent nests in the incubation period were watched on a morning early in April. Twice, one pair went to a nest tree other than their own. The first time, the pair quarreled on the roof of a chicken pen, with the male from this nest, and immediately the female flew down to join the group, but there was no more fighting. Next time, when the same three birds were on the ground directly under the nest, the brooding female flew toward them with many excited calls.

On another occasion the four birds from two nests in trees at opposite ends of a barn were together for about 5 minutes at one corner of the barn. Their loud calls indicated excitement. One, with raised tail, quivered its wings slightly, possibly indicating an early stage in the begging which the females develop. Then there was fighting in which all four birds took part, but which led to no decisive conclusion. Finally one took a stick to one nest, and its mate tried to carry one that was too heavy, so it went to the nest with no load. Both birds of the other pair then went to their nest. The purpose of the fight was obscure; possibly there was no immediate purpose, but only an indistinct urge to fight.

Study of numerous examples such as have been recounted brings the conclusion that the accounts reported so often as examples of courtship and pairing in the European magpie may have been encounters between groups of birds already paired. The pursuits could have been merely attempts to drive away intruders. And the congregation of small assemblies of excited birds could have been exhibitions of the common magpie trait of hurrying to investigate any disturbance. Also the reported examples have occurred too late in the nesting season to expect them to represent the very earliest stage of nesting. Recent observations indicate that pairs are well established in fall and that spring is the season for noisy squabbles incident to competition for nest sites or indicators of jealousy toward intruders. Until marked individuals can be traced through the whole cycle, it is not justifiable to consider this interpretation as conclusively established, but it now seems more reasonable than the traditional one.

There are other segments in the series of actions that amount almost to preparation for the behavior to come in the incubation period, and these are to be observed before, through, and after the period of nest building. Even though the birds are seen oftenest in small groups or flocks, it is probable that the units in the organization are pairs that remain together throughout the year. During and after the long period of nest building, or reconstruction, a large share of the time is spent at or near the nest. The two birds of each pair spend several hours of each fair day perched side by side on some limb close to the site. At such times one of them often utters a song that I have been able to hear as far away as 100 yards. There are other indications that during this preincubation time a magpie's attention is largely centered about its nest and its mate.

One of these attentions is the preening of one member of the pair by the other, presumably the male. I had opportunity to watch this at close range at the San Diego Zoo. In a cage of mixed kinds of corvids there were two yellow-billed magpies, considered by their keeper to be a mated pair. During most of the time I watched, one bird, apparently the male, was perched close to the other, and was working its opened bill through the feathers about, mostly on top of, the head of its mate. This is just what I have seen mated pairs do many times in the wild. The feathers were preened and worked over just as if the bird were searching for parasites, but the real significance of the behavior must be connected with mating. Most of the wild birds observed behaving in this manner have been perched directly on the nest or on a limb very close to it.

The most conspicuous habit in the series connected with courtship in the magpies is mate-feeding by the male. This begins to be developed at about the time the nest is completed and becomes well established by the time incubation begins. One demonstration of an early stage in the development of this habit was watched on a day early in March. One magpie was walking in a circle about 5 feet in diameter. It was fluttering its wings and walking around another magpie, which it seemed to keep in the circle. The second bird walked just a few inches ahead of the fluttering one, which kept its tail turned toward the center of the circle. When a third magpie lit nearby, the antics of the two birds stopped and they began to feed.

Another pair demonstrated the early stages in the establishment of the mate-feeding behavior. These two birds were foraging in a grainfield where the ground was nearly bare. The male walked about, paying little attention to its mate. The female at first ran after and put herself in front of the male, facing him with bill open, head lowered, and wings quivering. This bird seemed to hold its wings less widely opened and to move them more rapidly than did other individuals noted. The response of the male was merely to turn and walk in another direction. Once the female picked at some object on the ground, and immediately the wing-quivering reaction was aroused, and the bird hurried over to its mate, but again the response was negative. After about ten fruitless beggings the female began to pick up objects, presumably food, and for the next three or four minutes she was picking almost continuously, with only an occasional tendency to flutter the wings slightly. Next, the male flew to the top of a fence post. The female flew to the next post, and immediately upon alighting her wings were opened slightly. When the birds were on the ground, the female picked at objects much oftener than did the male. The supposition was that these actions were preliminary to actual incubation which was to begin shortly.

By late March, usually, food begging by females is conspicuous in yellow-billed magpie behavior. This is near the time of the start of incubation. In one such example the female of a pair was first noted perched just outside the nest, uttering the loud food-begging call. Later when the birds were back at the nest after an absence, they lit side by side, the female begging with widely spread wings, but no feeding was seen. Next it appeared that the male led the way to the nest, and when the female came near to it the male quickly left and the female entered and became quiet. It may be that the male thus coaxes the female back onto the nest to incubate when the urge is not strong, possibly in the early stages.

By the time all females are incubating, the late afternoon seems to be a normal hour for the female to come off and beg for food from the male. Males appear still reluctant to feed their mates, especially away from the nest, and they usually avoid the begging female, which crowds near with waving wings. At the nest they appear more tolerant, and it seems certain that feeding takes place at the nest as an inducement for the female to resume brooding. Often the feeding takes place wholly within the nest.

Late in the incubation period the brooding female appears restless and leaves the nest more often than in early stages. Then the feeding usually takes place in a tree but away from the nest itself. Usually the female flies out to meet the male on some nearby perch. It is possible also that the impulse to feed the female latterly becomes weaker in the male. They do not then, however, try to avoid the begging bird but will feed whenever she comes near. Sometimes, though, the female continues the pursuit and begging after it has received the food.

The forage range of individual male magpies during the incubation period varies from the near neighborhood of the nest to a place more than half a mile distant. Sometimes the bird hunts for food on the limbs and among the foliage of a tree, but oftener the foraging is done on the surface of the ground. A male from one nest tends to fly off in the same direction on all trips each day, but this direction may vary through the whole period. All the birds of a colony may forage over the same ground, or they may go in different directions, but, so far as I have been able to determine, they then ordinarily pay little attention to any magpie other than the brooding mate.

The mate-feeding is quickly terminated upon hatching of the young, and with the cessation of the persistent, loud calls, which are a part of the food-begging, the nesting colony becomes suddenly silent and the behavior is inconspicuous.

Presence of other birds about the nest is discouraged sometimes by magpies by vigorous pursuits. Two magpies watched in February were greatly concerned over a golden eagle that perched near their nest, and they tried repeatedly to drive it away. Special animosity is directed toward sparrow hawks in the vicinity of the nest, as if the magpies recognized the intent of these intruders. In one instance great effort was put forth to drive away two hawks from a nest, but when a western bluebird lit on a limb close to the nest, the magpie paid no attention to it. Once when a red-shafted flicker lit close to a magpie nest, the owner came immediately and drove it away. Similar effort has been watched with respect to California woodpeckers near occupied nests.

Other members of the Corvidae are treated with special enmity whenever one approaches a magpie nest. Magpies are able to drive away California jays ordinarily, but sometimes these smaller birds display an extra degree of persistence in refusing to leave the nest tree. Crows near a magpie nest arouse special activity, and they are driven away if it is possible. Sometimes they refuse to be driven and even turn on their pursuers and drive them from the site of their own nest. Once a nesting magpie attacked an intruding meadowlark with such fierceness that the two fell for 20 feet, or halfway to the ground, before the smaller bird escaped and fled with notes of alarm.

Defense of a nest site more than ordinarily effective against intrusion by a crow was noted on a morning early in February. A pair of magpies was discovered in a large valley oak, about a hundred feet from their nest site, where they had driven a crow to seek shelter in a thick clump of branches. One magpie would make a dash at the crow and retreat, and then the other would move toward it, but each took care to keep out of range of the crow's bill. Several times the crow dashed out after one or the other of the magpies, but always it retreated back to the protection of the limbs. This was kept up for 3 or 4 minutes, until four more magpies came, when the crow gave up, moved to the outer part of the tree, and then flew away. The magpies then dispersed.

Another incident in another year but in the same group of trees concerned a crow carrying a twig for its nest, which it had pulled from a blue oak about 80 feet from a new magpie nest. The pair of magpies kept such close watch of the crow and dashed toward it so frequently that it was unable to leave the tree with its stick. Finally, it placed the twig on a limb and tried to drive the magpies away, but this was not successful and it was forced to leave the tree without its twig.

Nearly all the examples of defense of nest site included in this section occurred in early stages of nest-building, before the start of incubation. When this stage is reached, the magpies seem too much occupied with their own program to pay much attention to other birds except on special occasion. Ordinarily this stage of nesting is reached nearly at the same time by almost all the pairs of magpies in a colony.

Nesting.—Habits connected with nesting in the yellow-billed magpie are, in general, like those of the black-billed species, but they contrast in several ways. Possibly the difference in nesting behavior between the two kinds seems more marked than in other types of behavior because the results of it, being definite objects, are more easily perceived by the human observer.

First, the nesting colonies are more compact and the nests are closer together in the yellow-billed form. This may have some connection with a more favorable foraging habitat that permits the yellow-billed form to live in a more gregarious society. Or it may reflect the result of some different need for group response to disturbance by intruders.

The actual position of the nest provides one major difference between the two kinds. In the yellow-billed one the nest is nearly always in some tall tree and far out on the limbs, or if in a medium-sized tree, it is likely to be in the periphery. Thus it is regularly at a different level and in a different site from the low, bushy one occupied by the American black-billed birds in their nesting.

Trees prominent among the ones nested in by yellow-billed magpies are sycamore, valley oak, live oak, blue oak, poplar, cottonwood, locust, and willow. One colony observed near Oroville by W. B. Davis tried repeatedly to nest in a clump of digger pines but with poor success, for these trees provided poor anchorage for the nests. They were easily dislodged by the wind, and sometimes the weight of the nest itself was enough to change the slope of the limb so that the structure would slide off to the ground. (See Linsdale, 1937.)

A curious item of yellow-billed magpie nesting, commented upon at length by Dawson (1923), is the resemblance of a nest to a clump of mistletoe. It happens that in California the area occupied by this bird and that occupied by two kinds of mistletoe are closely similar in their boundaries. Not only are the areas nearly the same, but the species of trees concerned are the same. Cottonwoods, sycamores, and valley oaks are the kinds of trees mainly involved in this peculiar relationship. The bird not only nests in trees having many clumps of this plant, but also it often builds actually within a mistletoe clump. Whether or not intentional selection is made by the bird for this purpose, it is obvious that the close resemblance of the two objects helps to screen the presence of nesting birds. Even a person experienced in detecting the nests is unable to distinguish one from a mistletoe cluster at a distance, so nearly alike are they in size, shape, and position. It seems possible that such a lot of decoys, even if the relation is the result of accident purely, might save some nests from discovery and destruction by persons not closely observant.

The inaccessibility of the nest because of its position on small limbs far above ground is sufficient to save it from the ordinary prospects for destruction by people climbing to it. Although the nests can be reached, the handicap is too great for most climbers. Forty to sixty feet is the normal height above ground for these nests. Thus the birds can live close to human habitation and nest with greater freedom from the kind of molestation ordinarily encountered by nesting colonies of black-billed magpies. This may account partly for the greater apparent lack of fear generally shown by the yellow-billed birds.

The mild winter climate in the range of the yellow-billed magpies seems to encourage, or rather to permit, an exceptionally early start at nest-building. In Monterey County, even in the mountains at the highest levels inhabited by the species, magpies regularly begin to build around the middle of December, before the shortest days come. Often a nest will reach a late stage of construction during a series of warm days at this season, but the coming of storms interrupts the progress, and the pairs rejoin the flock. With the return of warm, clear spells in February or March the building is quickly resumed. Normally the nests are completed soon enough for the laying to be completed before the end of March. Thus the egg-laying comes four to six weeks before the corresponding stage in the cycle of the black-billed birds living near the same latitude, but subjected to the more rigorous climate of the interior.

Warm, cloudless days late in January seem to arouse an extra amount of nesting activity, even when they follow as much as a month of nearly complete inactivity. Once, nest-building at lining stage on January 29 was carried on through a light rain. In three hours of morning watching, the pair carried 12 loads of material, sticks or mud, to the nest. Through a cloudy period the following morning the magpies were busy foraging, but when the sun broke through the clouds they went immediately to the nest. On the same date of the previous year, soon after a rain stopped and while it was still cloudy, several magpies were perched quietly in the vicinity of nest sites. One working pair left when the rain started again.

In this locality twigs for nests are regularly pulled from valley oaks, sycamores, and black locusts. Sometimes many in succession are tugged at before one can be loosened. The birds have some trouble in picking a route through the trees where the twigs they carry will not catch on the branches. During active nest-building, periods from 10 to 30 minutes elapse when no bird appears at the nest and when the pair is off foraging. In the active periods the whole effort of both the birds is directed to getting sticks to the nest and working them into place.

The urge for nest-building seems to grow stronger with the advance of the season. Also the member-pairs of a colony seem to approach such a synchronized program that they reach the stage of incubation at about the same time even though they do not start building together. However, some examples indicate that nests are sometimes begun long after the normal time. At the Hastings Reservation one season a pair of magpies was watched one morning, on May 11, busily carrying sticks to a nest in its early stages. Several sticks picked up from the ground were carried to it. A magpie was seen to enter this nest on July 24, but no actual use of it was made until the following year. Then a brood was brought off in the normal season.

Another late start at building a nest was noted in June. On the 19th a magpie carried a stick to a site in a fork about 50 feet up in a sycamore, and where 35 or more sticks already had been placed, many of them within the preceding 24 hours. No bird was seen about this nest again until the early morning of June 24, when each member of the pair brought a stick. Each bird then called and perched near the nest for about a minute and then they, separately, flew off to the ground. Again, a considerable quantity of material had been added since the preceding day. No more material was added. Two birds perched close to the site on June 27, but 5 days later all the magpies on the Reservation left for a period of days, thus apparently ending the story. But early in the morning of July 15 a noisy flock of magpies settled in this vicinity. One of the birds went three times to the partly built, late nest and crouching on it with wings aquiver uttered low, throaty, harsh calls, which sounded like *currow*, and turned around several times. Half an hour later the same performance was seen at the nest, but there was no further indication of its use at any later time.

Like the other kinds of magpie, the California one builds a nest that is sought by other birds as a home, but this use seems rather restricted to the sparrow hawk. Nearly every colony of yellow-billed magpies has at least one pair of nesting sparrow hawks. Although it may not be evident all through the season, there is considerable strife between these species when nest sites are being selected. After a given nest has been successfully defended and all the pairs are settled, the two species appear to take little notice of each other.

Pursuits of magpies by sparrow hawks are noted often in fall, beginning in September. Sometimes a magpie turns in a chase and pursues the hawk. These pursuits, however, may have little significance in the nesting activities. A different situation is present early in spring. An example was noted on an early day in March, when, just after 5:30 P.M., a sparrow hawk flew to a magpie nest in a sycamore. The two magpie owners came immediately and drove away the hawk. The possibility of the hawks taking permanent possession of a nest at this time of day, just before dark, may provide an explanation for the repeatedly observed circumstance that pairs of magpies keep an especially close guard in the tops of the trees over their nests for about half an hour each evening just before dark.

During the early part of the nesting season whenever sparrow hawks singly or in pairs approached and attempted to enter an occupied magpie nest, one or both of the rightful owners would come immediately and drive them from the vicinity. When smaller birds, for one example a flock of juncos, came near a magpie nest, the magpies paid no attention. They seemed to recognize the nature of the threat offered by sparrow hawks.

Eggs.—Seventy sets of eggs contained in the Museum of Vertebrate Zoology or collected by W. B. Davis made up a total of 455 eggs, or 6.5 eggs per set. Number in a set ranged from five to eight, and the modal number was seven. This indicates a slight tendency toward smaller sets than are laid by the black-billed kind.

Laying time for this bird is usually the latter part of March. Late nests are rare, and as already indicated they are almost certainly never completed or used in the season they are started. Extreme dates for nests with eggs are March 30 and June 2. All but six of the 70 sets mentioned above were collected in April. This no doubt indicates a shorter, and earlier, season suitable for nesting than that of the black-billed magpie.

Dawson (1923) described coloration of the eggs as yellowish glaucous or pale olive-buff, finely and rather uniformly speckled and spotted with buffy brown, or citrine drab, or grayish olive, or deep grayish olive. A considerable degree of variation in color was observed by Kaeding (1897) in the 30 or more sets of eggs he collected. Some were heavily blotched with lilac and buffy or purplish brown. Bendire (1895) observed that eggs with a greenish tinge in the ground color appeared more frequently in this than in the black-billed magpie.

Measurements of 195 eggs of this magpie were given by Dawson (1923) as follows: Average 30.8 by 22.4 millimeters (1.22 by 0.88 inches); index 72.1. Largest egg, 37 by 23.4 (1.46 by 0.92); smallest 26.7 by 20.3 (1.05 by 0.80). Measurements of 62 eggs in the United States National Museum were as follows (Bendire, 1895): Average 31.54 by 22.54 millimeters; largest 34.29 by 22.86; smallest 28.45 by 21.34. Kaeding (1897) has commented on the diversity in shape shown in his collection of over 30 sets; some eggs were short and rounded, while others were long and elliptical.

A set of eight fresh eggs ranged in weight from 7.6 to 8.8 grams, average 8.3, thus being considerably lighter than eggs of black-billed magpies.

Young.—Length of incubation period for the yellow-billed magpie is not definitely known but is assumed to be close to that of the black-billed one, or around 18 days.

Hatching of the young apparently changes the magpies to silent birds. One pair carried food to a nest at least 10 times in the 40 minutes it was watched on the morning of April 12, the visits thus averaging four minutes apart. Usually it was not possible to distinguish any objects in the bill, but once or twice supposed large insects could be seen projecting from the bill. The birds flew directly toward the site, but they perched on some limb 3 to 10 feet away and toward the main crown of the tree for a few seconds before going into the nest. The birds always entered the nest from the same side and left through the opposite one. Nearly always when one of the parents left the nest it flew directly to the ground and immediately began to search for food. On this date whenever a parent visited the nest, the young birds made calls that could be heard by a person 60 to 75 yards distant. These calls began when the parent entered the nest and they ceased as soon as it left. Once, both adults arrived at, and entered, the nest at about the same time. They flew away together a few seconds later, and one of them dropped a fecal sac from its bill when it had gone 50 yards.

These adults nearly always flew to a nearby orchard to forage, and four-fifths of their trips were to some freshly disked ground. On several trips one of the parents perched in the top of an orchard tree before going to the ground. They usually were silent, but once or twice a short series of notes was heard at the nest. The time spent at the nest on each visit averaged between 10 and 20 seconds.

Apparently typical behavior of a family of six young out of a nest and being fed by parents was watched for an hour near the time of nest-leaving. At first three of the young birds perched in a tree 30 yards from the nest. Later this tree held four young and two were in the nest.

The usual procedure in feeding on this day was for the young to keep a sharp lookout and, whenever an adult magpie came within sight, to start up a series of loud calls, higher in pitch than those ordinarily given by adults. If the approaching adult were not the parent of this brood, it continued on its way, and as soon as it had passed the cries would cease. If the approaching bird happened to be one of the parents, it would go to the group of young birds and the cries would be continued until a young one had been fed and the old one had left.

The cries of the young birds were accompanied by energetic flapping of the wings. The white on the wings helped to make the birds conspicuous while thus flapping, so that chances for the attention of the approaching bird being directed to them were increased. This may be an important function of this set of white markings. Whether developed because of its adaptive value or not, it certainly operates to disclose the locations of the young birds to the human observer and presumably likewise to the parent birds when they are approaching with food.

The brood of young magpies, being separated and in different trees, gave a good opportunity to see additional features of the response of parents to begging young. The destination of the approaching parent seemed to be controlled entirely by the amount of commotion made by the young. The group that began calling first and kept it up with greatest vigor was the one finally approached. Sometimes a parent abruptly changed its course when headed toward one group and went to another, apparently because of a greater persistence in begging there. Once, when an adult started to leave, it was attracted by calls from another group and turned back and went there. It was not determined whether any food was delivered to the second group.

The young magpies showed little ability to distinguish their own parents. Any magpie flying toward them aroused the cries and wing flapping. These ceased, however, as soon as the flying bird passed and took a course away from the young birds. Once the young birds begged when a California woodpecker flew over them. The amount of begging seemed to be a direct expression of the degree of hunger. Apparently the two young that stayed at the nest were not yet able to fly. The others could fly; hence they probably were larger and possibly they required more food than the ones at the nest. At least more trips were made to them by the parents. But the ones at the nest were not neglected. For a few trips after food was taken to the nest, the young ones there would remain almost quiet, giving only one or two notes during each visit.

When a parent landed in some tree other than the one holding the young birds, they would fly to that place, flap their wings, make loud calls, and attempt to get in front of the adult. The latter did not remain long after food was placed in the widely opened mouth of some one of the youngsters. Each parent, in turn, left and went to the nearby orchard to obtain another supply of food. The trips were about five minutes apart. The adults generally were quiet, but occasionally they uttered series of alarm notes, usually when away from the young ones.

Between feedings the young magpies moved about on the limbs of the tree by walking, hopping, and making short flights. Their legs appeared to be long, and their claws were the chief means of keeping a hold on the bark. When alighting they had difficulty in regaining a balance. Their tails were only 3 or 4 inches long. Not once was one seen on the ground. Their time was spent picking at the limbs, preening their feathers, or just drowsing. At frequent intervals a young magpie would raise both wings and stretch them over its back, partly folded, but would not extend them.

These young birds were quiet except when a flying bird approached them or they were being fed. During each feeding all the young kept their mouths opened widely while calling. Once one kept its bill widely opened as it flew from one tree to another. On another occasion a young bird flew out 15 or 20 feet to meet an approaching parent, which, however, paid no attention to it but continued on to the tree and fed another one. When a person tested their responses to disturbance by walking over to the tree, at first all the young uttered warning notes. After a few minutes they became quiet, and the ones at the nest withdrew into it. One of the other young ones flew to another part of the grove. Parents came, uttered alarm notes, and left. They returned in half an hour and there were more alarm notes.

In Monterey County a brood of bobtailed young at nest-leaving age was studied early in June. The parents came down to lower limbs within 6 or 8 feet of the intruding person to protest at his presence. Other adults came too. Notes of the adults and young differed considerably, the latter being weaker, softer, and higher. After being disturbed the young magpies moved into higher parts of the tree.

Another young bird, watched in early morning, was begging from an adult perched on a wire nearly 10 feet above ground. It crouched with wings quivering, held its bill near the adult's bill, and uttered low notes, but the old bird did not feed it. The youngster then flew down to the ground and foraged for itself.

An example of communication between adult and young was observed when a young magpie with tail two-thirds grown was handled in a room. Its cries attracted an adult (parent?), which perched on a limb outside the window and squawked long and loudly. Soon after the adult began to call, the young one stopped. The adult flew away. Again the young one called, the adult returned and squawked, and the young one became silent. When the young bird was released outside, the adult perched in a tree overhead, screaming loudly, but made no move. When a person caught the young bird again, however, the adult swooped down at him.

Plumages.—Plumages and changes of them in the yellow-billed magpie seem to be so nearly like those of the black-billed magpie, already described, as not to require separate treatment. Possibly the calendar of molt differs, but this has not been worked out.

On at least several of the magpies I noted on October 9, 1929, in the Sacramento Valley, bright yellow of the exposed skin could be seen extending back from the bill, below and nearly around the eye. On some, if not all, of the magpies observed on November 11, 1930, the yellow, bare area around the eye could be seen distinctly.

Four freshly killed birds from Santa Clara County were examined by me on October 19, 1929. All were in molt. In one, a female, the molt was nearly completed; the sheaths still showed on the contour feathers on the breast and around the head. The skin was yellowish, especially around the head, the base of the tail, and on the body at the bases of the feathers. The yellow bare space behind the eye was 10 by 10 millimeters in size. A male in the same stage of molt showed more yellow on the skin, especially on the under sides of the wings. Another female was farther along in its molt; it showed scarcely any yellow on the skin except around the head. All but the feathers of the throat and chin were free from sheaths. The fourth bird showed sheaths on the feathers about the head, those on the chin and throat being least developed.

Abnormal plumages in the magpie attract more than ordinary interest because of the possibility, which seems almost probability, that some of the existing geographic forms arose by the preservation of this kind of character. This preservation might have been accomplished by means of the kind of geographic isolation that now characterizes the yellow-billed kind. It is interesting, if not significant, that the yellow bill, which is the conspicuous mark of the Californian kind of magpie, has been discovered as an abnormality in other parts of the range of the genus.

Food.—Kalmbach's (1927) study of the food of this species made him conclude that it is somewhat more insectivorous than the black-billed species. At the same time, he pointed out, it is capable of committing practically all the offenses of which the latter is so frequently accused. He considered that its scarcity precluded the possibility of the yellow-billed magpie's doing serious damage. The stomachs examined indicated that 70 percent of the bird's food is obtained from animal matter and 30 percent from vegetable. Insects made up more than half the food. Conspicuous among these are grasshoppers, which appear to be most of the food after midsummer, until the cold weather of fall. Bees, ants, wasps, ground beetles, flies, carrion beetles, and true bugs ranked high. Carrion is consumed in winter and early in spring.

Observations on the manner of feeding have been recorded at length on the Hastings Reservation. Capture of flying insects on the wing is a habit regularly noted. During a period of warm sunshine on a January afternoon, magpies were making flights out and up from a dead oak on a knoll, presumably after insects. The flights ranged in length from 15 to 50 feet. At the end of each one the bird generally swooped and zigzagged up and down as if in pursuit of an insect. Return was to the same dead oak. Similar behavior was noted in the fall.

Another habit, characteristic of foraging magpies, is to search under objects such as chips of wood or cow dung. These sites are hiding places for a great variety of insects and other invertebrate animals. They are generally inaccessible to most foraging birds, which are too weak to uncover them, but magpies can get them with slight inconvenience. In addition, from the heaps of dung they often get grain.

Magpies keep a close watch for new food sources. They quickly find scraps of waste about houses, such as garbage or bits of food that may be thrown out on the ground. They watch other feeding animals, birds or mammals, and rush to retrieve any bit of food that may be lost or discarded. Other foraging magpies even are watched closely, and many pursuits occur when one attempts to carry some item of food. Any object too large to be swallowed immediately or to be carried away is likely to be the center of a contest so long as any of it remains. These encounters rarely reach the stage of actual combat.

To force others away from food, the magpies use a posture as a sort of bluff. Both parties in such a dispute stand very high, point their bills upward at nearly a 70° angle, throw out the breast and throat, and work the muscles as if producing calls. If any are made, they are inaudible to a person 50 feet away. The successful bluffer then walks toward the other bird as if to bump his chest against the other. This sometimes occurs, but oftener the gesture is sufficient to retire the opponent. This pose seems to leave its maker in extremely vulnerable position with sensitive throat exposed to attack, but no blows have been seen struck there. The retiring bird crouches down and shrinks aside, still holding its bill up.

Sometimes a bird resorts to blows instead of bluff when another becomes too persistent. Then a swift hop and peck at the head of the intruder drives it off. So long as the other birds do not try to feed, they are permitted to stand so close to the feeding bird as to bump it occasionally. When pursuits occur, they may be only for about twelve inches, but in the interval an onlooker usually takes possession, and frequently it then refuses to be driven away.

Some individuals merely wait their turn to feed. Occasionally one watches for a long time and then suddenly forces away the possessor. It may be that this period of watching serves to build up courage. Some individuals, which appear to the observer to be weaker ones, rush in to snatch morsels while the feeder tears loose another or pursues another bird. While feeding, these magpies often produce throaty songs, like pleasant squeals. Others squawk as they watch but do not sing.

Magpies exhibit the same sort of nervousness shown by smaller birds when they forage over open ground, and they commonly rush off at intervals to perch in trees or bushes. One morning 23 of them were foraging in a recently planted grainfield. When something disturbed them all flew nearly simultaneously into willows near a creek. If not badly frightened, some lit on fence posts instead. Usually they began to return to the field within 45 seconds. An occasional bird, however, dropped to forage in a pasture on the opposite side of the creek. Generally when two or more birds flew up, the whole flock rose, but one bird's leaving did not cause them to follow. Fewer and fewer returned to the field after each flight until the whole flock was foraging in the pasture.

Storage of food by magpies is noted regularly. It occurs oftenest in winter but has been observed also at other times of the year. Storage takes place usually in shallow pits dug by the magpie in the top layer of soil, but crevices among limbs of trees may be used too. The objects most often stored, at least in one locality, are acorns, but carcasses of small animals and left-over pieces of food may be stored also. Nearly always the cache is covered over with stems of grass or leaves and is so carefully hidden as to be nearly invisible. It may be found by a person only after marking the spot well. On occasion some object may be carried for a long time, as long as half an hour, and then buried, or it may be placed in the open on top of the ground and left. An item being carried may be laid on the ground while the bird examines or eats some other object. Acorns may be buried entire without the shell being opened or they may be partly eaten first. Whether magpies return to a cache once it has been hidden has not been observed, but they have been seen raiding the stores of other birds, always shortly after they were covered.

In some years acorns provide a large share of the yellow-billed magpie's food for several months beginning in mid-September. Live oaks and valley oaks provide most of the ones eaten on the Hastings Reservation. Sometimes the acorns are picked up from the ground, but oftener they are taken directly from the tree. The acorn is then carried to the top of some fence post or to a limb in another tree, where it is held underfoot and pounded until the shell is broken and the desired portion of the kernel is removed. Normally the acorn is carried with the small end pointing ahead and only one is taken, but sometimes it is turned, and the bird carries an extra one. Limbs of blue oak, black locust, and valley oak are used as an anvil for pounding the acorn, but the valley oak seems to be the favorite. Apparently the rougher bark and larger branches make this tree more satisfactory.

In hammering an acorn the bird does not strike with vertical blows; instead the bill comes down toward the feet at an angle of 80° from the horizontal. The bill is not lifted immediately on striking the acorn but is pushed into it. Sometimes hundreds of blows, in series of about 20, are required to open the shell. A magpie exhibits jealousy over an acorn in its possession and normally takes it to some secluded spot away from the other birds. Also it appears reluctant to leave one when disturbed by a person. Effort is likely to be made to retrieve an acorn that is dropped by accident. However, on their own accord, they commonly discard any acorn that proves difficult to open, and many partly eaten ones are left for another occasion or where some other bird can get them. These traits are exhibited with respect to other kinds of food, but they are more easily watched when large objects like acorns are concerned.

Additional mannerisms are shown by the following account of an incident watched on the Hastings Reservation. Four foraging magpies, on November 18, assembled under a live oak and looked up at the periphery of the tree where another one was clinging to an acorn-laden limb, which bent to a 60° slope under the weight of the bird. An acorn was wrenched free and carried to the ground. Another bird flew up and, after searching in three places, found a suitable acorn. This was seized at its base, transverse to the axis of the bird's bill, and wrenched free. The acorn fell to the ground, and the bird followed it swiftly, but too late, for one of two magpies on the ground had reached it and then the retriever refused to give it up. The dispossessed bird returned to the tree and obtained another acorn, which also fell. But this one it followed so swiftly that the other two birds could not reach it, and it retrieved and carried away the acorn with a squawk. The bird that had stolen the first acorn deserted it to try for the next one to fall and then returned to the first acorn and began to hammer it. As many as eight magpies were seen carrying acorns under this tree. Nearly all the acorns were obtained by loosening them and following them to the ground. If no other magpie was close, the bird got its acorn; otherwise a bird already on the ground got it. Each acorn finally retrieved was carried away, placed underfoot, and hammered open.

Other kinds of food obtained regularly from plants in this vicinity were figs and fruits of poisonoak, coffeeberry, and grapes. The fruits of coffeeberry ordinarily are carried to some perch where the pulp is extracted and swallowed and the skin is discarded. The seeds seem to be swallowed sometimes and discarded at other times.

When magpies are feeding broods of young they sometimes discover the nests of other birds and take the contents, but this happens on rare occasions. Once at the end of May, in the Sacramento Valley, I observed a commotion caused by a magpie in a large colony of nesting cliff swallows. At my close approach the magpie flew out from under the bridge that held the nests. It went back again and then out and away. All this time the magpie was being pursued by the large flock of adult swallows. Circumstances indicated clearly that the magpie was there to get young swallows, which at that time filled most of the nests. But no actual raid was seen. Many of the nests had long entrance tunnels, and they appeared too long to permit a magpie to reach into the main cavity of the nest.

Nearly 10 years elapsed before I again found a yellow-billed magpie molesting nests of small birds. In one tree where a pair of magpies and three pairs of Bullock orioles each had nests, several encounters were noticed between the species. One morning a magpie went to one of the oriole nests and poked its head into the cavity, but just what happened was not learned. It left and was pursued by the orioles. On other occasions the orioles were particularly bold in attacking young magpies just out of the nest.

In the same season and close to this spot one magpie acquired the habit of searching out and destroying linnet nests placed in crevices about the farm buildings. Apparently these raids were made to get the young linnets, but actual captures were not seen.

Behavior.—The yellow-billed magpie exhibits the general manner of the black-billed kind, but it appears less timid throughout its range and seems to live closer to human habitations. Possibly the black-billed magpie would respond just as quickly to the near presence of people if permitted to do so, but it regularly encounters a more aggressive disfavor. Whatever the cause, it appears obvious that the birds live closer about houses in California and they encounter less molestation. This means that it is easier to study them, for they willingly permit approach in many places to within 10 or 15 feet. At the same time they retain the capability of cautiously watching for danger, and they quickly slip away if disturbed by shooting. Much of the area occupied by the species is relatively free from this danger.

On the ground a magpie walks, hops, runs, dodges, or makes short leaps with the aid of its wings. The flight is usually short, and in the wind it is wavering, for the long tail then proves to be a hindrance, although ordinarily it gives the bird a graceful appearance. Types of perching places oftenest used when the birds are not on the ground are the larger-sized limbs of big trees. On telephone wires they appear to have no difficulty in balancing. When coming to a stop the bird may jerk its tail upward four or five times and then maintain its balance with the tail held close to 45° below the horizontal. Foraging birds may jerk their tails upward independently of need for balance. A slight jerk of the tail often accompanies a vocal note.

When starting to fly away after being disturbed the birds do not flap evenly. Several spurts may be distinguished. In one example the rhythm was three easy beats followed by two vigorous ones. The contrast was audible as well as visible in a bird flying 25 feet overhead. On another occasion this irregular flight consisted of five to eight rather weak wing beats and then four or five rapid, strong beats. The flight of the several individuals, however, was not synchronized. Another habit of flight observed often is that of gliding downward for long distances, extending the wings, and braking the speed at regular intervals.

Flocks of yellow-billed magpies may be seen all through the nesting period, but these are mainly temporary formations, probably accidental assemblages. When the young birds leave their homes, however, the flocking tendency soon becomes conspicuous again, and the birds form definite aggregations that have the quality of permanency. The unit of organization for this species then changes from the pair to the flock. For the next half year the behavior of the flock is our chief concern with this species.

A striking example of responses of members of a flock to a certain note of one of their species was observed in late November. At 4:30 P.M., after the birds had congregated and just before they started off to roost, they were perching quietly in some oaks. A dead snag broke under one of the magpies in the vicinity, causing it to fly away squawking in alarm. The flock, calling noisily, flew in a compact body to the scene of the disturbance. The birds perched on the remaining limbs of the snag and on nearby trees and then became silent. From there they finally flew off, 15 and 20 minutes later, to roost.

Roosting time is the occasion for congregation of the largest and most compact flocks. The birds assemble in tree tops and then seem to await the move of one to act as leader. Sometimes the group flies out to join one or more individuals flying overhead; sometimes they follow one of their number as it makes a start from the perching tree. If one bird turns back the whole group is apt to follow, whether the bird is ahead of or behind the others. On such occasions they are likely to come swooping back down to trees or thickets, swerving erratically, and wings whistling. In these assemblages there is apparent a great reluctance to make the first move, but once it is made the remaining birds rush to follow.

Daily routine of activity in the yellow-billed magpie is, perhaps, more easily traced than in most other birds, mainly because this bird is so conspicuous. Also at many places the magpies are so continuously and so easily accessible for observation in the vicinity of human living quarters that it is possible to trace the daily and seasonal changes year after year. Thus it becomes obvious that while the daily program is fairly rigid it is also highly variable.

The recorded program of the magpies on a normal day begins with the arrival of the flock, as a unit or in sections, from the roost. After a reassembly in trees, the birds scatter somewhat and begin to forage. Late in the morning they retire to trees for a midday rest, which is broken for another period of feeding before they form another group to fly off to roost. Through the main part of the nesting period this behavior is modified considerably by the necessity of keeping close to the nests.

Magpies regularly congregate to roost in a group through most of the year outside the nesting season. The site oftenest used on the Hastings Reservation is in a small ravine about a hundred feet wide, facing the west, and grown over with live oaks. The magpies roost in these dense trees. Each night, nearly an hour before finally settling at the roost, the birds assemble in sycamores or oaks in the canyon or the low hills a quarter to a half mile to the north of the roost. Half an hour after congregating, they fly to the roost. The assembling flock seems to become more and more unified, the longer the birds stay. Often there are squabbles with other birds, as crows or hawks, in which the whole group participates, at least vocally.

For 10 or 15 minutes preceding the flight across the canyon the magpies are rather silent. They perch mostly in tree tops, several in a tree, and there is little foraging. This period is followed by a series of outbursts of calls or a single one. Large groups make greater outbursts. The flight then begins. After a few outcries the birds become quiet as they fly more slowly and gain altitude steadily. The destination of this flight is a group of trees, usually leafless black oaks, used as gathering places both morning and night. Ten or fifteen minutes are required for all the birds to make the flight. Cold, strong wind, cloudiness, and rain tend to advance the time of flight across the canyon. Encounter with some other species tends to delay it. Change of roosting place from a common gathering place outside the normal daytime range of the colony to the nest site occurs usually in January.

Response to disturbance on the long flight to the roost is indicated sometimes by the birds when they suddenly change the course of their flight and dive back downward to the tops of thickets of trees. Their wings make a great roar, then all is silent. Within a short time, two minutes in one example, the flock rises and proceeds normally on its way.

Departure of the magpies from the roost on a fall morning is somewhat as follows: The birds fly up from the roost in live oaks in a ravine and enter large, leafless black oaks on a ridge. Calls are to be heard often before any birds appear and as early as 20 minutes before they finally leave the hill. The flight to the exposed trees is made singly or in pairs, and the birds perch silently, with heads drawn down close to their shoulders. They finally become vociferous, but the calls are short and quiet compared with ones made later in the day. Calls increase slightly or cease altogether before the birds leave. Gradually the single birds begin to move about and combine to make small groups. Sometimes all these groups unite into one big one, especially in disturbed weather, as rain, fog, strong wind, or low temperature. Finally the magpies leave the hill and fly, swiftly and low, to a wooded knoll on the canyon floor, where the long columns of birds converge in the tops of one or two trees. Calls become louder and more frequent than they have been on the hill or in the flight. If the groups are small, 10 or 15 minutes are required for all the magpies to leave the hill; if large, the time may be as little as 5 minutes. Next, the birds scatter from the trees to the ground.

Change in time of sunrise is followed closely by change in time of roost-leaving by the magpies, but there is considerable variation in the time of the actual flight from the hill. On many mornings, especially in December, the magpies, instead of all going to the lower part of the canyon, split up into groups, and some of them fly directly across the canyon to the top of another sunlit, warm hill. From there they move back later in the morning to the floor of the canyon.

Behavior of magpie flocks at the start of their day on the Hastings Reservation was observed on several occasions in the second half of July in 1938. Once, at 5:55 A.M., an observer suddenly became aware of raucous magpie calls. A flock in compact formation arrived near the center of the normal nesting area and split into three groups of six or seven birds each. Twenty were counted in all. Two groups lit in blue oaks, the third in a sycamore. All were calling loudly, harshly, and continuously. There was a constant interchange of individuals between the groups. The birds moved into many trees but stayed in the top branches. When they first arrived the din was terrible, as all 20 birds squawked loudly at the same time. Twenty-five minutes later the birds had spread out and the calls were more widely spaced.

Usually only two to four birds were calling at once. Two came down to drink at a water trough, but none had gone to the ground. The calling slowed down so that intervals of 10 to 15 seconds elapsed without a sound. By 6:30 the interchange from tree to tree had ceased.

In another 15 minutes small groups of two or three birds began to move off toward the north and the number of calls was still further reduced, so that as many as 45 seconds sometimes passed with no sound. Soon all the magpies had left the area where they had first settled.

Water troughs on the Hastings Reservation, originally set up for cattle, provided water for birds, but their shape proved to be a slight handicap in use. Magpies regularly bathe at these troughs by standing on the rim, tipping forward, and flapping their wings. They then fly to nearby trees to dry their plumage. A different procedure was noted on a cloudy, windy afternoon in January when a magpie drank twice from one of these troughs. The bird then jumped about 6 inches into the air and, with beating wings, gradually lowered itself until belly and feet touched the water. With the tail held up so as not to touch the water, it kept this position for about 15 seconds and then flew to the top of a nearby post to rest. After repeating this three times at one-minute intervals, the magpie flew to a locust tree where it ruffled and shook its plumage, picked at one or two breast feathers, and flew back to the trough. It perched beside another drinking magpie for two minutes and then took two more dips. The other magpie then drove this one off to a fence post where it dried its plumage and the two flew off to forage.

In many parts of California, for example in the Sacramento Valley, the planting of trees and the extension of cultivation have apparently favored the local spread of magpies. Study of present-day conditions in that region indicates that extension of human occupation of this land has, also, over a long period of years, resulted in increased numbers of this bird there. At an earlier time, from 1850 to 1890, there was a period of persistent destruction of magpies in California, which resulted in greater wariness of the birds and led to a disappearance of the species from some localities on the edge of its range. Besides the direct killing by shooting, magpies in this area, according to the testimony of numerous observers, have been killed by placing poisoned bait to prevent their taking of cultivated crops, by poisoned baits placed for coyotes, and by poison used in rodent-control campaigns. Despite the rather widespread notion, however, that the yellow-billed magpie is rare and that it is on the verge of extinction, there seems to be at present no reason for immediate concern over its welfare as a species. Extermination might conceivably be possible, but it would be so expensive and difficult that it would not occur under ordinary circumstances.

The yellow-billed magpie exhibits the tendency shown by magpies elsewhere to be attracted to herds of large mammals. The birds are seen regularly about sheep, cattle, and horses, but they seldom cause any actual injury to these animals. Mostly they are after insects and grain, which they find in greater abundance close to domestic stock. Horses seem, ordinarily, unconcerned at the attentions of magpies. Once I watched a cow that appeared much disturbed by magpies; several times it moved quickly toward one on a fence post and by shaking of its head caused the bird to move on to the next post.

Magpies in the vicinity of one house developed an active feud with a sheep dog, which continued for several months. The birds learned that a certain whistle by a person meant that food was to be given to the dog. They always came too and sometimes arrived even before the animal. The resulting quarrels over bits of food seemed to trouble the dog so much that it would become aroused over the mere presence of magpies or other birds of similar size and would rush after them with many barks whenever they came close to the ground. Late on a summer afternoon a magpie, perched on a clothesline, was watching the dog eat a bone. Three times the dog started to leave and each time the magpie flew down to the spot only to be driven off by the return of the dog. On the fourth departure the dog did not return when the magpie flew down to the ground, picked up the bone in its bill, and carried it away for at least 60 feet.

Yellow-billed magpies may be called resident wherever they occur even though there may be periods of several days each year when not a single individual can be found. Usually the longest period of absence from the Hastings Reservation each year lasts for about 10 days for the species and possibly longer for many individuals, and it comes in the early part of July. Apparently the helpless young hold the parents at the nesting site even after living conditions may become unsuitable in the long, dry days of midsummer. When the young are able to fly and care for themselves, small groups of magpies may be seen for a few days, and then some morning it is realized that not a single magpie is in sight. How far and just where and why they go are questions that have not been answered. Before the birds come back to stay regularly, single individuals may show up for a few hours or a flock may come to roost for a night or so, but obviously these birds are not following the regular schedule which is the routine for the species for most of the year.

This well-marked break in the regularity of seasonal occurrence may result from some local peculiarity in the environment that makes it intolerable for a few weeks. The annual extreme in dryness usually reached at this time when there is a minimum of green vegetation, of animal activity, and of moisture on the ground may drive away these birds, as soon as they are free of nesting duties, along with other species that disappear at about the same time. The exodus may last only until the shortening of the days reduces the evaporation and brings better conditions. Movement toward the coast of only a few miles would bring the birds to cooler and moister places.

The annual departure of the magpies, however, may reflect merely physiological change in the birds independent of the conditions in their habitat at this season. After the young are released from parental care, there may be no attraction in one spot of ground to hold any of the birds—nothing to keep them from wandering freely. But after a rest of a few weeks, faint beginnings of a new reproductive cycle may require a definite home area, and thus the birds come back to begin anew the cycle they have so recently completed.

No doubt a nucleus of the returning birds consists of individuals that left the same place, but the recorded observations indicate that a different number and presumably a slightly different set of individuals return. Some of the recently hatched birds may stay in new spots; they may join with other wandering groups and go to their home sites; they may come back with the parent colony. Thus the returning number may be smaller or larger than the colony at the end of the nesting, but it is rarely of the same size. The stabilization of numbers may require several months, and always it extends over a much longer period than the actual absence of the species from the area.

Thus may be seen, in a species ordinarily considered as strictly resident, behavior that appears exactly like that of migratory kinds which leave for long periods and which travel long distances. Once more it appears that migration is a characteristic of all birds and may be exhibited only in lesser or greater degree. It is not necessarily a basis for sharply limited classifications of species.

Voice.—Special attention to study of the voice of the yellow-billed magpie led Richard Hunt to conclude that in fall the birds had only one type of utterance, which, however, was varied. (See Linsdale, 1937.) The three variations or phases listed here he considered distinct enough to be described separately. His statements have been verified in field observations by me. The three phases are as follows:

1. *Qua-qua, qua-qua-qua,* etc., given in series from two to six *qua's.* The utterance is usually quite rapid when the *qua's* are more numerous. The note is loud and the expression is rather good natured or well disposed. The timber is raucous. It has more than a slight resemblance to that of the California woodpecker's "cracker" notes. When birds of the two species are heard calling at the same time, the timber

of the two calls is almost indistinguishable. An element in this utterance suggests the rich, harsh, scolding *chaack* note of Bullock's oriole.

2. *Quack?* A single note, rather mild in expression, yet querulous. This note has the same general timber as No. 1.

3. *Queck* or *kek.* Sometimes uttered alone and sometimes heightened from phase No. 2. The utterance has an almost absurdly weak tone. It reminds the observer of the call of the black-necked stilt. It is more piping than the other types of notes and is a little nuthatchlike. When first heard it was written down as *pêip.* The sounds in this note are less distinct and of the three types it is least spellable and least utterable by a human being.

Comparison of the voices of the two kinds of magpie in America is difficult when the birds must be studied separately. Some observers have been unable to distinguish them on this basis. Wheelock (1904), however, considered the call note of the yellow-billed form as less harsh and loud than that of the black-billed, and this is an observation that might be anticipated after studying other habits and the surroundings of the birds.

The primitive song similar to that of other corvids is produced often by yellow-billed magpies in December, January, and February, and apparently it is a normal part of the early segments of the nesting season. Indications are that it has significance in the nesting habits and possibly in other phases of magpie life.

Field marks.—The yellow bill provides an easy and certain means of distinguishing this species from the black-billed kind at ordinary distances. Its slightly smaller size is of little help because the two kinds do not occur together naturally. It may be separated from other birds by the general characters of the black-billed magpie.

Predators.—The horned owl may sometimes eat magpies, but so far the evidence is only circumstantial. At the Hasting Reservation, once in June, 20 minutes after three magpies had gone to the roosting place, a great horned owl flew up to a ridge near them where it hooted three times. One of the magpies moved over to investigate and became very excited, calling loudly for about four minutes before becoming quiet again. Other disturbances heard at magpie roosts were thought to be caused by the near presence of a horned owl. Most of these occurred before the birds settled for the night or after they left their roosting trees.

A long series of observations of accipiters and yellow-billed magpies brings the conclusion that magpies must sometimes be captured by sharp-shinned or Cooper's hawks. In summer and fall flocks of magpies carry on almost continuous squabbles with these hawks. Usually the hawks appear to take the initiative in the pursuits, but actual captures have not been seen, and many times the chases seem to be only a form of exercise. Many times when capture appeared certain, the hawk turned away just before reaching the magpie. Once two accipiters flew after the same magpie. Possibly these flights serve to make the magpies less cautious so that capture is easier when the hawk is really hungry.

One flock, in August, was flushed by a Cooper's hawk that flew into the tree. The hawk pursued the birds as they scattered and then re-formed a group, each one flying erratically and yelling in a sort of "confusion chorus." Termination of the chase was not seen, but it was thought that no capture was made.

Teamwork on the part of two magpies to drive away a sharp-shinned hawk was seen early in November. They took turns maneuvering into position to dive at the hawk as it pursued or fled from the other. The hawk soared upward, which it could do as fast as the magpies could fly. Thus it got above them and out of range of easy flight for them, but it was forced finally to leave the vicinity.

When a sharp-shinned hawk struck and carried off a *Zonotrichia*, one magpie fled before it, but four others immediately pursued it to a rose thicket, dived at it three times, and forced it to double back and seek refuge in a more dense thicket. The magpies then stationed themselves around the hawk, one above on a fence post, one in the bushes, and two on the ground, and they remained there watching silently for nearly five minutes.

In another example magpies in and close to a blue oak were excited over pursuits by a sharp-shinned hawk and a sparrow hawk, both with quarters in the same tree. The former was much more persistent in its drives, pursuing the magpies more swiftly and going closer to them. Each pursuit, however, was abandoned after about 50 feet, and the birds would return to the tree, the hawk in the lead. Behavior when the magpies were driven from the ground was much the same. On such occasions a harsh, growling note is uttered by each magpie just before the hawk reaches it, and this may have something to do with the latter's turning aside just before reaching the magpie. No magpie was touched, so far as could be seen.

Once three magpies, all making the growling note, joined in pursuit of a sharpshin, flying about 10 feet behind the hawk, and following it into a sycamore. Another time this hawk pursued a single magpie, which zigzagged to evade three attempts at capture and growled continuously. The mere presence of the hawk as well as its pursuits seemed to bother the magpies. After these disturbances the magpies moved to denser parts of the foliage of a tree and especially to places where there was a protective barrier of branches close overhead. Flocks pursued by hawks in the evening seemed to be harried more persistently than in chases at other times of the day.

DISTRIBUTION

Range.—Confined to California; nonmigratory.

The range of the yellow-billed magpie extends **north** in the Sacramento Valley to Redding and Anderson. To the **east** it is found to Smartsville, Placerville, Hornitos, casually Dunlap, and Three Rivers. The **southern** limits of its range are in the vicinity of Santa Paula, Santa Barbara, and the Santa Ynez River. **West** to San Luis Obispo, Paso Robles, and Monterey.

Egg dates.—California: 112 records, March 9 to June 2; 56 records, April 9 to 22, indicating the height of the season.

CORVUS CORAX PRINCIPALIS Ridgway

NORTHERN RAVEN

PLATES 27-34

HABITS

The raven is a cosmopolitan species, widely distributed over all the continents of the Northern Hemisphere; it has been separated into several subspecies, only two of which are recognized as North American. The range of this subspecies extends, according to the 1931 Check-list, from the Arctic regions in "northwestern Alaska, Melville Island, northern Ellesmere Island, and northern Greenland south to Washington, central Minnesota, Michigan, coast region of New Jersey (formerly), and Virginia, and in the higher Alleghanies to Georgia." This seems like a very extended and rather unusual distribution and suggests that Dr. H. C. Oberholser (1918) might be justified in giving the name *C. c. europhilus* to the bird he described from the Eastern United States.

The haunts of ravens and their behavior vary greatly in different portions of their range, depending on where they can most easily find their food and where they are least likely to be disturbed. In the far north they range widely over the open treeless tundra, along the seacoast and the banks of rivers, hunting for the carcasses of animals slain by hunters and the remains of sea-animal life washed up by the storms, or stealing the bait from fox traps; here they are also common about the Hudson's Bay Co. stations and the camps and villages of the natives, where they can usually find scraps and are not molested.

When we stopped for supplies at Ketchikan in southern Alaska, we found ravens very common and familiar. They came flocking down from the heavily forested mountains back of the town to compete with the gulls and crows for the garbage thrown out on the shore. They seemed much at home in the village also, perching unafraid on the roofs of houses or on the totem poles. They are useful as scavengers here and are not disturbed.

We found ravens common all through the treeless Aleutian Islands, frequenting the steep grassy hillsides and the rocky cliffs. In Iliuliuk Village on Unalaska Island they were especially abundant and absurdly tame. Along the beach and about the houses they were as tame as hens, sitting on the fences or on the roofs of houses like tame crows. They knew that they were safe here and made no attempt to fly away, unless we approached too near, when they merely hopped off to one side or flew a short distance to some low perch.

Farther south, along the Pacific coast and islands of Alaska, and on the Atlantic coast from Maine northward, ravens frequent the rocky cliffs along the shores and are especially apt to establish themselves in the vicinity of sea-bird colonies, where they can prey on the eggs and young of these birds. Here they are also found on the islands that are heavily wooded with spruces and other coniferous trees, sometimes placing their nests in the midst of heron colonies, which suffer from their depredations.

From Pennsylvania southward, ravens are mountain birds, living in the Carolinas usually above 3,000 feet, far above the range of the crows; from these heights they sometimes descend to the valleys, or even the islands along the coast, to forage among the colonies of sea birds. Dr. Samuel S. Dickey (MS.) tells me that in the mountains of Pennsylvania and the Virginias "most of them prefer to dwell among rocks, such as serpentine, quartzite, sandstone, and shale. They resort to perpendicular cliffs, to escarpments thrust above forests on the flanks of mountains, and to sloping talus beds in the valleys of streams. When such sites are molested by too frequent visits of mankind, or when blasting takes place for construction of railroads and highways, then ravens will move away. They will, if the case requires them to do so, take shelter in various species of evergreen growth." The Rev. J. J. Murray tells me that the raven is fairly common in Rockbridge County, in the center of the valley of Virginia. "It seems to me to be commoner now than it was 10 years ago. This is particularly true in the Blue Ridge Mountains."

Courtship.—Richard C. Harlow (1922), who has had considerable experience with ravens in Pennsylvania, believes that they remain mated for life. He bases this assumption on the fact that certain pairs that he has watched year after year show certain individual characteristics by which he can recognize them. The males show striking peculiarities in behavior and voice. Certain pairs always nest on cliffs, though suitable trees are readily available; other pairs always nest in trees, close to

available cliff sites; and the eggs of each female run true to form each
year. These are not, of course, positive proofs of his theory, but they
are at least suggestive.   He says of the courtship:

The pair return to the cliff together usually the first week in February, at
first only for a short visit each day, and later several visits. A heavy storm at
this time usually delays these visits for several days. At this time I have seen
them go to the nesting ledge, the female usually alighting on the ledge and the
male on a dead stub nearby and spend ten or fifteen minutes there. At this time
they often soar together high up in the air with wing tips touching, the male
always slightly above the female. At times he will give a wonderful display of
his prowess on the wing, either dropping like a meteor for several hundred
feet and fairly hissing through the air in the manner of the male Duck Hawk, or
tumbling like a pigeon over and over. During this period also, I have found
them perched together high up in an old dead tree caressing each other with
bills touching.

Nesting.—In the far north, beyond the tree limit, ravens have to nest
on the cliffs, usually near the coast, and often on the same cliffs with
gyrfalcons. But where suitable trees are to be found, they often nest in
them. MacFarlane (1891) found this species abundant at Fort Anderson and on the lower Lockhart and Anderson Rivers. "All but one of
the eight recorded nests were situated on tall pines, and composed of
dry willow sticks and twigs and thickly lined with either deer hair or
dry mosses, grasses, and more or less hair from various animals."
Major Bendire (1895) says: "While nesting sites on cliffs are generally
resorted to along the seashore, in the interior of Alaska on the Yukon
River, as well as on the numerous streams in British North America
flowing into the Arctic Ocean, they resort to some extent to trees, probably on account of the absence of the cliffs. Mr. James Lockhart found
a nest in a cleft of a poplar tree, 20 feet from the ground, at Fort Yukon,
Alaska, on May 29, 1862. * * * Their nests resemble those of the American Raven in construction. Near the seashore they are usually lined
with dry grasses, mosses, and seaweed, while hair of the musk ox and
moose is often used when procurable in the interior."

In the Magdalen Islands I saw three nests, quite inaccessible with any
means at our disposal, on high rocky promontories facing the Gulf of
St. Lawrence. In northern Ungava, Lucien M. Turner (MS.) found
the ravens nesting on cliffs near the seashore and about the mouths
of rivers.

The northern raven breeds regularly on the wooded islands on the
coast of Maine, where I have seen several nests in the vicinity of Penobscot and Jerico Bays. On June 10, 1899, I visited Bradbury Island in
Penobscot Bay for the purpose of investigating a breeding colony of
great blue herons. This is a high island, with steep rocky sides and open

pastures in the center, and is heavily wooded at both ends with firs and spruces and a few white birches. There was a rather large colony of the herons here, and in the midst of the colony was a raven's nest. A pair of ravens were flying about overhead, croaking and cawing angrily, and on the ground nearby were three of the herons' eggs, with big holes in the ends and the contents still fresh, as if recently taken from a nest; evidently we had disturbed the ravens at their stolen feast.

During the next two years I came earlier and found several old and new nests on Dumpling and Fog Islands in Penobscot Bay, and on some of the islands in Jerico Bay. The nests were all in spruce trees in thick woods, at heights varying from 25 to 32 feet above ground. Most of them were conspicuous from the ground below, but one was well hidden in a thick top. They were mostly huge structures, 2 to 3 feet in diameter and nearly as much in height; evidently they are sometimes added to from year to year, as we know that the Dumpling Island nest was used for three years in succession. They were made of crooked sticks as large as a man's thumb, mixed with smaller sticks and twigs. The nests were deeply hollowed and were warmly lined with sheep's wool and sometimes a little usnea; in one nest there was a patchwork lining of black-and-white wool. These ravens evidently lay at irregular intervals, for on April 23, 1900, I examined a nest containing two fresh eggs and another with two lusty young.

But Maine ravens do not all nest on the coastal islands. Paul F. Eckstorm writes to me: "On March 28, 1940, I took a set of five eggs from a nest situated in a crevice of a cliff 27 feet above the highest shelf accessible on foot. The nest was about 4 feet in diameter from front to rear, with the cavity at the rear 6 inches from the back wall. The nest was made roughly of course sticks, largely hemlock, and lined entirely with deer hair taken from some old carcass. The top of the cliff bulges into a great overhang extending like a roof about 6 feet above and, in places, 10 feet beyond the nest site. The nest was located near a large pond and about 15 miles back from the general coastal trend. The cliff has a southern exposure and warms up in advance of the adjacent flatter country."

Mr. Harlow (1922) has given us a most comprehensive account of the nesting of the raven in Pennsylvania, from which I quote the following extracts:

There are two distinct types of nesting site chosen here in Pennsylvania—the cliff site and the tree site (the cliff nests outnumbering the tree nests in the proportion of about eight to one). * * *

One feature is almost invariably demanded by the cliff nesting Ravens and that is that the location be dark and well shaded. Usually the darkest available section of the cliff is selected where the ledges are shaded by hemlocks which often grow on the smallest ledge on the face of the cliff. Very frequently the nest will be placed under an overhanging tongue of rock so that it will be protected from above and I have yet to see a nest in use that is not sheltered either by trees or by an overhang. * * * The height of the cliff seems to be a secondary consideration to the shade, though rarely is a cliff with a straight drop of less than fifty feet chosen and they run from there up to two hundred feet.

In the case of the tree nesting Ravens, the first requisite they seem to demand is the highest available tree, and the second is good cover in the very top of the tree. The tree nests are giant structures over four feet across and yet the birds conceal them so well in the very top of the tree that they are frequently very hard to see from the ground. * * * The tree nests are usually placed in a double or triple vertical crotch from seventy to over one hundred feet up, nearly always the highest available strong crotch but in one instance a horizontal crotch four feet out on a large limb was used.

The base of the nest varies from little more than three feet to five feet in the largest nests with an average of almost four feet. The cavity averages a foot in diameter and six inches in depth, the depth varying considerably. * * * The base is composed almost entirely of dead branches and sticks, freshly broken by the birds themselves. When built upon last year's nest, the freshly broken sticks make a sharp line of contrast where they are built upon the old excrement bespattered rim of the previous year. Some of these dead branches are over three quarters of an inch in diameter and over three feet long. * * *

The two most constant features of the cup lining are bark shreds and deer hair, the latter predominating when available. * * * The bark strips, shreds and fibres are obtained from dead trees, underneath the rough outer bark and they frequently use grapevine shreds as well. Some nests are lined almost entirely with white hair from the belly of the deer and some with red from the back, the birds using just what is available from the carcass. Outside of these main features the lining varies according to material available in the various localities. Tufts of hair from domestic cattle or from dogs as well as horse hair are frequently found. Bits of fur from the skunk, opossum and wildcat, sheep wool, bits of green moss scraped from the sides of rocks are all used by various pairs. I have found one nest heavily felted with material which the birds had been picking from an old felt hat and in another lining were bits of rope. Perhaps the most striking nest was one containing a heavy lining of deer hair and flourishing on one side of the cavity was the entire tail of a deer. * * *

The Raven is essentially a solitary bird and the nests of different pairs are usually a considerable distance apart. The only pairs I know of which nest at all near to one another are six miles apart. I know of no bird which comes into direct contact with the Raven during the breeding season but the Duck Hawk. * * * There seems to be a mutual respect between the two species and though they have occasional disagreements I have known them to nest on ledges only forty feet apart, the Raven having young while the Duck Hawk had eggs.

Dr. Samuel S. Dickey has sent me some extensive and interesting notes on ravens in various parts of the country, but space will allow only a few extracts. His notes on Pennsylvania ravens agree very closely with the foregoing quotations from Mr. Harlow's published paper. Out of 17 nests recorded in his notes, 13 were on cliffs or ledges, 3 were in hemlocks at heights varying from 45 to 80 feet, and one was

in a white pine at 85 feet above ground. He says that when the season
is not too backward and cold ravens may begin carrying sticks for the
nest during the middle of February, and he has seen nests finished as
early as February 25, but that most nests are not ready for eggs until
March. "Ravens usually take what they want rather near at hand,
although they may move off some miles for substances suitable for the
lining. I have known them to enter the forest, remain either in undergrowth or low branches, and shortly arise with a sizable stick in the
bill. After the bird had arisen to a height of several hundred feet, it
would begin to circle, the stick in its bill visible through a field glass.
Then it would toss the stick loose from its hold, would snap at it,
thrust forth the body and take it again, and even drop the stick and
plunge admirably downward, taking the stick from the air before it
struck the ground."

Walter B. Barrows (1912) mentions a nest, found by Dr. Max M.
Peet on Isle Royale, Lake Superior, that was in a very unusual location. He quotes Dr. Peet as follows: "While exploring the ruins of
the deserted town near the head of Siskowit Bay, on September 10,
a nest of the Northern Raven was found in the old stamp mill. It
was placed in the small hollow formerly occupied by the metal plate
upon which the head of the stamp fell. The side walls of the stamp
mill are broken down in places so that the entrance to the interior
was simple."

Eggs.—The northern raven is said to lay two to eight eggs to a full
set. Four and five are the commonest numbers. The largest sets seem
to be found in the far north; MacFarlane (1891) says six to eight,
but Harlow (1922) reports that in Pennsylvania sets of three are rather
common and that certain pairs never lay more than three or four; he
says that six is very rare and that he has only one record each of
seven and two. Dr. Dickey records only one set of seven. The eggs
are indistinguishable from those of the American raven, which are
more fully described hereinafter.

Some eggs of eastern ravens are so heavily marked that the ground
color is nearly obscured, and others are so faintly and sparingly marked
that they resemble some types of eggs of the white-necked raven. Often
dark and light types occur in the same set, suggesting that the pigment
may have become nearly exhausted before the last egg was laid. The
measurements of 50 eggs average 50.2 by 34.3 millimeters; the eggs
showing the four extremes measure **56.0** by 33.3, 50.8 by **36.1, 41.7**
by 33.0, and 44.5 by **30.2** millimeters.

Young.—Several observers agree that the period of incubation is
about three weeks and that the young remain in the nest about four

weeks. Harlow (1922) says that during incubation "the male feeds the female upon the nest but does not as a rule sit upon the eggs except in cold stormy weather when the female leaves the nest for food."

Dr. Dickey says in his notes: "When the eggshells curl and burst, the infants squirm in the cup of the nest. They are weak organisms, streaked with orange, yellow, and dusky and having areas of dusky gray down upon them. They are gorged with food periodically, about every half hour during daylight hours, by both male and female parents. Thus they grow fast and within five days disclose bands of slate-blue pinfeather shafts upon their wings and tails, as well as in stripes on their breast and sides. The pinfeather shafts then disintegrate and scatter the deciduous scales; these fall into the nest and spot its rim. Thus the dull, lusterless first feathers appear; they almost conceal the bare skin coloration. Gradually the entire pinfeather scabbards disintegrate and the first plumage dominates."

W. Bryant Tyrrell has sent me some notes on two nests, found on ledges in Shenandoah National Park, Va. In one of these the young were just hatched on March 26, 1939, and were well feathered on April 23. He says that the young "stay in the nest four or five weeks, though the adults have to look after and feed them for some time after their leaving the nest."

Young ravens, during their first summer at least, are often absurdly tame. Langdon Gibson (1922) says that young birds that he observed in Greenland "were trusting and inquisitive. At our boat camp in August, 1891, on Hakluyt Island, some young birds alighting on the flat shelving rocks on which we were cooking our evening meal, literally walked into camp, and at distances of no more than fifteen feet, ate the entrails of Guillemots that we tossed to them. We found them playful and at the expense of 'Jack,' a Newfoundland dog, amused themselves by leading him a chase. The birds would allow 'Jack' to approach within a few feet and then with a flop or a hop, would keep just out of his reach."

Theed Pearse tells me that on Vancouver Island young ravens are very unsuspicious, settling on nearby low trees and gurgling, and allowing approach to within 25 or 30 feet, or flying over within easy gunshot range.

Young birds, taken from the nest shortly before they reach the flight stage and reared in captivity, make interesting and amusing pets, much like young crows in behavior.

Plumages.—Two young ravens that I found in a nest on the coast of Maine were about as large as pigeons; they had evidently been hatched blind and naked, for their eyes were not quite open, and they had developed only a scanty growth of grayish-brown down on the dorsal tract. They were not attractive objects; their abdomens were fat and distended, as if they had been well fed; and their great, gaping, red mouths were wide open, as they stretched up their heavy heads on their weak and shaky necks.

The development of the plumage is referred to above. The juvenal plumage is practically fully acquired, with most of the natal down rubbed off, before the young bird leaves the nest. In the full juvenal plumage, the wings and the tail are much like those of the adult, clear lustrous black with greenish and purplish reflections, but the contour plumage of the head and body is dull brownish black, without any metallic luster. A partial postjuvenal molt, involving the contour plumage and the lesser wing coverts, but not the rest of the wings and tail, occurs in summer, beginning in July or earlier; this molt is sometimes completed in July and sometimes not until August or early in September. This produces a first winter plumage that is practically adult, full lustrous black, with the peculiar shaggy and attenuated feathers on the throat. Adults have a complete postnuptial molt in summer and early in fall, apparently at about the same time as the young birds; I have two adults in my collection, from Alaska, that were in full molt, body, wings, and tail, on June 11; and in some cases the molt is not completed until October. The sexes are alike in all plumages. Spring birds show some signs of wear and fading, but apparently no molt. Theed Pearse tells me that molting is very irregular and that he has seen adults molting their primaries as early as May 15 and others still molting as late as October 21.

Food.—The northern raven is one of our most omnivorous birds and a filthy feeder. Almost any kind of animal food that it can catch, kill, or find is grist to its mill. In the far north, especially in winter, it must live largely on carrion, the carcasses of various animals that it finds cast up on the shores. Dr. Dickey was told that in the dead of winter, when hard pressed for food, ravens will follow the dog teams and fight for the steaming dung as soon as dropped by the dogs.

Dr. George M. Sutton (1932) writes of their winter feeding habits on Southampton Island: "Rarely, if ever, did they prey upon ptarmigans or Arctic Hares, though they were known to pursue and even occasionally to wait for lemmings; but their principal food appeared to be the carcasses of walruses, seals, or whales, which were located and regularly fed upon before the winter set in. A dead whale thus sometimes furnishes a flock of ravens sustenance for the winter, after the gulls have departed and the Polar Bears gone to sleep. In patrolling their range they keep an eye open for all seals killed at the edge of the floe, or for caribou, freshly dragged down by the Arctic Wolves. And there are always, of course, the fox-traps, where they can steal bait, pull the foxes to pieces, or tear up the Snowy Owls, which have been caught."

Lucien M. Turner says in his unpublished notes on Ungava: "In the fall of the year they eat great quantities of berries and, after having satiated their hunger, repair to the rocks on a point of land and digest them. Their stains are everywhere visible on the rocks." In October he found scores of ravens along the Koksoak River, "where they had collected to feed upon the refuse of the hundreds of carcasses of reindeer, which had been speared by the Indians and Eskimos, and decomposing along the banks, whither the winds and currents had drifted them."

In the summer time along the coast and islands of southern Alaska, where the ravens live in the vicinity of sea-bird colonies, they make an easy living by robbing the nests of gulls, murres, and cormorants. As soon as a nest is left unprotected, the ravens dash in, seize an egg or young bird, and fly off with it. When a man invades the colonies all the sea birds leave their nests, and the ravens make repeated raids, returning again and again to carry off egg after egg, concealing for future use what they cannot eat at once.

Theed Pearse mentions (MS.) some feeding habits of ravens on Vancouver Island. He has seen them while feeding on the tidal flats "fly up and down with a bump on wet sand and search the surrounding area, presumably for sandworms disturbed." One was seen following a grazing heifer, keeping close behind it and picking up the insects disturbed; one appeared to follow right at the heels of the beast, "kept looking up, then periodically it would fly up to the flank, either picking up an insect or, as the action suggested, impatient at the slowness of progress." He has also seen ravens following a plow, as the gulls do; and once he saw one "feeding on Saskatoon berries, from which it drove away a flock of crows." He says, in its favor: "During the years of great abundance here, there were broods of ducks raised close to where there would be 20 or more ravens each day. I never saw any sign of ravens attacking young birds or chickens, and the congested area was all farm land with chickens running about the fields."

J. A. Munro reports to me that the stomach of a raven, taken on July 25 in British Columbia, contained a mass of blackberry pulp and seeds, 90 percent, and the fragments of a shore crab; another, taken August 14, held 10 fly maggots, 70 percent, fragments of an amphibian, two winged ants, and 3 seeds.

Dr. Dickey (MS.) writes of the feeding habits of ravens in Pennsylvania: "Along the major rivers of the Appalachians I have noticed that they frequent the banks of streams to procure dead animal matter cast up by freshets. They take crayfish, mussels, minnows, fish, tadpoles, and frogs. Where mountain folk haul carcasses of horses and cows into lonely recesses of the uplands, I have known ravens to appear and cleanse the bones. They contend with crows, starlings, blue jays, and turkey buzzards for morsels of food. I have found that, when they have gorged themselves on organic matter, they will dig circular holes in sod, about 3 inches wide and 4 inches deep, in which they bury pieces of meat that they desire to have properly seasoned. They return and utilize it at an opportune time."

Reid McManus (1935) tells of a raven that entered a henhouse in New Brunswick, when food was scarce in the winter, and killed a sickly hen; it escaped when surprised but returned to feed and was killed; its stomach contained only a piece of skin from the hen and a few feathers. Mr. Harlow (1922) has known ravens "to eat the buds of various trees when hard pressed for food." There are several other items, not referred to above, that have been mentioned as included in the raven's varied diet, viz: Mice, rats, lizards, snakes, various insects such as beetles, grasshoppers, and crickets, several forms of marine invertebrates picked up along the seashore, and mollusks. The raven seems to have learned from the gulls, or perhaps the gulls have learned from the raven, the trick of breaking the shells of mollusks by dropping them on the rocks.

A study of the food of the raven would seem to indicate that it is not a serious menace to man's interests. The harm that it does to young lambs, poultry, or wild birds' eggs is probably overestimated and more than offset by the good it does in the destruction of injurious rodents and insects. Most ravens live far away from human habitations, and where they do come in contact with villages, trading posts, and camps in the north they are useful as scavengers.

Charles Macnamara tells me that in the lumber camps of Ontario in winter "the shanty men, working too far from their camp to return for dinner, were always careful to bury in the snow the flour bag containing their frugal meal of bread and pork, so as to hide it; for the raven, if he found it, would promptly tear it open with his powerful beak and devour the contents. The French Canadians interpreted the birds' hoarse cry as 'Poch! Poche!' ('Bag! Bag!'), and said he was calling for the lunch bag."

Francis Zirrer says in his notes that ravens "often frequent those parts of heavy timber where the waters of the spring thaw and later rains remain longest. Very little vegetation develops in such places; the ground remains mostly bare. In the rich, black humus, however, an enormous number of larvae of various species of Diptera live.

And the ravens, besides many other birds, take full advantage of the abundant and nutritious food. Sedately they walk, swinging their bodies from side to side, or jump awkwardly back and forth, turning the leaves and pieces of bark and decayed wood or boring after the small but juicy morsels. During this period, and only then, one hears their metallic click, sounding like the stroke of a light hammer on a piece of heavy tin—one of the most remarkable sounds in the north Wisconsin woodlands. No one lucky enough to hear it will pass it without marvel, comment and inquiry as to the origin of it."

Behavior.—The flight of the raven is so fully described under the following subspecies that it is hardly necessary to say anything further about it here. It shows great mastery of the air in its majestic flight; it can stand almost motionless in the teeth of a gale, hover in the air like a sparrow hawk, or take advantage of the upward current on a steep hillside to rise and circle like a large hawk. Mr. Pearse tells me that when these birds were so abundant there, there was a regular flight line night and morning to and from their feeding grounds toward the mountains in the interior of Vancouver Island; they always passed over sometime before dark and would return in the morning at a corresponding period after sunrise. They never went by in a flock, but in small parties of eight or more, once as many as 40. They probably had some roost in the interior. Baird, Brewer, and Ridgway (1874) mention a roost discovered by Captain Blakiston near Fort Carlton; his "attention was first drawn to it by noticing that about sunset all the Ravens, from all quarters, were flying towards this point. Returning to the fort in the evening by that quarter, he found a clump of aspen-trees, none of them more than twenty-five feet high, filled with Ravens, who, at his approach, took wing and flew round and round. He also noted the wonderful regularity with which they repaired to their roosting-place in the evening and left it again in the morning, by pairs, on their day's hunt. They always left in the morning, within a minute or two of the same time, earlier and earlier as the days grew longer, on cold or cloudy mornings a little later, usually just half an hour before sunrise."

The raven is one of our most sagacious birds, crafty, resourceful, adaptable, and quick to learn and profit by experience. Throughout most of its range and under ordinary circumstances it is exceedingly shy and wary; it is almost impossible to get within gunshot of one in the open; one is seldom seen flying from its nest, as it hears the intruder coming and departs; I have never seen one return to even the vicinity of its nest while I was near it. Yet it knows full well where and when it is safe; about the northern villages and stations, where it is appreciated as a scavenger and not molested, it is as tame as any dooryard bird; but even here it is always on the alert, and, if one picks up a stone or makes any other suspicious move, it is off in an instant. Kumlien (1879) relates the following, to illustrate its resourcefulness:

I have, on different occasions, witnessed them capture a young seal that lay basking in the sun near its hole. The first manoeuvre of the ravens was to sail leisurely over the seal, gradually lowering with each circle, till at last one of them dropped directly *into* the seal's hole, thus cutting off its retreat from the water. Its mate would then attack the seal, and endeavor to drag or drive it as far away from the hole as possible. The attacking raven seemed to *strike* the seal on the top of the head with its powerful bill, and thus break the tender skull. * * *

I witnessed a very amusing chase after a *Lepus glacialis.* There were two ravens, and they gave alternate chase to the hare. Sometimes the raven would catch the hare by the ears, and hare and raven would roll down the mountain side together thirty or forty feet, till the raven lost his hold, and then its companion would be on hand and renew the attack. They killed the hare in a short time, and immediately began devouring it. * * *

Young reindeer fall an easy prey to them. When they attack a young deer, there are generally six or seven in company, and about one-half the number act as relays, so that the deer is given no rest. The eyes are the first parts attacked, and are generally speedily plucked out, when the poor animal will thrash and flounder about till it kills itself.

C. J. Maynard (1896) writes: "Dr. E. L. Sturtevant informed me that he was at one time standing on a beach at Grand Menan, when he saw a Gannet soaring very high in the air with, what appeared to be, a black spot above and below it. The bird seemed distressed and continued to mount upwards until both dark spots were seen to be above it, when suddenly it fell from that immense height, struck the ground, and was actually dashed to pieces by the force of the shock. Dr. Sturtevant approached it, when a Raven sprang from the body and flew away."

The behavior of ravens with other species of birds, notably crows, hawks, and vultures, has been commented on by many observers, but it is not always apparent which is the aggressor. Emerson Tuttle writes to me that the raven "rarely, if ever, leaves a sentinel on watch, relying on the crow or the herring gull to give the alarm. Once on the ground among crows and herring gulls, the raven dominates. I have seen a gull utter his screaming challenge in the face of a raven, but once the raven moved forward, the gull gave way. I have seen the raven and the goshawk together on several occasions. I watched a young goshawk chase and strike at a raven. The raven did not seem disturbed, though each time the goshawk rolled to one side and struck at him, the raven let out an oath and avoided the touch by rolling and dropping. It would be reading too much into the episode to suggest that the raven enjoyed the chase, but such was the impression he gave. The goshawk tired first and gave it up."

Crows often mob ravens, as they do owls or hawks, but seldom seriously attack one. Dr. Dickey (MS.) writes: "Two ravens emerged from a gigantic cliff. All at once a turkey buzzard flew down near the cliff and acted as if it were searching for something. One raven pursued this buzzard and actually struck it; the raven continued to pursue the slow-moving buzzard until four crows drew near. The crows harassed the raven, but it was too nimble on the wing to be actually hit by them. The raven would move speedily to the right or left every time the crows struck. Then the mate of the first raven appeared from trees, and two of them were attacked by the four crows. The buzzard in the meantime joined the throng; all the birds ended up in an apparent playful manner on the wing near the crags; they continued, while I watched, to make dives and sallies at one another."

He tells, also, of a "vicious combat" that took place between a red-tailed hawk and a raven. "The raven, ired as it was about its molested nest, whipped and drove off the predator."

Rev. J. J. Murray writes to me: "I have reported in *The Auk*, on the word of a very trustworthy mountaineer, the amazing habit in the raven of worrying the turkey vulture until the vulture disgorges the food and then eating the vulture's vomit. I have commonly seen crows chase ravens, but have only once noted the reverse."

Ravens sometimes display decidedly playful tendencies (some of which are mentioned under the American raven), aerial acrobatic feats, and spectacular dives. Theed Pearse tells me that he has seen similar behavior. One bird was seen carrying a fir cone in its beak. The tall Douglas firs "seemed alive with ravens," and it was interesting to see them picking the cones; "the bird would fly up to the branch and hang onto the cone with its beak, with the wings partially extended, and get the cone off by tugging at it. There was one particular branch that seemed to attract them, at the top of a tall ragged tree open at the top. Birds would come to this same branch and clip off a small twig with the beak, sometimes holding it in the beak for a minute or so, even flying away with it, but usually the twig would be dropped when detached." He saw another playing in the air with what looked like a piece of dried skin, dropping it and catching it in its claws.

Dr. Nelson (1887) says: "They have a common habit of rising high overhead with a sea-urchin (*Echinus*) in their beaks, and after reaching an elevation of several hundred feet of allowing the shell-fish to fall. As a consequence, it is common to find the shells of these radiates scattered all over the hill-sides in the vicinity of the sea; apparently the ravens do not do this with the intention of gaining readier access to the contents of the shell, and I do not recall a single instance where a raven followed the shell to the ground, although on several instances I have seen the birds dive hastily after the falling shell and capture it in their beaks before it reached the ground, apparently in sport."

Team play often enters into the raven's activities. B. J. Bretherton wrote to Major Bendire (1895): "I saw a native dog one day with a bone which he vainly endeavored to eat. While so engaged he was espied by a Raven, who flew down and tried to scare the dog by loud cawing, in which he was shortly afterwards assisted by another, both birds sidling up to the dog's head until they were barely out of his reach. Just at this time a third Raven appeared on the scene and surveyed the situation from an adjacent fence, but soon flew down behind the dog and advanced until within reach of his tail, which he seized so roughly that the dog turned for an instant to snap at him, and at the same moment the bone was snatched away by one of the Ravens at his head."

Lucien M. Turner (MS.) relates the following performance that he witnessed on the banks of the Koksoak River: "A few miles below the falls on the river I saw at one time over a hundred of these birds. The banks of the river at this locality were very high and crumbling with the process of freezing during the night and thawing during the day. Here the birds resorted to have the fun of coasting down this hillside. A dozen at a time would stand, either sidewise or with their heads upward, and start down with the rolling pebbles and clay, each bird constantly uttering its harsh croak, which reverberated among the hills until the air was filled with their coarse notes. This noise was heard over a mile before we paddled up to the birds, where we stopped to witness their amusement. The trees in the vicinity contained numbers of ravens aiding the sport with their cries of approval, or taking their turns as the others became tired."

Referring to its character, he says: "The raven is bold and fearless when able to cope with an adversary and rarely fails to drive any intruder but man from the locality. I have seen a single bird successfully attack a white gyrfalcon and cause it to forsake the hillside adopted by the raven for its home. On the other hand, the raven is one of the most cowardly birds, rarely attacking without certainty of superiority in itself, or trusting to its harsh notes to call assistance from its comrades."

Mr. Zirrer (MS.) adds the following notes on the behavior of ravens: "Although they brave storms when no other bird ventures in the open, they are, especially the young, much afraid of the heavy summer thunderstorms. Again and again I have noticed several of them, young birds I presume, sitting during the thunder, lightning, and heavy downpour on a strong, horizontal, lower branch of a big tree, under the protective canopy of densely leaved branches above, expressing all their fear and anxiety by stretching and craning their necks in all directions, and emitting many peculiar, plaintive sounds. At such times it was possible to approach them very closely without taking much precaution; the birds were plainly too frightened to be watchful.

"In localities where ravens are not persecuted and feel secure, they come very near the buildings and become bold. They will attack a cat within sight of a dwelling and take away whatever little game the cat might be carrying at the time. Early one morning, when our big white tomcat was on the way home with something, apparently a meadow mouse, and no more than 100 feet away, he was suddenly attacked by a pair of ravens with a design on that mouse. Without warning the two big birds, which must have been sitting on a tree nearby, dived at the big cat, which, surprised and not knowing what was happening, gave an enormous leap, dropped the mouse and ran. This, of course, was what the birds wanted. Picking up the mouse and disappearing among the tree tops, they went faster than I was able to tell."

Voice.—The raven has a variety of notes. As I recorded it many years ago, on the coast of Maine, its commonest note seemed to be a loud croak, deep-toned, and audible at a great distance, *croake-croake*; we also heard occasionally a hoarse *croo-croo*, not so loud or so penetrating; on one occasion we recorded a richer, more musical note, *croang-croang*, with the resonance of a deep-toned bell, on a lower key than the other notes, less harsh and rather pleasing, and sometimes ending in a loud cluck. Then there was the short, guttural *cur-ruk* or *cruk*, with the rolling r's, given mostly on the wing, singly or repeated several times.

Mr. Harlow (1922) has heard "a very distinct hollow, sepulchral laugh 'haw-haw-haw-haw.'" And he says: "During the period of courtship and incubation there are two distinct notes that I have not heard at any other time. One is a soft 'crawk,' which the male gives to the female when he is sitting near her while she is on the nest ledge or incubating. The other is a series of 'crawks' given while on the wing and with rarely a note best expressed by the syllables 'ge-lick-ge-lee' given either between the 'crawks' or still more rarely as a single note."

J. Dewey Soper (1928) gives the raven credit for considerable musical versatility. He writes: "The northern raven possesses a musical, guttural note with a slightly bell-like quality. This note is employed at times throughout the year. The raven at any time may, also, utter a strange call like *thung-thung-thung*, which bears a remarkable resemblance to the mellow twang of a tuning-fork, being, like it, rich, full, vibrant, and musical. Another expression has a metallic, liquid-like quality after the style of the red-winged blackbird, though greatly magnified in volume.

The ravens possess a great range of notes, from their customary melancholy croaks, through numerous performances in striking imitation of other birds such as geese and gulls, up to the melodious accomplishment first mentioned."

Mr. Tuttle took his flight pictures of ravens by imitating their rallying cry, which, he says "is rather like the second note of the peacock's raucous call—'harraowh'. More than any bird I know, the raven will converse with himself for hours at a time, a curious gargling, strongly inflected talk. It is not very hard to steal up on him when he is so engaged." He adds the following, as the raven's conversation: *Cáhonk-cáhonk; cwaanh; cwahonk; onk-onk; craaounk;* and *koeh, koeh.*

Dr. Dickey says in his notes that "they rarely give evidence of what may also be called a song, so ardently do they vent a long, drawn-out strain, such as *spor-spree-spruck-spur-per-rick-rur-ruck*. Lisps, croaks, buzzing sounds, and gulps may be heard at odd intervals from ravens in the breeding season."

Mr. Zirrer writes in his notes: "From the middle of August to about the end of September, and as a rule in the afternoons only, they congregate in a secluded spot of heavy timber and hold their daily concerts. For this purpose they select one or two of the tallest trees, sit facing one another and sing, mostly solo, but sometimes more at once. The song is a musical warble, not very loud and, considering their size and otherwise rough, croaking call, extremely attractive. The birds, however, are very alert throughout the performance and when frightened once will not return to the same spot again, but otherwise they will return daily."

Field marks.—To the casual observer a raven may look like just a large crow and so not be recognized. But to the trained eye of an ornithologist there are several points of difference. The raven is decidedly larger, with a wing expanse of over 4 feet, against less than 3 feet in the crow; its tail is also proportionately longer and more rounded. But size alone is not a safe guide unless there is direct comparison at the same distance.

Its voice is quite distinctive, as explained above, though young ravens sometimes "caw" like crows. And its flight is very different from that of the crow, swifter and less steady, with frequent turnings from side to side, accompanied by two or three rapid wing beats and with occasional attempts at tumbling; its sailing or soaring flight is majestic and often used.

Mr. Tuttle says (MS.): "The four field marks by which one can most easily distinguish the raven from the crow, lacking the presence of both birds for size comparison, are the heavy, triangular head, with apparent bulges at the base of the jaws as seen from below (probably the part where the brow joins the beak), the sharp break of the wings at the shoulders, the openings between the primaries, and the large fan-shaped tail. All these features can be clearly seen in the flight pictures submitted." See plate 33.

Enemies.—Ravens have few natural enemies. They have been known to have occasional squabbles with gyrfalcons, duck hawks, red-tailed and red-shouldered hawks, and crows, but such encounters generally result in favor of the ravens, with little damage inflicted on either party. J. Southgate Y. Hoyt writes to me of such an incident that he witnessed near Lexington, Va., on April 7, 1939:

Just as we located this year's nest, I heard the cry of a duck hawk. From around the end of the range came the raven with the duck hawk flying high above it, calling loudly. The raven croaked a few notes of protest, but continued its slow and deliberate flight along the range. As I watched this unusual sight, I saw something at which I still marvel.

The duck hawk stooped at the raven, calling faster. Just at the point when I expected to see the raven get a hard blow, it flipped over on its back with its feet up in the air and warded off the blow. I could not see whether it used its feet or just assumed an attitude of guard. The raven did not seem to use its wings in turning over but was upside down in a small fraction of a minute. At this the falcon swooped up in the air again, still screaming loudly. The raven turned over again just as quickly as it had turned onto its back and resumed its course slowly and steadily along the face of the mountain.

The duck hawk, having again reached its position over the raven, stooped as it had before. Again the raven turned over onto its back to ward off the blow. This performance was repeated eight times as the raven crossed before me and finally settled in a pine tree at the end of the cliff. The duck hawk swooped up to a tall dead tree nearby and sat there motionless. The next I saw of the raven was the pair of them flying back along the top of the mountain, and the duck hawk was nowhere to be seen.

Visiting this same mountain again this spring (1940), I witnessed a similar performance between the raven and the duck hawk. This time the fight continued for several minutes high in the air over the edge of the mountain.

The raven's worst and most effective enemy is man, because of the damage it does, or is supposed to do, to domestic animals, some wild animals, poultry, and nesting wildfowl and other game. Fortunately for the ravens, they are so sagacious and wary that very few can be shot, but many have been killed in crow traps and in various animal traps. Theed Pearse (1938) tells of large numbers that were trapped near Comox on Vancouver Island; 76 were trapped in 1933, 120 in 1934, 62 in 1935; "in 1936 the number taken was sixty-three, and in January alone of 1937, seventy were killed. Thus, during these years of abundance, four hundred Ravens were destroyed in the Crow traps alone, and it would be safe to add another hundred as having been shot or otherwise destroyed (I came across one caught in a trap set for Mink). The local Game Warden gives the huge total of 535 accounted for up-to-date."

He says that he "never saw a Raven doing anything that could be described as harmful to the farmer, the sportsman, or other bird life." Ducks, mallard and teal, bred in the sloughs between the two slaughter houses where the ravens fed, and no diminution in the broods of these ducks was noted during the periods of greatest abundance of the ravens. One of their chief feeding grounds "appeared to be recently ploughed land and grass-fields where the birds could be seen picking up food, probably noxious insects such as cutworms, etc."

Fall.—Most ravens are apparently resident throughout the year over the greater part of their range, but there is some evidence of at least a partial migration from extreme northern habitats. Hagerup (1891) referred to what seemed to be migrations at Ivigtut, Greenland, as follows: "I frequently noticed that when a strong wind blew from the north they migrated in great numbers toward the south. The largest of these migrations took place August 30, 1887, when one hundred to two hundred crossed the valley. They were seen through the entire day coming from the north side of the fjord, flying low over it, stopping a little at the south shore, then crossing the valley until they reached the mountains. At the base of the hills they first began to rise in the air, working upwards in spiral curves without any flapping of wings, until abreast of the summit, when they sailed away to the south."

Mr. Pearse's notes state that "there appears to be a regular line of migration along the east coast of Vancouver Island." Near Courtenay, on August 16, 1923, 30 were seen going south, following one another in a scattered formation and flying parallel to the shore line. Eleven were seen on October 12, 1936, and a flock of 25 or 30 on December 8, 1935, all flying along the shore in the same direction. He says: "Where the cliff makes an abrupt turn west, the birds seemed undecided and stopped, wheeling around and some even playing. After a short time the greater part of the flock moved seawards, going in an easterly direction away from Vancouver Island and in the direction of another island between there and the mainland, about 15 miles away. The other birds remained wheeling around above the cliff, evidently not liking to face the sea, though the day was fine and the sea calm. It was rather amusing to follow the actions of these birds; first a party of ten started off; others followed, so that more than half of the flock was on its way; there were some faint hearts among these, which straggled back to join the birds that were still wheeling over the land. Twelve at least continued the journey, and, shortly afterward, the faint hearts were out of sight around the bend of the cliff, following the shore line."

Winter.—At least a few ravens remain all winter, even at the northern limits of their breeding range. Donald B. MacMillan (1918) records it as a winter resident at Etah, northern Greenland, but says that the "majority migrate south about September 15th." And Langdon Gibson (1922), referring to McCormick Bay, in latitude 77° 40′ N. in northern Greenland, says: "I am fully satisfied that these birds do not all migrate in the fall because, after the sun had disappeared for the winter, we heard their hoarse croaking and five days before the sun reappeared, February 7, 1892, I saw in the dim twilight on the beach near our house a Raven lazily flopping along."

Other explorers have recorded ravens in winter on Baffin Island, Southampton Island, in Ungava and northern Labrador, and along the Arctic coast of Canada, where the few that remain must eke out a meagre living, with deep snows covering the ground and hiding all the familiar feeding places; then, driven desperate with the pangs of hunger, they risk their lives in attempts to steal the baits from fox traps, which often results fatally, as they are either killed outright or left to freeze under a pall of drifted snow.

CORVUS CORAX SINUATUS Wagler

AMERICAN RAVEN

PLATES 35, 36

HABITS

The ravens of the Western United States have long been called by the above scientific name and the rather inappropriate common name for a bird that is so decidedly western. George Willett (1941) has recently shown that we might well recognize two western races within the United States. The measurements that he has accumulated "appear to indicate a large race *(principalis)*, with heavy bill and tarsus, in Alaska and British Columbia; another large race *(sinuatus)*, with slender bill and tarsus, in the Rocky Mountains and Great Basin region; and a small race *(clarionensis)* ranging from interior valleys of California to Clarion Island, Mexico."

The subject of this sketch might well have been called the western raven, as it occupies the western half of the United States and much of Central America. It is smaller than the northern raven, with a relatively smaller and narrower bill and a longer and slenderer tarsus. It is a wide-ranging species, with a scattered distribution, and seems to have no especially favored haunts. It is at home alike in the mountains and on the plains or deserts, in the forests or on the open ranges; it may be seen flying from its nest on some high cliff in a deep rocky canyon, or perched on some tall pine high up in the mountains.

M. P. Skinner tells me that in Yellowstone National Park ravens are seen almost anywhere and at all seasons, perched on the ground or on some prominent rock, or about the geyser basins, and they are common in the lodgepole and fir forests from the lowest altitudes to the highest peaks. "In spring ravens are on the edge of the lake ice about to break up; they are rather frequent about the buffalo herds and often visit garbage dumps and old camp sites; they even visit occupied camps."

Mrs. Nice (1931) states that the raven was formerly an abundant resident in Oklahoma in the days of the buffalo, but that with the disappearance of the bison the ravens have gone. Many ravens were killed by eating poisoned baits and the viscera of wolves that had been poisoned. "Here seems to lie the explanation of the practically complete disappearance of this once abundant bird from Kansas and Oklahoma—the extermination of the buffalo on whose carcasses it fed, and the unintentional, yet wholesale, poisoning by cattlemen."

According to Dickey and van Rossem (1938) this raven is a "fairly common resident of the interior mountains and foothills" of El Salvador "from Los Esesmiles eastward. * * * The raven occurs principally in the pine regions of the Arid Upper Tropical Zone, but in late fall and winter descends to the foothills. Extremes of altitude are 800 to 8,500 feet. * * * In the pines on Los Esesmiles ravens were decidedly more numerous than anywhere else in El Salvador and were seen almost daily. There were at least a dozen pairs within a radius of five miles from camp at 6,400 feet, and these were scattered at elevations of from 6,000 to 8,500 feet. Below 6,000 feet ravens were less numerous, but nevertheless were distributed generally all over the pine country down to about 3,000 feet."

Bendire (1895) says of the haunts of ravens: "It seems to make little difference to these birds how desolate the country which they inhabit may be, as long as it furnishes sufficient food to sustain life, and they are not hard to please in such matters. One is liable to meet with them singly or in pairs, and occasionally in considerable numbers, along the cliffs of the seashore, and on the adjacent islands of the Pacific coast, from Washington south to Lower California, as well as in the mountains and arid plains of the interior, even in the hottest and most barren wastes of the Colorado Desert, as the Death Valley region, and through all the States and Territories west of the Rocky Mountains. * * * I have met with them at every Post at which I have been stationed in the West."

Courtship.—With the springtime urge of love-making, the otherwise sedate and dignified ravens let themselves go and indulge in most interesting and thrilling flight maneuvers and vocal performances. Chasing each other about in rapid flight, they dive, tumble, twist, turn somersaults, roll over sidewise, or mount high in the air and soar in great circles on their broad, black wings. Their powers of flight shown in these playful antics are no less surprising than the variety of their melodious love notes, soft modulations of their well-known croaks, varied with many clucking and gurgling sounds. Their exuberant spirits seem to be overflowing at this season.

Other forms of playful springtime antics are described by Dawson (1923). One he called a "game of tag," in which several birds took part, chasing each other about and playing with a "yellow something," passing it from bill to claws, or from one bird to another. "After this I witnessed an aerial minuet by two gifted performers,—a tumbling contest, wherein touching hands (wing-tips), with one bird upside down, was varied with simultaneous somersaults and graceful upright, or stalling, presentations."

Nesting.—I have seen a few nests of the American raven in Arizona and in California. At the northern end of the Huachuca Mountains, Ariz., on April 14, 1922, we saw a pair of ravens building their nest on a steep rocky declivity; they were flying about, carrying nesting material and croaking, but the nest was not finished. On April 20, my companion, Frank Willard, climbed to an almost inaccessible nest on a high perpendicular cliff in Apache Canyon in the Catalina Mountains; it was located on a ledge under an overhanging rock, but by the skillful manipulation of a long rope he managed to reach the nest and collect a set of five eggs. In a neighboring canyon, on the same day, we found a big nest in a large cottonwood tree that a pair of ravens were repairing; this was the only tree nest that we saw in Arizona. We saw some other old ravens' nests on high, precipitous, rocky eminences, some of which were occupied by western red-tailed hawks.

In California, J. R. Pemberton gave me two very interesting days with the ravens, March 19 and 20, 1929, in Kern County, driving for many miles among the abandoned oil wells in the valley between the Kettleman Hills and Wheeler Ridge. We collected five sets of eggs, two of six, two of five, and one of four eggs. The ravens were nesting in the abandoned oil derricks, usually near the tops, at heights ranging from 58 to 104 feet above the ground. The nests were securely built on the framework, either in a corner or against the ladder, which made it a simple matter to climb to them. They were made mainly of the stems and branches of sagebrush, mixed with other sticks and rubbish, deeply hollowed and warmly lined with a profusion of wool of various colors, which, judging from the smell, must have been taken from dead sheep; this was mixed with such matter as cows' hair, bits of hemp rope, and pieces of cloth. They varied in height from 18 to 24 inches; a typical nest measured 24 inches in outside diameter, 14 inches across the lining, and the inner cavity was 8 inches in diameter and 5 or 6 inches deep. The birds usually left the nests as we approached, but some remained on until we were part way up the ladder. One bird did not fly until my head was nearly on a level with the nest. Some birds departed at once, but others flew around close by, croaking.

A few other nests were noted in various parts of California, mostly inaccessible, on rocky cliffs or in potholes in sandstone cliffs. One that I saw when I was out with the Peyton brothers in Ventura County, on April 7, 1929, was in a pothole in a perpendicular sandstone cliff, about 50 feet high; the nest was about 30 feet from the bottom of the cliff and was reached with the aid of a rope ladder. The nest was made of sagebrush and other sticks and was lined with cows' hair of various colors, bits of rag, and strips of yucca fiber. It contained five eggs.

I saw no tree nests in California, except one shown to me by Wright M. Pierce; this was in a Joshua tree on the Mohave Desert (pl. 36). And Major Bendire (1895) says that about Camp Harney, Oreg., where ravens were very common, "out of some twenty nests examined only one was placed in a tree. It was in a good-sized dead willow, 20 feet from the ground, on an island in Sylvies River, Oregon, and easily reached; it contained five fresh eggs on April 13, 1875." Dawson (1923) mentions a California nest in the top of a white oak, and remarks: "In an experience covering some scores of nests, this was the only example of a tree-nesting Raven. I am told, however, that they do nest in trees in Mendocino and Del Norte Counties, where they are also exceptionally common."

In central Lower California, according to Griffing Bancroft (1930), "they build in a normal manner on cliffs or more often in tree yucca or multifingered cardón." And Dickey and van Rossem (1938) say, referring to El Salvador: "On February 8, 1927, a pair of ravens was found working on a nest in the topmost branches of a forty-foot pine at an elevation of about 7,000 feet on Los Esesmiles. The tree was one of a group of half a dozen growing on a bare ridge and was directly above a trail over which a dozen or more people traveled daily. This nest could be seen half a mile away and would have been conspicuous even without the presence of the builders, both of which were constantly arriving and departing."

James B. Dixon (MS.) tells me that in San Diego County the ravens nest in trees nearly as much as on cliffs; out of the ten records of nests that he sent me, five of the nests were in trees. Ravens often build their nests on cliffs overhanging the seacoast. A. D. DuBois sends me the data on a nest in San Diego County that was 20 feet below the top of a bluff overlooking the Pacific Ocean; it was in a pothole, about 18 inches back and well protected from the weather; it was composed of sticks and lined with cow's and skunk's hair. W. E. Griffee (MS.) says that "occasional pairs of ravens nest in the western Oregon valleys and they occur more frequently along the coast, but the heavy timber west of the Cascade Mountains makes these occasional pairs decidedly inconspicuous." Most of his experience with ravens has been east of the Cascades, where, he says, a large majority of them "nest in rimrocks, usually low rims not over 50 feet high." But he has also found their nests in boxelder, locust, juniper, poplar, and willow trees.

Old ravens' nests are often used by hawks and owls. I took my first and only set of prairie-falcon eggs from the remains of an old raven's nest, and I have found red-tailed hawks and horned owls appropriating ravens' nests that were still in good condition. It is a common occurrence to find the ravens and falcons nesting in the same canyon, or on the same cliff, and not far apart; it seems to be a sort of tradition that where one is found the other will be found in the vicinity, but Mr. Griffee, who has had considerable experience with both, thinks that they nest together only where suitable sites are scarce; and he mentions two cases where ravens formerly nested close to falcons and are now nesting in trees at some distance. This community of interest is not due to any affection between the species, as is shown by the spirited encounters that sometimes occur between them.

Bowles and Decker (1930) give an interesting account of the nesting of this raven on, or in, man-made structures in a deserted agricultural community between the Yakima and Columbia Rivers in Washington:

In travelling over many miles of this country we have seen the following varieties of nesting sites: Several different parts of windmills; rafters in small one-room shacks; in barns; in various places in houses; one a few feet up in a small tree; and one on top of a bookcase in a school house. * * * Only one nest was built on the outside of a house, this being placed on a porch directly above a small bay window. * * * In the low, river country, where natural sites are scarce, we have found the nests on high tension poles, oil derricks, telegraph poles, and on the beam of a railroad bridge. One of the last mentioned was only twelve feet from the ground and two feet below the rails. [A freight train rumbling over this bridge did not cause the bird to leave its nest.]

Another interesting proof that these birds do not mind disturbance in and around the nest was where a windmill had been used as a site. For some strange reason the nest had been built around the plunging rod, which, the mill being in working order, went up and down through the outer wall of the nest whenever the wind happened to be blowing. * * *

The material used is almost literally anything that strikes the fancy of the birds, although the common types are composed outwardly of coarse sticks and twigs for the most part. However, we have several times found them built almost altogether of different kinds of wire, while at other times the ribs of sheep and the smaller bones of cattle form a large percentage. One nest contained a large jawbone, with most of the teeth intact. * * *

"A curious feature of their nest building," they say, "is that they never pick up a piece of material that has fallen from the nest, even though they may have to fly for miles to get more." They cite a case where a pair of ravens had made a number of unsuccessful attempts to build a nest on an insecure board inside of a small building, but finally succeeded. "As a result the floor beneath the nest was one great mass of almost every imaginable sort of material that could be found for miles around, there being included dozens upon dozens of bones of many kinds. In all we estimated that there must have been between twelve and fifteen bushels of material, showing how pertinacious these birds are when they have decided upon a site for their nest."

The raven's nest is often filthy and unsanitary; the wool and hair used for the lining are often taken from dead animals and so are highly offensive to the human nostrils; and Dawson (1923) says that "as if this were not enough, the sitting bird drenches the whole recklessly with its own excrement, making it a veritable abode of harpies." And Bendire (1895) found that "when the nest was occupied the lining was always alive with fleas."

He says further: "The American Raven becomes attached to a site when once chosen, and although its eggs or young may be taken for successive seasons, it will return and use the same nest from year to year. I have taken three sets of eggs (evidently laid by the same bird) from the same nest for successive years; they were readily recognizable by their large size and style of markings." Bowles and Decker (1930) say: "Should their first set of eggs be taken another is laid, usually in the same nest; and in some cases three sets have been laid in the same nest, with intervals of from seventeen to twenty-two days between sets. Sometimes the same number of eggs is produced in each set, but often the second and third sets will contain one egg less than the first. We have found that one egg is deposited daily until the set is complete." Bendire (1895), on the contrary, says that the eggs are laid on alternate days, or even at longer intervals.

Eggs.—The American raven lays from four to seven eggs to a set, but five and six are the commonest numbers, and as many as eight have been recorded. Mr. Griffee (MS.) says that larger sets are laid by northern birds than by those breeding farther south; he estimates that of all complete sets in northern Oregon 30 percent would have five eggs or less, 35 percent six eggs, 30 percent seven eggs, and 5 percent eight eggs.

The eggs vary in shape from ovate to elongate-ovate, or rarely cylindrical-ovate. They are merely large editions of crows' eggs and not so much larger as one might expect; some of the smallest ravens' eggs are not much larger than large crows' eggs. The colors and markings of ravens' eggs have nearly all the variation shown in crows' eggs, though I have never seen the darkest types of crows' eggs quite matched. The ground color varies from "glaucous," through various shades of "greenish glaucous," to "pea green"; Bendire (1895) adds greenish olive and drab to the list of variations. The markings, in shades of dull, dark browns, drab and olive, show considerable variation in pattern; some eggs are sparingly marked with small spots, and some are profusely covered with small spots and fine dots; others are unevenly marked with irregular blotches and scrawls.

The measurements of 54 eggs in the United States National Museum average 49.53 by 32.76 millimeters; the eggs showing the four extremes measure **60.5** by 37.6, 51.8 by **48.3**, **40.9** by 31.6, and 48.3 by **30.5** millimeters.

Incubation.—Major Bendire (1895) writes: "Only one brood is raised in a season. Incubation lasts about three weeks, commencing when the set is completed, and I believe both sexes assist in this labor. When the female is sitting on the nest the male may frequently be seen perched on some small bush or a dead branch of a tree on the opposite side of the canyon from where the nest is situated, uttering an occasional 'klunk-klunk' and keeping a sharp lookout. Should anyone approach in that direction, though some distance off, he will warn his mate, uttering a low alarm note while flying past the nest, when she will usually slip off and try to keep out of sight, while he endeavors to draw attention to himself, acting at the same time as utterly unconcerned as if he had no interest whatever in that particular locality."

Young.—The young are well cared for, fed, and guarded by both parents. When the birds are four weeks or a month old their wings are sufficiently developed for flight and they are ready to leave the nest. Attended by their parents for some time after that, they are taught to forage for themselves. Soon after they have learned how to hunt for their food they all disappear from the vicinity of their nesting site and resort to the valleys where food is more easily obtained. After a few weeks the family party breaks up, and the young, now able to shift for themselves, are deserted by their parents.

Plumages.—The plumages and molts are the same as in the northern raven, to the account of which the reader is referred.

Food.—Ravens are not at all particular about their choice of food; almost anything edible will do, from carrion to freshly killed small mammals and birds or birds' eggs, other small vertebrates, insects, and other small forms of animal life; garbage and various forms of vegetable material are also welcome.

No thorough analysis of the year-round food of the raven seems to have been made, but A. L. Nelson (1934) has published a thorough study of the early summer food of this raven in southeastern Oregon, "based on examination of the stomach contents of 18 adult and 66 nestling birds, the latter representing 18 broods." Bird remains occurred in 21 stomachs, the bulk percentage amounting to 6.37 for nestlings and 7.72 for adults. "Shell fragments of birds' eggs were noted in 14 stomachs, forming by volume 2.03 percent of the bulk." But this probably does not come anywhere near representing the number of eggs destroyed, for bits of shell are seldom eaten and egg contents are not easily detected. Small mammals formed the largest percentage of the food, 34.26 percent for adults and young combined. "Examination showed that thirty-five of the sixty-six nestlings, or 53 percent, were fed on rabbits, while eight of the eighteen adults, or 44 percent, had fed on these animals." These were mostly young rabbits, and probably some of them were carrion.

Amphibians formed 7.40 percent of the food of the young and 3.62 percent of the food of the adults. "The total percentage of reptile food for nestlings amounted to 6.43, for adults 0.84, and for all birds 5.23. * * * Insects, as a group, stand next in importance to the rabbit as a food item, amounting to about 33 percent of the total. The adults had a greater percentage of insects in their diet than did the nestlings, the percentage for the former being 48.56, and for the latter 29.74. * * * In the order of their importance in the diet, from the percentage standpoint, representatives of the following seven orders of insects were identified: Homoptera, Diptera, Hymenoptera, Coleoptera, Lepidoptera, Orthoptera, and Heteroptera. The orders Orthoptera and Heteroptera were so sparsely represented as to be insignificant, together amounting to less than ¼ of 1 percent of the total diet.

"* * * The only vegetable item taken by adults was corn. It was present in two stomachs, being recorded to the extent of 35 percent in one and 2 percent in the other. Of the nestlings, eight stomachs contained vegetable material, two stomachs containing corn to the extent of 42 and 33 percent, respectively, and three, containing oats in percentages of 62, 15, and 8, respectively." Thus it appears that, although some of these stomachs contained rather high percentages of these grains, the number of birds involved was so small that "the determined vegetable material, corn and oats, amounted to" "only "2.35 percent of the total diet" of all the birds involved. And probably, at other seasons of the year, vegetable matter forms a larger proportion of the raven's diet.

Mr. Skinner says in his notes from Yellowstone Park: "Ravens habitually feed on such carrion as dead elk, deer, and small animals; and I believe they follow bears and coyotes at times to benefit by anything they may find or kill. They frequent the garbage piles for scraps and they show little fear of the bears. I have seen a raven on marshy ground eating a frog; and I was once greatly surprised to see a raven on a tree limb reach up three inches and grab a fly that attempted to fly over."

Bendire (1895) writes: "Among various misdeeds it is charged with killing young lambs, chickens, and turkeys, as well as with destroying the eggs and young of different species of wild fowl; and while this is true to some extent, yet where these birds can get a reasonable amount of food from other sources they rarely disturb domestic animals of any sort. I have more than once seen a Raven feeding among my poultry, apparently on friendly terms with both young and old; they never molested any to my knowledge; nor have I ever heard complaints of shepherds that their lambs were troubled, much less killed, by them. Their food consists principally of carrion, dead fish, and frogs, varied with insects of different kinds, including grasshoppers and the large black crickets so abundant at some seasons in the West; they also eat worms, mussels, snails, small rodents, including some young rabbits, as well as refuse from the kitchen and slaughterhouse."

Charles A. Allen wrote to him, however, that "in the interior of California the Raven destroys many young chickens and turkeys around the ranches. In the spring months I have frequently seen one of these birds flying overhead with a young fowl or an egg in its bill." The chances are that most of the damage complained of is done by comparatively few individuals and that the species as a whole probably does more good in the destruction of injurious rodents and insects than it does harm. Ravens must also be credited with their usefulness as scavengers.

Behavior.—The flight of the raven is sometimes slow and measured, like that of the crow, with which it is often confused by casual observers, but it is more majestic, grander, stronger and swifter, varied with sailing or soaring in a manner that would rival a *Buteo,* or with spectacular dives and plunges.

Mr. Skinner says in his notes: "I have seen them high above a snowy ridge, apparently 'riding the gale' seemingly for the mere pleasure of it.

They are given to circling in spirals above carrion quite after the manner of vultures. I have seen ravens fly by, croaking, and at occasional intervals turn back to make a circle before going on. One day I noted four ravens performing various evolutions in the wind, one had some prey that he would carry a distance in his claws and then transfer it to his bill for a short distance before changing back."

W. W. Rubey (1933) witnessed some remarkable flight behavior of ravens at the summit of Wyoming Peak (elevation 11,363 feet), Wyo., of which he writes:

Shortly after noon [October 5] we reached the summit. Immediately we were set upon by a flock of Ravens that dropped down upon us most unexpectedly. The birds, about thirty of them, rushed at us in long, nearly vertical dives, croaking, snarling, and almost barking out their harsh notes. So real did their "attack" appear that we threw rocks in an effort to drive them off. On the first dive, each bird veered off from us at distances of 25 to 100 feet, fell past the peak, then swerved back up and dived again. For a moment, retreat seemed not a bad idea; but soon the Ravens tired of their sport with us and took to another game in which they exhibited a type of bird-flight entirely new to me.

To the west, Wyoming Peak falls off rather abruptly 3500 feet to the valley of Greys River. The Ravens rose perhaps 500 feet above us, then plunged suddenly into a remarkable series of dives, spins, and coasts which eventually carried them almost out of sight to the forests far below. Their maneuver was carried out somewhat as follows: At the top of the preliminary climb each bird turned sharply straight down and fell a short distance with closed (or at least closely cupped) wings. Then, as the speed of fall increased, the wings seemed to open part way and the dive was deflected somewhat from the vertical. Promptly, the Raven began to spin or 'barrel-roll' about its longitudinal or bill-to-tail axis, slowly at first, then more and more rapidly. This rotating fall continued at an accelerating velocity through a vertical distance of several hundred feet. At length, perhaps because the speed could no longer be endured, the wings were opened wider, the angle of dive began to level off, and the axial spinning gradually slowed down until, when the coasting flight became horizontal, rotation ceased. Each bird immediately swerved back up as far as its momentum would carry it and, from an elevation about 500 feet below that of the start, dived again. Thus, the entire performance was repeated over and over again, each successive dive leveling off farther and farther down the steep mountainside. * * *

In the two hours that we remained on the peak, the Ravens were there no less than eight times. Each time, if I remember correctly, they displayed their trick of spinning dives, and on four of their visits they made their mock attacks on us.

Three times, however, they found better game than men for their bullying. Once they put up a Golden Eagle from some ledge on the cirque wall north of the peak, and the majestic bird fled shamelessly and with all speed for a peak two miles to the east, with the whole pack in noisy pursuit. Soon they returned and quickly routed another Eagle from near the same ledge. This bird fared worse than the first one because he failed to get started far enough in advance of the Ravens. The entire flock surrounded and badgered him relentlessly for some time as he literally fought his way toward the Salt River Range, miles to the west.

Yet it was on the last one of their visits—while we were there—that the Ravens found their greatest sport. In the intervals between Raven raids, I had noticed that we were in the midst of a large but (at first) widely scattered flock of Leucostictes. These small birds were industriously feeding on the snow and among the rocks of the peak, and they seemed not at all disturbed by our presence. * * * Finally, the Ravens on one of their swooping raids somehow managed to frighten the Leucostictes into flight; and the entire flock of approximately 200 individuals took to the air almost simultaneously. Immediately, the Ravens were at them, dashing swiftly and noisily through the thickest of the compact flock, scattering it, then charging again each time that it reformed. Not content with merely this disruption of the flock, the Ravens began following up the separated groups of apparently panic-stricken Leucostictes, diving into them viciously. We saw no actual casualties but it seems probable that some of the Leucostictes, despite their expert dodging, must have been struck down during the repeated dives of the larger birds.

Mr. Skinner writes to me: "Frequently I find an eagle on the ground or feeding on carrion with a circle of ravens about him; they are there, I think, more for the sport of mobbing the eagle than anything else. When so tormented the larger bird seeks refuge in a tree top; we often flush an eagle and his attendant ravens from the thick top of a cedar. But, if the ravens mob the eagles, they have their own tormentors in the little Brewer's blackbirds. I have seen a big raven fly past closely pursued by eight blackbirds, and he seemed unable to defend himself in the air or to escape from his more agile pursuers by flight, so he alighted in the top of a small pine, where by constant snapping and several fierce lunges at his tormentors, he managed to keep them at a distance. Soon he tried to escape by flight, but was forced back to his tree again. I have often noticed that, if a raven happens to pass within sight of a Brewer's blackbird nest, all the blackbirds within sight and hearing take after him, and do it every time a raven passes. By late August and September the blackbirds give up this sport of mobbing ravens."

On the ground the raven is sedate and dignified, walking easily with a stately tread, when not hurried, or hopping less gracefully forward or a little sidewise. As a rule it is a very shy and wary bird, difficult to approach, but it is sagacious enough to know when and where it is safe and is often quite tame under such circumstances. Mr. Skinner tells me that he once rode his horse to within 15 feet of one on the ground, and then it merely hopped away. About ranches and farmyards, where it knows that it will not be molested, it is often quite tame, but in the wilderness it seems to know how far a gun will shoot.

Voice.—The various notes of the western ravens do not seem to differ much from those of the eastern or northern birds, which are more fully treated under the northern raven. Major Bendire (1895) says:

"Their ordinary call note is a loud 'craack-craak,' varied sometimes by a deep, grunting 'koerr-koerr,' and again by a clucking, a sort of self-satisfied sound, difficult to reproduce on paper; in fact, they utter a variety of notes when at ease and undisturbed, among others a metallic-sounding 'klunk,' which seems to cost them considerable effort." Bowles and Decker (1930) mention a pair that "gave an endless variety of creaks and croaks, quacked very much like a Mallard duck, bawled like a cat, and, in short, made it exceedingly easy to believe that there would be little difficulty in teaching them to talk." I have never heard western birds give the deep-toned, bell-like note that I have heard from birds on the coast of Maine, nor have I seen it mentioned in print, unless Bendire's "metallic-sounding 'klunk' " or Dawson's (1923) "curious, mellow, *hunger-ó ope*" may refer to it.

Winter.—Mrs. Bailey (1928) says that, in New Mexico, "in the fall the Ravens with their grown young ascend high into the mountains, even to the tops of the highest peaks," 11,000 to 13,000 feet, but that "the snows of early fall drive the birds from these extreme altitudes. The winter is spent in the valley and foothills for the most part below 7,500 feet. * * * In the winter, Mr. Ligon has found the Ravens going about in pairs and he thinks that they remain mated. At this season they come into towns and are far less shy than during the early summer. At Albuquerque one was seen by Mr. Loring perched on a cow's back, and at Deming they were found feeding in the streets acting as important scavengers, while a dozen seen in a hogyard, feeding with the hogs, allowed a person to pass within twenty feet of them without their flying away."

Fred Mallery Packard writes to me from Estes Park, Colo. "Small flocks of ravens may be seen almost daily in winter near Estes Park between 7,500 and 8,500 elevations, occasionally flying higher. After February fewer are to be seen, and these usually in pairs flying near the tops of the mountains above 10,000 feet. In September and October they are to be seen in small numbers flying over the alpine meadows and southward down the canyons at 12,000 feet or higher, descending to the lower valleys when the snows fall on the heights."

John E. Cushing, Jr. (1941), summarizes his report on the winter behavior of ravens by saying: "The ravens in the vicinity of Tomales Bay, Marin County, California, roost together in a brushy canyon on a small hill near Valley Ford, Sonoma County, during the fall and winter months. During the day the birds disperse over the surrounding country, some of them travelling at least forty miles a day. The colony numbered about 200 birds on the two times that counts were made. It has probably been in existence for at least nine years, quite possibly much longer."

Range.—Circumpolar; North America, northern Central America, Europe and northern and central Asia; nonmigratory.

The American races of the raven are found **north** to northern Alaska (Cape Lisburne, Cape Beaufort, Meade River, and probably Demarcation Point); northwestern Mackenzie (Fort Anderson and Fort Pierce); northern Franklin (Winter Harbor, King Oscar Land, Cape Sabine, and Cape Lupton); and northern Greenland (Polaris Bay and Navy Cliff). **East** to eastern Greenland (Navy Cliff, Renet, Cape Wynn, and Ivigtut); Labrador (Port Manners and Gready Island); Newfoundland (Lewis Hills and Base Camp); Cape Breton Island (Englishtown); Nova Scotia (Wolfville, Halifax, and Grand Manan); formerly western New York (Canandaigua Lake and Ithaca); central Pennsylvania (State College and Chesteroak); West Virginia (Coppers Rock); southwestern Virginia (White Top Mountain); western North Carolina (Grandfather Mountain and formerly Craggy Mountain); formerly northwestern South Carolina (Caesars Head and Mount Pinnacle); northeastern Georgia (Brasstown Dome and formerly Toccoa); formerly central Texas (San Angelo); Veracruz (Jalapa); central Guatemala (Chanquejelve and Barrillos); northern Honduras (between Opotelma and Siguatepeque); and northwestern Nicaragua (San Rafael del Norte). **South** to northwestern Nicaragua (San Rafael del Norte); El Salvador (Volcán de San Miguel and La Reina); southern Guatemala (Volcán de Fuego and Quezaltenango); southern Oaxaca (Tapana); the Revillagigedo Islands; and Clarion Island. **West** to Clarion Island; Baja California (Natividad Island, Cerros Island, Guadalupe Island, and Todos Santos Island); California (San Clemente Island, Anacapa Island, Farallon Islands, and Tuscan Buttes); western Oregon (Prospect and Cape Foulweather); western Washington (Grays Harbor and Bellingham); western British Columbia (Friendly Cove, Nootka Sound, and the Queen Charlotte Islands); and Alaska (Forrester Island, Near Islands, St. Matthew Island, St. Lawrence Island, and Cape Lisburne). The raven is now uncommon or rare over most of its range in the United States.

The range as outlined is for the two races that are currently recognized as North American. The northern raven *(Corvus corax principalis)* is found across the northern part of the continent from Alaska to Greenland south to the northern United States and, in the Allegheny Mountains, to northern Georgia; the American raven *(C. c. sinuatus)* is found chiefly in the West, from southern British Columbia and North Dakota south to Nicaragua, formerly east to Missouri and Indiana.

Casual records.—One was reported as seen at St. Georges, Bermuda, on December 23, 1918. There are also several records, chiefly for the Northeastern States, that appear to be outside the normal range. Among these are: New Hampshire, one was recorded from Sutton on December 20, 1878, and one was taken at Warner on February 18, 1879; Vermont, one was obtained at Bennington on November 7, 1909, and another at Hartland on November 19, 1912; Massachusetts, one was taken at Tyngsborough prior to 1859, one was taken in the fall of that year at Springfield, two were taken about the same time at Dedham, and one was taken at Northampton prior to 1901; Connecticut, one was taken on September 18, 1890 at South Manchester, and one was seen at Norwalk on May 25, 1919; eastern New York, one was taken in 1848 at what is now Brooklyn; New Jersey, one was reported from Morristown during the winter of 1881-82, and individuals were recorded as seen at Barnegat Inlet on April 13, 1924, and January 17, 1932, with several other observations between these dates; and Maryland, a specimen was collected at Sunnybrook on November 8, 1929.

Egg dates.—Alaska: 3 records, April 26 to May 29.

Arctic Canada: 9 records, May 1 to June 16.

California: 96 records, March 2 to May 19; 48 records, April 1 to 16, indicating the height of the season.

Labrador: 8 records, April 15 to May 12.

Maine: 9 records, March 24 to April 29.

Nova Scotia: 27 records, March 23 to May 11; 13 records, April 3 to 13.

Pennsylvania: 24 records, March 3 to April 10; 12 records, March 13 to 20.

Washington: 65 records, March 6 to May 26; 33 records, April 1 to 23.

CORVUS CORAX CLARIONENSIS Rothschild and Hartert

SOUTHWESTERN RAVEN

As suggested by George Willett (1941), the A. O. U. (1945) committee has decided to admit to our Check-list the "small race *(clarionensis)* ranging from interior valleys of California to Clarion Island, Mexico."

Mr. Willett's study of this species indicates that there are three subspecies recognizable within the limits of the United States, as mentioned under the preceding form *(sinuatus)*.

What information we have about the habits of the southwestern raven will be found under the preceding form, as that account was written before this subspecies was recognized.

Plates

PLATE 1

Alberta.

S. S. S. Stansell.

CANADA JAY.

PLATE 2

March 28, 1940.

A. D. Henderson.

Belvedere, Alberta, March 22, 1939.

TWO NESTS OF THE CANADA JAY.

PLATE 3

Isle Royale, Mich., April 1935.   J. V. Coevering.

Ben East.

Adult.

CANADA JAYS.

PLATE 4

Bear Lodge, Wyo., March 24, 1905. P. B. Peabody.

Nesting site.

Colorado, May 30, 1931. A. M. Bailey and R. J. Niedrach.

Adult on nest.

ROCKY MOUNTAIN JAY

PLATE 5

Crater Lake Park, Oreg., August 10, 1926.   J. E. Patterson.

ADULT GRAY JAY

PLATE 6

Near Toronto, Ontario.   W. V. Critch.

NORTHERN BLUE JAY.

PLATE 7

W. V. Critch.

YOUNG NORTHERN BLUE JAYS.

Toronto, Ontario, June 15, 1939.

PLATE 8

NORTHERN BLUE JAYS IN WINTER.

Near Toronto, Ontario.

PLATE 9

Eliot Porter

NORTHERN BLUE JAYS

Illinois, June 8, 1942.

PLATE 10

W. V. Critch.

Near Toronto, Ontario.

NORTHERN BLUE JAY. MALE FEEDING

PLATE 11

Duval County, Fla., June 24, 1931.   S. A. Grimes.

St. Cloud, Fla., April 9, 1924.   A. A. Allen.

SOUTHERN BLUE JAYS ON NESTS.

PLATE 12

Sequoia National Park, Calif.   Gayle Pickwell.

San Bernardino Mountains, Calif., April 21, 1940.   J. G. Suthard.

BLUE-FRONTED JAY.

PLATE 13

Huachuca Mountains.Ariz., May 30, 1922.   A. C. Bent.

NEST OF LONG-CRESTED JAY.

San Bernardino County, Calif., June 23, 1935.   J. S. Rowley.

NEST OF BLUE-FRONTED JAY.

PLATE 14

FLORIDA JAYS.

St. Johns County, Fla., April 1932.

PLATE 15

St. Johns County, Fla. May 1931.   S. A. Grimes.

Fledgling.

Duval County, Fla., April 1932.   S. A. Grimes.

FLORIDA JAY.

PLATE 16

Englewood, Fla.   W. F. Smith.

Feeding tame adults.

St. Johns County, Fla. April 1931.   S. A. Grimes.

FLORIDA JAYS

PLATE 17

California.   H. D. and Ruth Wheeler.

LONG-TAILED JAYS

PLATE 18

Jackson County, Oreg., May 6, 1924.   J. E. Patterson.

NEST OF LONG-TAILED JAY.

PLATE 19

W. L. Finley and H. T. Bohlman.

Near Los Angeles, Calif.

CALIFORNIA JAY

PLATE 20

San Bernardino County, Calif., May 16, 1916.   W. M. Pierce.

Azusa, Calif., November 15 1939.   R. S. Woods.

CALIFORNIA JAYS

PLATE 21

Cochise County, Ariz.   F. C. Willard.

WOODHOUSE'S JAY.

Oracle, Ariz., December 1903.   E. R. Forrest.

ARIZONA JAYS.

PLATE 22

Huachuca Mountains, Ariz., May 1, 1922.   A. C. Bent.

Two nests in one tree.

Huachuca Mountains, Ariz.   .   F. C. Willard.

ARIZONA JAY.

PLATE 23

Russell Reid.

Bismarck, N. Dak., May 22, 1923.

W. L. and Irene Finley.

Eastern Oregon.

AMERICAN MAGPIE.

PLATE 24

Klamath Lake, Oreg., April 25, 1925.

J. E. Patterson.

J. E. Patterson.

Swan Lake, Oreg., April 1925.

NESTS OF THE AMERICAN MAGPIE.

PLATE 25

Burley, Idaho, June 1935.   E. C. Aldrich.

W. L. and Irene Finley.

YOUNG AMERICAN MAGPIES.

PLATE 26

J. E. Patterson.

Merced County, Calif., March 25, 1934.

NEST OF YELLOW-BILLED MAGPIE.

PLATE 27

S. S. Dickey.

Centre County, Pa., March 16, 1919.

S. S. Dickey.

Centre County

NESTING SITES OF THE NORTHERN RAVEN.

PLATE 28

B. P. Tyler.

TYPICAL SET OF EGGS OF THE NORTHERN RAVEN.
(Natural size.)

PLATE 29

May 27, 1940.

Nesting site.

Near Bar Harbor, Maine, May 14, 1940.

Maurice Sullivan.
Courtesy National Park Service

NESTING OF THE NORTHERN RAVEN.

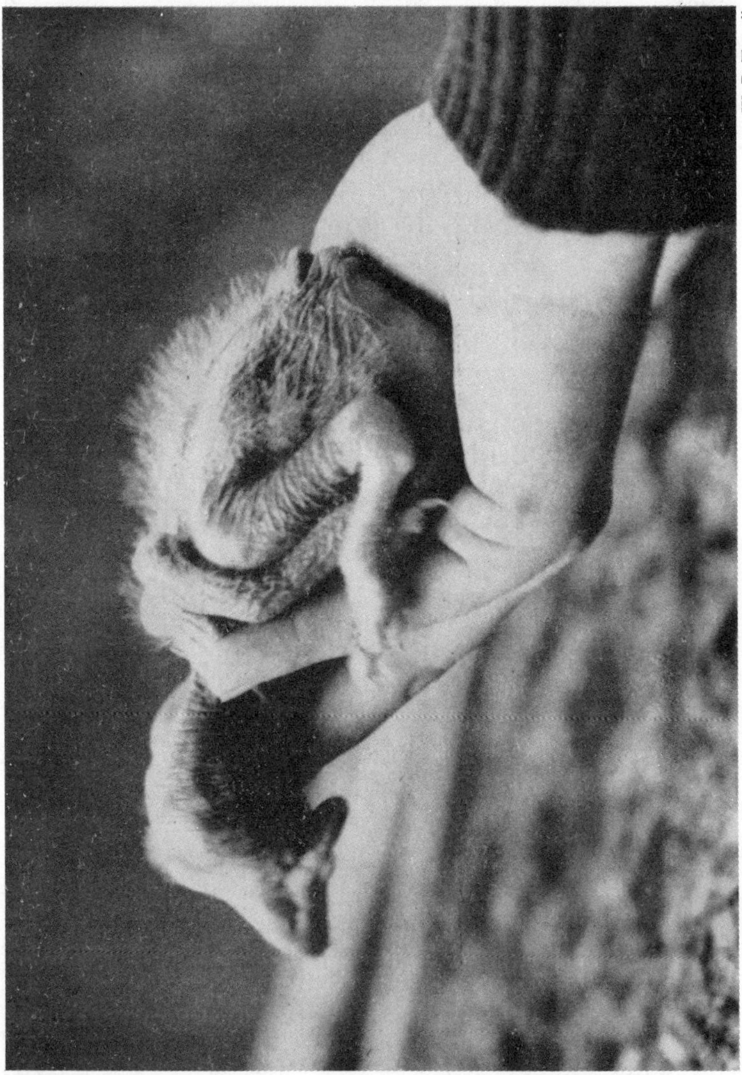

PLATE 30

W. B. Tyrrell.

Shenandoah National Park, Va., March 31, 1940.

NORTHERN RAVEN LESS THAN A WEEK OLD.

PLATE 31

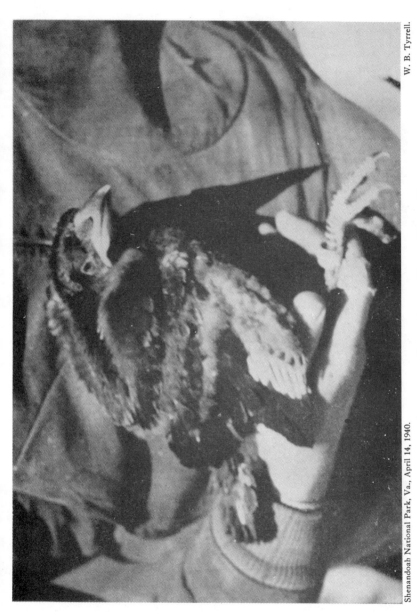

W. B. Tyrrell.

Shenandoah National Park, Va., April 14, 1940.

NORTHERN RAVEN ABOUT THREE WEEKS OLD.

PLATE 32

Shenandoah National Park, Va., 1937.   W. B. Tyrrell.

NESTING SITE OF THE NORTHERN RAVEN.

PLATE 33

Emerson Tuttle.

Marquette County, Mich.

NORTHERN RAVENS.

A. D. Cruickshank.

Emerson Tuttle.

PLATE 34

Marquette County, Mich.

NORTHERN RAVEN.

Emerson Tuttle.

PLATE 35

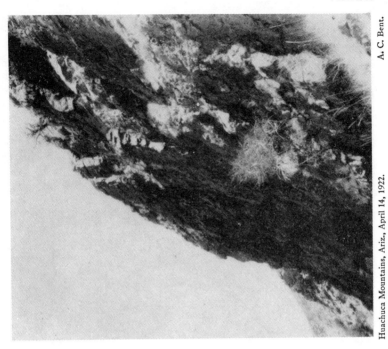

A. C. Bent.

Huachuca Mountains, Ariz., April 14, 1922.

NESTING SITES OF THE AMERICAN RAVEN.

A. C. Bent.

Kern County, Calif., March 19, 1929.

PLATE 36

Nest in A. M. Ingersoll collection.

Mohave Desert, Calif.   W. M. Pierce.

Nest in a Joshua tree.

PART TWO

CONTENTS.

CORVUS CRYPTOLEUCUS Couch

WHITE-NECKED RAVEN

PLATES 37, 38

HABITS

The white-necked raven is smaller than the American raven, but larger than any of the crows; it has a relatively shorter and deeper bill than the larger raven; and it derives its name from the fact that the feathers of the neck and upper breast are pure white for at least their basal half. The name *cryptoleucus* is well chosen, for the white bases are well hidden; they can be seen, with the specimen in hand, by lifting the feathers; but in life they are seldom seen, except when the wind ruffles the plumage or when the bird bends its neck far downward in feeding.

This raven is essentially a bird of the deserts and open plains of the Southwestern States and Mexico. It formerly occupied a wider range in Colorado, western Kansas, and western Nebraska, but, with changing conditions, it has practically disappeared from these regions. Aiken and Warren (1914) have this to say about the withdrawal of the white-necked raven from its range in Colorado, where it was formerly abundant:

Some strong incentive was necessary to have induced these birds to wander northward from their native range in western Texas and New Mexico. This was offered by the slaughter and extermination of the buffalo herds on the western plains which was going on during the late sixties and early seventies. Pioneer settlers were pushing ahead of the railroads; transportation was by teams, and travelers camped along the road and fed grain to their stock. The Ravens, probably first attracted by the buffalo carcasses that strewed the northern plains later followed along the routes of team travel and fed on scattered grain left by campers. By 1874 the buffalo were nearly gone; completed railroads had put the wagon freighters out of business; frequent houses along most roads provided shelter for travelers and camping became unnecessary; the food supply of the White-necked Raven was curtailed and the bird presently retired to its former habitat.

The same thing happened to a less extent in New Mexico, for Mrs. Bailey (1928) says that "before the buffalo disappeared the birds occurred much farther north. * * * In New Mexico, at the present time, they breed from the lowest, hottest valleys of the State up to about 5,000 feet, and less commonly a thousand feet higher to 6,000 feet at Silver City."

As we drove westward from the valley of the San Pedro River toward the Huachuca Mountains, in southern Arizona, we crossed a wide, unbroken plain, a steady, gradual rise of gently sloping land; for the first 10 miles it was covered with a scanty growth of mesquite, creosote bushes, yuccas, and various cacti, typical of the arid plains of that region; but, as we drew near the mountains, approaching 4,000 feet in altitude, the plain gradually changed to an open grassy prairie, broken only by the rows of scattered trees that grew along the washes extending outward from the canyons and by an occasional solitary mesquite of medium size. On the grassy prairie horned larks, meadowlarks, lark buntings, and lark sparrows were common; and everywhere the white-necked ravens were in evidence, and their bulky nests were conspicuous even in the most distant trees. Such an environment as this seems to be the typical habitat of this raven in other portions of its range.

Courtship.—The springtime activities of this raven are thus described by Herbert Brandt (1940) as observed by him in Texas: "During early April the raven begins in the broad mesquite area to make this his bridal bower, to engage in his courtship, and to select the site of his future home. The building process is carried on leisurely because at that season there are many social affairs and quick nest-building is unnecessary since egg-laying is a May urge. It is then that the community takes to the sky, and the male especially is wont to perform in the air—soaring, side-slipping, wheeling, and tumbling, thus distinguishing himself as an aerialist extraordinary. At that time his snowy-lined neck-piece becomes so enlarged that the feathers stand straight out like a fluffy boa, while those on his chin upturn at an acute angle, and the over-weening, black-bewhiskered rogue is then the picture, to his ebony admirer, no doubt, of a handsome, chivalrous swain."

Nesting.—We found it a simple matter to locate the nests of the white-necked raven on the open plains of southern Arizona as they were usually in solitary trees and conspicuous at a long distance. One of our nests was in a large sycamore along a wash, 30 feet from the ground. Another was 30 feet from the ground in an ash on the open plain. The other five nests examined were all in small mesquites on the open plain, 9 or 10 feet from the ground. Frank Willard's notes for the same region record one nest 40 feet up in a sycamore, one 10 feet up in a willow, and one 12 feet from the ground in a mesquite in a wash.

Major Bendire (1895) says that "the favorite nesting sites in southern Arizona are low, scrubby mesquite trees, next oak, ash, desert willow, and yucca, and in southern and western Texas ebony and hackberry bushes are likewise not infrequently used for this purpose.

"The nests are usually poorly constructed affairs, and are a trifle **larger than** those of the common Crow. Outwardly they are mainly composed of thorny twigs, while the inner parts are lined with cattle hair, rabbit fur, and frequently with pieces of rabbit skin, wool, dry cottonwood bark, grass, or tree moss, according to locality. This lining is frequently well quilted and again apparently thrown in loose. They are extremely filthy and smell horribly. Old nests are repaired from year to year, some of them being, as Lieutenant Benson expresses it, seven or eight stories high, showing use for as many years."

The nests that we examined were rather loosely built of large sticks externally, but the inner cup was deeply hollowed, compactly made, and smoothly lined with strips of inner bark, cow's hair, wool, and occasionally a few rags. A typical nest measured 20 inches in outside diameter, and the inner cavity was 8 inches in diameter and 5 inches deep.

In addition to the sites mentioned above, nests have been found in low mesquite bushes 4 feet from the ground, in walnut trees, cottonwoods, palo verdes, tall tree yuccas, and giant cactus, as well as on telegraph poles, or windmill towers, or on almost any structure that will hold them.

Mr. Brandt (1940) writes of their nest-building:

As nearly as we could ascertain, the female does all the carpentry, but her glossy mate escorts her back and forth, strutting, full-chested, about her, puffing out his throat and uttering purring croaks of encouragement. She seems to pay not the least bit of attention to him, but hurries on with her building, interlacing the sticks and then adding thereto, in the base, a binding mat of grasses, rootlets, pieces of rope, newspapers, or other handy trash. She proceeds then to elevate the outer wall with well chosen sticks and at the same time raises the soft inner lining until a deep cup is formed, usually finished with cow or horse hair, though rabbit fur likewise is favored. The bird molds the basin of the nest with her breast, pushing, prodding, and pounding with sharp movements, all the while snuggling down into the bowl. * * * In one case, in Arizona, a nest was found in the process of being colorfully decorated with the black and white fur of the skunk, and the very air was redolent of that fact. A few hundred feet away we came upon the odoriferous carcass of the former owner of that fur with its back cleanly plucked. In the More museum are three ravens' nests made entirely of rusty wire strands instead of sticks, and these have been wound into a rather neat, presentable, wire basket, proving the dexterity of this ingenious bird.

The white-necked raven is a late breeder. We found our first eggs on May 29, and some new nests were still empty at that date. Out of 66 records mentioned by Bendire (1895) the earliest is May 6. "Only twelve other sets were recorded for May, and these usually in the latter part of the month. All the remaining sets were taken in June, and fully half of these after the middle of that month. * * * I can only account for the remarkably late nesting of this species by the fact that insects and small reptiles, which probably furnish the larger portion of the food of these birds, are much more abundant in southern Arizona after the rainy season commences, about the last of May, than before, and these birds seemingly understand this and act accordingly."

Shaler E. Aldous (1942), in his report on this raven, says: "Activity around old nests begins in April, and sometimes the ravens stay constantly in the vicinity of chosen nests as if maintaining claim to them."

Eggs.—The white-necked raven lays three to seven eggs, rarely eight, but the commonest numbers are five, six, or seven. I cannot improve on Major Bendire's (1895) fine description of them, which is based on a series of 288 eggs in the United States National Museum, so I shall quote it here:

The eggs of the White-necked Raven are, in nearly every instance, readily distinguishable from those of the other species of the *Corvinae* found in North America, and this is due to the characteristic style of their markings. The ground color varies from pale green to grayish green, and only very rarely to a light bluish green. Two distinct types of markings are found among these eggs, the principal but usually not the most notable one consisting of a mass of longitudinal streaks and blotches of different shades of lilac, lavender grey, and drab, running from pole to pole of the egg, and these are again more or less hidden and partly obliterated by heavier and more regularly defined spots and blotches of different shades of brown. In not a few sets these lighter and more subdued shades are wanting, and are replaced by a more conspicuous brown; but almost all of the eggs show the peculiar longitudinal streaks and hair lines so prominently characteristic of the eggs of the genus *Myiarchus*. Besides the more regularly shaped markings common to the balance of the eggs of our *Corvinae*, they are on an average also decidedly lighter colored, and a few eggs are almost unspotted. Scarcely any two sets are exactly alike. The shell is strong and compact. In shape they are mostly ovate; a few are elliptical and elongate ovate.

The measurements of 288 eggs in the United States National Museum average 44.20 by 30.22 millimeters; the eggs showing the four extremes measure 48.8 by 33.8 and 38.1 by 27.9 millimeters.

Young.—According to Bendire (1895), "only one brood is raised in a season. Both sexes assist in incubation, which lasts about twenty-one days; this usually begins only after the set is completed; but young birds varying in size are sometimes found in the same nest." Young birds apparently remain in the nest about a month, though I have no definite information on this; they probably hatch late in June or early in July, and young birds of various ages have been found in the nests all through July. "Early in August the young birds begin leaving the nests, and when they have attained their growth young and old gather together in enormous flocks" (Swarth, 1904).

Plumages.—Nestlings are like other young ravens or crows, naked at first but soon scantily covered with brownish-gray down. They are fully fledged in the juvenal plumage before they leave the nest. The juvenal body plumage is dull black, without any of the purplish gloss of the adults; but the bases of the feathers of the neck, chest, and breast are pure white; the lanceolate feathers of the throat, so prominent in the adult, are lacking; the wings and tail are as in the adult; the basal half of the lower mandible is light colored, probably flesh-colored in life. Young birds that Mr. Swarth (1904) raised in captivity began to molt about the first of October and were in full winter plumage by the first of November, having renewed all the contour plumage, but not the wings and tail.

I have seen no molting adults, but probably their molts are similar to those of the young birds.

Food.—Ralph H. Imler (1939) has made a comparative study of the winter food of these ravens and crows in Oklahoma. He concluded that the crows were apparently more beneficial than the ravens, as they ate many more insects and weed seeds. The percentages of the different kinds of food found in the stomachs of 20 ravens killed in December were as follows: Beetles, 0.1; grasshoppers, 1.8; mammals, 4.5; sorghums, 29.8; corn, 17.3; melons and citron seeds, 3.0; hackberries, 37.5; sunflowers, 4.5; and debris, 1.5 percent.

It seems to be quite as omnivorous as other ravens and crows and quite as useful as a scavenger, picking up whatever scraps of food are thrown out from camps and kitchens and carrying off and hiding what it does not eat. Major Bendire (1895) saw one dig a trench and bury a salmon croquette in it, covering it up and marking it for future reference; the Major dug it up, and when the raven returned for it he was disappointed and flew away in disgust.

Mrs. Bailey (1928) lists its food as "principally animal matter, including carrion (as dead jack rabbits), cottontails and cotton rats, field mice, lizards, cicadas, alfalfa caterpillars and 'conchuela'; also cactus, wild fruit, and probably waste grain. Stomachs of five young about ten days old examined by Ligon contained three small nestlings, probably horned larks, birds' eggs, a small lizard, beetles, grasshoppers, and 'jar flies'."

Vernon Bailey (1903) says: "The abundant and juicy fruit of the cactus, Opuntia, Cereus, and Mammalaria, supplies part and probably a large part of their food during July, August, and September, enabling the ravens as well as some of the mammals and even men to make long journeys into waterless valleys with comparative comfort."

Since the above was written, an extensive research report on the white-necked raven has been published by the Fish and Wildlife Service (Aldous, 1942), in which some 35 pages are devoted to a study of the food of this species, to which the reader is referred for details. The summary contains the following general statement: "Laboratory examinations of 707 adult and 120 nestling stomachs of the white-necked raven and field examinations covering almost every month show the bird to be an omnivorous and resourceful feeder and demonstrate that its seasonal food is governed largely by the factor of availability. In all, 288 different items (214 of animal and 74 of vegetable origin) were identified (table 5, p. 47), and if all material found in the stomachs could have been specifically identified no doubt the number would have been increased. Although the animal items far outnumbered the vegetable, the total volumes of the two kinds of food were about equal in the adult diet. The nestlings, though, were almost entirely carnivorous."

Insects made up most of the bulk of the food; grasshoppers (51.21 percent) were the largest item, beetles being second in volume, and Lepidoptera (mostly cutworms and other injurious larvae) third. Hemiptera, such as stink bugs and leafhoppers, were consumed in small quantities, mainly by the nestlings.

Spiders, earthworms, myriapods, and snails were eaten sparingly.

"Mammalian food was important in the diet, ranking second in the animal food of the adults and third in quantity in the nestling food. Most of it consisted of carrion, which was obtained chiefly from carcasses of horses, cows, sheep, and rabbits. * * * Small rodents were eaten sparingly. * * * Birds, including domestic poultry, and their eggs were found in but a small proportion of the stomachs. * * * Reptiles and amphibians formed about 6 percent of the food of the nestlings but only slightly more than 2 percent of that of the adults. * * *

"Cultivated crops offer the greatest supply of food to the white-necked ravens and so are somewhat responsible for the sporadic concentrations of these birds. Grain sorghums were the most important plant food item found in the stomachs examined and made up more than a fourth of the adult birds' subsistence. * * * Cultivated crops of less importance that are attacked by the raven and may be severely damaged locally are corn, peanuts, melons, tomatoes, castor-beans, sunflower seeds, and pears. * * * Wheat, oats, barley, and rye are minor crops in the raven territory and were not fed on excessively by the birds examined.

"Wild fruits were consumed in large quantities during the summer and early fall and therefore played an important part in helping reduce the amount of feeding done in cultivated crops at that time."

Behavior.—In a general way the behavior of the white-necked raven is much like that of its smaller relative, the western crow, though its flight is rather more like that of the larger ravens. Mr. Swarth (1904) writes:

They are usually quite tame and unsuspicious, paying little or no attention to a man on horseback or a wagon passing by; but after being shot at a few times soon become very wary and hard to approach, and as they are usually out on the open prairie it is an easy matter for them to keep out of the way. On one occasion I approached a flock of thirty or forty busily engaged in catching grasshoppers, and as they began to leave long before I arrived within gunshot, I thought to try an experiment; wondering if an appeal to their curiosity might not be as successful as it usually was with jays. Tying a stone in the corner of a red bandana handkerchief, I tossed it high into the air, and the result far exceeded my expectations; for though standing in plain sight, they came headlong to see what it was that had fluttered to the ground, and from that time on I had no difficulty in securing White-necked Ravens. When one or more were shot out of a flock the remainder did not fly off and alight again, but usually circled about, keeping in rather a compact body and ascending higher and higher; not descending to the ground for a considerable length of time, and usually a long ways off. * * *

In the spring of 1903, I noticed a place on the plains some eight or ten miles from the mountains, where some species of bird was evidently roosting in large numbers. The plains are covered with brush at this point, mostly scrubby mesquite, and for a space some two hundred yards long and twenty-five or thirty yards wide the trees were almost destroyed by the use to which they had been put. The ground beneath was inches deep with excreta, and the trunks and branches of the trees were white with the same; while they were almost totally denuded of leaves, except at the extreme top where a little green still lingered. In many cases the limbs were broken down by the weight of the birds. From the appearance of the excreta it was evidently a large species of bird that was roosting there, and as on a careful examination none but raven feathers could be found lying about, I came to the conclusion that it was they that were using the place, though I never found them roosting in such large numbers in any one place before.

William Beebe (1905) thus describes the coming of a vast horde of these ravens to roost in a canyon in Mexico:

And now as the sun's disk silhouettes the upraised arms of an organ cactus on the opposite summit, scattered squads of another army of birds appear and focus to their nightly rendezvous—the White-necked Ravens of the whole world seem to be passing, so great are their numbers. As far as the eye can see, each side of the canyon gives up its complement of black forms; one straggling ahead uttering now and then a deep, hoarse-voiced croak. From all the neighbouring country they pour in, passing low before us, one and all disappearing in the black depths of a narrow, boulder-framed gorge. A raven comes circling down from above and instantly draws our eye to what we have not noticed before, a vast black cloud of the birds soaring above the *barranca* with all the grace of flight of vultures. The cloud descends, draws in upon itself, and, becoming funnel-shaped, sifts slowly through the twilight into the gorge where the great brotherhood of ravens is united and at rest.

Bradford Torrey (1904) writes amusingly of being "mobbed" by a flock of these ravens near Tucson, Ariz. As he approached a lonely ranch a flock of these birds "rose from the scrub not far in advance, with the invariable hoarse chorus of *quark, quark.*" He continues:

I thought nothing of it, the sight being so much an every-day matter, till after a little I began to be aware that the whole flock seemed to be concentrating its attention upon my unsuspecting, inoffensive self. There must have been fifty of the big black birds. Round and round they went in circles, just above my head, moving forward as I moved, vociferating every one as he came near, "quark, quark."

At first I was amused; it was something new and interesting. * * * But before very long the novelty of the thing wore off; the persecution grew tiresome. Enough is as good as a feast; and I had had enough. "Quark, quark," they yelled, all the while settling nearer,—or so I fancied,—till it seemed as if they actually meant violence. They were doing precisely what a flock of crows does to an owl or a hawk; they were mobbing me. * * * The commotion lasted for at least half a mile. Then the birds wearied of it, and went off about their business. All but one of them, I mean to say. He had no such notion. For ten minutes longer he stayed by. His persistency was devilish. It became almost unbearable. The single voice was more exasperating than the chorus. * * * "Quark, quark!" the black villain cried, wagging his impish head, and swooping low to spit the insult into my ear.

On another occasion he watched the playful antics of a flock of ravens going to their roost, of which he says: "Again and again, in the course of their doublings and duckings, I saw the birds turn what looked to be a complete sidewise somersault. * * * Sure I am that more than once I saw a bird flat on his back in the air, * * * and to all appearance, as I say, he did not turn back, but came up like a flash on the other side." On subsequent occasions he concluded "that the birds turned but halfway over; that is to say, they lay on their backs for an instant, and then, as by the recoil of a spring, recovered themselves."

While the above play was going on, "another and a larger flock were sailing in mazy circles after the manner of sea-gulls. * * * More than once I have watched hundreds of the birds thus engaged, not all at the same elevation, be it understood, but circle above circle—* * * till the top ones were almost at heaven's gate."

Field marks.—"The field character that best distinguishes the white-necked raven from the crow, with which it intermingles on the north and east boundaries of its range in Texas and Oklahoma, is without question the raven's less open-throated and distinctly lower-pitched and guttural voice. Other distinguishing field characteristics are the raven's slightly larger size; its longer and coarser beak; its slightly more rounded tail silhouette when in flight; and its tendency to soar, at which time the tips of the primaries are separated and upturned. Occasionally, also, the white bases of its neck feathers can be seen when the plumage is ruffled by the wind." (Aldous, 1942).

Economic status.—Mr. Aldous (1942) says about this: "It is extremely difficult to arrive at a generally applicable verdict with respect to a bird with such varied habits and such an adaptable nature as the white-necked raven. The occurrence of ravens in large numbers makes them potentially capable of doing either severe damage or much good, and during the season their habits may vary from one extreme to the other. If the birds were evenly distributed throughout the year and did not congregate they probably would be more beneficial than detrimental. In judging the economic status of the ravens examined in this study their yearly food habits may be segregated roughly into beneficial, 37 percent; detrimental, 33 percent; and of neutral significance, 30 percent. * * *

"In order to obtain the farmer's point of view regarding the raven, a farm-to-farm canvass covering 100 farms was made in Howard County, Tex., in April 1936. * * *

"The 100 farmers interviewed estimated that they grew about 13,644 acres of sorghum each year, and every farmer but one considered that the greatest loss the ravens caused him was to this crop. The estimated annual loss per acre ranged from nothing to $3 and averaged $0.66. At this average rate the annual loss from the total acreage of grain sorghums grown in Howard County would be $49,500. * * *

"The following opinions comparing the ravens with other pests were volunteered. Six farmers considered small birds—including lark buntings, English sparrows, and blackbirds—more detrimental than ravens to the grain sorghum crop; two thought that rabbits were as bad as ravens and two thought them worse; two believed that ducks consumed more grain sorghums than ravens; and two said that coyotes were more destructive than ravens to their melons."

The ravens are, also, accused of spreading contagious diseases of livestock and poultry, such as hog cholera, blackleg and roup, through their carrion feeding habits; but this has not been proved.

In some treeless regions, the ravens have formed the habit of building their nests on the cross arms of telephone poles. As they often use old haywire and cast-off barbed wire in their nests, these cause short circuits; this has cost one telephone company $2,500 to $5,500 annually to patrol the line and keep it clear. "There have been as many as 202 instances of wire trouble that called for special investigation in a year (1934), and between 700 and 800 pounds of scrap wire have been removed annually from the nests and the ground beneath the lines. Shooting, poisoning, and trapping have accounted for 1,500 to 2,000 ravens yearly, but the trouble persists."

Various control measures, such as shooting to kill or frighten away the birds, catching them in steel traps, poisoning them, or destroying their nests, eggs and young, have been tried with varying success, but none of these is very satisfactory. "The most selective and safest means of reducing the numbers of white-necked ravens is by catching them alive in large cage traps of the type known as the Australian crow trap that have demonstrated their efficiency on various occasions (fig. 12). One trap that was operated for 12 days in November caught 512 white-necked ravens; and 4 traps, used at one place from September 1934 until the following spring, caught 10,000."

Enemies.—According to Mrs. Bailey (1928) this raven has some friends and some enemies among the agriculturalists in New Mexico. One man stated that every raven was worth a dollar to him, as without the ravens it would be impossible to raise a crop of alfalfa seed, for they are the only control they have for the "conchuela," an insect of the stink-bug family; any one of his hands found shooting a raven was fired then and there. In another place the ravens were reported as saving the hay crop by feeding on the alfalfa caterpillar. Still another man was down on the ravens because during the melon season they destroyed $25 worth of cantaloupes and truck crops a day. The chances are that after balancing all the evidence it will be found that the ravens do more good than harm and should not be molested, except in a few special cases. Vernon Bailey (1903) writes:

> Out in one of the driest, hottest valleys of the Great Bend country of western Texas a pair of big Mexican ravens came beating over the valley ahead of our outfit one day, when they were suddenly attacked by two pair of the smaller, quicker, white-necked ravens. The attack was vigorous, not to say vicious, with quick repeated blows and pecks till the feathers flew. From start to finish the big birds sought only to escape, but this seemed impossible. They pounded the air in vain effort to out-fly their tormentors, dove to the ground but were forced to take wing again, circled and beat and tacked to no purpose, and finally began mounting steadily in big circles, taking their punishment as they went, the smaller birds keeping above and beating down on them in succession till all were specks in the sky, and finally lost to view. Such a drubbing I never saw a smaller bird inflict on a larger, before or since, and it was probably well deserved. The nests of the white-necked ravens are unprotected from above and eggs are said to be a delicacy to any raven.

Fall.—After the young birds are strong on the wing, these ravens gather in immense flocks and travel about over the country, visiting the most likely feeding places and gradually drifting southward. F. C. Willard (1912) witnessed a heavy migration early in November in Cochise County, Ariz.; this happened just before a very severe winter, during which these ravens were entirely absent from that section. They migrated in one immense flock, which "extended over a distance of nearly three miles along the foot hills of the Dragoon Mountains near Gleason in this county. There did not seem to be any regular flight, but a sort of general slow movement to the south. The birds were present in many thousands and it was two days before the last stragglers disappeared."

Winter.—In its winter resorts in Texas the white-necked ravens are highly gregarious. Mr. Brandt (1940) writes:

When the winds of winter roar down from the north this black clan then gathers into large communities and moves about the countryside in active, restless flocks, often numbering thousands of individuals. They may then be seen feeding forward on the ground in the great open pastures, the rear birds eddying over those ahead and alighting, imparting to the flock the effect of rolling along. They then visit the cities and villages of the region, making themselves perfectly at home, and are less afraid of man than ever. The encroachment of civilization seems to have little or no effect on their numbers and they may be found perched in the trees and on the roofs of the houses, and feeding in the streets and yards. * * * To tour over these bare high prairies in January would be bleak indeed were it not for the two typical lively objects of the region—the White-necked Raven and the tumbleweed.

DISTRIBUTION

Range.—Southwestern United States and northern Mexico; nonmigratory.

The range of the white-necked raven extends **north** to southern Arizona (Baboquivari Mountains, Papago Indian Reservation, and Oracle); New Mexico (Cactus Flat, Cutter, and Fort Summer); rarely east-central Colorado (Hugo); and Oklahoma (Arnett). **East** to western Oklahoma (Arnett); central Texas (Haskell, Albany, probably Turtle Creek, and probably Brownsville); and Tamaulipas (Charco Escondido). **South** to northern Tamaulipas (Charco Escondido); Nuevo León (Monterrey); Coahuila (Saltillo); Chihuahua (probably near Chihuahua City and San Pedro); and southern Sonora (Hermosillo). **West** to central Sonora (Hermosillo and Magdalena); and southeastern Arizona (Fort Huachuca and the Baboquivari Mountains). The range is said to extend south to the Mexican state of Guanajuato, but the supporting evidence is not known. Formerly the species was common along the foothills in eastern Colorado north to the Wyoming line.

Casual records.—It is probable that white-necked ravens formerly were not uncommon in western Kansas and Nebraska. In the latter State one was recorded from the Republican River region in April 1877, and it was noted near Sidney sometime prior to 1904. Several were noted at Wallace, Kans., October 12-16, 1833, and one was taken at Ellinwood on November 8, 1934. Recorded occurrences in California and Montana are not considered properly authenticated.

Egg dates.—Arizona: 94 records, May 6 to June 27; 48 records, June 6 to 17, indicating the height of the season.

Texas: 58 records, March 15 to June 16; 30 records, May 12 to 20.

CORVUS BRACHYRHYNCHOS BRACHYRHYNCHOS Brehm

EASTERN CROW

PLATES 39-42

CONTRIBUTED BY ALFRED OTTO GROSS

HABITS

It has been aptly stated that if a person knows only three birds one of them will be the crow. The crow, if we include all the five subspecies, is widely distributed over the greater part of the North American Continent. Throughout this area this familiar bird is instantly recognized by anyone who sees it. Because of its striking coal-black plumage, its large size, its unusual adaptability, its extreme cunning and apparent intelligence, its harsh garrulous notes, and its habit of frequently appearing in the open, it has become one of the best known of our American birds. The common name crow is universally applied, and I know of no English local synonyms for it. Even before white man came to America it was well known to the Indians and every tribe had its name for this bird, which was such a conspicuous creature of their environment.

Unfortunately the crow has a questionable record as far as his relations to human interests are concerned. No bird has been the subject of more heated controversy than the crow, and none of our birds have been more violently persecuted by man. In spite of incessant persecution the crow has been able to outwit his human adversaries by its unusual intelligence and instinct of self-preservation, to the extent that it has been able to maintain its existence in all parts of its wide and diversified range. For this the crow commands our admiration.

Spring.—A few crows winter in northern New England, but the majority of them are found farther south during the season of extreme cold weather. The first arrivals of the spring migration reach Maine during February, but it is not until the latter part of the month or the first week of March that they become common. Low (1934), in connection with banding operations at the Austin Ornithological Research Station on Cape Cod, Mass., has collected data that suggest that three populations of crows may be found there as follows—permanent residents, breeding birds that winter to the south, and northern breeders that either winter or migrate through the region.

Determinations of sex ratios at roosts by Hicks and Dambach (1935) indicate that the migration of the sexes may differ in range and extent. Certain of our populations of crows undergo a relatively short migration, but banding operations conducted in Oklahoma by Kalmbach and Aldous (1940) prove that many of the crows wintering in that State migrate to the Prairie Provinces of Alberta, Saskatchewan, and Manitoba, a flight of more than a thousand miles. One crow shot at Meadow Lake, Saskatchewan, at latitude 54° N. had traveled 1,480 miles and another at Camrose, Alberta, 1,435 miles from their winter home in Oklahoma. Out of 714 crows banded, 143 recoveries were obtained. Of 65 crows recovered during the nesting season, 49 were from the Prairie Provinces. It is obvious that many of the returns recorded in the States north of Oklahoma were on their way to or from the Canadian breeding grounds. The results obtained by Kalmbach and Aldous not only give us definite information concerning the extent of crow migration but are important in their relation to the value of the extensive control measures undertaken in Oklahoma.

Crows have been used for important experimental work concerned with different phases of migration. William Rowan, proceeding on the hypothesis that the migrating stimulus is a physiological one originating in the gonads or sexual organs, experimented on various birds, but chiefly the crow. The crows were confined in outdoor aviaries at Edmonton, Alberta, and exposed to temperatures as low as 44° F. below zero, but from the first of November until early January they were subjected to an ever-increasing amount of light, supplied by electric bulbs. In this way they were artificially subjected to light conditions that approximated those of spring. At the close of this period it was found that the gonads had actually attained the maximum development normally associated with the spring season. Control crows not subjected to the light treatment showed no development of the gonads. The birds, both the light-treated individuals and the controls, were marked, banded, and then liberated. By means of radio and other publicity, the cooperation of hunters was solicited for the return of the bands. While bands from eight of the experimental crows were returned from the north and northwest (two of them from a point 100 miles northwest of the point of liberation), an equal number were recovered from the south and southeast, thus to some extent nullifying the experiment. This work does indicate that the stimulus that initiates migration is a physiological one, and it is assumed to be a hormone produced by the interstitial tissue of the reproductive organs.

Courtship.—Edward J. Reimann in correspondence writes of the early courtship of crows he observed in the vicinity of Philadelphia, Pa. On March 8, 1940, he saw crows paired at most of the nesting localities along the Pennypack Creek. In some of these places two or three birds and at times four or five, what he supposed to be males, were seen chasing a female in courtship. Late in March the crows were rather noisy as he passed through each prospective territory. At some places courting was still going on where small groups of crows milled about the trees. Males chased the females, courting them while performing aerial gyrations of diving and wheeling. It was apparent to Reimann that the birds were pairing off, were claiming their nesting territory, and were about to drive their unwanted rivals from the scene.

Charles W. Townsend gave the subject of courtship of many birds serious and careful study, and no one is better qualified than he in the recording and interpretations of their performances. The following account is based on his observations of crows at Ipswich, Mass. His published account (1923) in part is as follows:

Courtship in birds is expressed in three ways, namely in display, dance and song. * * * The courtship song of the Crow consists of a rattle, a quick succession of sharp notes which have been likened to the gritting of teeth. That this is a courtship song and not merely one of the bizarre expressions of this versatile bird, is shown conclusively by its association with courtship display and dance. Like all bird songs it is commonest in the spring, but may occasionally, as in the case with many bird songs, be heard at other times, especially in the fall of the year, when it is explained by the "autumnal recrudescence of the amatory instinct." Although the song is generally given from a perch, it may also be given on the wing, constituting a flight song, although there is no other difference in the character of the two songs.

The whole courtship of the Crow varies somewhat, but the following description of this act, seen under favorable circumstances, is fairly typical. A Crow, presumably the male, perched on a limb of an oak tree, walked towards another and smaller Crow, presumably the female, that seemed to regard him with indifference. Facing the smaller one, the male bowed low, slightly spreading his wings and tail and puffing out his body feathers. After two bows, he sang his rattling song, beginning with his head up and finishing it with his head lower than his feet. The whole performance was repeated several times. The song, such as it was, issued forth during the lowering of the head. * * *

During the love season, fights by rival Crows are common. Each bird tries to rise above the other in the air, and, with noisy outcry, each attacks the rival. Sometimes their struggles are so violent that the birds come to the ground, where they continue their fight and sometimes roll over together in their efforts, all the time voicing their wrath.

On the other hand, one may sometimes chance upon the loving actions of affianced couples. More than once I have seen one of a pair that were sitting close together in a tree, caress the other with its beak and pick gently at its head. The mate would put up her head to be caressed, and I have been reminded of billing doves.

Later Townsend (1927) made further observations which he elaborated upon as follows:

Spending the nights in an open lean-to in my "forest," at Ipswich, I found myself listening every morning to the courtship song of the Crow close at hand, and, on May 3, 1926, I discovered from my bed that a pair had their nest in a white spruce twenty-five yards from me, so that I was able to watch them closely. At about four-thirty every morning I awoke to the rattling song of the Crow, and I often saw one flying about in irregular circles, singing and chasing another. Both alighted on trees, especially on a spruce, from time to time. The song was given in the air and from a perch, and once I heard it given as a whisper song. I also heard for the first time at the end of the rattle a pleasing sound which suggested the cooing of a Pigeon or the note of a cuckoo clock, but softer and more liquid. It was usually double—I wrote it down *coi-ou* or a single *cou*—and generally repeated several times, although sometimes given only once. These soft sounds, which I heard many times when the bird was near, generally followed the rattle, but were often given independently. When the bird was perched, he bowed and puffed out his feathers at the time of their delivery as during the rattling song. The cooing was also given in the air and on one occasion, I saw a bird drop slowly down with wings tilted up at an angle of forty-five degrees, singing as he fell. The rattle song was once given fifty-four times in succession, followed by a series of *cous*.

The female was at times very importunate, calling slowly *car car* like a young bird begging for food. If the male approached, the calling would become more and more rapid and end exactly as in the case of a young bird in a gurgle or gargle—*car, car, car, cowkle, cowkle, cowkle.* After mating the male would fly to the next tree and call loudly *caw-caw* several times. Occasionally the loud *wa-ha-ha-ha* was given. An examination of the nest made at this time showed three heavily incubated eggs.

Nesting.—In northern New England and the Maritime Provinces the vast majority of the crows nest in coniferous trees and those that I have examined have ranged 18 to 60 feet from the ground. Of 22 nests observed in Maine, 12 were in pines, 6 in spruces, 3 in firs, and only 1 in a hardwood tree, an oak. A nest containing six eggs found on May 20, 1936, near Brunswick, Maine, is typical. It was in a large pine located near the center of a 10-acre grove. The nest was built close to the trunk of the tree and was supported by three good-sized horizontal branches at a point 42 feet from the ground and approximately 30 feet from the top of the tree. The foundation of the nest was made up of branches and twigs of oak, beech, and pine, the largest ones were one-fourth to three-eighths of an inch in diameter and 10 to 16 inches in length. The nesting bowl was made up of smaller twigs interwoven with strands of bark. The soft compact lining was entirely of finely separated fibrils of bark, which apparently were shredded by the birds before being placed in position. The foundation of nesting materials measured 22 by 26 inches, the depth of the nest from the upper rim to the base was 9 inches, and the rim of the nest proper was 12 inches in diameter. The interior of the nesting cup occupied by the bird was 6 by 7 inches and its depth 4½ inches.

All the nests of the crow are substantial and well built; they are crude in general external appearance but always delicately and warmly lined. The main departure from the type described above is the nature of the materials used in lining the nesting bowl, a difference somewhat dependent on their availability. Different nests may be lined with moss, reed fibers, grass, feathers, twine, rags, wool, fur, hair, roots, seaweed, leaves, and similar materials.

The crow seems to prefer coniferous trees not only in the northern sections of its range but even in the south where such trees abound. In States where hardwood trees predominate, they are more frequently selected as nesting sites. T. E. McMullen, who has made extensive observations on the nesting sites of 227 crows in Pennsylvania, Delaware, and New Jersey, reports finding 112 nests in oak trees, 62 in other species of hardwood including 13 in maple, and 11 in beech trees. The remaining 43 were in coniferous trees, 24 in pine, 17 in cedars, and 2 in hemlocks. The above nests varied from 10 to 70 feet in height from the ground, but the majority exceeded 25 feet. Edmund J. Reimann writes that he has found nests in Pennsylvania that were built at a height of 100 feet from the ground.

In the agricultural areas of the Middle West, where there is a lack of large trees, crows resort to second-growth timber and shrubs of various kinds. In central Illinois favorite nesting sites are the Osage-orange fences. These hedges, abundantly armed with thorns, offer excellent protection, even against the prowling naturalist who may wish to examine the nests.

The crow is adaptable in the choice of its nesting site. In the western Canadian provinces there are numerous instances where the crow has nested on the ground either from choice or because of the lack of trees. Ferry (1910) found a crow's nest at Quill Lake, Saskatchewan, that was situated on the ground at the forks of the dead branches of a fallen and nearly burned up weather-bleached poplar tree. At Regina, Saskatchewan, Mitchell (1915) found a crow's nest on the ground between wild-rose bushes; others were placed on clusters of rose and low bushes just a few inches above the ground. On June 13, 1935, Aldous (1937) found two crow's nests built on the ground along the shore of Lake Manitoba. Another nest containing three eggs was found in the tules over the water, and a fourth nest was built on marshy ground among the reeds. In the latter two cases there were trees and brush in the vicinity, and apparently these situations were a matter of choice on the part of the birds.

Horning (1923) cites an unusual experience with a nest that he found at Luscar, Alberta. "We found a crow's nest in a willow thicket about ten feet from the ground, on May 28, 1922. The situation surprised us, as the Crow usually builds very high, and there were high trees within a few hundred yards. We thought that the presence of an abundant food supply, in the shape of a dead cow, within twenty-five yards may have been the reason for the choice of nesting site. We cut down the nest, which contained three eggs, newly laid, and photographed it, leaving it not more than two feet from the ground, and inclined at an angle of about 55 degrees. We removed the eggs. * * * Judge of our surprise, on re-visiting the nest on June 1 to find four new eggs. * * * It seemed to us very unusual for the Crows to re-occupy the nest especially when so close to the ground and at such an angle."

Occasionally crows select sites that are an extreme departure from the usual situations. Harold M. Holland in correspondence states that a pair nested in the hollow of an old stub located in a wooded tract in Knox County, Ill. They nested in this place for at least three seasons in preference to other numerous apparently suitable locations offered by the surrounding woods. Potter (1932) states that a pair of crows remodeled the top of a disused magpie habitation.

Bradshaw (1930) comments on unusual nesting sites he found in Saskatchewan as follows: "In many treeless sections of the prairie, such as Big Quill Lake, crows have been found nesting on the cross-arms of telephone poles. In such cases one usually finds nearby a marsh well-stocked with ducks, coots, rails, grebes, and other marsh-loving birds. Probably the easy available food supply is the principal factor for the crow locating in such areas. * * *

"The most unique nesting site of the crow encountered was one found on the top of a chimney of a country church, between the towns of Pense and Lumsden." On the same road a pair of crows built their nest in a chimney of an abandoned house. In both cases, however, there were plenty of trees that the crows might have chosen for their nests.

Dr. S. S. Dickey, who has made extensive observations of crows in Pennsylvania during the nesting season, contributes the following observations made of the procedure of nest-building: "The female descended into the underwoods or would move along branches of the trees to masses of twigs. She would take one of them into her beak, twist it loose from its fastenings, and hurry with it to the site she had chosen for her nest. At first she tended to drop sticks en route, or else would proceed awkwardly in placing them in a fork or crotch. She dropped many sticks, causing a veritable heap of rubbish near the base of the nesting tree. Finally after many trials she managed to arrange a loose array of sticks in the base of the fork. Most of the work was done in the morning hours between 7 and 11 o'clock. Thereafter she appeared to weary and would fly away in company with the male in search of food. Late in the afternoon and shortly before dusk she proceeded again to work on her nest. The walls grew consecutively from coarse sticks and twigs to finer materials. She added mud, strands of rope, rags, corn husks, mats of dry grass, roots, moss and weed stems, and strips of bark from various kinds of trees. The rim was nicely rounded off with strips of grapevine bark. The interior of the deep wide cup was tightly lined with inner bark fibers, pads of hair, fur, wool, and green moss. It required approximately 12 days to complete the nest after the first sticks had been placed.

"If bad weather conditions prevailed, several days would elapse before the first egg was deposited, although in one nest an egg had been laid in spite of the fact that the edge of the nest was encrusted in snow. During fair warm weather eggs were found in the nest a day or two after the nest had been completed."

Although not mentioned by Dr. Dickey, it has been noted by many observers that both male and female take an active part in the building of the nest as well as sharing in the incubation of the eggs.

Eggs.—The number in a complete set of crow eggs is usually four to six, but in some cases there are only three and in others as many as eight or nine. Macoun (1909) reports an unusual set of ten eggs. In the latter instances it is probable the large number of eggs are the product of two birds, as it has been observed that two females in addition to the male have shared a single nest. Bendire (1895) has given us an excellent description of the eggs of the crow based on a wide experience and the study of large numbers of specimens. His account is as follows:

"Crows' eggs are rather handsome, and vary greatly in shape, size, color, and markings; the majority may be called ovate, but both short and rounded ovates, and elliptical and elongated ovates are also found in a good series. The ground-color varies from malachite and pale bluish green to olive green, and occasionally to an olive buff. The markings usually consist of irregularly shaped blotches and spots of different shades of browns and grays. In some specimens these are large, and irregularly distributed over the egg, usually predominating about the larger end, leaving the ground color clearly visible. In others again the markings are fine, profuse, and evenly distributed, giving the egg a uniform dark olive-green color throughout."

Bendire gives the average measurements of 292 eggs in the United States National Museum as 41.40 by 29.13 millimeters or about 1.63 by 1.15 inches. The largest egg of the series was 46.74 by 30.78 millimeters, or 1.84 by 1.21 inches; the smallest 36.07 by 25.91 millimeters, or 1.42 by 1.02 inches.

Sometimes eggs of abnormal size have been found. G. Ralph Meyer collected a set of eggs in which one egg measured 2.00 by 1.25 (50.8 by 31.8 millimeters), much larger than the largest egg in the large National

Museum series. One or more eggs dwarfed in size have been found in sets in which the other eggs are normal, but these usually prove to be sterile.

There are a number of reported cases of erythristic crow eggs, in which there is present an excessive amount of red pigment. In correspondence William Rowan, of Edmonton, Alberta, informs me that he has two sets of erythristic eggs that he obtained from central Alberta. They were laid by the same bird in successive years, and he states further that this same type of egg has been found in the same nest for seven successive years. Mr. Rowan believed that these eggs were unique and represented the first recorded case of erythrism in crow's eggs. However, there are published descriptions of so-called abnormal red-colored eggs that are undoubtedly cases of erythrism. Following are a few that have come to my attention. Bendire (1895) states as follows: In an abnormal set of five eggs, presented by Dr. A. K. Fisher to the United States National Museum collection, four have a *pinkish* buff ground color, and are minutely speckled with fine dots of ecru drab resembling somewhat in general appearance a heavily marked egg of the American Coot. * * * In another specimen, presented by Dr. Louis B. Bishop, the ground color is salmon buff and this is blotched with pinkish vinaceous. The entire set of six eggs was similarly colored. Sage, Bishop, and Bliss (1913) mention six pinkish eggs of a set obtained near New Haven, Conn., on May 8, 1884. Jacobs (1935) describes a set of five eggs he found May 1, 1934, in a nest located in a willow tree near Waynesburg, Pa., as follows:

Throughout the whole set there is not the slightest suggestion of the usual greenish-drab shades. The shell, held to the light, appears a rich cream-white such as seen in the eggs of the Eastern Sparrow Hawk, and on the whole, resembles in coloration eggs of the latter collected on the same day. The smallest egg is less thickly marked and contain sparingly scattered hold patches of mauve and maroon purple, which tints are brought out by the brick-red laid over varying shades of lilac and lavender, the majority of them all on the smaller half of the shell. It is a beautifully spotted egg with *brick-red,* mauve and maroon purple about equally apportioned and equalling the amount of lilac and lavender shades which are untouched by the reddish pigment.

The ground color of the four eggs originally rich creamy-white with lavender blendings in paler underlays is heavily mottled over with brick-red giving the shells a uniform rich vinaceous appearance, over which are diffused blotches of strong vinaceous-cinnamon blending into the underlays. Thus we have, in these five crow eggs, specimens appearing like huge Cactus Wren eggs but the general red shade is really stronger than that of the wren's eggs.

Incubation.—The incubation period of the crow is 18 days. One brood is reared each year, but in the southern part of the nesting range two broods each season are not unusual. Both male and female may take part in incubation and both share in the care of the young. Macoun (1909) reports a nest in which both birds were sitting on the eggs at the same time. The cavity of the nest was much larger than usual. There were five nearly incubated eggs in the same stage of development, indicating that these birds were male and female rather than two females. Occasionally three crows may be seen about the nest, but because of lack of sexual differences of plumage it is difficult to determine whether they represent cases of polygamy or polyandry. There is indirect evidence, however, that two females may be concerned. There are a number of cases on record where two sets of eggs were found in a nest that hatched on different dates. Jung (1930) found a crow's nest on June 15, 1928, in Alberta, Canada, that contained three eggs and one young about a week old. When the nest was visited the next day a fourth egg had been added. Three crows were seen about the nest and it is apparent that two of them were females, both of which were contributing eggs to this communal nest. In other cases three crows were concerned with a single nest, which contained a normal set of eggs hatching on the same date. Here, it is probable, two males were involved.

Young.—The young when first hatched are pink or flesh color and scantily clad with tufts of grayish clove-brown on the head, back, and wings. At five days of age the eyes are open and the exposed parts of the skin have acquired a brownish-gray color. At 10 days the principal feather tracts are established by the rapidly growing feather papillae. At this stage they assert themselves by loud clamorings for food, and the presence of a nest may be revealed by their incessant calls, especially as they grow older. When the young crows reach the age of 20 days many of the contour feathers are unsheathed, presenting a dull black color. Tufts of down still cling to the tips of these juvenal feathers, especially in the region of the crown. The eyes are a dark blue-gray, the scales of tarsus and toes are grayish black, the upper mandible or maxilla is black, and the lower mandible is pale yellow or horn color streaked with gray. The lateral basal portions of the gape are yellowish orange. At this age tufts at the base of the bill are developed.

After four weeks most of the feathers are completely unsheathed. The young at this stage also show a marked change in behavior especially in regard to a human visitor. Before this time they were passive but now exhibit fear and offer resistance at being handled or lifted from the nest. At this time they may stand on the rim of the nest or even leave to nearby branches of the tree where they are fed by the adults. In the course of another week they are capable of leaving the nest and making their initial flight. If disturbed they may leave the nest before reaching the age of five weeks.

Plumages.—The young in the completed juvenal plumage are dark grayish black above, with the underparts somewhat duller in tone; the wings and tail are black with violet and greenish reflections; iris bluish and the bill and feet grayish black.

The first winter plumage is acquired by a partial postjuvenal molt, which involves the body plumage and wing coverts but not the rest of the wings or the tail. The young in this plumage are similar to the adults, but the feathers show less gloss and the majority of the specimens have a greenish hue. The underparts are of a duller black, the belly with a dull slaty cast. The first nuptial plumage is acquired by wear, the feathers becoming brownish and worn by the end of the breeding season. The adult winter plumage is acquired by a complete postnuptial molt. The sexes are alike in plumages and molts. All parts, including bill, legs, feet, and claws, are deep black. The plumage of the body has a distinct metallic gloss of violet, and the wings are glossed with bluish violet and greenish blue; iris brown.

Albinism is common in the crow, judged from the more than 25 reported cases that have come to my attention. Since an albino crow offers such a striking contrast to the normal plumage, and because crows are more readily observed than the more secretive species, there are many reports of albinism. A few of the more interesting cases are cited below.

In the Bowdoin College collection there is a female crow collected at Yarmouth, Maine, that is pure white, including the bill, feet, and claws. The iris of this specimen was pink and so the bird was a pure albinistic type. Two albino crows taken from a nest near Portland, Maine, in 1910 were mounted by J. A. Lord, a taxidermist in Portland. An albino crow was seen at South China, Maine, for a period of several weeks during August 1930. F. A. Stuhr, of Portland, Oreg., reported having four live crows that were taken from a nest in Lane County, Oreg. Three of them are almost entirely white, showing only slight black colorations on the primaries and secondaries and at the base of the bill. The iris of these birds is brown, but the feet and tarsus are nearly white. Fleming and Lloyd (1920) report that two albino crows were taken from a nest 9 miles north of Toronto on June 29, 1908. Both birds were grayish white, the eyes blue-gray, the feet lead black, and the beak horn color. Harry Piers (1898) reported a partial albino collected near Halifax, Nova Scotia. His description is as follows: "Its general color was brown, darker on the throat, cheeks and belly; scapulars and feathers of back margined obscurely with whitish; primaries mostly whitish; tertials white; tail feathers light reddish brown margined with whitish on outer edge; legs, bill and iris brown." Several crows similar in coloration to the one described by Piers but with certain variations have been reported by other observers.

Warne (1926) cites a very unusual case of a pet crow that after five years suddenly acquired white feathers in each of its wings; when the wings were spread, about half of the area was white. Previous to this time they were black. Albinism is a hereditary character, and why white feathers would replace black feathers after five years is difficult to explain.

Longevity.—We have relatively few records on the longevity of the crow. Banding of the birds has not been conducted in sufficient numbers or for a long enough time to yield definite results, but the following four banding returns are of interest: A crow banded as a nestling in Saskatchewan in July 1924 was shot five years later in July 1929 only a mile and a half from the place of banding; one banded at Garden Prairie, Ill., was shot five years later at Marengo, Ill., on March 25, 1934; one banded at Richmond, Ill., on May 28, 1927, was shot seven years later in Kenosha County, Wis., on March 13, 1934; and one banded at Lundar, Manitoba, on May 1, 1926, was shot seven years later in Grant County, S. Dak., on April 2, 1933.

Kalmbach and Aldous (1940) are of the opinion that relatively few crows in the plains area live more than four years. This supposition is based on the rapid decrease in the number of returns during the years following the release of the birds. Out of 143 returns of 714 crows banded, 76 were received the first year and 47, 12, and 8 (first six months) in the successive years. All were reported killed, which emphasizes the intense persecution the crow receives from the hands of the gunner. It is possible, state these authors, that the number of returns for the crows banded might have been greater were it not for the fact that, in their winter home, many are killed in bombings under conditions not conducive to the recovery of the bands.

Crows kept in captivity have lived spans of life exceeding 20 years, but it is doubtful if many individuals in nature ever approach that age.

Food.—Few ornithological problems have been of greater widespread controversy than the economic status of the crow. It is an omnivorous feeder and readily adapts its food habits to the changing seasons and available food supply. Its food varies so greatly that isolated observations may be very misleading unless the food habits are considered from the standpoint of the entire population through all seasons of the year. If one is biased it is relatively easy to find abundant evidence either for or against the crow. It is no great wonder that this bird has been the subject of heated debate between the conflicting interests of those who wish to destroy and those who would protect this species with no thought of control. The advocates of either side of this question are probably sincere, but what we need is a common-sense solution of the problem, combining the interests of both factions. Only the thoughtless short-sighted person desires to have the crow completely exterminated, and the overzealous conservationist should submit to a reasonable control of a species when large numbers prove destructive to man's best interests.

The resourcefulness of the crow is vividly indicated by the fact that the Biological Survey identified 650 different items in the food eaten by 2,118 crows collected in 40 States and several Canadian provinces. According to Kalmbach (1939), "about 28 per cent of the yearly food of the adult crow is animal matter and consists of insects, spider, milli-peds, crustaceans, snails, the remains of reptiles, amphibians, wild birds and their eggs, poultry and their eggs, small mammals and carrion." About two-thirds of the animal food consists of insects, chief among which are beetles and their larvae and Orthoptera (grasshoppers, locusts, and crickets), each group constituting more than 7 percent of the food of the crow, and comprises the essential beneficial feature of the food habits of the species.

The numbers of insects eaten vary with the season. For example, few May beetles are eaten early in spring, but by April they constitute 5 percent of the food and in May, at the peak of abundance of May beetles, they comprise nearly 21 percent of the bird's diet. Likewise, the monthly increase in grasshoppers from May to September is shown in the crow's food, in which these insects constitute respectively by month 4, 6, 14, 19, and 19 percent of the food taken.

At the time of outbreaks of such insect pests the crow becomes a valuable agent in their control and herein lies the chief benefit to the farmer. Examples of isolated cases revealed the presence of 85 May beetles in one stomach, 72 wireworms in another, 123 grasshoppers in another, and 438 small caterpillars in a single crow's stomach collected in Michigan. In central Illinois I have seen large flocks of crows follow-ing the plow, where they were devouring great numbers of grubs of the destructive May beetle. It is also a common experience to see them digging up the grubs in the pasturelands where these pests were abun-dant. Alexander (1930) states that in Kansas the early spring crows eat enormous numbers of grubs and cutworms, which are very destruc-tive to wheat in that State.

Nestling crows require even greater quantities of insect food than do the adults. One brood of four examined by the Biological Survey had eaten 418 grasshoppers and another brood of seven had eaten 585 of these insects; one individual had taken the record number of 143 grass-hoppers. Of 157 nestlings obtained in Kansas, 151 had been fed grasshoppers. Caterpillars, always a favorite source of food for nestling birds, were present in more than a third of the 778 nestling crow stomachs examined.

The insect food of the crow is one of the strongest points in its favor and should be given proper consideration in judging the economic status of the species. The crow is an enemy of gypsy and browntail moths, but it has been observed that new colonies of moths often form about the nests of crows, indicating that these birds may serve as an agent in the spread of these pests.

Unfortunately the food of the crow is by no means restricted to insects, and among the bird's less admirable traits is its destruction of eggs and young of other species of birds, a habit that has placed the crow on the black list of many both sportsmen and bird lovers. However, these depredations, in many instances, have been greatly and perhaps willfully exaggerated in articles advocating the destruction of the crow, which have appeared in many sporting columns of newspapers and magazines. The examinations by the U. S. Biological Survey reveal that only about a third of 1 percent of the animal food of the adults and 1.5 per cent of the food of nestlings is derived from wild birds and their eggs, and only about one in every 28 crows and one in every 11 nestlings had eaten such food.

The percentage of such food, as would be expected, runs higher in crows that inhabit the proximity of nesting waterfowl. Examinations of adult crows collected in such situations in the prairie provinces of Canada show that they had eaten four times the quantity of other birds and their eggs, and the young six times the quantity eaten by crows collected in the United States. On the basis of frequency of such predation in Canada the adult crow is ten times and the nestling crow six times as bad as their fellows in the United States. This pronounced record of bird and egg destruction in Canada was due primarily to the fact that the birds collected were taken in close proximity to nesting waterfowl, almost to the exclusion of any obtained in agricultural sections.

Observations on the Lower Souris Refuge in North Dakota in 1936 and 1937 showed that the crow is not an outstanding hazard to waterfowl there. Only 1.7 percent of the 351 nests studied in 1936 were destroyed by crows, while in 1937 the birds preyed upon 3.4 percent of the 566 nests under observation. Even with the latter rate of loss, the crow on this refuge is at present considered to be a minor hazard to waterfowl.

Many independent observers have reported the destruction of eggs and young of both game and song birds, and there is no doubt that the crow at times is guilty of serious depredations. Baker (1940) reports the destruction of a colony of 1,500 little blue herons and 3,000 snowy egrets nesting in an island of timber known as "Live Oaks" on the coastal prairie, 9 miles south of Waller, Tex., by about 40 crows that inhabited the section. On Great Duck Island, off the coast of Maine, where I studied a colony of black-crowned night herons for an entire season, the crows destroyed 27 of the 125 nests under observation. During the season of 1940 crows proved to be a serious menace to the eider ducks nesting on Kent Island, Bay of Fundy, where Bowdoin College has established a bird sanctuary and scientific station. On this island the crows had the habit of carrying their booty to certain convenient places to be devoured. At one such rendezvous I counted 37 eider-duck eggs and 24 herring-gull eggs and in another 22 eider-duck eggs and 28 of the gull eggs. Crows have been reported as carrying an entire egg in their beaks, but at Kent Island the egg was usually punctured by a thrust of the beak. On several occasions we observed them carrying off the downy nestlings. In August I found a place where there were more than a dozen juvenal gulls that had been killed and partially eaten, presumably by the crows. Certainly in sanctuaries such as "Live Oaks" and Kent Island, where a special effort is being made to preserve certain species of birds, the control of the crows is necessary, as it is when they become too abundant in the vicinity of nesting fowl, such as in the Prairie Provinces of Canada.

Depredations on poultry have been reported. For example, Mousley (1924) states that he saw 16 young chickens carried away by crows. Numerous reports have been published citing instances where crows have killed and eaten various species of small birds, and even birds as large as the partridge have been killed and eaten.

Such depredations, though they may call for certain measures of control, in no way warrant the total destruction of a species that has been shown to be beneficial to man's interests at other times. In the case of poultry means of protection can be readily improvised.

Of interest but of lesser economic importance is the consumption of small mammals, crustaceans, mollusks, amphibians, snakes, and carrion. Eifrig (1905) found the crop of a crow filled with earthworms. Along the seacoast, especially during the winter months, mollusks constitute a most important element of the food. It is a common practice of the crow to carry clams, scallops, mussels, or sea-urchins to a considerable height to let them fall on the rocks to be broken and thus enable them to secure the edible contents, a habit shared by other birds, notably the herring gull.

Along the New England coast, especially in Maine, I have seen groups of crows on the mud flats at low tide, where they were feeding on the myriads of invertebrates that abound there. I have also seen them feeding on dead fish left behind by the tide, and at one time seven crows were taking their turns at the carcass of a dead seal. It is not unusual to see them thrusting their beaks into the mud to secure what seemed to be a *Nereis,* a marine annelid worm, much after the fashion that robins retrieve earthworms from our lawns. F. H. Kennard (MS.) on July 15, 1923, saw a young crow foraging on his lawn for earthworms. For over an hour he and others observed the crow pulling up the worms. After they were pulled out the crow would stand on the worm and cleanse it with its bill before swallowing it. Brewster (1883) relates an experience of crows eating 20 good-sized trout that had been hidden in a spring. The farmers along the Maine coast complain that crows as well as gulls are a nuisance in removing fish placed on their fields as fertilizer. Ball (1938) reports similar damage in the Gaspé region, and other complaints have come from New Brunswick and Nova Scotia.

When hard pressed crows may resort to all manner of means to obtain food. For example, Isel (1912) has seen crows enter the business district of Wichita, Kans., to feed from garbage pails back of restaurants; Crook (1936) has observed crows feeding on car-killed animals, including dogs, cats, chickens, opossums, pigs, and even skunks; Guthrie (1932) states that crows prize a dead snake as much as a living one; Anderson (1907) reports that in Iowa crows frequent the slaughterhouses to feed upon the waste of slaughtered animals; Scott (1884) observed crows feeding on a carcass of a dog while the temperature registered 14° below zero. These cases serve to emphasize the role played by crows as scavengers. They also attest the omnivorous feeding of crows and their extreme resourcefulness in securing a livelihood under adverse conditions. Such adaptability insures the success of any species in spite of persecution.

According to Kalmbach (1920) vegetable matter forms nearly 72 percent of the adult crow's yearly food, and over half of it consists of corn. Of 1,340 adult crows collected in every month of the year, 824 (over 61 percent) had fed on corn. During April and May, when the corn is sprouting, corn constitutes about a third of the food, and at the harvest in October it supplies over half of the crow's diet. The damage by the crow is chiefly to sprouting corn, corn "in the milk," or when the ripened grain has been stacked in shocks. Of the three, the second seems to be the most serious. It is not so much the corn the crow actually eats at this time but the subsequent injury resulting from water entering the ears from which the husks have been partially torn that makes the loss so important.

In 1938 the United States Biological Survey made a special investigation of the crow damage to grain, sorghums, and Indian corn growing on 210 farms comprising 39,797 acres in Grady County, Okla. The results reveal that Oklahoma has a winter crow population of between three and four million. The damage to grain sorghums was appraised at 3.8 percent and to Indian corn 1.7 percent. The loss of these crops in Grady County alone for the year was estimated to be $18,370.

On the basis of this investigation the Biological Survey concluded that in southwestern Oklahoma there may be need of measures of control. This situation is now being met by the systematic bombing of the roosts. The case of the crow in Oklahoma is qualified by the statement that in some of the wheat-raising sections of Oklahoma the wintering crows are a benefit.

It was concluded that the crow problem, though serious to sorghums and corn in some counties, is not of sufficient magnitude in the State as a whole to demand combined State and Federal action for its solution. Kalmbach (1920) writes that in the Northwest States, where corn is not raised extensively, wheat replaces corn in the crow's diet. The damage is especially severe at the time wheat is sown or is sprouting. Oats and buckwheat are also occasionally eaten, but the larger part of these grains represents a waste product.

"Apples and almonds are less frequently injured; while the aggregate losses to beans, peas, figs, oranges, grapes and cherries are not important. Fruits of the various sumachs, poison-ivy and poison-oak, bayberry, dogwood, sour gum, wild cherries, grapes, Virginia creeper and poke-berry" are also common ingredients of the food. "The mere consumption of wild fruit by the crow involves nothing of economic importance," but the "digestive processes destroy practically none of the embryos of the seeds, and crows act as important distributors of certain plants, some of which, as poison-ivy and poison-oak, are particularly noxious."

The indigestible parts of the crow's food, such as bones, teeth, fur, and hard seeds, are regurgitated in the form of pellets as is customary with such birds as hawks and owls. An examination of these pellets gathered at crow roosts reveal interesting elements of the food eaten by any such crow population. Townsend (1918) collected several hundred pellets from a crow roost located in Essex County, Mass. These pellets amounted in bulk to 662 cubic centimeters of material after they were broken up into their composite parts. The examination of this material by E. R. Kalmbach, of the Biological Survey, revealed 13 kinds of insects and 7 other invertebrates including *Melampus, Nereis, Mytilus,* and *Littorina.* Among the vertebrates there were fish, bones and scales of a snake, shells of hen's eggs, four meadow mice, a star-nosed mole, two short-tailed shrews, and large fragments of bone. There were seeds and parts of no less than 20 plants, of which the following are of special interest—10,000 seeds of bayberry, 2,300 seeds of poison-ivy and species of sumac, 360 seeds of cranberry, and varying numbers of seeds of juniper, smilax, winterberry, grape, and nightshade. There were also very small quantities of wheat, barley, corn, buckwheat, and seeds of pumpkin or squash, apple, and pear. These results again emphasize the omnivorous feeding habits of the crow as well as its resourcefulness during adverse winter conditions.

It is important to know not only what the crow eats but also how much it eats to enable us to form a complete picture of the economic status of the species. Forbush (1907) made careful records of the food eaten by captive crows, which throw considerable light on this problem. He found that two well-grown crows fed 20 to 25 ounces of food a day just maintained their own weight, but less than that amount was not sufficient. When the quantity of food given the birds was largely reduced there was a corresponding reduction in their weight. He concluded that young crows, when fledged absolutely, require a daily quantity of food equal to about half their own weight and will consume much more than this to their advantage if they can get it. When this amount is multiplied by the number of crows in an entire population the results are impressive. Experiments on the time required for assimilation of food revealed that from the time of eating to that when the undigested parts of the food were emitted average 1½ hours. It is not only what they eat at a single time, but it must be remembered that the average crow gorges no less than eight to ten full meals a day. Hicks and Dambach (1935) found that the average weight of the filled stomachs of 75 adult crows was 36.6 grams and that their food contents averaged 11 grams.

The following interesting experience submitted in correspondence by R. Bruce Horsfall reveals how we may unwittingly condemn the crow when the facts are not clearly understood. Mr. Horsfall bought a farm near Redbank, N. J., where he planted five acres in corn and ten acres in asparagus. He noted that the lower end of his field, where the crows were present each day during the early morning hours, yielded no harvest. Mr. Horsfall immediately jumped to the conclusion from published accounts of crow depredations on farm crops that these birds were responsible for his loss. Without further investigation the crows were shot and the bodies left there as a warning to others. After a number of crows were killed an examination of the stomach contents revealed a mass of greenish liquid filled with cutworm heads, black beetles, and other undigested materials. On the following day a visit was made to the fields in the early morning hours at about the time the crows were accustomed to be present. Great numbers of cutworms were found before they dug into shelter for the day. Mr. Horsfall thereupon decided to welcome his much-maligned friends and he had reason to regret his past hasty judgment. He placed ears of corn on the ground and left the fields to the crows. They recognized the change of attitude, returned in numbers, cleared the field of cutworms, and rewarded the owner by giving him a full yield. Since this experience Mr. Horsfall has been a staunch friend of the crow.

Forbush (1927) relates a similar experience of Gardner Hammond, of Marthas Vineyard, Mass. "Mr. Hammond owned great pastures where many sheep grazed. He told me once that he had offered a bounty of fifty cents each for Crows, as the birds had already killed about 200 of his newly born lambs, and that the native hunters under the stimulus of this bounty had killed nearly all the crows about the Squibnocket region. Notwithstanding my objection he continued to offer the bounty, although he expressed some fear that the expense would leave him bankrupt. About three years later he hailed me one day to see if I could determine what had destroyed the grass in his pastures. The grass was dead, having been cut at the roots by white grubs which had increased so rapidly after the destruction of the crows that they had already ruined a large part of the pastures. The offer of a bounty was withdrawn and the pastures gradually recovered."

Charles P. Shoffner, associate editor of the *Farm Journal* of Philadelphia, sent a questionnaire regarding the economic status of the crow to the readers of the journal who are scattered all over the agricultural districts of the United States. The results of this questionnaire are interesting, since they present a cross section of public opinion of a group of citizens most vitally concerned in the problem. Some of the replies were copied directly from the reports of the Department of Agriculture or other sources, but 9,731 were selected as being apparently based on personal observation or opinion. Among these 1,801 were in favor of the crow and 7,829 against him. Of the latter, 7,573 replies charged damage to crops, 6,937 to poultry, 4,112 to young pigs, sheep, rabbits, etc., 6,796 to song birds, and 6,493 to game birds. As Mr. Shoffner truly says, due weight must be given to the fact that reports were solicited by mail and it would be natural for farmers who had suffered serious damage to write their disapproval, while those who had suffered little or no loss would not trouble to do so. The interesting point is that so many persons defended the crow.

The conclusions of the *Journal* were:

1. The Crow wherever found in large numbers is injurious to farmers from March to December.

2. Where Crows are numerous they should be reduced in numbers and this should be done under active cooperation of State or National Agricultural Authorities. The Crow need not be exterminated.

3. The good Crows do by eating insects does not compensate for the damage done by eating eggs and young of other birds.

4. In acting as scavengers, Crows carry disease; farmers should bury or burn at once all dead animals.

There is a great difference in local conditions. In the West crows are a serious menace, while in parts of the East they are neutral or actually beneficial. Of the conclusions arrived at by the *Farm Journal*, those who have studied the economic relations of the crow will take exception to conclusion No. 3.

(For further comment on this questionnaire, see The Auk, vol. 43, pp. 140-141, 1926.)

Behavior.—The crows return to their roosting place early in the afternoon. The flight then is high and quite direct. Various estimates have been made of their speed in flight. Crows have been known to keep up with trains running at the rate of 60 miles an hour, but the speed determined by Townsend and others indicates that under ordinary conditions it seldom exceeds 20 to 30 miles an hour. If this rate is correct then a crow during sustained flight of a 10-hour day would cover only about 250 miles. Ducks, according to Lincoln (1939), travel 400 to 500 miles in the same period.

Townsend (1905) in writing of his experiences with crows in winter at Ipswich, Mass., records some interesting incidents as follows: "Hearing a great outcry among a party of Crows one day at Ipswich, I saw several swooping down to within a few feet of a fox. Reynard seemed not a whit disturbed, and carried his brush straight out behind as he sauntered along. * * * I have heard them make a virtuous outcry over a couple of innocent hares that were running through the dunes.

"Tracks show that it is a common habit for Crows to drag their middle toe in walking and sometimes all three front toes are dragged. Again, tracks of the same or other Crows show that the toes are lifted up without any dragging. I have seen Crows hop, and have found evidence of that in the sand. In landing from the air, their tracks show it is often their habit, to bound or hop forward once with feet together, before beginning to walk."

Some observers have stated that the crow in flight carries its feet extended backward, but F. H. Kennard, in some unpublished notes, records an observation made under favorable conditions. He was very near a crow that was silhouetted against the snow as it took flight. It raised its legs, after dangling them straight, directly up under and flew off with the closed claws showing as two lumps barely projecting from the feathers of the lower breast.

The feet of the crow are not well adapted for grasping, and their appearance would not at all suggest that they are prehensile, yet these birds do at times carry fairly large objects by means of their feet. At Kent Island I saw a startled crow grasp an eider duckling in its claws and transport it to cover in a thick growth of spruces. Chamberlain (1884) observed crows carrying two young of a brood of robins in their claws, and Kneeland (1883) has seen crows carrying fish heads and other objects too large and too heavy to be conveniently carried in the bill yet too precious to be left behind when food is scarce, as it often is during the winter. Chamberlain also saw a pet crow seize a partially eaten ear of boiled corn in its claws and fly away with it when accosted by a barking dog. Fred J. Pierce (1923), in an article entitled "A Crow that Nearly Looped the Loop," presents the following interesting observation: "I noticed a Crow flying overhead carrying an article in his feet that looked like a mouse or something of that sort. This Crow wanted to transfer the morsel to his bill, and in trying to do so bent his head underneath him so far that he lost his balance and barely escaped overturning in the air. This must have surprised him considerably, but he was a determined Crow and shortly tried it again with no better success. He was continuing his vain efforts when lost to view, but as his unsteady flight had brought him very near the ground, he doubtless alighted, where his object was accomplished with much less danger to his equilibration."

The adult crow is very wary and suspicious of man, an instinctive behavior for self-preservation that has been acquired through generations of experience. Yet crows taken from the nest at the proper time have become pets that have exhibited the greatest confidence in their companionship with human beings. There are innumerable instances on record in which crows have proved to be interesting and entertaining pets.

Lorenz (1937) has pointed out that a young bird, when reared under artificial conditions, will invariably react to its human keeper in exactly the same way that it would have reacted, under natural conditions, to birds of its own species. He has also stated that the period of acquiring this imprinting is confined, in some species, to a very definite and often astonishingly brief period, and that certain actions of the bird for the remainder of its life depend on the imprinting during this crucial period.

Cruickshank (1939) in testing out the statements of Lorenz, contrasts the behavior of two crows that he kept as pets. The first was taken from the nest when it was only two weeks old. It was raised in his home

with great attention and soon reacted to him and his wife as it would
have to its own parents if left in the wild. The crow followed them
about, fluttering its wings and excitably begging for food. After it had
learned to fly it paid no attention to local wild crows or to other human
beings about the camp, but would single out Mr. Cruickshank or his
wife and follow them everywhere. The food-begging act was performed
for them only. The appearance of either of them or the sound of their
voices was sufficient to start its begging. In various other ways this pet
crow showed that it had thoroughly accepted its human foster parents
and rejected all others. The other crow, obtained a few years before,
had been taken on the day it left the nest. Though kept in isolation for
the ensuing two weeks this crow never accepted Mr. Cruickshank in any
way. His appearance never released the begging act, and the bird was
always interested in the calls of nearby crows. At the first opportunity
it flew off into the woods and never returned. This individual evidently
had been obtained at too late a period. The imprinting had already
taken place, and even close attention and strict isolation did not initiate
a reverse.

The above experience readily explains the varying success persons
have had in attempting to make pets of crows. Many have written about
their pet crows, but one of the most detailed accounts is presented by
Norman Criddle (1927), who had four crows that he obtained near
Treesbank, Manitoba, on June 19, 1926. These birds exhibited considerable fear when first obtained, and it was necessary to feed them
by force, but after a day they strongly exhibited the begging reactions.
They greeted his approach by enthusiastic cries for food and their fear
of man had vanished. Later, when able to fly, they were allowed to
roost among the trees, but in the morning they collected around the
feeding cage and his approach was always greeted with enthusiasm.
They would alight on his head and shoulders as readily as on any other
perch. During the day the crows devoted much of their time collecting
and hiding objects of various kinds. As they grew older, berries and
other food were hidden with the definite object of using it later when
hungry. One of the crows would alight on its foster parent's shoulder,
pull out the pocket handkerchief, deposit a throatful of berries, and then
carefully shove the hankerchief back into place on top of them. The
love of destructiveness became a dominant trait. Newspapers and
brightly colored flowers in the garden were pulled to small bits, and
other objects were similarly treated. When a pan of water was provided they soon took to bathing, although they had never experienced
water before. Bathing and playing in the water became a regular pastime.
Sometimes, after flying to Mr. Criddle's shoulder, they would playfully

pinch his ear or run their beaks through his hair. One of them repeatedly tried to dislodge the button from his cap. Each of the crows was different in its personality and details of behavior. Mr. Criddle presents a multitude of experiences indicating that these crows performed just as they might have done toward their own kind if left in nature.

Pet crows are known to possess unusual ability to articulate the words and imitate the sounds of the human voice. They readily master such simple words as "mama," "papa," "hello," "howdydo," and others, and human laughter is often imitated to perfection. This presents the question as to how the crow is able to articulate and imitate so well. We should not expect to find the tracheal syrinx and its controlling muscles to be well developed in a bird that is not recognized for its ability to sing. To the contrary, the crow has a complete set of voice muscles. It is these muscles that give parrots and certain passerine birds such a variety of vocal modulations, so that they can mimic other birds or even the human voice. Hence it is not at all surprising that crows exhibit this unusual ability of imitation.

Pet crows are known to be very adept at learning and to meet new and previously inexperienced conditions. Coburn (1914) has proved this ability experimentally. He found that crows learn very quickly to distinguish the correct exit door when placed in a dark box from which there were translucent and lighted exits, each of the same area and light intensity but of different shapes. In this way it was shown that they distinguished with very little practice between a circle, a triangle, a square, and a hexagon. In this and other tests the experimenter was convinced that the crow's reputation for brains is quite deserved, and that Henry Ward Beecher was correct when he said that if men could be feathered and provided with wings, very few would be clever enough to be crows!

Voice.—The crow does not excel in its musical ability, but it has a great versatility in its voice. It has an interesting repertoire of many calls and notes, which serves it well in its interrelations with its fellows. It also has superior imitative faculties, and captive crows have exhibited unusual aptitude in learning new calls. Even human laughter is imitated, at times so appropriately uttered that it is difficult to think of it as mere coincidence.

The calls and notes have been subject to diverse interpretations; hence several representative authors have been quoted to present a better-rounded concept of them. Hoffmann (1904) states: "Besides the ordinary *caw,* and the many modifications of which it is capable, the crow utters commonly two other striking notes. One is a high-pitched laugh, *hă-ă-ă-ă-ă-ă;* the other a more guttural sound like the gobble of a turkey, *cŏw cŏw cŏw."* Knight (1908) interprets the various calls of the crow as portraying signals that have a distinct meaning to their fellows:

When a band of crows is feeding one or two are generally posted as sentinels and a *caw c-a-a-w* of warning from these is sufficient to make all seek safety. Their call *caw-caw* is uttered in varied tones and different accents so that it is capable of meaning a great many things from alarm to satisfaction, and one acquainted with their ways can usually tell just what they are saying in a general way. For instance I have never failed to correctly judge from their excited and confused cries that they had an owl penned up somewhere and were engaged in "mobbing" it to their satisfaction. The alarm *"caw"* uttered sharply and quickly, which means "look out" is well known to about everybody who has ever seen a Crow. Their prolonged cries of distress when their home is menaced should be easily recognizable. The prolonged *car-r———a———c———k* of a love sick individual in spring, uttered in various tones and drawn out into prolonged gurglings, though somewhat like the call of the young for food is still quite different.

Forbush (1927) writes:

Some Crows, if not all, are capable of producing unusual, tuneful or pleasing sounds. As an example of the unusual let me refer to an individual that I heard early one morning on Cape Cod repeating for over an hour syllables like *clockity-clock, clockity-clock;* while as showing the musical attainments of the species mention may be made of a Crow that I saw on the banks of the Musketaquid, August 10, 1906, which uttered a series of exceedingly melodious, soft, cooing notes unlike any others within my experience. In the same locality on July 14 a young Crow remarked very plainly *aaaou, cou, cou, cou, aaaou, coucoo.* On October 20, 1903, I heard and saw a Crow give an excellent imitation of a whine of a dog. * * * I have heard from Crows a varied assortment of notes, some of which apparently were imitations, such as the cry of a child, the squawk of a hen, or the crow of a young rooster. The cooing notes mentioned above were similar to sounds uttered by the male in courtship. At this season, also, the male has a peculiar cry which may be an attempt at song and has been represented by the syllables *hollow-ollo-ollo.*

Townsend (1923) gives the following account of the calls of the crow:

There are many other words in the Crow vocabulary than the simple *caw,* and I find a number of them recorded in my notes. Many are common and familiar sounds of the countryside, and their recognition is always a pleasure. First, one may consider the modifications of the *caw.* Of these, *orr, orr,* are common, as well as *ah, ah,* the latter delivered at times as with a great feeling of relief. Again, the note may sound like *gnaw, gnaw,* delivered with a nasal inflection and in a taunting manner.

On the other hand the notes may lose all semblance of the typical *caws,* and rapidly repeated and wailing *kaa, wha, wha, wha, kaa, wha, wha, wha,* may be heard, or, as I have written at other times, *ou, ahh, ahh, ahh.* Again, a loud and cheerful *ha, ha, ha,* may be heard, suggestive of one of the calls of the Herring Gull. A despairing *nevah, nevah,* is not uncommon. Occasionally one may hear a loud *cluck.* One of the most extraordinary combinations of Crow notes that I have ever heard was emitted near my house at Ipswich early one

April morning. The bird called *chuck-chuck, whoo-oo,* and then *cawed* in the ordinary manner, repeating the formula in this order several times. Its significance was hidden.

The conversational notes of a small group or family of Crows are always entertaining, and the observer is impressed with the extensiveness of their vocabulary and with the variations in their feelings. At times the notes are low and confidential, pleasant and almost melodious, if I may use that word here; again they are raucous and scolding, bursting at times into a veritable torrent of abuse. In the same way, in human conversations, one may, even without understanding the words, be able to interpret the meanings and motives involved. [See under Courtship for additional notes.]

Allen (1919) has called our attention to the time rhythm, which he attributes to a well-developed esthetic sense of the crow. He has noted that the *caw* notes are not only in triplets but at times they give four *caws* in groups of two (2-2); again he noted that the bird cawed 2-1 a large number of times in succession and on other occasions 2-1-1. The time was so regular that he could detect no variations. The length of the several notes and their pitch and quality were uniform, the rhythm being all that differentiated the phrase from other performances of the crow.

Allen does not believe the series of combination of calls represents a code of signals, nor does he believe them to be purely mechanical and involuntary, but he thinks the crow takes delight in the rhythm and variety of his utterances. He asks the question, "Is he not, in a limited way, a true artist, a composer as well as a performer?"

Wright (1912), in a study conducted at Jefferson Highlands in the White Mountains, N. H., determined the order and manner in which summer resident birds within range of hearing awoke and voiced themselves. According to Wright the crow is a comparatively late riser, as it ranks twenty-fourth among the common birds in time of voicing itself. Fourteen records show that the earliest times at which a crow was heard to call were 3:35 and 3:36 A.M. The average time of the first call was 3:44 A.M. The variations of the crow's awakening was only 21 minutes on 14 occasions, ranging in date from May 27 to July 9, and covering ten seasons. Wright concluded that the crow was one of the most regular in awakening of the common birds he observed.

Enemies.—The crow is recognized as an enemy of certain species of birds, especially in the destruction of their eggs and young, but it is itself in turn preyed upon by hawks and owls. Horned and snowy owls have been seen to capture and kill crows, and the remains have been found in the stomach contents of others. Likewise remains of crows have been found in the stomach contents of red-shouldered and red-tailed hawks and goshawks, and probably the crow falls a victim to other species of the larger hawks. Even the smaller species of hawks may sometimes exhibit a daring inclination to tackle a crow. White (1893) relates an experience in which the small sharp-shinned hawk was seen to attack successfully a crow on Mackinac Island, Mich. Sutton (1929) found the crop of a Cooper's hawk, killed near Shippensburg, Pa., packed with feathers and flesh of a crow. According to the observer a second Cooper's hawk was seen to fly up from the spot where the first was killed, and nearby among the weeds was a partly eaten and fairly well plucked crow, the flesh of which was still warm. Dr. Sutton, although admitting he is unable to prove the case, believes that one or both of the hawks killed the crow.

It is well known that even smaller birds, notably the kingbird, may harass a crow and make its existence very uncomfortable. Currier (1904) in an account of crows observed at Leech Lake, Minn., writes: "One pair in particular had our sympathy. They had a nest full of young in a scrub oak standing alone out on the marsh, where several pairs of Kingbirds, and thousands of Redwings were breeding. Every time a Crow made a move it was pounced upon by from two to a dozen of the smaller birds and forced to light for a time. The Yellow-heads would also join in at times, but they were not so persistent. The Redwings seemed to be the worst."

I have seen crows that have chanced to enter sea-bird colonies viciously and violently attacked by terns.

That crows are never on good terms with predaceous birds, especially owls, is evidenced by the great commotion aroused among the crows whenever an owl is discovered. Fortified by numbers, they exhibit great audacity and may harass an owl for hours at a time. In fact, the presence of an owl may frequently be revealed by the cawing and behavior of the crows at such times. Their antipathy for owls is so great that they may be lured by a stuffed owl placed by a gunner who wishes to destroy them. For the past 15 years I have had several live horned owls in a large flight cage in the backyard of my home in Brunswick, Maine. Almost every morning during the spring migration, flocks of crows ranging from a dozen to 25 or 30 alight in the surrounding trees and awaken the entire neighborhood by their haranguing calls. The crows alight on top of the cage but the least movement on the part of the owls sends them scampering to the tree tops under loud protests. Seeking renewed courage the crows descend again and again to repeat the performance. This goes on in spite of the fact that it is in the midst of a thickly settled portion of the town.

Crows, as well as other birds, fall as victims of flesh-eating mammals. Errington (1935) in his study of the food of midwest foxes, reports that crows are eaten by them.

Wilson (1923) reports a case in which a crow was attacked by a large snake, but such instances are probably rare.

Among mammals, the crow's greatest enemy is man. Since the economic status has been questioned thousands of crows have been killed by poisoning, shooting, and especially by bombing the populous roosts. A few are killed on the highways by automobiles.

No comprehensive study of the diseases of the crow has been made to my knowledge, but as has been shown in the case of other species of birds, disease is probably an important factor in the life of the species.

Mitchell (1929) reported an epidemic of tuberculosis in crows of western Ontario, where he conducted experiments to see if infection is likely to be carried to other animals. Eaton (1903) has given us a detailed report of an epidemic of roup in the Canandaigua crow roost in Ontario County, N. Y., during the winter of 1901-2. Eaton estimates that at least a thousand crows succumbed to the disease in that region alone. Dr. Fox (1923) in a pathological examination of 16 crows, found cases of tropidocerca, occasional intestinal cestodes, and a few filaria.

Dr. E. B. Cram (1927) lists five internal nematode parasites found in the crow. Three of these are found in the proventriculus, the glandular part of the stomach, and the other two in the trachea, or lungs. The parasites are as follows: *Acuaria cordata* (Mueller), found in the wall of the proventriculus; the males range from 10 to 11 and the females 22.5 to 40 millimeters in length. *Microtetrameres helix* Cram, found in the walls of the proventriculus; the males of this small worm are 4.9 and the females 1.2 to 1.3 millimeters in length. *Tetrameres imispina* (Diesing), known only from the female, which is 3 millimeters in length; this parasite is found also in the proventriculus. *Syngamus trachea* (Montague), occurring as adults in the trachea and bronchi and as larvae in the lungs; immature worms have been found in the peritracheal tissue and air sacs; the males are 2 to 6 and the females 5 to 20 millimeters in length. *Syngamus gracilis* Chapin, found only in the trachea; the males are 3 to 3.3 and the females 8 to 11 millimeters long.

Harold Peters (1936) lists three lice and one tick as common external parasites of the crow. The lice are *Degeeriella rotundata* (Osborn), *Myrsidea americana* (Kellogg), and *Philopterus corvi* (Osborn) and the tick is *Haemaphysalis leporis-palustris* Packard. A different species of tick and two species of mites have been found on the southern crow. The tick is *Amblyomma americanum* (Linnaeus), found in a crow from South Carolina, and the mites are *Liponyssus sylviarum* (Canestrini and Fanzago) and *Trouessartia corvina* (Koch), found on crows collected in South Carolina and Florida, respectively.

But by far the worst enemy of the crow is man. Where crows are numerous, especially in their winter roosts, enormous numbers are killed by bombing with dynamite. As one example of this, Dr. Walter P. Taylor (MS.) tells us that in Collingsworth County, Tex., on April 7, 1937, bombs were exposed in a shinnery clump to kill crows. There was one stick of dynamite to each bomb, and the bombs were connected with wires, so that they could be fired simultaneously. Sixty bombs were set off at the first discharge, at which it was estimated that 40,000 crows were killed; at the second shot, 120 bombs were set off, killing nearly as many more. Other bombing operations are mentioned under "Roosts."

Roosts.—During the summer crows associate only in pairs at their isolated breeding places, but in fall they exhibit a marked gregarious inclination, and birds from many miles of territory congregate in immense roosts comprising thousands, sometimes tens and even hundreds of thousands, of individuals. These roosts are not only made up of the birds breeding in the region but the flocks are augmented by birds that have migrated from nesting grounds located farther to the north. In New England there is a marked tendency for the crows to move from inland areas to roosts established near the coast. Food is the primary factor involved in this shift; whereas the feeding grounds in the interior become covered with snow and ice, the seacoast provides an uninterrupted food supply that is replenished with every flow of the tide. Even the severe winter weather does not drive the hardy members from the roosts established in the dense coniferous forests that fringe the coast. Most of the roosts in northern New England are comparatively small, however, and one must go farther to the southward before meeting with aggregations of unusual size.

Townsend (1918) has presented a vivid account of a crow roost that contained approximately 12,000 individuals, located in the thickets and hardwoods on Castle Hill near Ipswich beach, Mass. Following are extracts from Dr. Townsend's paper, which portray in detail scenes similar to those many others have experienced.

In the short winter afternoons the Crows begin their flight to the roost long before sunset. By three o'clock or even as early as one o'clock, especially in dark weather and in the short December days, this bed-time journey begins, while in the latter part of February the flight is postponed until half past four or a quarter of five. From every direction but the seaward side the Crows direct their course towards the roost. Three main streams of flight can be distinguished: one from the north, from the region of the Ipswich and Rowley "hundreds,"— the great stretches of salt marshes that extend to the Merrimac River,—a second from the west and a third,—apparently the largest of all, broad and deep and highly concentrated,—from the south.

It was the last of these rivers that on a cold December afternoon with a biting wind from the northwest I first studied. * * * It was an impressive sight. About 3 o'clock the Crows began to appear, singly and in small groups, beating their way in the teeth of the wind towards the north. In flying over the estuary of the Castle Neck River they kept close to the water as if to take advantage of the lee behind the waves; over the land they clung to the contour of the dunes. As we walked among these waves of sand the Crows often appeared suddenly and unexpectedly over the crest of a dune within a few feet of us. Silently for the most part, except for the silken rustle of their wings, they flew over in increasing numbers until it was evident that they were to be counted, not by hundreds, but by thousands. Many of them alighted on the dunes to the south of the roosting place; sand, bushes and stunted bare trees were alike black with them. Others assembled on the bare hillside to the east. About sunset a great tumult of corvine voices issued from the multitude,—a loud cawing with occasional wailing notes,— and a black cloud rose into the air and settled in the branches of the bare trees to the west of the roost. From here as it was growing dusk they glided into the evergreens for the night.

The last day of the year 1916, I spent with Dr. W. M. Tyler in the dunes. The wind was fresh from the northwest,—the temperature was 15° Far. at 6:30 A.M., 18° at noon and 20° at 6 P.M. As early as one o'clock in the afternoon a few Crows were seen struggling north over and close to the surface of the dunes. Others were noticed flying high and towards the south. This southerly flight came from over Castle Hill to the north, passed the roost and continued on over the dunes. At half-past three some of these birds, which were apparently turning their backs on their usual night's lodging place, met with a large company coming from the south and all settled together in the dunes about two miles south of the roost. Some of the birds coming from the north, however, settled in the bare fields by the roost, and their numbers here were augmented by a stream from the west. This concourse on the hillside set up a great tumult of cawings just before four o'clock. At five minutes after four the united multitude of northerners and southerners rose from their meeting place in the dunes and flew low to join their noisy brethren on the hillside. This river of black wings from the south was a continuous one and it was joined just before its debouch on the hillside by the stream from the west. The river from the north had split into two layers: the lower flying birds came to rest on the hill,—the higher flying ones favored by the strong northwest wind, continued on their way south, notwithstanding the great current that was sweeping north below them. They joined their comrades in the dunes and retraced their steps. No signs of starvation and impaired vigor in these unnecessary flights, or in the game of tag in which two or more of the birds at times indulge!

The pace is now fast and furious. The birds are anxious to get within touch of the roost before it is dark but none have yet entered it. At 4:15 P.M., 135 birds pass in a minute from the south alone on their way to join the concourse on the hillside. A little later this southern river becomes so choked with birds that it is impossible to count them. From our point of vantage in a spruce thicket on the hill we can see that this flock stretches for two miles into the dunes and it takes them four minutes to pass. The speed of flight, therefore, must be roughly about thirty miles an hour. At 4:15 P.M. the sun sets, but in the yellow glow of the cloudless sky the birds can be seen pouring by from the west and south. The bulk of the stream from the north now comes to rest on the hillside for only occasionally can a crow be seen flying to the south over the heads of the southern stream.

At 4:35 P.M. Dr. Tyler and I again counted the southern stream for a minute as they flew silently between us and the lighthouse. One of us counted 160 the other 157 birds, so it is probable that our counts are fairly accurate. This constant watching of the black stream from the south against the white lighthouse produced in both of us a peculiar optical illusion. The lighthouse and dunes seemed to be moving smoothly and swiftly from north to south!

At 4:37 P.M. a great cawing arose from the hillside and a black cloud of birds rose up, some to enter the roost, others to subside on the hillside. It was evident that the birds from time to time had been diving into the roost. At 4:40 P.M. it was rapidly growing dark and the tributary streams were evidently dwindling. Only 50 went by the lighthouse in a minute. Five minutes later it was nearly dark and only a few belated stragglers were hurrying to the concourse on the hill.

At 4:45 P.M. Dr. Tyler and I walked around to the north of the roost and although we could see nothing in the darkness we could hear the silken rustle of wings and feathers as the Crows were composing themselves for the night's rest among the branches of the trees. The babble of low conversational notes that went up from the company suggested the sounds of a Night Heronry although *cawings* and *carrings* were interspersed with the *kis* and *uks* and *ahhs*. * * *

In the dim light we could make out that the hillside field between the roost and the sea was still blackened with birds that were continually rising up and entering the trees. Some of them perched temporarily on the bare tops of the hard woods where they were visible against the sky. The noise and confusion were great. It would seem as if the roost was so crowded that the birds had to wait their time for a chance to get in and that a constant shifting of places and crowding was necessary before the Crows could settle in peace for the night. Hence the prolonged varied conversation; hence the profanity.

It was an intensely interesting experience, this observation of the return of the Crows to their night's lodgings, and one wished for eyes all about the head, well sharpened wits to interpret and a trained assistant to take down notes. * * *

At the full of the moon on the sixth of January I visited the roost at 9 P.M., a time when all well regulated Crows should, I had supposed, be sound asleep. As I approached the roost much to my surprise I heard distinct sleepy cries like those of young herons, and when I reached the edge of the roosting trees there was a tumultuous rush and bustle of Crows flying from tree to tree and overhead. Strain my eyes as I would only occasionally could I catch sight of a black form, although the air was brilliant with the moonlight and the reflection from the snow. I turned back at once as I had no desire to disturb the birds' slumbers but it was evident that many, even at this late hour, had not settled down for the night.

The morning flight from the roost takes less time than the evening return. As I approached it in the semi-darkness at 6:25 A.M. on January 7, a distant cawing could be heard and a minute later nine Crows were seen flying off to the south, and three minutes later, nine went off to the west. At half past six, after a great uproar of *caws* and *uks*, occasional rattles and wailing *ahhhs*, a broad stream boiled up from the roosting trees and spread off towards the west, obscurely seen in the dim light except when the birds stood out against the beginning red glow in the east or against the light of the setting moon in the west. As I stood concealed on the hillside among a grove of spruces, the Crows passed over my head, noiselessly except for the silken swish of their wings, fully a thousand strong. Then no more for over five minutes although the tumult in the roost continued in increasing volume. At 6:40 the roost boiled over again, but the birds spreading in all directions soon united into a black river that flowed over the dunes to the south. The settings of this black stream were the white sand dunes and the luminous glow in the east which had become a brilliant crimson fading to orange and yellow and cut by a broad band of pink haze that streamed up to the zenith. The morning star glowed brightly until almost broad daylight. The sun rose at 7:14. At 7 I entered the roost and hurried away the few hundred remaining birds some of whom were in the bare tops of the hardwoods ready to depart, while others were still dozing in the evergreens below.

Nuttall's Ornithology (1832) gives an account of two roosts on the Delaware River in Pennsylvania. One of them was on an island, near Newcastle, called the Pea Patch, a low flat alluvial spot, just elevated above high-water mark, and thickly covered with reeds. The crows took shelter in the reeds and at one time during the prevalence of a sudden and violent northeast storm accompanied by heavy rains, the Pea Patch Island was wholly inundated in the night. The crows apparently made no attempt to escape, and were drowned by thousands. The following day the shores for a distance of several miles were blackened by their bodies.

Stone (1899) states that the crows that inhabited Pea Patch and the neighboring Reedy Island were estimated at 500,000.

Another famous crow roost is one located in Brookland, near Washington, D. C., which accommodates practically all the crows that feed in the vicinity. Oberholser (1920) estimated that this roost contained 200,000 birds. A very large crow roost was located at Arlington, Va., across the Potomac from Washington. Dr. W. B. Barrows estimated that 150,000 to 200,000 crows came to it every night during the winter of 1886-87.

Widmann (1880), in connection with an account of a crow roost located on Arsenal Island opposite the southern part of St. Louis, writes: "As early as August they begin to flock in, first by hundreds, then by thousands, and in December hundreds of thousands sleep there every night. The roar they make in the morning and evening can be heard for miles around, and the sight of the influx of these multitudes in the evening is something really imposing." Later Widmann (1907) in writing about this roost stated: "All through fall and in moderately cold weather in winter, the Crows spent the nights perched ten to fifteen feet above the ground in the willow thicket of the island, but when the cold weather became intense they deserted the willows entirely and spent the nights on the snow-covered sand bank in front of the willow thicket and exposed to the fierce northwest and north wind. When they had gone in the early morning, every bird had left an imprint of its body in the form of a light depression in the snow with a hole in front made by the bill and a few heaps of excreta on the opposite side, showing the bird had spent all night in that position, always with the head turned toward the wind, letting the wind sweep over its back, but keeping the feet from freezing."

Although crows are very resourceful in combating the adverse weather conditions of winter, extreme subzero temperatures have been known to play havoc with them at the roosts. J. W. Preston wrote Bendire (1895) of a roost of 40,000 crows located near Baxter, Iowa, in which many of the birds died of starvation during the cold winter of 1891-92 because they were blinded from the freezing of the corneas of their eyes. Likewise, Ridgway in *Science*, February 10, 1893, p. 77, mentions the sufferings of the crows in a roost near Washington, D. C. He states that many had their eyes frozen, which was followed by the bursting of the organs and the consequent death of the birds from starvation.

Crows have probably evolved the habit of congregating in roosts for mutual protection, but in the present day, since the verdict concerning their relations to man's interests in certain States has been pronounced against them, thousands of individuals are killed by man at the very roosts where they sought refuge against danger. Imler (1939) states that 26,000 birds were killed by the bombing of a large crow roost near Dempsey, Okla., on December 10, 1937. The Game and Fish Commission bombed another roost at Binger, Okla., on December 6, 1938, killing 18,000 crows. Frank S. Davis, inspector for the Illinois State Department of Conservation, killed 328,000 crows in roosts near Rockford, Ill., with the use of festoons of dynamite bombs. This wholesale slaughter was given great publicity, appearing with photographs in the issue of *Life* for March 25, 1940. Numerous roosts throughout the winter range of the crow in the Middlewest and South have been dealt with in a similar manner. In addition to shooting and bombing, poisons also have been employed. This unprecedented destruction of bird life has been received with both commendation and violent criticism. Some of the larger roosts numbering hundreds of thousands of individuals provide us with one of the most spectacular scenes of bird life. It is indeed unfortunate that departments of conservation find it necessary to destroy them.

Winter.—Along the New England coast winter is one of the most interesting seasons for a study of crows. At this time they are more numerous than during summer, since the snow-bound conditions of the interior bring them to the tidal shores, where there is a more accessible and constant food supply. They may be seen leaving the roosts early in the morning, often before sunrise, in groups of two or three to a dozen or 20. At this season they seem to lead an aimless kind of existence, meandering here and there, flying low over the mud flats or open fields in a persistent search for food. Sometimes their wanderings take them long distances, going hither and thither until a carcass or other food supply is located. At all times they are alert and suspicious, always proceeding toward food with caution, often alighting on convenient vantage points to carefully inspect the surroundings and to make sure no harm is in store for them. Finally an individual more audacious, perhaps hungrier than the others, approaches to test out the situation. If he succeeds in escaping harm the others quickly join him in active competition to gorge themselves. At such times one bird may act as a sentinel to give warning in the event of approaching danger.

Edward J. Reimann in correspondence concerning crows seen in winter in the vicinity of Philadelphia, Pa., writes: "Crows in winter, especially when ice has formed in the waterways, will be found frequenting the low flats of streams and creeks left bare by the low tide. They can be seen congregated in immense flocks feeding on the seeds of arrow-arum (*Peltandra virginica*). When the rivers are full of drift ice, crows seem to take a particular delight in perching on the cakes and traveling up and down stream with the tide. On some occasions crows were seen to be eating fish frozen in the ice."

DISTRIBUTION

Range.—Most of North America; migratory in the northern regions.

Breeding range.—The crow breeds **north** to Alaska (probably Kodiak Island, Seldonia, and Hinchinbrook Island); southern Mackenzie (Fort Simpson and Grandin River); southern Keewatin (50 miles south of Cape Eskimo); southern Quebec (Onigamis, Godbout, and Mingan Island); and Newfoundland (probably Port au Port and St. John's). The **eastern** limits of the range extend southward along the Atlantic coast from Newfoundland (St. John's) to southern Florida (Royal Palm Park and East Cape). The **southern** limits extend westward along the coast of the Gulf of Mexico from southern Florida (East Cape) to Texas (Houston and Kerrville); northern New Mexico (Glorieta, Santa Fe Canyon, and Santa Clara); central Arizona (White Mountains and the Salt River Bird Reservation); and probably northern Baja California (Guadalupe). **West** to probably Baja California (Guadalupe); California (San Diego, Buena Vista Lake, Stockton, and Red Bluff); Oregon (Warner Valley and Silver Lake); Washington (Camas, Westport, and Everett); British Columbia (Stanley Park, Gull Island, and Massett); and Alaska (Forrester Island, Sitka, and probably Kodiak Island). The crow is also resident in Bermuda.

The range as outlined is for the entire species, which is now separated into four geographical races. The eastern crow *(Corvus brachyrhynchos brachyrhynchos)* is found from Maryland, the northern parts of the Gulf States, and northern Texas north to Newfoundland, Quebec, northern Manitoba, and southwestern Mackenzie; the southern crow *(Corvus b. paulus)* occupies the southeastern part of the range (except Florida) west to eastern Texas; the Florida crow *(Corvus b. pascuus)* is found only in the Florida Peninsula; the western crow *(Corvus b. hesperis)* occupies the western part of the range north to southern Saskatchewan and central British Columbia.

Winter range.—Resident throughout the southern part of the breeding range and north to southern British Columbia (Comox, Chilliwack, and Okanagan Landing); southern Saskatchewan (East End); southern Manitoba (Portage la Prairie); southern Ontario (Sault Ste. Marie, North Bay, and Ottawa); northern Vermont (St. Johnsbury); Maine (Avon and Ellsworth); New Brunswick (Scotch Lake and Fredericton); Nova Scotia (Wolfville and Pictou); and southeastern Newfoundland (St. John's).

Migration.—Since in winter the crow is found in limited numbers, as far north as southern Canada, dates of arrival and departure do not convey a true picture of its migrations. Extensive migratory flight is, in fact, confined to the birds of the Great Plains region. In the files of the Fish and Wildlife Service there are several hundred instances of crows banded during the summer in Saskatchewan that were subsequently recovered during the following fall and winter at well-connected series of localities south through North and South Dakota, Nebraska, Kansas, Oklahoma, and Texas. Similarly, crows banded during winter in Oklahoma and Kansas were recovered during the following spring and summer north through Nebraska and the Dakotas, to Manitoba, Saskatchewan, and Alberta. Crows banded at northern points on the Atlantic coast show very little movement but are generally recovered within 100 miles of the point of banding.

Egg dates.—Alaska: 15 records, May 10 to June 11; 8 records, May 17 to June 6, indicating the height of the season.

Alberta: 21 records, May 2 to June 9; 10 records, May 13 to 24.

British Columbia: 10 records, April 17 to June 4.

California: 112 records, March 21 to June 12; 56 records, April 9 to 21.

Florida: 52 records, January 21 to May 27; 26 records, February 26 to April 2.

Illinois: 39 records, March 27 to May 22; 19 records, April 9 to 27.

Kansas: 22 records, March 3 to May 15; 11 records, April 3 to 19.

Maine: 60 records, April 16 to May 27; 30 records, April 28 to May 8.

New Jersey: 146 records, March 30 to June 12; 74 records, April 15 to May 7.

Ontario: 12 records, April 14 to July 5; 6 records, April 25 to 30.

Oregon: 17 records, April 16 to May 27; 9 records, April 29 to May 11.

Texas: 9 records, February 28 to April 26.

West Virginia: 27 records, April 5 to May 28; 13 records, April 12 to 19.

Washington: 6 records, April 22 to May 22.

CORVUS BRACHYRHYNCHOS PAULUS Howell

SOUTHERN CROW

HABITS

This southern race was named by Arthur H. Howell (1913) from a type collected in Alabama and described as "decidedly smaller than *Corvus b. brachyrhynchos,* with a much slenderer bill. Nearest to *Corvus b. hesperis* but with shorter wing and slightly larger bill." He says further: "Although the bird is nearest to *C. b. hesperis* in size, its range apparently is separated from the range of *hesperis* by a strip of country in central Texas in which no crows breed." He gives as its range "Alabama, Mississippi, Louisiana, southeastern Texas, Georgia (?), South Carolina, and north to the District of Columbia and southern Illinois." It evidently intergrades with *brachyrhynchos* at the northern and western limits of this range. He also says that it is decidedly smaller than the Florida crow, *pascuus,* which is rather remarkable, as Florida races of other birds are generally smaller than the more northern races.

I cannot find that the habits of the southern crow are materially different from those of the eastern crow on the one hand, or the Florida crow on the other hand, depending on the conditions in which it lives. It builds similar nests in many different kinds of trees and feeds on similar classes of food. It has some of the bad habits of the northern race but is nowhere so abundant as to do much damage, and it destroys so many injurious rodents and noxious insects that it probably does more good than harm and should not be molested.

M. G. Vaiden writes to me: "These birds are not so plentiful in the Yazoo-Mississippi Delta of the Mississippi as they are in the hill section lying some 70 miles to the east of the Mississippi River. However, large numbers are found during the latter part July and in August and September, feeding along the mud bars of the river and about the great number of barrow pits left from levee construction, where they feed along with the herons, wood ibises, egrets, and buzzards on the dead and dying small fry left as the pools dry up. They are destructive to other birds' nests, especially killdeers and terns, where they nest along the river on sand bars. I have seen the crow destroying the nests of terns, as described in a short article in *The Oologist* for February 1939, page 24."

Albert J. Kirn tells me that in Texas he has seen crows harassed by a marsh hawk and has also seen a crow attacking a marsh hawk. In southern and eastern Texas he finds them fairly common, mostly in river bottom woods, but thinks that the resident birds do not range very far west in the State.

The measurements of 40 eggs of the southern crow in the United States National Museum average 41.4 by 28.9 millimeters; the eggs showing the four extremes measure **47.3** by 30.2, 42.9 by **31.2,** and **35.8** by **20.2** millimeters.

Dr. Walter P. Taylor tells me that he called up a crow that came within 150 feet of him and sat in the top of a tree. To his surprise "it began to sing. One would never suspect that such tender notes could come from the raucous throat of a crow. The notes resemble somewhat the clucking of a rooster when he is calling hens to some dainty morsel he has found. The crow's song is more varied, however; sometimes he adds a high-pitched tone, and then again continues with his clucking and gurgling noises."

CORVUS BRACHYRHYNCHOS PASCUUS Coues

FLORIDA CROW

HABITS

The Florida crow is smaller than the eastern crow, except for its bill and feet, which are relatively larger and heavier. It is somewhat larger than the southern crow *(paulus)*, which has a smaller and slenderer bill.

The Florida crow *(pascuus)* is generally distributed over peninsular Florida, except in the northwestern part, west of the Aucilla River; it apparently intergrades with *paulus* somewhere in the region of St. Marks. I have met with it in various parts of the State but found it nowhere especially abundant, much less common, in fact, than crows are in many other parts of the United States. We found it most commonly in the flat pine woods, especially about the small cypress swamps, but saw it also in the mixed oak and palmetto hammocks and on the prairies.

Nesting.—What few nests we saw were in the thick groves of tall, slender longleaf pines. Arthur H. Howell (1932) says: "The nests are frequently placed in oak trees in the hammocks, sometimes in pine trees near the border of a cypress swamp, or in a lone tree on the prairie, 7 to 40 (rarely 60) feet above the ground. They are composed of oak twigs and Spanish moss, and lined with horse hair, cabbage-palm fiber, and small pieces of bark."

Bendire (1895) says of some nests collected by Dr. Ralph: "Several nests were found by him in tall, slender pine trees in low, flat pine woods, usually bordering on swamps. The nests were located in the tops of trees, on horizontal limbs, and close to the trunk, at distances varying from 45 to 70 feet from the ground. They are usually composed of small sticks, lined first with Spanish moss and then with strips of cypress bark; occasionally a few feathers from the sitting bird, hair from cows' tails, bunches of fine grass, and grass with the rootlets attached entered into the composition of the linings, and in one instance the eggs were laid on about half a pint of fine rotten wood. The nests average in measurement about 24 by 9 inches in outer diameter, the inner cup being about 16 inches in width by 5 inches in depth' [the 9 inches probably refers to the height and not the diameter].

Harold H. Bailey (1925) says that they nest in "almost every kind of tree from mangrove, gumbo limbo, and cabbage palm of southern keys to pine, oak, and hardwood trees" farther north.

Eggs.—This crow lays three to six eggs, most commonly four or five, which are indistinguishable, except for a slight average difference in size, from the eggs of other crows. The measurements of 40 eggs in the United States National Museum average 41.1 by 28.7 millimeters; the eggs showing the four extremes measure 45.5 by 20.2, 43.2 by 30.9, 37.7 by 29.0, and 40.6 by 20.2 millimeters.

Food.—Mr. Bailey (1925) says: "They have little chance [in southern Florida] to feed on or destroy the farmers' grain, or work in the plowed areas for insect life, but resort to food such as frogs, lizards, small snakes, large grasshoppers, and snails; while during the breeding season the colonies of herons, ibis, anhinga, and other water birds are robbed of eggs or even small young. In every part of the country where I have studied them, I have found them very destructive birds as a whole."

Behavior.—The habits of the Florida crow are much like those of crows elsewhere, though it seems to be tamer and more sociable, less shy, probably because it does less damage to crops and is not persecuted so much as are the more northern subspecies. D. Mortimer (1890) writes:

It is common to see it feeding about the streets and vacant lots of Sanford, especially when the palmetto fruit is ripe enough to eat. It associates freely with the Boat-tailed and Florida Grackles, and also with the Red-winged Blackbird and the Rice-bird, and I have seen flocks including all these species enjoying themselves about the town. It always retreats before any small bird that undertakes to chase it, though it does so apparently because it is too indolent to drive off its assailant, and not on account of timidity. Omnivorous in the fullest sense, it is always on the lookout for any edible morsel. I have seen Florida Crows attach themselves to the Osprey as soon as the latter captured a fish, and tag it about as if to secure any scraps that might fall during the meal. The Osprey is disturbed by this intrusion and tries to strike the Crows with its wings if they come too close.

Voice.—In addition to the ordinary crow notes, two observers have noted some notes that seem to be peculiar to this subspecies. Mr. Mortimer (1890) says: "The Florida Crow has a peculiar note that I never heard uttered by any crow at the North. It is a loud, rattling sound something like the cry of the Cuckoo, and puzzled me much as to its source until I detected the bird in the act of producing it." And Mr. Howell (1932) refers to what is perhaps the same note: "It suggested the rattling call of the Sandhill Crane, though not so loud, and resembled, also, the 'churring' note of the Red-bellied Woodpecker."

CORVUS BRACHYRHYNCHOS HESPERIS Ridgway

WESTERN CROW

PLATES 43-45

HABITS

The western crow is decidedly smaller than the eastern crow, with a relatively smaller and slenderer bill. It is very close to the southern crow *(paulus)* in size, but it has a longer wing and a slightly smaller bill; furthermore, it is widely separated from it in its breeding range. Its range, as given in the 1931 Check-list, is "western North America, from central British Columbia, southern Saskatchewan, and Montana south to northern Lower California and central New Mexico." It apparently breeds very sparingly, if at all, in the extreme southern portions of its range, where it occurs mainly as a winter visitor. Just where it intergrades with the eastern crow on the north and east does not seem to be definitely established. We listed the birds we found in southwestern Saskatchewan as *hesperis,* and a bird taken at Walsh, in southern Alberta, was referred by Dr. Bishop to this race. Frank L. Farley tells me that the birds around Camrose, central Alberta, are of the eastern race. The western crow is very irregularly distributed throughout its range, being rare or entirely lacking in many regions and exceedingly abundant in others; in the regions where it is abundant it is much more numerous than the eastern crow is anywhere, often nesting in colonies.

Dawson (1923) says: "The crow in California is no such constant factor of bird life as he is in the East. He is, instead, very local and sharply restricted in his distribution, so that to a traveller the appearance of Crows is rather a novelty, something to be jotted down in the field-book; and Crow country can scarcely comprise more than a twentieth part of the total area of the State."

Nesting.—About Crane Lake, Saskatchewan, we found a few pairs of crows nesting in the small willows; seven nests were recorded in my notes during the latter half of June, all of which contained young. I found crows nesting at two quite different localities in California. On April 10, 1929, in San Diego County, while exploring a long row of sycamores along the banks of a dry stream, I saw a number of crows' nests and many of the birds. The nests were all well up near the tops of rather large sycamores; many of them were apparently occupied, as the birds were very solicitous; but, as the trees were hard to climb and as I did not flush any birds off any of the nests, I did not disturb them. Again, in Ventura County, on April 27, M. C. Badger and I found crows very common and noisy in a large tract of small cottonwoods and willows along the Santa Clara River; we saw a number of nests, mostly in small live cottonwoods, but some in dead cottonwoods or in willows and 15 to 20 feet from the ground; those that we examined were empty. They did not differ materially from other crows' nests.

It seems to be characteristic of the western crow to build its nest at low elevations, as well as in groups or colonies; small trees or bushes seem to be satisfactory as nesting sites, and nests have been found even on the ground. Major Bendire (1895) says that at Fort Lapwai, Idaho, he "occasionally found them breeding in what might be called small colonies, and this was not due to scarcity of timber for nesting purposes; in fact, I once saw here three occupied nests in a single small birch tree, where a number of good-sized cottonwood trees were to be found close by and equally suitable. * * * Cottonwoods, junipers, and willows are most frequently used. Nests are usually placed at heights varying from 20 to 60 feet; but I have found some barely 6 feet from the ground, and in many localities in the West they are rarely placed over 20 feet up. Here also they are said to occasionally nest on the ground, but I have never observed this personally."

Capt. L. R. Wolfe (1931), however, did find a nest on the ground in Albany County, Wyo., and published a photograph of it. "It was in the short grass, on flat, open prairie. The nest was placed in a depression which at one time might have been the entrance of a badger hole. The depression was such that the rim of the nest was just level with the surface of the ground, and it was well filled with grass, weeds and small pieces of sage and willow. The nest had an outside rim of weeds, sage and willow twigs and a few small sticks and was lined with strips from weeds and with cow's hair. The general construction and bulk of the nest was about the same as any tree nest of the species. A few larger sticks and twigs were scattered around on the ground surrounding the nest. This nest contained seven eggs."

W. E. Griffee writes to me from Oregon that nests he found near Portland were mostly in "trees along the sloughs on the Inverness Golf Course, where the crows, like other birds, are protected." On April 25 and 26, 1940, in Lake County, east of the Cascades, he collected seven sets "from scrubby willow clumps out in the Chewaucan Marsh," and thinks he could have taken 30 or 40 more sets there in those two days, if he had concentrated on them. There must have been a large concentration of breeding crows in that region. A still larger concentration is mentioned by Leon L. Gardner (1926) in Klickitat County, Wash. On Rock Creek, crows were found at their noon siesta, and many old nests were located with an abundance of evidence in feathers and droppings that for miles up and down the creek a vast rookery existed. Although this was in August, when the crows had congregated to feast in the almond and apricot orchards, many of the vast hordes referred to later probably nested in that rookery where the nests were seen.

J. A. Munro has sent me the data for 14 nests, found in the brush along the creek near Okanagan Landing, British Columbia; 10 of these were in willows, and one each in an alder, a blackhaw, and a poplar; the heights from the ground varied from 10 to 20 feet.

Eggs.—Five or six eggs form the usual set for the western crow, but four or seven are often laid, and as many as eight or even nine have been recorded. J. G. Suthard tells me that he once found a set of nine eggs, eight intact and one broken. "The ninth egg had been broken and the shell telescoped onto another egg, to which it was firmly attached by the dried yolk and albumen. All the eggs were similar in size and shape." Three other nests in the vicinity were examined, in which the "eggs were definitely different from this set."

The eggs of the western crow are similar to those of the eastern crow, except in average size. The measurements of 40 eggs in the United States National Museum average 41.1 by 28.8 millimeters; the eggs showing the four extremes measure **46.2** by 31.2, 44.7 by **31.5, 35.8** by 26.4, and 37.7 by **24.4** millimeters.

Food.—Like other crows, the western subspecies is omnivorous; it probably does some good in the destruction of harmful insects and rodents, and as a scavenger; but in many places it has formed the habit of congregating in enormous numbers to feast on cultivated nuts and fruits, which makes it a serious menace and requires effective control measures to save any of the crops.

Mrs. Wheelock (1904) says that "in California, acorns, beechnuts, berries of various shrubs and trees, seeds and all kinds of fruit, with insects such as locusts, black beetles, crickets, grasshoppers, spiders, cutworms, angleworms, and injurious larvae form a large part of its daily menu. In addition small mammals and snakes, frogs, lizards, snails, crawfish, fish, all kinds of dead flesh, and the eggs or nestlings of other birds are his victims. * * * The fact that all feathered creatures are arrayed against him is proof to me that, from the bird-lover's standpoint, he does more harm than good."

Ralph H. Imler (1939), in his comparative study of the food of crows and white-necked ravens in Oklahoma, found that the crows were apparently more beneficial in their feeding habits than the ravens since they ate many more insects and weed seeds. His analysis of 14 crow stomachs showed the following proportions: "Beetles, 4.6 percent; grasshoppers, 9.4 percent; mammals, 1.2 percent; grain, sorghums, 24.7 percent; corn, 24.9 percent; melons and citron seeds, 18.9 percent; sunflowers, 16.3 percent."

Some of the items mentioned by Mrs. Wheelock may be placed to the credit of the crow, and probably the large percentages of grain in Mr. Imler's report may be waste grain and therefore neutral; but there is another side to the picture, which is as black as the crow's plumage. Mrs. Nice (1931) says that "in northern Oklahoma crows have become a serious pest in pecan orchards" and describes their crafty methods of work and the none too effective methods of control. "Vast numbers have been killed in Oklahoma: 10,000 in one week near Chickasha in February, 1926, 11,000 in Payne County during the winter 1927-28, 3692 in two nights by means of dynamite near Vinita in January, 1929, according to the newspapers; but vast numbers remain."

Maj. Leon L. Gardner (1926) tells an interesting and most remarkable story of an immense congregation of crows and the devastation that they wrought: "In the region of Goodnoe Hills, Klickitat County, Washington, a very promising enterprise in raising almonds and apricots was developing. It was reported, however, by the farmers of that region that in fall enormous flocks, amounting to 'millions of crows,' came into this region and destroyed practically the entire crop of fruit and nuts, together with considerable acreages of watermelons." When he arrived, about the middle of August, "the almonds were ripening fast and crows had assembled in mass from the surrounding country for the annual feast." He continues:

They were in flocks of hundreds to thousands. The usual formation was in one enormous flock which worked as a unit in some selected orchard with a few outlying and unimportant groups feeding at random from other orchards. When in the air, this large host looked at a distance like a huge cloud of gnats.

It was difficult to estimate the exact numbers but an approximation could be made with the larger group when they settled to work in an orchard. In a 20-acre almond orchard there were about 1500 trees. To each one of these trees one could count from ten to thirty Crows with an average of fifteen to the tree. The flock on our arrival numbered perhaps 15,000 which, in a short time, nearly doubled in size until we estimated fully 30,000 members in the various groups. * * * With regard to the almond damage there was no argument. The destruction of an $800 crop was complete in two days after which the Crows moved on to a new orchard. * * *

It was the practice of the Crows, after a hot afternoon's work, to spare themselves the trouble of flying any considerable distance to water by feeding on watermelons. They were never seen to attack melons in the morning but always after a dry day's work. Furthermore green melons were sought and eaten as well as ripe ones indicating it was the moisture they were seeking. In any event the damage amounted to practically 100% of the crop, no melon being spared unless it happened to be concealed in the vines.

Various methods of control or protection of the crops were tried, but with indifferent success. Shooting served only to drive the crows from one orchard to another, and they paid no attention to scarecrows or to belling or stringing the trees. Poisoned carcasses of rabbits were scattered about, but after a few crows had been poisoned the others soon learned to avoid these baits. Poisoned watermelons proved more effective in protecting this crop, for, after a few crows had been killed, no further visits were made to the melon patches. Almonds, slit open, poisoned, and scattered in conspicuous places in the orchards, finally succeeded in driving away the crows, though "the actual number of crows poisoned was extremely small not exceeding 1% of the flock."

He says in conclusion: "The use of poisoned almonds, when properly conducted, proved successful in protecting the crops but demonstrated anew the exceptional sagacity of this bird. The first reaction was one of extreme panic at some of their number being fatally affected by their chief article of diet. This was manifested by tumultuous clamoring and confusion of the flock while sudden sallys and forays were made into distant parts of the Hills only to be met by the same fatal consequences. The flock then rapidly reacted to the changed environment by abandoning attempts at feeding from the almonds and indeed, by departing from the entire region."

There is another score against the crows of western North America, the destruction of the eggs of waterfowl in the great breeding grounds of ducks in the Middle West. We noticed that the nests of crows that we saw in Saskatchewan were largely congregated around the shores of the lakes and sloughs where ducks were breeding. On one small island in Crane Lake we counted 61 ducks' nests in a few hours' search; the following year this island was practically deserted by the ducks; we charged the damage largely to a coyote and a family of minks, but very likely the crows found here an abundant food supply.

Frank L. Farley writes to me that crows have so increased in central Alberta as to become a serious menace to waterfowl. He (1932) writes:

It is significant that as the crows gained in numbers there was a corresponding decrease in the number of ducks, particularly of the marsh-nesting species. In the choice of nesting sites it was noted that the crows favored wooded areas adjacent to lakes and marshlands where ducks nested, this no doubt for the purpose of being close to a rich food supply during the nesting season. Mr. Francois Adam, a former prominent farmer of the Edberg district, and on whose farm are extensive marshes, told the writer of the discovery of twenty-two ducks' nests on his place one Sunday in May. The following Sunday, being suspicious all was not well with the nesting ducks, he again visited the marsh and was surprised to find every nest empty, and many crows busily engaged in every part of the marsh, searching for nests that had been overlooked. He stated that there were a dozen crows' nests in the willows surrounding the lake, and others nearby in scattered clumps on the prairie. It is evident that in such places ducks could not carry on nesting operations successfully.

Eastern crows, which are usually not too abundant anywhere, may not do enough damage to offset the good that they do, making their economic status at least neutral. But the above samples, of which there are probably plenty more, show what damage they can do when concentrated in large numbers.

The western crow, also, has its good traits. S. F. Rathbun writes to me from Seattle: "In the cultivated and more or less open sections lying along the eastern side of the Sound, on occasions early in spring, I have seen numbers of western crows on their northward movement. Wherever the locality had freshly plowed fields, hundreds of crows were on the ground, gleaning every kind of animal food exposed by the plow. The feeding birds never were disturbed by the farmer, who usually regarded them with favor because of the good work the crows were doing.

"One food item, of which this crow is fond, is the fruit of the red-berried elder, *Sambucus callicarpa,* a more or less common shrub of the bottomlands of western Washington. When this bush hangs heavy with its brilliantly colored berries, one often finds the crows eating them. While doing this they are noisy, for the birds will try to alight on the tops of the fruit-laden bushes, and, as these fail to support them, many birds fall fluttering to the ground amid much excitement and commotion. Once I shot a crow when it was feeding on the berries. I found its gullet packed with the fruit and its stomach also; this organ with the digestive tract proved to be so deeply dyed a dark red that it appeared to show that, for some time at least, elderberries had been the chief food of the bird."

Behavior.—There is little to be said about the behavior of the western crow, which does not differ materially from that of its eastern relative. It is the same clever, sagacious bird, wary when in danger and tame where it feels secure. Ridgway (1877) saw three individuals at a stage station in Nevada that "walked unconcernedly about the door-yard with the familiarity of tame pigeons, merely hopping to one side when approached too closely." On the other hand, Henshaw (1875), in Arizona, found them "quite numerous, associating freely with, and apparently the boon companions of the ravens. Yet, even here, I found that they had lost little of their traditional shyness, and it was some time ere I procured a specimen. Gun in hand, I found no difficulty in approaching the trees where sat the ravens, looking down upon me with a comical glance of wonder, tinged with a slight suspicion that all was not just as it should be. But the crows had long before taken the alarm, and made themselves scarce, and from some secure perch sent back their warning *caws,* given, as it appeared to me, with more than the usual earnestness, as though deprecating the stupidity of their big cousins."

Enemies.—Probably western crows occasionally attack hawks or owls, just as the eastern birds do; and sometimes the tables are turned. Joseph Mailliard (1908) tells of a pair of Cooper's hawks attacking a flock of crows that were quietly perched in the tops of some dead trees. The hawks did some good team work, one attacking from above and one from below, but the crows were too alert and no serious damage was done before Mr. Mailliard shot both of the hawks.

Fall.—Western crows wander about more or less in fall and winter, and there is a limited migration from some parts of the breeding range, a gradual southward drift that extends the range of the subspecies somewhat south of its summer range. There is probably, also, a retreat from some of the higher altitudes down into the valleys and about the ranches in search of food, and perhaps a coastwise trend. Theed Pearse tells me that some cross over from the mainland to Vancouver Island, where they associate with the northwest crow *(caurinus).* The two can be easily recognized by their voices.

Joseph Mailliard (1927) mentions a migratory movement in Modoc County, Calif.: "In September, 1925, the crows gathered as they had in the previous year. The number seemed to·reach the maximum about September 23, when I estimated the size of the band to be in the neighborhood of 1,000 individuals, all feeding in the stubble field back of our quarters. Ten days after this, small numbers were noted mov-

ing toward the south, which movement continued daily until comparatively few remained on October 15, when our party left the field."

Winter.—Some western crows spend the winter in the northern parts
of their range, for J. A. Munro tells me that sometimes large flocks are
seen in the vicinity of Okanagan Landing, British Columbia, in winter.
"They are extremely local in their feeding habits during winter, remaining in the vicinity of slaughterhouses, or other places where food is
easily obtainable, and returning to the same roost, often several miles
away, every evening."

The winter population of crows in Oklahoma, on which so much destructive bombing has been done, seems to consist largely of crows
from farther north.

Kalmbach and Aldous (1940) report that "the banding of 714 Crows
in south central Oklahoma during the winter of 1935-36 has yielded,
during the three and one-half years following their release, 143 returns,
slightly more than 20 percent of the birds banded." They say:

Analysis of these returns shows that, of the 65 Crows recovered during the
breeding and rearing season (April 1 to August 31), 49 (75 per cent) were killed
in the Prairie Provinces of Canada. The dates and locations of numerous other
returns recorded in the states north of Oklahoma indicate that many others of this
group of Crows may have been on their way to or from Canadian breeding
grounds. During this same period of the year not one of the winter-banded
Crows was recovered in Oklahoma, clearly indicating that winter Crow control
in Oklahoma can have little or no effect on nesting upland game or insectivorous
birds of that state.

Although winter Crow control in Oklahoma is destined to remove some birds
that would enter the problem of Crow-waterfowl relationships in the Canadian
provinces, the effect of this control is certain to be much 'diluted' if the results
are to be judged in a continental perspective. This comes about because only a
portion of the Crows in Canada can be classed as duck-egg predators, and because
the Crow, in what might be termed destructive abundance, occupies possibly only
a sixth of the duck-nesting area of Canada and Alaska.

CORVUS CAURINUS Baird

NORTHWESTERN CROW

HABITS

The crows of the Northwest coast from the Alaska Peninsula and
Kodiak Island southward as far as the Puget Sound region of Washington were originally described as a distinct species, largely on account
of their coastwise habitat and a somewhat different voice. Many modern
writers, including the framers of the 1910 Check-list, have listed *Corvus
caurinus* as a species. But it was reduced to the rank of a subspecies
in the 1931 Check-list; it will now be restored to full specific rank.

It is similar in appearance to *hesperis* but is smaller, with relatively
smaller feet.

Its favorite haunts are on the seashore, from which it seldom strays very far; it is a common resident bird about the wooded shores of the bays and on the beaches, where it feeds with the gull on shellfish and refuse thrown up by the waves. Theed Pearse writes to me that on Vancouver Island it is now showing signs of spreading out into many square miles of logged-over and burnt-over hillsides near the shore but that its main habitat is on the shores, especially where there are small coniferous trees.

A. M. Bailey (1927) says that in southeastern Alaska these "crows are especially numerous about the towns and villages, hanging about the camps for food. At low tide, the flocks repair to the flats, where they secure an easy living among the mussel beds."

Nesting.—J. H. Bowles (1900) writes:

On the Tacoma Flats, at the head of Commencement Bay, is a small cluster of Siwash Indian houses, which are bordered by a line of scrubby apple and cherry trees. In these trees six or seven pairs of this sociable little crow band together in a colony during the nesting season. The nest is placed in a crotch at a distance from ten to eighteen feet above the ground, the same one being made over each returning season. On one occasion I saw two occupied nests in an apple tree only twenty feet high. Its appearance differs greatly from that of *americanus*, as it closely resembles a round basket having a very slight projecting rim of sticks. The average rim of projecting sticks in a series of *americanus* I have found to be 9.78 inches, while that of *caurinus* is only a trifle over 4 inches. The inner dimensions average about 7 inches in diameter by 4 inches in depth. The composition also is nearly the same, only the material used is much less coarse, being a foundation of fine sticks and mud, lined with cedar bark.

Mr. Bailey (1927) "found a nest in Patterson's Bay, Hooniah Sound, May 17, which was about twenty feet from the ground in a small hemlock. The nest was a rather bulky affair of spruce twigs, lined with dried grass, while the interior cup was composed entirely of deer hair. There were four eggs in the nest. Crows were abundant on Forrester Island, and it was there that Willett called my attention to a peculiar habit of theirs, that of nesting under boulders on the beach. They placed their nests far back in rather inaccessible places."

S. J. Darcus (1930) says that on the Queen Charlotte Islands "many nests were found, all built on the ground beneath bushes or windfalls close to sea shore." Mr. Pearse tells me that on Vancouver Island most of these crows resort to the vicinity of the sea for nesting and nests will be found in quite low bushes and even in the side of a sandy bank. On June 16, 1940, a nest containing four eggs was found about 8 feet from the ground in a small fir in the logged-over area. Earlier reports of nests roofed over, like those of magpies, were probably based on incorrect identification. Mr. Rathbun writes to me: "For several years my home was near the crest of a high bluff along the Sound, its base bounded by the beach. The abrupt side of the bluff was thickly covered by a second growth of evergreen and deciduous trees, some of good size, and in several of the former a number of pairs of northwestern crows nested each spring."

Sidney B. Peyton writes to me that these crows are numerous on Forrester Island, where the majority of the nests were not over 8 feet from the ground in the thick spruce trees. One was "in a hole in a cliff about 15 feet up," and three others were under boulders on the beach "about 100 feet above the high tide mark."

Richard M. Bond has sent me some notes on this crow; he says that on Bainbridge Island in Puget Sound, "where the timber had only been gone over lightly for the best trees, the forest was almost in its virgin state. Here I was able to locate only two occupied nests, both about two-thirds of the way up fairly large Douglas firs—about 70 feet from the ground. In the San Juan Islands, even in virgin stands, the trees are in most places very small, and nests are easy to locate from the ground. Douglas fir is here the commonest tree, and though hemlock, western red cedar, alder, and others are also present in some numbers, I have never found a crow nest in any tree but the fir. The crows nest on the main islands rather like the western crow does; that is, in scattered groups, but they also nest on some of the small islets of an acre or less, where there are only two or three scraggly firs. I do not remember finding more than one nest on such an islet."

Eggs.—This crow lays ordinarily four or five eggs. They are indistinguishable, except in size, from the eggs of other crows and probably a large series would show most of the variations common to the species.

The measurements of 40 eggs average 40.4 by 28.2 millimeters; the eggs showing the four extremes measure 44.2 by 28.8, 40.8 by 29.8, 36.7 by 27.9, and 40.9 by 25.9 millimeters.

Young.—In the locality mentioned above, S. F. Rathbun had a good chance to watch the behavior of the young crows, which, early in July, were still being fed to some extent by their parents; "now and then one of them would sidle up to an adult and stroke the old bird's beak, evidently coaxing to be fed." He says in his notes: "It is interesting to watch the old and young birds; there are upwards of 50 of them. The other day, during a strong wind, many of the crows played about in it; some of their aerial evolutions were most graceful and reminded one of the raven's flight ability. Those I watched seemed to battle the wind for the pure love of the sport, old and young birds alike indulging in it.

"All the crows were drifting around in the wind just before sunset. There are now almost 70 of the birds. The young among them are still practicing flying, and some fly gracefully for they rise and fall with a floating motion, apparently without effort. The old birds are easy to distinguish, for they sit quietly in the trees and gravely watch their young at play."

Mr. Pearse tells me that he has seen young on the wing as early as June 10 and has seen young just out of the nest as late as August 23. He seems to think that two, or possibly three, broods may be raised in a season.

Food.—The main feeding grounds of these small crows are on the beaches, where they are useful as scavengers, picking up the refuse thrown out by the fishermen, and where they show no fear of the natives, who never molest them, but they are shy of strangers. There they feed also on shellfish, crabs, and any edible refuse thrown up by the waves. In winter they find an ample food supply on the extensive beds of mussels on the tidal flats. In summer they frequent the salmon streams to feed on the dead fish, and are welcome as scavengers about the salmon canneries. Mr. Pearse tells me that late in summer and in fall berries such as wild cherry and saskatoon form a large part of their food; they eat fruit also and are especially fond of pears and apples, though they ruin more than they eat; in the fall of 1935, after an unusually early frost, they fed on the frost-rotted apples that still hung on the trees.

Mr. Bailey (1927) writes: "At low tide, the flocks repair to the flats, where they secure an easy living among the mussel beds. It is a common sight to see Crows darting in the air, as they drop mussels upon rocks, to break them. If the wind is blowing, they allow for the curve, and usually do not make many misses in their endeavor to hit a certain boulder. * * * These birds, too, are especially bad about plundering the nests of their neighbors and no species is safe from them, for they are continually hunting, possessing a boldness even greater than the Raven. They rob the sea birds nesting under boulders as well as the Murres upon the cliffs. They are not so conspicuous in their plundering however, as the Ravens, for they eat their eggs where they find them, and so probably put their time in to better advantage."

The presence of a human being in a sea-bird colony sends all the gulls, cormorants, murres, and pigeon guillemots off their nests, which is the signal for the crows to rush in, grab an egg from an unprotected nest, and fly off with it; the crows return again and again as long as the rightful owners are kept off their nests; this results in great destruction among the eggs and young of these colonial birds, for which some overzealous bird photographer may be unwillingly responsible. The eggs and young of land birds probably suffer to a less extent. Mr.

Pearse tells me that some young birds are taken, but the crows do not seem to be persistent in hunting for them, though some individuals may get the habit. He once saw a crow that was apparently watching to locate a robin's nest and was being mobbed by a lot of flycatchers, warblers, and chickadees. He next heard the agonized cries of a young robin, which the crow had captured and was carrying off; the old robins attacked the crow and made it drop the young bird. "The other birds in some way recognized the crow as dangerous and kept up their mobbing. Though I have seen the northwestern crow working through wooded areas, evidently looking for young birds or nests, there is none of the systematic beating of the ground that may be seen done by the western crow; moreover, I have seen very few cases of nests that may have been destroyed by crows."

R. M. Bond sends me the following notes on the feeding habits of the northwestern crow: "I have seen crows foraging in fields, where they were seen to capture grasshoppers. They also, in one case, were observed eating ripe wild blackberries (*Rubus vitifolius*). I have also seen them feeding behind a plow in company with Brewer's blackbirds and gulls." He says that they had favorite times for feeding on the beaches: "One was near high tide, when the incoming water separated the house garbage from the ashes, with which it was frequently dumped, and when the innumerable amphipod 'sand fleas' were retreating before the water to the shelter of windrows of kelp and seaweed. I had a blind in a hollow log of driftwood and could watch the crows eat enormous numbers of sand fleas, as the flock worked by me. The other favorite period was near low tide, when cockles and gastropods were exposed.

"Carrion was a favorite food. This consisted mainly of dead fish that washed up on the shore. Dead dogs, cats, horses, etc., were also eagerly eaten, though they amounted to a very small percentage of the available food. Once a dead porpoise washed ashore on Blake Island, and for two or three days there was a stream of crows passing across about three miles of open water to this feast and back, apparently to feed their young."

Stomach analysis of three specimens by J. A. Munro (MS.) showed that the bulk of the food consisted of shore crabs and small mollusks; crab remains amounted to 80 percent in one case, and small mollusks to 80 and 60 percent in the other two cases. Mollusks and crab remains were found in all three stomachs, and the remainder of the food included a few insect remains, fish eggs (probably sculpins), oat husks, and miscellaneous vegetable matter.

Behavior.—As northwestern crows apparently do little damage to human interests, they are much tamer than crows elsewhere and pay but little attention to human beings. They make themselves at home about the Indian villages, where they are almost as tame as chickens and hardly move out of the way of the children playing on the beaches.

Mr. Pearse's notes contain several references to the behavior of these crows. He has seen 12 crows "vigorously chasing a raven"; again he has seen the two species feeding side by side, though the crows recognized the "superiority of the raven and would not contest the feeding grounds"; crows feeding on berries hurried away when a raven came into the bush. He has seen nesting crows drive away a bald eagle that had settled on a tree nearby, making stoops at it, the eagle squealing as though at least annoyed; and he says that they will frequently attack a flying eagle, if it comes near where they are feeding; sometimes the eagle will turn at the crow, to strike it with its talons, which drives away the latter.

He tells of one of a pair of crows that had lost part of its beak, perhaps in a trap; it was being fed by its mate, "which regurgitated the food. The healthy bird was crooning to it and stroking it with its beak, a really touching sight."

One day, in July, he saw "two diving and swooping around in a stiff breeze, stooping at each other and turning over, a kind of game of tag. At times a flock will plane-dive down to the ground, and this usually presages a change of weather, usually wind."

Mr. Munro tells me that on four occasions in January and February he has seen a heavy flight, totaling about 800 birds, passing over Departure Bay about dusk and apparently settling in their winter roost in some thick woods.

Voice.—Ralph Hoffmann (1927) says: "A trained ear can usually detect the difference in their notes from those of the Western Crow; they are usually slightly hoarser and lower in pitch but vary in pitch and quality and are at times very close to the Western Crow's. In the mating season they have a 'gargling' note similar to that of the Western Crow."

Mr. Bailey (1927) writes: "They are probably the best imitators of their family in Alaska, and the variety of their notes is unusually large. Their most characteristic one is noted when the old bird is feeling especially foolish, for they duck their heads toward their feet, and then give an upward tug, at the same time emitting a sound like the pulling of a cork from a bottle."

R. H. Lawrence wrote to Major Bendire (1895) of a vocal performance that may have been part of a courtship display: "A flock of about one hundred and twenty were noticed February 7, 1892; a few were perched apart on a tree or snag, uttering strange sounds, like 'koo-wow,

kow-wow, koo-wow,' the last syllable drawled and accented or emphasized; then, with a slight spreading of the shoulders and the tail, the
head being down and the tail drooped, they produced by a curious
chattering of the bill a sound (not made in the throat, I judged) which
resembled that of horny plates struck together, and causing an odd
shuddering of the head and even of the body. This was repeated a few
times, varied with a noisy 'caw, caw.'"

Fall.—Mr. Pearse writes to me that on Vancouver Island "the locally
bred crows commence to flock as early as the middle of August, but it
would be at least a month before the outside birds would arrive. Most
falls there is a considerable immigration to the farmlands from outside
areas, probably from inlets farther up the coast or from the mainland.
At times during winter 1,000 birds will be feeding together on the
cultivated areas. Banding operations show that there is a southerly
movement of the locally breeding crows in fall and that birds banded
in winter go to the mainland the following spring. Each year the crows
disappear from here the end of August, and very few are seen until
a month later." He also has seen some evidence of a northward migration in spring.

CORVUS OSSIFRAGUS Wilson

FISH CROW

PLATE 46

HABITS

This small and well-marked species of crow is widely distributed
along the Atlantic and Gulf coasts, as well as in the lower valleys of
some of the larger rivers. It reaches its northeastern limit in southern
Massachusetts, where it is a rare and local straggler, mainly in spring.
I am quite sure that I have seen it on two occasions in Bristol, R. I.
It is a fairly common summer resident on the coast of Connecticut,
chiefly in the western part, and on certain parts of Long Island, N. Y.
From New Jersey southward it is an abundant bird and practically
resident.

Its favorite haunts are the coastal marshes and beaches, the banks
of streams, and to some extent the shores of inland bodies of water.
In Florida, the numerous streams, lakes, and marshes furnish suitable
haunts for them over a large part of the inland country, especially
where they can prey upon the breeding colonies of herons and other
water birds. Dr. Samuel S. Dickey tells me that he has seen them

along the banks of inland rivers in central Pennsylvania, as far west as Harrisburg and Columbia.

Courtship.—Dr. Dickey (MS.) writes: "During the first two weeks of April fish crows become especially animated and proceed with mating impulses. Generally two males are seen to bicker over a single female. The three of them then hurry through the high canopies of crack willows, elms, oaks, and even some evergreens. They will half unfold the wings, lean back against boughs, and open their red beaks in a seeming defensive attitude. Then away they glide, from the trees of the stream banks, across wide plantations of truck gardeners. They will, on breezy days, dally with one another, and even touch wings and heads. In all, they have a playful, captivating manner in midair at this time of year."

Nesting.—Aretas A. Saunders writes to me that in the vicinity of Fairfield, Conn., where the fish crow is a "regular, but rare summer resident," nesting takes place late in April or in May." The nests are generally in small colonies, two or three pairs with their nests not far apart in a certain locality. I have found such colonies in two types of localities—swampy woodlands where the trees are tall and the nests high up and rocky places on the edge of a salt marsh, where the rocks stand up like islands in a salt marsh sea and are clothed with red cedars and pitch pines. Nests in such places are in the pitch pines and not very far up. Such localities are used year after year if conditions are not disturbed. At present I know of but one nesting locality of the swampy woods type."

I find in Owen Durfee's notes for May 10 and 11, 1903, the records of two Connecticut nests, also near Fairfield. The first was 64 feet from the ground in an 11-inch black oak. "The nest was composed of ˙small, dry sticks, well mixed in with old cornstalk strings. It was placed in the topmost crotch of the tree, where the diameter was only 2 inches, the tree being only about 4 feet higher. It was 14 inches in diameter and built up 14 inches high. Inside it was 7½ inches in diameter and hollowed 5½ inches. It was lined, but not felted, with strips of grapevine bark." The second nest was similar, 61 feet up in a 13-inch chestnut and about 9 feet from the top of the tree. It was 18 inches in diameter and built up 12 inches in a three-pronged crotch. "The lining was principally of strips of inner bark of the chestnut, with two large clumps of white horsehair and a little grapevine bark."

The only fish crow's nest I have ever examined was found near Little Egg Harbor, N. J., on May 27, 1927. In a large patch of baccharis bushes on a low sand dune on the edge of Little Sheepshead, I found a colony of seven or eight nests of the green heron. The crows had robbed three or four of the heron's nests. The crow's nest was located 7 feet up in one of the largest of the baccharis bushes. It was made of dead sticks and twigs and lined with strips of inner bark and a few feathers. It measured 14 by 15 inches in outside and 7 inches in inside diameter; it was hollowed to a depth of 6 inches.

T. E. McMullen has sent me his data for 138 nests of the fish crow, found in New Jersey and vicinity. Most of these, 80, were in holly trees, 12 to 30 feet from the ground; 34 were in cedars, 5½ to 25 feet up; 9 were located in oaks, at heights varying from 18 to 50 feet; 9 were placed in pines, 17 to 90 feet from the ground; and there was one each in a beech, a gum, a sassafras, and a wild cherry, and two in maples, all at intermediate heights.

Two nests have been reported at heights far above those already mentioned. The Rev. H. E. Wheeler (1922) mentions a nest, found on the bank of the Arkansas River, that was "well toward the top of a huge sycamore 110 feet from the ground." This nest "now contained no rootlets, but was lined with a mass of sycamore balls and horse hair!" (This was after the eggs hatched. Before that the nest was "lined with leaves and rootlets.") But Arthur T. Wayne (1910) reports the loftiest nest of which I can find any record: "About twenty-five years ago this species used to breed regularly in St. Paul's churchyard, in the city of Charleston, where it placed its nest in the topmost branches of a gigantic sycamore tree fully one hundred and fifty feet from the ground, and it also bred in later years in private yards along East Battery."

Major Bendire (1895) writes: "A nest taken by Dr. Ralph near San Mateo, Florida, was composed of sticks with a little Spanish moss attached to them, and was lined with pine needles, strips of cypress bark, and old Spanish moss. It was placed in the top of a slender pine tree, in low, flat pine woods, 81 feet from the ground. Some nests are lined with dry cow and horse dung, cattle or horse hair, dry leaves, eelgrass, and shreds of cedar bark, while pine needles seem to be present to some extent in most of them. They are mostly placed in evergreens, such as pines and cedars, and generally in the tops, either in natural forks or on horizontal limbs, close to the trunk, usually 20 to 50 feet from the ground. They prefer to nest near water, but occasionally a pair will be found making an exception to this rule, and nests have been found fully 2 miles away from the nearest stream or swamp."

In the nesting site photographed by Mr. Grimes (pl. 46), two or three pairs nest every spring in the tall slash pines; the highest fork that will support the nest is usually selected.

Eggs.—The fish crow lays ordinarily four or five eggs to a set, rarely more. These are exactly like the eggs of the other crows, except in size. They show all the ordinary variations in color, pattern, and markings that are to be found in the eggs of the eastern crow. The measurements of 46 eggs, in the United States National Museum, average 37.17 by 26.97 millimeters; the eggs showing the four extremes measure 42.9 by 27.4, 37.8 by 28.7, 34.5 by 27.5, and 37.7 by 25.2 millimeters.

Young.—Bendire (1895) says that "both sexes assist in incubation, which lasts from sixteen to eighteen days, while the young remain in the nest about three weeks. Only one brood is raised in a season, but if the first set of eggs is taken they will lay another, and not infrequently in the same nest."

Plumages.—The young fish crow is hatched naked and blind, but it soon acquires a scanty growth of grayish-brown natal down. This, in turn, is replaced by the juvenal plumage, which is practically completed before the young bird leaves the nest. The juvenal body plumage is dull brownish black, blacker above and browner below; the wings, except the lesser coverts, are much like those of the adult and so is the tail, but they are somewhat less lustrous black with greenish reflections; the bill and feet are grayish black.

The postjuvenal molt, which involves the contour plumage and the lesser wing coverts, but not the rest of the wings and tail, begins in July and is completed by September or earlier. This produces a first winter plumage, which is much like that of the adult, but somewhat duller. At the first postnuptial molt, which is complete, during the following summer, the young bird becomes fully adult. Adults have one complete, annual molt during summer and early fall. The sexes are alike in all plumages.

Food.—Like other crows the fish crow is largely omnivorous, with a long list of acceptable material available. As it spends most of its time along the seashore, the banks of streams, and the shores of inland bodies of water, its food consists largely of various kinds of marine or aquatic life, or other material washed up on such shores. It may often be seen hovering over the water, like a gull, looking for floating objects that it can pick up. On the beaches and salt marshes these crows feed on small crabs, especially fiddlers, shrimps, crawfish, dead fish and perhaps some live fish, and any kind of carrion or offal that they can find. They steal the eggs from the nests of terns, willets, Wilson's plovers, and clapper rails. William G. Fargo (1927) says that they have regular feeding stations where they bring their food to eat it; under a small yellow pine at Wakulla Beach, Fla., in a space about 4 by 6 feet, he found the remains of 79 or more clapper rails' eggs, one willet's egg, two Wilson's plovers' eggs, seven hens' eggs, several turtles' eggs, one fish head, and one rock crab.

Fish crows do immense damage in the heron colonies in Florida; wherever I have been in the many breeding colonies, fish crows have always been flying about, looking for a chance to steal the eggs from an unguarded nest. As the herons all leave their nests as soon as a man approaches, the crows have plenty of chances to enjoy a good feast, and they make the most of it. They rob the nests of all the herons, large and small, as well as the ibises, spoonbills, anhingas, and even cormorants. While we were photographing for parts of two days in the great Cuthbert rookery, nearly all the nests within sight of our blinds were completely emptied; and our experience was similar elsewhere in Florida. Howell (1932) says that they "perch on the bushes, watching for a sitting bird to leave its nest, whereupon they immediately swoop down and carry off an egg. In the large rookery at Orange Lake, it is estimated that two-thirds of the nests are robbed by the crows, which are there very abundant."

Fish crows are sometimes seen in the plowed fields, picking up grubs; they are also said to eat ants, and several observers have mentioned grasshoppers in their food. N. B. Moore says in his notes, made many years ago, that these crows alight on the backs of cattle, to pick up the ticks that are burrowing into the skin and sucking the life blood from, as well as annoying, these animals; this may be an ancient habit, as it does not seem to have been recently observed.

Bendire (1895) states that on the Smithsonian grounds in Washington they "have been noticed repeatedly carrying off and eating the young of the English Sparrows." Wilson (1832) writes: "There is in many of the ponds there [Georgia], a singular kind of lizard, that swims about with its head above the surface, making a loud sound, not unlike the harsh jarring of a door. These the Crow now before us would frequently seize with his claws, as he flew along the surface, and retire to the summit of a dead tree to enjoy his repast."

The vegetable food includes a variety of berries, fruits, and seeds, such as pokeberries, mulberries, hackberries, huckleberries, the fruits of red cedar, sour gum, palmetto, magnolia, holly, dogwood, papaw, red bay, catbrier, and mistletoe, and the seeds of locust, wildrice, etc. Some grain is eaten, such as corn and oats, but most of this is probably waste grain picked up in the fields after harvesting. Probably some cultivated fruits are taken, but not enough to be of great economic importance.

Audubon (1842) mentions the berries of the dahoon *(Ilex cassine)*; "they are seen feeding on them in flocks often amounting to more than a hundred individuals." They are also fond of the berries of the Chinese tallowtree *(Sapium sebiferum)*. "The seeds of this tree, which is originally from China, are of a white colour when ripe, and contain a considerable quantity of an oily substance. In the months of January and February these trees are covered by the Crows, which greedily devour the berries." He adds that they eat pears, and are very fond of ripe figs; they do considerable damage to the latter and have to be driven away from the fig trees with a gun.

According to Mr. Howell (1932), "Scott says that in October the birds congregate in enormous flocks and feed extensively on palmetto berries. Nehrling states that they eat the fruits of the cocos palms. Oranges and tomatoes are sometimes eaten, but apparently the habit is not sufficiently prevalent to result in much damage."

Harold H. Bailey (1913) says that in Virginia considerable damage is done to the peanut crop. "As the farmers turn their hogs into the peanut fields to fatten on the nuts left in the ground after taking off the vines, the Fish Crows thus rob the hogs of a great amount of food, while many pounds of nuts are taken from the stacks while the peanuts are still on the vines drying."

Behavior.—The fish crow does not differ materially in its habits from its better-known and larger relative. Its flight is similar, but it is quicker and more given to sailing, giving a few flaps of its wings and then sailing along for a short distance. It often poises in the air, hovering on rapidly beating wings, as it scans the ground or water beneath it for possible food. When a number of these crows are together, they often indulge in circling maneuvers, flying around in a confusing formation and then straightening out and proceeding on their way. Audubon (1842) writes:

While on the St. John's river in Florida, during the month of February, I saw flocks of Fish-Crows, consisting of several hundred individuals, sailing high in the air, somewhat in the manner of the Raven, when the whole appeared paired, for I could see that, although in such numbers, each pair moved distinctly apart. These aerial excursions would last for hours, during the calm of a fine morning, after which the whole would descend toward the water, to pursue their more usual avocations in all the sociability of their nature. When their fishing, which lasted about half an hour, was over, they would alight in flocks on the live oaks and other trees near the shores, and there keep up their gabbling, pluming themselves for hours. Once more they returned to their fishing-grounds, where they remained until about an hour from sunset, when they made for the interior, often proceeding thirty or forty miles, to roost together in the trees of the loblolly pine.

Fish crows are more sociable and more nearly gregarious in their habits at all seasons than are their northern relatives. They are seldom seen singly; they often nest in small colonies or groups; and wherever there is food to be obtained, especially in the vicinity of heron rookeries, they are always to be found in large numbers. But the biggest aggregations are to be found in the winter crow roosts. M. N. Gist, the warden at the Orange Lake rookery, estimated the winter crow population at that locality as 50,000, some of which may have been Florida crows, according to Mr. Howell (1932), who adds: "At Goose Creek, Wakulla County, in January, 1920, we observed long lines of Fish Crows every morning shortly after sunrise, flying westward along the beach from the direction of St. Marks Light. Several residents of the neighborhood told us that the birds roosted on beaten down tracts of rushes and drift in the marshes along the lower course of the St. Marks River. At Panasoffkee Lake, Crows are said to roost in large numbers in willow bushes in the marsh at the edge of the lake. At Lake Monroe, February 18, 1897, Worthington saw a flock of about 2,000 Fish Crows going to roost in rushes."

At North Island, S. C., early in December 1876, Maynard (1896) saw a great flight of fish crows that he thought were migrating. "They were evidently migrating for they came down the coast in an almost unbroken stream and continued to fly all day. I think I saw more pass the island than I ever saw before. It did not seem possible that there could have been so many of these Crows in existence for they could be counted by tens of thousands." This may have been merely a local movement, for the birds might have been seeking shelter from the hard, cold northeast wind that was blowing at the time; and fish crows are known to spend the winter much farther north.

Voice.—The note of the fish crow is quite different from that of our common crow, shorter, less prolonged, more nasal, staccato, and not so loud; it is hoarser, as if the bird had a sore throat or a cold. I wrote it in my notes as *cor,* or as an exact pronouncing of the word "car." Mr. Wheeler (1922) writes it *caa-ah,* and refers to a two-syllabled note, *ah-uk.* Bendire (1895) says: "Their call notes appear to be less harsh and are uttered in a more drawling manner than those of the Common Crow; they are also more variable. They consist of a clear 'cah' or 'cahk,' repeated at intervals of about thirty seconds, and are usually uttered while the bird is perched on the extreme top of a tree. They also utter a querulous 'maah, maah' or 'whaw, whaw,' varied occasionally to 'aack, aack,' or 'waak, waak.' It is almost impossible to reproduce such sounds accurately on paper, and no two persons would render them alike."

Field marks.—The most reliable field mark for the fish crow is its voice; and this can usually be counted upon to identify it; there is, however, a chance for confusion when young common crows are first on the wing and giving their weak calls. The appearance on the wing is slightly different; the wing of the fish crow seems to be more pointed at the tip of the primaries, and broader at the base, where the secondaries are relatively longer than in the common crow, but the difference is not easily detected. Fish crows are inclined to soar or to hover and are often more gregarious than the common species. The difference in size is an unsafe character, unless the two species can be closely compared.

Enemies.—All small birds hate crows and will drive them away from the vicinity of their nests, for the protection of their eggs and young. I have twice seen red-winged blackbirds attacking fish crows, just as kingbirds attack the larger crows. The herons are the chief sufferers from the depredations of fish crows. Perhaps some of the larger herons may destroy the young of the crows. The following incident is suggestive. Mrs. C. W. Melcher, of Homosassa Springs, Fla., tells me the following story: "One day in spring I heard the raucous cry of the Ward's heron, but with it was mingled an unusual note of distress. I ran to the porch just in time to see the heron fly into the river, where he sank to his body. Close behind him came a fish crow, and, as the heron sank into the water, the crow flew about his head and delivered several telling strokes, the heron meantime emitting loud cries of fright and distress. At last the crow ceased his chastisement and flew away. Then the heron laboriously lifted himself out of the water and flew away squawking. A day or two later I again heard the distress note and ran to look. This time they were in the air, the heron squawking as he flew, with the crow in full pursuit. At intervals for about ten days I saw the same performance."

Harold S. Peters (1936) lists two species of lice, *Myrsidea americana* and *Philopterus corvi,* that have been found on fish crows as external parasites.

DISTRIBUTION

Range.—Atlantic and Gulf coast regions of the United States; not regularly migratory.

The range of the fish crow extends **north** to northwestern Louisiana (Caddo Lake); central Arkansas (Little Rock); central Alabama (Coosada); northwestern South Carolina (Greenwood); central Virginia (Charlottesville); southeastern New York (Rhinebeck); and Rhode Island (Warren). **East** to the Atlantic coast from Rhode Island (Warren) south to southern Florida (Royal Palm Park). **South** along the Gulf coast from southern Florida (Royal Palm Park, Fort Myers, and Apalachicola) to southeastern Texas (Orange). **West to** southeastern Texas (Orange) and northwestern Louisiana (Caddo Lake).

During some winter seasons the species may withdraw from the northern parts of its range, but it is usually found at this season north to Long Island.

Casual records.—There are several records for Massachusetts, both in coastal areas and in the Connecticut Valley. Most of these are in March and April. Reported occurrences north of this State lack confirmation.

Egg dates.—Connecticut: 23 records, May 5 to June 6; 11 records May 10 to 15, indicating the height of the season.

Florida: 7 records, April 6 to May 13.

New Jersey: 106 records, April 29 to June 21; 54 records, May 12 to 22.

Virginia: 50 records, May 4 to June 10; 25 records, May 14 to June 10.

CORVUS FRUGILEGUS FRUGILEGUS Linnaeus

ROOK

CONTRIBUTED BY BERNARD WILLIAM TUCKER

HABITS

The rook is admitted to the American list on the basis of an accidental occurrence in Greenland. Schalow (1904) states that a male was shot by Petersen on March 23, 1901, at Kungarsik, near Cape Dan, on the east coast, and points out that as young birds are not rare as vagrants in Iceland this example may be supposed to have reached Greenland from there. This bird is stated to be preserved in the Museum at Copenhagen.

Few birds are more familiar in England than the rook, as might, indeed, be inferred from the fact that the term rookery, applied to its nesting colonies, is an everyday word in the English language and has been adopted as the most natural term to apply to populous breeding colonies of such different birds as penguins and seafowl and even, rather oddly, of seals. Rooks are a part of the English landscape. Their social habits and predilection for nesting close to human habitations both help to make them familiar, and the bulky nests high in the leafless tree tops of copse or hedgerow make the deserted rookery in winter as conspicuous as it is in spring by reason of the bustle and clamor of its inhabitants.

The rook is primarily a bird of agricultural country with sufficient trees to afford it nesting sites, but not too heavily wooded, for it feeds mainly on open ground, resorting to both pasture and arable land. Moorlands, heaths, marshes, and unreclaimed land are not so much favored, though rooks may be met with at times on all such types of ground and in some districts not rarely.

Rookeries being easily located and the nests large and easy to count with only a small margin of error, the rook is an excellent subject for census work, as a result of which the breeding population of the species over large areas of Great Britain has been accurately determined in recent years. The average density in typical agricultural country is found to be about 16 nests (or 32 breeding birds) to the square mile, with a variation, in cases of areas over 100 square miles, from 5 nests (North Wales) to 33.5 nests (Upper Thames Valley) a square mile.

Courtship.—A fair amount has been written about the sex behavior of the rook. It has been especially studied by Edmund Selous (1927) and G. K. Yeates (1934), and others have contributed miscellaneous observations, but the subject would still repay further attention. The ordinary courtship display is of a simple kind and may often be observed in the opening months of the year. The male bird, either in a tree or on the ground, droops his wings and bows several times to the female with outstretched neck, accompanying the movement with cawing and fanning out of the tail. The female may or may not be disposed to respond to these advances, but, if she is, her reaction usually takes the form of fluttering her wings with the body depressed in a crouching position. She may even make a slight answering bow from time to time, and sometimes a display takes place in which the two sexes behave similarly, both birds fanning their tails and bowing to one another and at least on occasions both fluttering their wings.

Wing-fluttering accompanied by elevation of the tail is the female's normal expression of readiness for coition, and she may take the initiative in soliciting without any preliminary advances by the male. The elevation of the tail is the diagnostic action in this connection, for the wing-fluttering should perhaps be regarded as primarily a food-begging action. Ceremonial feeding and coition are, however, so closely connected that it is doubtful how far actions related to one or the other can be properly dissociated. The display commonly leads up to the presentation of food by the male, and this is generally followed by coition, which in the breeding season normally takes place only on the nest. Sometimes, however (though the contrary has been stated), it may occur on a branch or on the ground. Coition and ceremonial feeding occur well into the incubation period, but the associated display is then hurried or absent. Often the male "simply flies straight down from a bough above on to the female's back, gives her food and simultaneously amidst much wing-flapping mating is accomplished" (Yeates, 1934). It should be observed that during incubation, which is normally performed by the female only, the male habitually supplies his mate with food, but the feeding has then an obvious function to perform as such and is no longer of a purely ceremonial character.

Recrudescence of sexual behavior takes place long before actual breeding commences. Sporadic displays, and occasionally even coition, may be observed in the fall, and both have been recorded as early as October, though at least as regards coition this must probably be regarded as exceptional. From late in November to December, when the weather is mild, coition seems to be not uncommon, and ceremonial feeding has been observed from the end of December. A mutual fondling of bills, which is often observed in mated birds, has also been recorded in fall. These expressions of affection outside the breeding season are probably in the main not merely promiscuous. There is good reason for supposing that rooks pair for life, as is known to be the case in such solitary Corvidae as the raven, and that "married couples" maintain their association in the flocks. Nevertheless a certain amount of promiscuity does take place at the rookeries, in the form of sudden assaults on incubating females by intruding males in the absence of the regular mate. In this connection reference must be made to the violent scuffles between several or a number of birds which occur from time to time in every rookery, centering round a nest. Selous (1927) was the first to recognize that these are, at least generally, attacks by neighbors on a pair engaged in the sexual act. But Yeates (1934), who had the advantage of making his observations from a hide in the tree tops, obtained fairly strong evidence for the unexpected conclusion that the attacks are generally confined to promiscuous mating where the male is an interloper. Although he observed coition repeatedly he found that such mobbings occurred only occasionally, and in three instances observed while he was specially concentrating on this point there was a very strong presumption, amounting to virtual certainty in two of them, that the male involved was not the mate of the female. He writes:

"On these three occasions referred to the evidence in two of them was, to my mind, quite certain. At one nest a male came down, mated and was mobbed and driven off. The very next minute the rightful male came with food. He could not possibly have collected it and returned in the interval. At the second nest mating took place and a very violent mobbing scene resulted (there were four birds attacking). Within a few minutes the rightful male came and again mated—and he was left in peace. The third instance was a little more uncertain—but was of a similar nature to the first."

The subject of these mobbings deserves further study. If Yeates's conclusion is correct—and his evidence cannot be lightly dismissed—it would indicate a very singular state of affairs and one presenting an intriguing problem to students of bird behavior and psychology, since it is certainly not to be explained by attributing a sense of morality to the birds! It should not be overlooked, however, that in various other species of birds the sight of a pair engaged in the sexual act has been observed to have a very provocative effect on other individuals, which are stimulated thereby to interfere in very much the same way that has been described in rooks—and this quite irrespective of the "legitimacy" or otherwise of the nuptial behavior.

What may be regarded as a kind of courtship—or display flight, though its full significance is not altogether clear, may sometimes be noticed, several birds following one another with peculiar slow wing beats. Headlong dives from a height and other aerial evolutions may sometimes be observed at the beginning of the breeding season, but such performances have apparently no sexual significance (except indirectly as an expression of excitement) and are much more characteristic of the fall, under which heading they are further described.

Nesting.—Rooks are colonial birds, normally building their nests in tall trees, and it appears to be a matter of indifference to them whether the trees are in compact groups or scattered. Rookeries may be built on the one hand in trees dispersed over open parkland, fields, or lawns, irregularly spaced along hedgerows or in closer rows forming windbreaks or avenues, or on the other hand may occupy more compact groups of trees forming copses, spinneys, plantations, or small woods or in the shrubberies of gardens. More rarely a rookery may be situated on the borders of a large wood, but it is probably safe to say that no rookery is ever found far inside any extensive and uninterrupted tract of woodland. A great many rookeries are near farms or buildings or in the grounds of country houses. No doubt the tendency to nest near farms may be partially accounted for by the fact that very often a farm has a spinney or a windbreak planted near it, as well as by the better food supply near such settlements, and no doubt the well-timbered character of the grounds of so many country houses is sufficient to account for the occupation of many such sites. But the observer who critically notes the surroundings of a large number of rookeries will probably find it difficult to resist the impression that the birds have often deliberately selected a site near a building though others were available.

Rooks are not averse to nesting in trees in towns provided suitable feeding grounds are available fairly close at hand. On account of its parks Inner London preserved a few rooks until comparatively recently. A small rookery that existed for many years in the garden of Gray's Inn in the City lasted till 1915, its abandonment being variously attributed to disturbance by recruits drilling under the trees or to the depredations of carrion crows. Other details about London rookeries are given by Macpherson (1929).

In hilly or partly hilly districts the main concentration of rookeries will be found in the lowlands, the density of rook population beginning to fall off at a comparatively inconsiderable elevation. In southern and midland England, at any rate, the decrease above about 400 feet is quite definite, though it is evident that the determining factor is the more extensive cultivation of the lowlands, which provide better feeding grounds than the hills, and not altitude as such. In the writer's home county of Somerset quite large rookeries of 200 nests or more occur up to 1,000 feet, and the highest site (with 38 nests in 1933) is at 1,350 feet. On the Carboniferous limestone plateau of Derbyshire nearly all rookeries are between 1,000 and 1,200 feet.

Rooks nest in both broad-leaved trees and conifers, and, provided the trees are of substantial size and of a habit of growth such as will provide suitable lodgement for the nests, they seem to have no special preferences, their choice of trees in any area depending primarily on the relative abundance of the species of tree available. Occasionally rookeries are found with the nests built in saplings of only some 25 to 35 feet in a hedgerow, but such sites are not common. Quite exceptionally nests have been built on buildings such as church spires. Up until 1835 a pair occupied such a nest on the weather-vane of the steeple of Bow Church, in London, and other cases are recorded by Yarrell (1882).

The size of rookeries varies a great deal, from half-a-dozen nests, or even fewer, to hundreds. Rookeries of over a hundred nests are common, but small colonies are most numerous and in most districts the commonest size for a rookery is something under 25 nests. The largest English rookery, to the best of the writer's knowledge, is one of over 600 nests in a small wood in Oxfordshire, but in parts of Scotland there are immense rookeries of over 1,000 and even over 2,000 nests, which have no parallel south of the border. Several colonies of the same order of size (the largest with 2,400 nests) exist in Holland. In both countries it seems possible that the great size of the colonies may be correlated with a certain scarcity of suitable sites in areas not so well timbered as many in which rooks are found.

It should be observed that while some rookeries are compact others may be dispersed in irregular groups over a considerable area, and in dealing with these latter the student is soon confronted with the question, What constitutes "a rookery"? Groups situated close to one another are obviously essentially parts of a single colony, but what degree of separation justifies us in speaking of two groups as two distinct rookeries? Many factors have a bearing on the problem, which is too complex to discuss here, but it may be said that in the present state of knowledge any criteria adopted must necessarily be somewhat arbitrary, and the point is mentioned mainly in order to stress that statements as to the dimensions attained by individual rookeries will be affected in a certain proportion of cases by the criteria employed. But for the most part the large colonies mentioned would be reckoned as single rookeries on any reasonable scheme.

At the opposite extreme to these giant colonies, isolated nests are not very rare, but as a rule they have little permanency, and it has been repeatedly stated that isolated nests situated at all close to a regular rookery are liable to be raided by the occupants of the latter, such individualistic tendencies being regarded with disfavor in rook society. But more critical observation on this point seems desirable.

Rooks are early breeders, but, although the rookeries are visited from time to time in the winter and more frequently as the season advances, the serious business of nest building or repair rarely begins before late February. The nests are generally built in the topmost, slender branches of trees, often several or even a considerable number in one tree, and are frequently very difficult or almost impossible for even a good climber to reach, though an enterprising egg collector or photographer will not as a rule have great difficulty in finding a rookery where the nests are more accessible. They are built of sticks, solidified with earth and lined with grasses, dead leaves, moss, roots, straw, and the like. Wool and hair, beloved of the carrion crow, are only quite exceptionally employed and then only in small quantities. Exceptionally the use of feathers in quantity has recently been recorded, where a plentiful supply happened to be available close at hand. Both sexes build, the cock doing most of the collecting and the hen most of the arranging. The sticks are broken off from trees or pillaged from other nests and are not collected from the ground. Even a stick accidentally dropped is not ordinarily retrieved, though rarely one may be picked up again. Eggs are usually found in the latter half of March or early April. The task of incubation is performed solely by the hen, to whom the cock brings food at the nest, and sometimes, but not always, begins with the first egg laid. As an exception a case of a male taking a turn on the eggs has been recorded (Nethersole-Thompson and Musselwhite, 1940). The Rev. F. C. R. Jourdain (1938), from whom some of the above details are also quoted, states that the incubation period is 16-18 days. Not all the nests are built at once; a few may go on being added far into April, perhaps by young birds coming into breeding condition for the first time.

Eggs.—The eggs are of the usual *Corvus* type, but rather small. The following condensed description is given by the Rev. F. C. R. Jourdain in the work already quoted:

"Usually 3-5, sometimes 6, rarely 7; 8 and 9 recorded: vary much as Crow's: ground-colour from light bluish-green to green and greyish-green but never so blue, and rarely show much of ground-colour, being more uniformly marked with shades of ashy-grey and brown. Erythristic varieties have occurred abroad and also Northern Ireland (H. T. Malcomson). Average size of 100 British eggs, 40.0 x 28.3, Max.: 47.1 x 26.2 and 39.5 x 30.5, Min.: 35.1 x 27.1 and 44.2 x 25.8 mm., averaging smaller than Crow's."

Young.—The fledging period of the young is given by S. E. Brock (1910) as 29 to 30 days. Yeates (1934) found that for the first 9 or 10 days the male collects all the food, but after this the female also takes a share in the work. The food is brought in the capacious pouch under the tongue, which produces a prominent swelling under the chin, conspicuous in birds coming into the rookery after a foraging excursion. When the time comes for fledging the young birds, according to Yeates's observations, are enticed out of the nest by the female, who refuses to bring food to the nest but waits with it in the branches, calling to the chicks until they venture out. They remain for some time in the branches until their wings are strong enough to support them and finally join the old birds in the fields, where the parents continue to feed them at first. Mortality before leaving the nest is severe. Four young hatched may be taken as a fair average, but counts quoted by Yeates show that on an average at most only two chicks to a nest are fledged. Only one brood is normally reared, but cases of breeding in fall and even as late as November have been recorded.

Plumages.—The plumages and molts are fully described by H. F. Witherby in the *Handbook of British Birds* (1938).

Food.—The food of the rook is thus summarized by Jourdain (1938): "In agricultural areas corn is staple food, but potatoes, roots, fruit, acorns, walnuts, peas, berries, and seeds also recorded. Animal food includes insects: Coleoptera (many injurious species), also larvae, larvae of Lepidoptera, Dermaptera, Orthoptera, Hymenoptera, Hemiptera, Diptera, and larvae, etc.: also earthworms, Mollusca (snails and slugs) : millipedes, spiders. Carrion (dead lambs, etc.) occasional: birds killed in hard weather and young or eggs frequently taken (game-birds, duck, Stone-Curlew, Lapwing, and many species of small birds) : small mammals (mice, shrew, young rabbit): small fish also recorded."

W. E. Collinge (1924), who analyzed the contents of 1,306 stomachs, found that the food was made up of 59 percent vegetable matter and 41 percent animal matter. The vegetable matter was made up of 35.1 percent cereals and 13.4 percent potatoes and roots, and only 6.1 percent of items considered "neutral" from an agricultural standpoint. Of the animal total 28.5 percent consisted of forms considered injurious to agriculture (the great bulk being injurious insects), while 3.5 percent consisted of beneficial insects and 9 percent of "neutral" insects and earthworms. On this basis 52 percent of the rook's food consumption inflicts damage to agriculture, and only 28.5 percent is beneficial. The influence of the rook on British agriculture has been much discussed, but until recently all such discussions suffered from the vital defect that the crucial information on the actual density of the rook population was lacking. But thanks to recent accurate census work over large tracts of agricultural country in England, taken in conjunction with the requisite statistics on crop production and Collinge's data already quoted, it can now be stated with considerable confidence that in spite of widely held opinions to the contrary the numbers of rooks are nowhere sufficient to exert any important effect on the crop production of the country, or of any considerable district, as a whole, and this may be stressed as a good illustration of how census work undertaken primarily for its biological interest may have a very real economic value.

Behavior.—The ordinary gait of the rook, like that of other crows, is a sedate walk, which may be interrupted by one or two less dignified hops if the bird is bent on securing some tasty morsel ahead of a competitor or otherwise feels the need of hurry. The flight, when making for a definite objective, is direct and deliberate, with regular wing beats varied only by gliding where air conditions are specially favorable, as when crossing a valley or when about to pitch. But on more desultory flights about the feeding grounds and around or above the rookery more irregular wing action and a good deal of gliding may be observed. Eminently gregarious, rooks are usually seen in parties or flocks, which may be of large size, differing therein from the carrion crow, which is much more of an individualist, though it must not be supposed that either single rooks or flocks of carrion crows are very unusual.

Voice.—The cawing of rooks, though it cannot be called melodious, is to most lovers of nature a pleasant sound, redolent of the country. The ordinary note may be described as a hoarse *kaah*, and typically it is of somewhat lower pitch than the more raucous croak of the carrion crow, but actually the rook's voice has a much greater range of pitch than the crow's. According to Nicholson and Koch (1936), who prepared gramophone records of the voices of many British birds, the rook's vocabulary has a range of about 425-1,800 cycles a second, while the crow's range is only about 600-750. The ordinary caw is subject to considerable modulations and variations, as anyone may observe who listens to the sounds in a rookery in spring or at a roost in winter. There are also other more or less distinct notes, and Edmund Selous (1901) has recorded over 30 sounds, but his list is unsatisfactory, as it includes too many sounds that seem to be little more than variants of the same essential note. I have shortly mentioned some of the more distinct notes in the *Handbook of British Birds* (1938), but a systematic and critical study of the rook's vocabulary has yet to be made. On occasions of exuberance in spring rooks are sometimes moved to a kind of uncouth attempt at song. As I have noted in the work referred to, one such performance is described as resembling a bass or guttural reproduction of the varied and spluttering song of the starling, and other more or less discordant variations have been noted.

Field marks.—The adult rook is easily told from other black-plumaged Corvidae by its bare grayish-white face. But this characteristic is not acquired until the first summer after hatching, and young birds with fully feathered faces are not easily distinguished from carrion crows. The bill of the rook is rather slenderer and the culmen usually less curved, but this is not a very good character, as it is easy to find young birds of the two species whose bills are so alike that they could not possibly be distinguished by this feature in the field. A better character is the loose baggy appearance of the feathering of the thighs, which is noticeable at all ages.

Enemies.—The rook has no enemies serious enough to constitute in any way a menace to the species. The chief enemy is man, who destroys a certain number because of depredations on his fields and at many of the larger rookeries organizes regular rook shoots in spring, at which considerable numbers of young birds recently out of the nest are detroyed. These find their graves in rook pies, which are generally agreed to be excellent eating, though the writer has not tried them. The peregrine falcon, and on the Continent the goshawk, will sometimes take a rook, and the golden eagle has also been recorded as doing so, but this can happen only rarely, for the haunts of eagle and rook do not overlap to any great extent and the eagle is nowhere common. The common buzzard *(Buteo buteo)* and European sparrow hawk *(Accipiter nisus)* are recorded as taking young birds occasionally, and if the opportunity offers the carrion and hooded crows are not above raiding the nests of their relatives. As shown by Elton and Buckland (1928) a very high percentage of young rooks may be infected with gapeworms morphologically identical with *Syngamus trachea* of poultry. Ninety-four percent of 33 young birds examined were infested, mostly rather heavily, but adults appear to be only slightly parasitized, so that it is probable that infected young birds gradually rid themselves of the parasites as they grow older, as happens in chickens. Cram (1927) records *Corvus frugilegus* as host to the following nematode parasites: *Syngamus trachea, Acuaria anthuris, A. cordata, Oxyspirura sygmoidea*, and *Microtetrameres inermis*. G. Niethammer in the *Handbuch der deutschen Vogelkunde* (1937) gives a list of ecto- and endoparasites obtained from the rook.

Fall and winter.—Outside the breeding season rooks are as gregarious as when nesting, feeding together in flocks by day and roosting collectively at night, and their habits during this period of the year are of much interest. The general outlines of their behavior are tolerably constant, but details vary considerably according to local conditions and other factors. Their foraging excursions are not confined to the fields, for after breeding they will resort regularly to woods, more especially of oak, to feed upon caterpillars, as well as to the outskirts of the moors in hilly districts.

As the season advances the scattered parties tend to fuse together into larger flocks, but in most districts it is noticeable that these flocks contain remarkably few young birds. It is clear from several lines of evidence that from causes still not precisely determined there must be a very heavy mortality among the young during July and August. Careful observations by J. P. Burkitt (1936) in Ireland led him to place the number of young birds surviving in winter at as small a figure as about 10 percent of the number of adults, and this figure seems to be reached as early as August. But the proportion of young to adults in flocks seen in the fields is commonly even less, and many flocks consist of adults exclusively. Some observations of the same author suggest that the birds of the year independently of the adults (though commonly in company with jackdaws) tend to form parties that wander far away from the rookeries, but the subject of the habits of young rooks is one in need of further careful study.

At least in some districts from early in summer until fall parties of rooks are apt to roost in any convenient wood or copse near which they happen to find themselves, but in other cases the birds may continue to roost at their own rookery or may join up into larger roosting assemblies comprising the occupants of more than one breeding colony.

The really large winter roosts are not as a rule at full strength before September or early in October, though varying numbers may roost at these sites from the close of the breeding season. Indeed, the winter roost is normally also the site of an existing large rookery, and when it is not there is reason to believe that it is usually, if not always, the site of a former colony, as is known to have been the case in various recorded instances. [1] These large roosts are often of many years' standing and comprise thousands of birds drawn from a large area. They are commonly shared with jackdaws *(Corvus monedula)*, and collectively they account for the bulk of the winter population of rooks, but in most districts there are also smaller subsidiary roosts, and here and there the members of a single rookery may roost there throughout the winter. The feeding territories corresponding to individual roosts are generally fairly constant in a broad way, but in the absence of any natural boundaries like the sea or a range of hills have apparently no very clear-cut frontiers as a rule, and there is considerable evidence that in some places flocks from different roosts may mingle amicably on the same feeding grounds.

The dispersal from the roost at daybreak is rapid, and the flocks proceed in a direct and purposeful manner to their feeding grounds, though often with a pause to visit their own rookery for a short time. The return journey, on the other hand, is usually of a markedly leisurely character. The birds begin to move in toward the roost from the more outlying areas at a comparatively early hour, and in Northumberland Philipson (1933) found that by about 1 o'clock the peripheral areas of the feeding territory were usually deserted. In the less remote areas flocks may be seen feeding in the fields considerably later, but there is a gradual move toward the roost, often in two or more stages with an interval of feeding between. Very commonly the adult birds assemble at their own rookery before moving off, and as the flocks and parties converge toward the roost they frequently combine into larger bodies. In the neighborhood of the roost there are often regularly used collecting grounds at which large numbers of birds gather and feed for a time before the final move, and in the fields immediately adjacent to the dormitory the earlier arrivals resume their feeding in a restless way for a time before retiring. As more and more birds stream in excitement rises, and the observer is treated to astonishing displays of acrobatics in the air before the fresh arrivals settle. The fields are now black with birds and at last about dusk the whole body rises as if at a word of command, and with much cawing the procession to bed begins. Above the wood the

---

[1] It will be understood that this applies to the British Isles and cannot be the case in southern Europe, where the rook occurs only as a migrant.

stream of birds breaks up again into groups and parties, and after one or two preliminary circlings the birds pour down into the trees with a renewed crescendo of cawings and other strange noises before finally settling down for the night.

The aerial evolutions referred to afford as impressive and exciting a spectacle as an ornithologist can wish to see and reveal an agility and a capacity for frolicsome behavior that are not a little surprising in these usually sedate birds. They plunge and dive headlong from a great height, roll, tumble, sideslip, and chase one another in swerving and switchback flights like a veritable avian circus troupe. Such displays, in which immature as well as adult birds participate, are not confined to roosting time, but may be seen also earlier in the day, especially in September. As mentioned under "Courtship" something of the same kind may also be observed in spring, but it is much more characteristic of the fall. It is an expression of a sort of general *joie de vivre* without sexual significance, and if the scientist needs to classify it somehow it must be reckoned as a form of social, not sexual, behavior.

Allusion has been made above to visits to the rookery during the period under discussion. Though occurring, especially, early in the morning soon after daybreak and late in the afternoon they are by no means confined to these times, for the birds often feed close to their breeding places and pay visits to the trees during the day. As the season advances these visits become more frequent and more prolonged, and early in the year signs of a renewed interest in building and nest repair may be observed, as already described, but the winter roosts are not generally abandoned till about mid-March. The mode of abandonment seems to vary; it may apparently take place all at once or be spread over some days up to ten or more, but sooner or later the birds are all roosting again at their own rookeries and the annual cycle begins once more.

DISTRIBUTION

[Acknowledgment is made of assistance derived in the compilation of this section from the sections on "Distribution Abroad" by F. C. R. Jourdain and on "Migration" by N. F. Ticehurst in the *Handbook of British Birds* (1938).]

Breeding range.—Occurs throughout the British Isles wherever there are trees for it to breed in, but not in the Shetlands. Breeds in greater part of Continental Europe from latitude 63½° N. in Norway, 60° in Sweden, 62° in Finland and northern Russia, east to Perm (eastern Russia) and south to mid-France, northern Italy, Serbia, Bulgaria, and southern Russia. Allied races are found in west temperate and eastern Asia.

Winter range.—Breeding birds are sedentary or mainly so in temperate parts of range, but migrate from Scandinavia, north-central Europe, and colder parts of Russia. The winter range extends south to Spain and Portugal, central and south Italy, Sicily, Corsica, and Sardinia, the Balkan Peninsula, Cyprus, and southwestern Asia.

Spring migration.—Birds from northern and eastern Europe that have wintered in England leave the east coast between mid-February and the third week of April, the movement being at its height on the coast of Suffolk at the end of March and early in April. Banding records show that these birds come from as far away as northwestern Russia and as near as Holland, as well as from intermediate regions, including northern Germany, east Prussia, and Lithuania. Many birds, and at times very large numbers indeed, pass over Heligoland on the spring migration, the earliest date recorded by Gätke being February 4, and many pass through Germany in March and April on their way to the Baltic Provinces and Russia. Migrants in Italy and Greece leave in March and early April, the latest date recorded by C. J. Alexander for the Rome district being April 7. In central Spain rooks are stated to be (at least locally) common until March and in the north a flock has been observed as late as April 18.

Fall migration.—In fall (from mid-September) the main movement of rooks on the Continent of Europe is westward from the northeastern, eastern, and central regions toward France, Belgium, and eastern England. The birds that winter in the last-named area mostly arrive from the end of September to the third week of November. Alexander's (1927) earliest date for arrival in the Rome district was October 28. In western Greece, according to some observers, rooks arrive in October, and Lord Lilford (1860) says toward the end of the month, but about mid-November seems more usual. In central Spain rooks are described as being present from November.

Casual records.—Recorded occasionally or casually from east Greenland, Iceland, the Faeroes, Novaya Zemlya, northwestern Siberia, the Azores, Madeira, Balearic Islands, Malta, Algeria, and Egypt.

CORVUS CORNIX CORNIX Linnaeus

HOODED CROW

CONTRIBUTED BY BERNARD WILLIAM TUCKER

HABITS

The claim of the hooded or gray crow to a place on the American list rests on its casual occurrence in east Greenland, where Schalow (1904) states that several examples have been collected by Danish zoologists at Angmagsalik, having presumably wandered from the Faeroes by way of Iceland.

The hooded crow and the black carrion crow *(Corvus corone)* are exceedingly closely allied; indeed they appear to differ in no structural feature whatever, but only in coloring. Moreover, where their ranges overlap they regularly interbreed, producing fertile hybrids showing every intergradation of color. These considerations have led some ornithologists to argue that they ought to be treated as races of one species, and there is undoubtedly a great deal to be said for this view. At the same time, no competent biologist nowadays can doubt that racial differences frequently provide the material out of which specific differences evolve, and, this being so, the decision as to the precise point at which two diverging races become deserving of specific rank must necessarily be somewhat arbitrary. In the present case the two forms seem to be so near the border line that considerations of convenience may be legitimately allowed to carry some weight, and as it is distinctly more convenient for descriptive purposes to consider them as separate species, this practice, as followed by both the A.O.U. Check-list and the new *Handbook of British Birds,* may be the more readily accepted.

The distribution of the two birds is curious, for the range of the hooded crow cuts that of the carrion crow completely into two parts separated by many hundreds of miles, the carrion being the crow of southwestern Europe, and absent from the rest of that Continent and from western Asia, but reappearing in central, eastern, and northeastern Asia. The ranges only overlap along relatively narrow zones of contact, a circumstance justifiably stressed by those who would regard the two forms as races of one species, and in these areas, as mentioned above, they interbreed more or less freely. The hooded crow might be regarded as a geologically more recent form which arose in the intermediate area and has replaced the black form over a part of its range.

The hooded crow frequents both cultivated and uncultivated country, which may be fairly well timbered, but not densely wooded, or quite devoid of trees. It is often seen on the seashore or along the borders of lakes and estuaries.

Courtship.—The following account has been given by C. and D. Nethersole-Thompson (1940):

While watching a flock of some fourteen birds on December 9th, 1938, the trait observed by Miss E. V. Baxter and Miss L. J. Rintoul was noted, viz., repeated jumps into the air and descent to the same place. Chough-like dives with wings nearly closed, fluttering with dangling legs and upturned wings of males(?) over females(?), and low skimming flights by flying over perched birds were also observed. During these performances there was much croaking from time to time. Display-flight of the male (20.3.38) is a series of short dives with semiclosed wings, and sometimes a half-roll, bird croaking continuously. Aerial display in spring also includes "corkscrew" or figure-of-eight flight high in the air by a pair of birds. This is also accompanied by much calling. Another form of posturing several times watched by us shows the male bird, while perched on a tree bowing to the female. In this display, carried out several times in succession, wings are spread and tail expanded. In April, 1940, we watched coition of Hooded Crows take place on a bare, burned patch of moorland. Prior to the act the male for over a minute violently shook his raised wings and this he also continued after coition.

It appears from this description, as might be expected, that the courtship and sex behavior of the hooded crow differ in no respect from those of the carrion crow.

Nesting.—The hooded crow will nest either in trees or on some ledge of a coastal cliff, generally protected by an overhang of the rock. Unlike those of the rook, which likes to build in the slender topmost branches, tree nests are generally placed in a comparatively stout fork. Inland, even in localities where trees are rare or absent, it rarely builds in cliffs or rocks but quite frequently places its nest in bushes, sometimes quite low. Ussher and Warren (1900) state that in the west of Ireland, where it habitually breeds in bushes in this way, it usually selects one on an island in a lake, no doubt for greater security. Occasionally a nest may be built on the ground among heather on low islets in lakes.

The nest, which is built by both sexes, the male doing most of the collecting and the female most of the building, is described by Jourdain (1938) as "strongly constructed of sticks, heather-twigs, seaweed-stems, moss, and earth, lined wool, hair, and sometimes feathers." These crows usually employ a quantity of wool in lining their nests, in contrast to the rook. Jourdain also mentions that in the Shetland Islands and northern Ireland large bones have been found in the foundations of nests.

Eggs.—The following condensed particulars are given by Jourdain (1938): "Usually 4-6, very rarely 7; much resemble Carrion-Crow's; ground-colour varying from light blue to deep green, generally blotched and spotted over whole surface with shades of umber-brown, and underlying ashy shell-marks. Some have only few markings on blue ground or are entirely devoid of them: one light-coloured egg in clutch not uncommon. Erythristic variety recorded once or twice. Average size of 100 British eggs 43.5 x 30.3. Max.: 52.0 x 32.0 and 43.4 x 33.0. Min. 39.2 x 29.8 and 44.3 x 28.4 mm."

Eggs may be found at the end of March and in April in the British Isles.

Young.—The incubation period is 19 days and is performed solely by the hen bird, who is fed by the cock. The feeding may take place at the nest or the female may leave the nest to meet her mate and stretch her wings a little. Quite exceptionally, it would seem, the male may take a turn on the nest; a change-over at the nest has been observed in the British Isles by J. Walpole-Bond, quoted by Jourdain (1938). Incubation normally begins with the first egg. The nestlings are fed by both parents, but chiefly by the hen and are fledged in four to five weeks.

Plumages.—A full account of plumages and molts is given by H. F. Witherby in the Handbook of British Birds (1938).

Food.—The hooded crow is a rapacious and destructive rascal, hated by farmer and gamekeeper alike. Hardly any kind of animal food comes amiss to it, and it is an inveterate egg stealer. Carrion and refuse of all sorts are greedily devoured. It will forage diligently on rubbish tips and among tidal refuse on the shore or feast more sumptuously on a carcass when it can find one. Wounded or sickly animals and birds fall victims to it, and occasionally some healthy bird may be pounced on and laid low by a vicious blow from the bill, or even pursued and captured in fair chase. The young of ducks and other birds form a comparatively easy prey, and it will range the moors systematically in search of nests with eggs. Occasionally small mammals such as mice and voles, and even young rabbits and hares, are captured. Shepherds accuse it of attacking lambing ewes and new-born lambs, and even larger lambs or other animals that it can take at a disadvantage may succumb to its attacks. A contributor to the journal British Birds some years ago described how two hoodies set upon a lamb some weeks old, which had got into difficulties in a marsh, and quickly killed it (Simpson, 1926). When the observer rushed up from a distance of about 400 yards both eyes of the unfortunate animal had been pecked out and it was dying, apparently from injuries inflicted on the brain through the eye sockets. Such tragedies are probably not rare. Frogs are also sometimes killed and even small fish are recorded, so that no class of vertebrates is immune from its attentions. Among invertebrates, Mollusca, including Cardium, Mytilus, Tellina, Patella, Buccinum, Littorina, and Purpura, among marine forms, as well as the fresh-water Anodonta and land snails (Helix), insects (chiefly Coleoptera, but also moths, Trichoptera, and larvae of Odonata, Plecoptera, and Diptera), spiders, sea-urchins, small Crustacea (sandhoppers and crabs), and earthworms have been recorded, and some vegetable matter, such as grain, potatoes, and turnips, is also taken. A number of the above details are quoted from Jourdain's summary of published data in the Handbook of British Birds (1938).

A special habit of the hooded and carrion crows, well known to modern ornithologists but recorded of the present species in Ireland more than 800 years ago by the twelfth-century theologian, traveler, and writer Giraldus Cambrensis, is that of carrying up mollusks and crabs and dropping them on the shore or on rocks or shingle in order to smash the shells. Critical observers have not generally considered that they exercise any intelligent selection of hard as opposed to softer surfaces for this purpose; nevertheless there is evidence that in some places they have learned to utilize masonry or walls for their operations. The subject is discussed by Oldham (1930).

Voice.—The ordinary note of the hooded crow is a raucous croak. It is the writer's considered opinion that there is no constant difference between the voices of the hooded and carrion crows, and in this he is supported by H. G. Alexander (MS.), a most careful and experienced ornithologist with an excellent ear. Various observers have professed to be able to detect a difference, but as to its nature their accounts are not consistent. The question was also discussed some years ago in German ornithological journals and the conclusion reached after critical investigation was the same as that just stated. Reference should be made to Meise (1928) and Kramer (1930).

The note of the crow is not only more raucous but typically rather less deep than the typical caw of the rook, but the range of pitch of the crow's vocabulary is very much less than the rook's, namely about 600-750 cycles a second as against 425-1,800 (Nicholson and Koch, 1936). Nicholson further stresses a greater resonance, more deliberate timing, and tendency to repeat three to four times with a long pause before the next repetition. The common note might be rendered as *kraah,* with an exaggeration of the nasal sound heard in such words as "twang," and also *kraarrr.* A higher-pitched *keerk, keerk, keerk, konk, konk* irresistibly suggesting a distant motor horn, a short, rattling note of aggression, and others may also be heard from the carrion crow, and there can be little doubt that they are common to both species. The croak and variants are sometimes repeated a great number of times from a tree top or some other elevated position, and at times a succession of croaking and bubbling notes functioning as a kind of song may be heard. The brooding female on the nest sometimes produces a very soft "crackling" sound, which may continue uninterrupted for several minutes and has been called a "nest-song" (R. Zimmermann, 1931).

Behavior.—The behavior of the hooded crow is much like that of other typical corvids. Its gait is a sedate walk, varied by perhaps two or three rather ungainly hops if hurried. The flight is direct, with rather deliberate wing beats. Where air conditions are favorable, as in crossing a valley or about coastal cliffs, gliding flight may be employed to some extent, but typically the wing beats are regular and uninterrupted. It does not soar like the raven.

Though breeding in isolated pairs and usually reckoned a solitary species, the hooded crow may be seen in parties outside the breeding season and even at times in flocks, while it shares with many other Corvidae the habit of occupying communal roosts, more fully referred to under "Winter," which may attain large dimensions where the species is numerous. Even in the more solitary Corvidae distinct social tendencies are observable, and assemblies of crows are sometimes clearly something more than merely fortuitous gatherings for feeding purposes. In the case of the carrion crow gatherings have been observed in which in addition to chasing or hopping among the branches of trees there are pursuits and maneuverings on the ground, sometimes with a distinct formality about them. Such behavior does not seem to have been expressly recorded in the case of the hooded crow, but in view of what has already been said as to the identity of the habits and behavior of the two species in other respects it can hardly be doubted that the same sort of thing occurs. It is difficult to assign any function to these gatherings beyond the quite general one of affording some outlet to the social urge that seems to be common in varying degrees to practically all corvine birds.

For the rest the general habits and behavior of this species are evidently closely similar to those of the American crow. Like that species, it is a wary, cunning, and intelligent bird.

Field marks.—The gray body contrasting with the black head, wings, and tail at once separates the hooded crow from the carrion and American crows.

Enemies.—Man is the chief enemy of the hooded crow, and it has been mentioned that it is very unpopular with gamekeepers, shepherds, and poultry farmers. It seems to be little interfered with by birds of prey, though no doubt the more powerful species like the peregrine will take one occasionally, and the golden eagle has also been recorded as doing so. Cram (1927) records the following nematodes parasitic in *Corvus cornix: Porrocaecum semiteres, Acuaria anthuris, A. cordata, A. depressa, Tetrameres unispina,* and *Physaloptera malleus.* A list of ectoparasites and endoparasites recorded from the species is given by Niethammer (1937).

Fall and winter.—It has been mentioned above that outside the breeding season hooded crows may be seen in parties and flocks as well as in pairs or singly, and where the species is plentiful it gathers at night in large communal roosts. The writer knows of no record of such gigantic roosts as have been described in the case of the American crow, but nevertheless they may attain considerable dimensions. Thus, roosts of many hundreds are recorded in the Outer Hebrides, where the birds generally sleep on the ground among heather on islands in lochs. Elsewhere they roost on trees and on the Continent roosts on high buildings are also recorded.

DISTRIBUTION

[Acknowledgment is made of assistance derived in the compilation of this section from the sections on "Distribution Abroad" by F. C. R. Jourdain and on "Migration" by N. F. Ticehurst in the *Handbook of British Birds* (1938).]

Breeding range.—Breeds in the more northerly part of Scotland and in Ireland; in England only occasionally and chiefly in the eastern counties. Outside the British Isles it breeds from the Faeroes, Scandinavia, Finland, and north Russia south to Denmark, Germany east of the Elbe, Czechoslovakia, western Yugoslavia, Italy, Sicily, Austria, Hungary, southern Russia to the Caucasus, and north Persia. Allied races breed in Corsica, Sardinia, the Balkans to Palestine and Egypt, southeastern Russia (?), Cyprus, Crete, Iraq, Iran, western Siberia, etc.

Winter range.—Birds from the colder parts of the range migrate at the approach of winter and spread southwestward into western Germany, the Netherlands, Belgium, and France and to eastern and southeastern England. These winter visitors in western Europe are commonest in the coastal districts. Birds in the more southern parts of the range are sedentary.

Spring migration.—Hooded crows that have wintered in England leave the east coast between mid-March and the third week of April, with extreme dates as early as mid-February or as late as May 24. The Atlantic coasts of France are forsaken about the same time. Great numbers pass over Helgoland and the Kurische Nehrung, referred to more fully in the next section. At the latter place the movement begins in February or March according to weather conditions and lasts well into April, while the passage of smaller numbers continues until May.

Fall migration.—The most famous locality for the passage of hooded crows is the Kurische Nehrung, the narrow strip of land about 50 miles long separating the Baltic from the Kurisches Haff in east Prussia and on which is situated the ornithological station of Rossitten. Here almost incredible numbers pass on the fall migration, and as the birds travel by day and commonly quite low down the migration is very spectacular. According to Niethammer (1937) it begins about September 22 and lasts to the end of November, with the peak from October 14 to 25. The young birds appear first and then the adults. Frequently according to weather conditions later waves may pass even until January. Large numbers also pass over Helgoland in autumn. On the east coast of England hooded crows arrive from about the second week in October to the third week in November, but occasionally as early as August 5.

Casual records.—In addition to the Greenland occurrences hooded crows are recorded casually from Iceland, Novaya Zemlya, and Spitsbergen among the northern countries and from southern Spain, northwestern Africa, and Egypt in the south.

CYANOCEPHALUS CYANOCEPHALUS (Wied)

PINYON JAY

HABITS

"Maximilian's jay," as it was called by some of the earlier writers, "was discovered and first described by that eminent naturalist Maximilian, Prince of Wied, in his book of travels in North America, published in 1841," according to Baird, Brewer, and Ridgway (1874). "Mr. Edward Kern, who was connected with Colonel Fremont's exploring expedition in 1846, was the first to bring specimens of this interesting and remarkable bird to the notice of American naturalists, transmitting them to the Philadelphia Academy." The bird is now known to have a wide range in the Rocky Mountain region and in the Sierras and Cascades, farther west, breeding chiefly in the pinyon and juniper belt, but wandering erratically over much of the intervening and adjacent regions at other seasons.

Blue crow, a common local name, seems most appropriate for this bird, as it greatly resembles these birds in many of its actions and habits. Its blue color gives it its closest resemblance to a jay, but its short tail and its highly gregarious habits, together with its nomadic tendencies, are hardly jaylike. Its systematic position seems to ally it more closely with the crows than with the jays.

The foothills and lower mountain ranges, where the slopes are covered with a scattered growth of nut pines or pinyons *(Pinus edulis)* and junipers *(Juniperus occidentalis),* with perhaps an undergrowth of sagebrush, are the favorite haunts of the pinyon jay, especially during the breeding season. Here large straggling flocks of these short-tailed blue crows may be seen trailing over these stunted open forests, looking for their favorite food in the nut pines, or building their nests in the low trees. But they may be here today and gone tomorrow in their restless wanderings.

Migration.—The pinyon jay throughout most of its range is not really a migratory species, though it has been reported in flights that looked like migrations. These flights, which occur in both spring and fall, seem to have no definite north or south direction at either season but are quite as often seen moving either east or west, or in other directions. These are probably not migrations in the strict sense of the word but rather mass movements to or from breeding grounds or from one feeding ground to another. The fact that the birds fly in large flocks, and often for a long time in one direction, leads to the impression that they are migrating. There is, however, a limited migration in the northern portion of its range, where Bendire (1895) says that it is "only a summer visitor, migrating regularly."

Nesting.—Pinyon jays prefer to breed in large or small colonies on the foothills or mountain slopes below 9,000 feet, placing their nests in the pinyons, where they straggle down the slopes toward the desert scrub areas, or in junipers, which grow in such places, or in scrub oaks; the nests are usually not more than 10 or 12 feet from the ground and often lower, though James B. Dixon tells me that "one colony had nests high in the pines, 25 to 50 feet off the ground." J. C. Braly (1931) found a large colony nesting near Grandview, Oreg., of which he says: "These nests were all in small junipers from three to seven feet above the ground. During our investigations [elsewhere?] we found over fifty nests of these birds, the great majority in juniper trees from three to eighteen feet up, while a few nests were found in yellow pine trees up to eighty-five feet." Sometimes three occupied nests were found in one tree.

In Dawson County, Mont., E. S. Cameron (1907) saw no evidence of these birds breeding in colonies and found only two nests, one of which he describes as follows:

The pair were first noted to be carrying twigs on May 19, at which date the nest was about half-finished, both birds assisting in its construction. Without the guidance of the birds it is unlikely that I should have found the nest at all, placed, as it was, near the extremity of a thick pine bough and completely screened from observation except from above within the tree. The nest was of large size with a smaller interior cup, the whole of the exterior, together with a platform on which the cup rested, being composed entirely of dead greasewood sticks and a few rootlets. The width across the sticks was 14 inches, and the height of the nest 8 inches. The cup was very strongly made of dead grass, pulled by the birds into a material like tow, and so thickly matted together, that it remained intact when nearly all the surrounding sticks had been blown away. Some dead thistle leaves were woven into the rim. The inner cup was 5½ inches in diameter and 2½ inches deep.

Mrs. Bailey (1928) says that the nests are "deep, bulky, and compactly built, with a framework of twigs and shreds of bark supporting the deep, well felted cup; made variously of finer shreds of bark, plant fibers, fine rootlets, weeds, wool, hair, dry grass, and a few feathers."

She mentions a few nests found by J. Stokley Ligon in New Mexico: "He says they nest generally from March 1-31, in gray live oaks among the pinyons, though occasionally in pinyons, even where the oaks can be had. * * * On February 17, while the ground was still half covered with snow, on the southwest side of Black Mountain in the Datil Forest, at about 7,500 feet he found one nest about complete and others under construction, in scattered scrub oaks on a steep grassy canyon side. There were more than fifty birds in pairs and flocks mingling and scattering and flying about noisily. On March 3, he returned to the colony and found nests in almost all the scrub oaks of sufficient size, but never more than one in a tree. One, half completed, was in a juniper. The birds, slow to leave their nests, finally did so noisily. As it had snowed many times since his first visit, the nests were damp from melted snow. Nearly all contained four eggs, but one had five." In the same general region, and at about the same elevation, he discovered a second, smaller nesting colony on March 4; and a third colony of perhaps 150 birds was found on March 28; many of these nests held eggs.

Major Bendire (1895) tells us that "the first nests and eggs of this species were found by Mr. Charles E. Aiken, near Colorado Springs, Colorado, on May 13, 1874. * * * The first naturalist, however, who observed the nests and young of this species was Mr. Robert Ridgway, who found a colony nesting in a low range of piñon-covered hills in the vicinity of Carson City, Nevada, on April 21, 1868."

Col. N. S. Goss (1891) reports that his brother, Capt. B. F. Goss, took nine sets of eggs of the pinyon jay in May 1879 near Fort Garland, Colo. "The nests were all in high, open situations, two of them well up the steep mountain sides, and none in valleys or thick timber. All were in small piñon pines, from five to ten feet up, out some distance from the body of the tree, and not particularly well concealed."

J. K. Jensen (1923) says that "the nesting season extends from February to June, during which time fresh eggs may be found," in New Mexico. He found a set of four fresh eggs on May 18, and on March 19, "a colony of thirteen nests each containing four young, some full grown." On March 15 he located a colony of 17 nests, all with fresh sets, two of three, eleven of four, and four of five. "All the nests found were placed from two to eight feet from the ground—average height five feet, and all but one were built in piñon pines, this one being placed four feet up in a juniper."

Eggs.—Four or five eggs, oftener the former, constitute the ordinary set for the pinyon jay; as few as three incubated eggs have been found and rarely as many as six eggs or young have been recorded. The eggs vary in shape from short-ovate to elliptical-ovate, but ovate is the prevailing shape. They are only slightly glossy. The ground color is bluish white, greenish white or grayish white. This is usually evenly covered all over with minute dots or small spots in various shades of brown, from reddish brown to purplish brown; sometimes there are larger spots or even small blotches, which are apt to be concentrated about the larger end. Bendire (1895) says that "an occasional set is blotched heavily enough to nearly hide the ground color, but this appears to be rarely the case." The measurements of 50 eggs in the United States National Museum average 29.2 by 21.7 millimeters; the eggs showing the four extremes measure **32.0** by 22.6, 31.7 by **23.4**, **26.3** by 20.7, and 28.0 by **20.0** millimeters.

Young.—Major Bendire (1895) writes: "Incubation lasts about sixteen days. The Piñon Jays are close sitters and, like Clarke's Nutcracker, are devoted parents. The young are able to leave the nest in about three weeks, and may easily be distinguished by their somewhat duller plumbeous blue color. They at once form in flocks and rove about from place to place in search of food."

Mr. Cameron (1907) says: "To the best of my belief, both birds share the duties of incubation. * * * It is an interesting sight in June, to watch a flock of some hundred or more Piñon Jays which contains a large proportion of the newly fledged young. After the latter can fly well they still expect the parents to feed them, and clamor incessantly to be fed, repeating their shrill monotonous cry of *wauck* on a single note, whether on the ground or in the pine branches, voracious, openmouthed fledglings walk towards the parents, flapping their newly acquired wings to attract attention. The old birds may then be seen supplying them with grubs and insects. I observed one female feed a single offspring on the ground several times in a few minutes."

Mrs. Wheelock (1904) says that the young "learn to extract the sweet kernels of the piñon nuts before they leave" the nest. "They are also fed quite as fully on grasshoppers from which legs and wings have been carefully removed."

Plumages.—Mr. Cameron (1907) writes: "The naked slate-colored young were hatched on June 15, so that the time of incubation was about 18 days. They are fully feathered at two weeks old, being then a uniform lavender of exactly the same color as the flower of that name, with bill, legs, and feet to match. The hue is darkest on the quills and lightest on the crissum. After leaving the nest they become more ash gray, lighter below; the tail is then dark slate with a light tip, and the ends of the primaries almost black. Until after the fall moult the birds show no real blue."

The postjuvenal molt seems to be very variable as to date, on account of the variation in the date of hatching, but it occurs before fall and apparently involves everything but the wings. I have seen two in this molt on August 15.

The first winter plumage, which is worn until the following summer, is much like that of the adult female, but much duller throughout, the general color being bluish gray rather than grayish blue, with brownish rather than blackish primaries and with pale-gray under parts, more whitish in the anal region. Adults apparently have their postnuptial molt before the middle of September, as after that date they all seem to be in fresh plumage.

Food.—The chief food of the pinyon jay, or rather its favorite food, is the sweet nut of the pinyon pine *(Pinus edulis)*, but it also eats to some extent the nuts, seeds, or young tender cones of other pines, particularly the yellow pine *(Pinus scopulorum)*. Other vegetable food includes various wild fruits, such as the fruits of the red cedar and the boxelder, various seeds, and some grain; these jays are said to do considerable damage to grain crops, but probably most of the grain is picked up as waste grain.

The animal food consists of grasshoppers, beetles, and other insects and, to some extent, the eggs and young of small birds. J. B. Dixon tells me that "these birds are relentless in their search through the desert scrub for other birds' nests and destroy their eggs and young when found." Other observers do not seem to emphasize this habit.

H. W. Henshaw (1875) says: "A large flock of these birds were seen near Silver City, N. Mex., October, busily engaged on the ground feeding upon grass seeds. Those in the rear kept flying up and alighting in the front rank, the whole flock thus keeping in continual motion." Near Tularosa, late in November, he saw "a large flock engaged in catching insects on the wing, and in this novel occupation they displayed no little dexterity. From the tops of the pine trees, they ascended to a considerable height, when, hovering for an instant, they would snap up an insect and return to near the former position, remain for a moment, and again make an essay."

Mr. Cameron (1907) adds: "Like Magpies, however, they are practically omnivorous, and a Piñon Jay has been known to meet its fate in a wolf trap by which destructive instrument so many of the former have perished. Like Magpies, too, Piñon Jays come about the ranch house in the hope of receiving scraps from the table, alighting but two or three yards from the door, or on the hitching post where the horses are tied. They are also very fond of insect food, and may be seen walking about as they turn over dried cattle manure in search of coleoptera. Mr. Dan Bowman informs me that in his locality (Knowlton) soft corn on the cob has a great attraction for them."

Mr. Braly (1931) saw one of these birds feed its mate while the latter was engaged in incubation: "On coming in with food, a male usually perched on the top of a tree forty or fifty feet distant from the nest and called the female off to be fed. While being fed, she made a screeching series of calls similar to those of a young bird and continually fluttered her wings, and if the male flew to another tree, she followed, begging for more food. Having finished feeding, the male flew back to the feeding ground and the female flew directly to the nest, making it very easy to find. The feeding was closely observed and was solely by regurgitation, an unusual procedure for any of the crow or jay family."

Behavior.—Pinyon jays are among our most highly gregarious birds at all seasons, and, except in their nesting colonies, they are always restless and erratic nomads and almost always noisy, making their presence known as they sweep over the foothills in flocks of hundreds. In many of its movements this bird is more like a crow than a jay; its flight is crowlike, though considerably swifter, and has been likened to that of the robin or Clark's nutcracker; on long flights from one feeding ground to another, it often travels in compact flocks. It spends much time on the ground, where it often feeds in rolling flocks; its gait is a dignified walk or easy run, with its body more or less erect and its head held high, more like that of the starling than like the bouncing hops of the jays. If a flock is disturbed while feeding in the pines, the first one to leave gives a warning cry and the others follow in a leisurely manner, one at a time, until all are on the wing.

John T. Zimmer (1911) writes: "The birds are not particularly wary but are somewhat difficult to approach at times owing to their restless nature which keeps them constantly moving. I have been standing in the line of approach of a flock of Pinion Jays and had them settle all around and within a foot or two of me and not show the least sign of fear when I moved around among them. They would turn and peer at me and were full of curiosity. Even when I shot they would merely rise, wheel around with loud outcries for a moment or so and then settle down and continue their activities as if nothing had happened to disturb them."

Voice.—Mr. Braly (1931) says that "the female has a call given when near her nest, that closely resembles *krook, krook.* The male has a peculiar whistle-like note when one is near a completed nest and a very jay-like note when the female is disturbed from her nest." According to Mr. Cameron (1907), "their presence is always proclaimed

by their shrill cry of *wī-ār whǎck, wī-ār whǎck;* the last note short, but
the first two notes long and high pitched like the caterwaul of a cat."
Ralph Hoffmann (1927) writes it differently: "Besides the mewing call
queh-a-eh, given in flight, they utter, when perched, a continual *queh,
queh, queh."* Mr. Zimmer (1911) describes their note as a "high,
nasal 'kree-kree-' or 'karee-karee-', repeated rapidly many times in succession or long drawn out." Bendire (1895) says that some of the
notes "are almost as harsh as the 'chaar' of the Clarke's Nutcracker,
others partake much of the gabble of the Magpie, and still others resemble more those of the jays."

Field marks.—A dull-blue, crowlike bird, with a short tail and a long,
slender bill, could hardly be anything else but a pinyon jay, especially
if seen flying about, or feeding on the ground, in flocks. The voice
is also distinctive, if one is familiar with it; and it is very noisy.

Enemies.—The pinyon jay may be something of a nest robber, but
it also has been preyed upon itself occasionally. Mr. Cameron (1907)
saw the young disappear at intervals, one after the other, soon after
hatching, from a nest that he was watching. They were hatched on
June 16, and by July 2 only one fully fledged bird remained in the
nest. Being at a loss to account for the disappearance of the young,
he sat down to watch, and after a long wait he saw a pair of northern
shrikes fly straight to the nest tree. Fortunately for the surviving
youngster, the parent jays were at home; they attacked the shrikes and
drove them away.

Fall.—As soon as the young are strong on the wing they begin to
gather into larger flocks than ever and start on their erratic fall and
winter wanderings. These huge flocks, numbering hundreds and sometimes a thousand or more, swoop over the foothills and open country
in a rolling mass, the birds in the rear overtaking the leaders and all
screaming their loudest. Their movements are not governed by climatic
conditions but by the scarcity or abundance of the food supply. Even
where pine nuts and cones and cedar berries are abundant, the supply
is soon exhausted by the many mouths to be fed, and the flocks move
on to seek new fields, with hundreds of eyes on the alert to detect the
presence of any available food supply. When there is a scarcity of
pinyon nuts, or other normal food, or when they have exhausted the
supply, the flocks sweep down on the grainfields and may do considerable
damage to late crops of beans, corn, or other cereals. All through the
fall and winter, flocks may be seen coming and going, but they may be
very abundant during one season and entirely absent from the same
region the next season; or they may be there in hundreds one day and
all gone the next.

These fall flocks sometimes indulge in interesting flight maneuvers; Henshaw (1875) quotes C. E. Aiken as follows: "At Fort Garland, Colo., in October, 1874, I saw probably a hundred of these birds in a dense, rounded mass, performing evolutions high in the air, which I have never before known them to do; sweeping in wide circles, shooting straight ahead, and wildly diving and whirling about, in precisely the same manner that our common wild pigeons do when pursued by a hawk. This singular performance, with intervals of rest in the piñons behind the fort, was kept up for about two hours, apparently for no other purpose than exercise."

Laurence B. Potter tells me that he had an excellent sight record of a pinyon jay, at short range, at Eastend, Saskatchewan, on September 16, 1910.

Winter.—At least some pinyon jays spend the winter as far north as Montana, where, according to Mr. Cameron (1907), "in midwinter, Piñon Jays seek deep ravines and love to sun themselves either on a bank or in the branches of low cedars which grow there. When thus sheltered these noisy, restless birds will sit motionless for some time without calling to each other. At this season their food seems to consist entirely of cedar berries."

At the other end of the line, in Brewster County, Tex., Van Tyne and Sutton (1937) say that "they are a characteristic species of the region in winter, but are practically never seen in summer." In intermediate regions throughout their range they are locally abundant or scarce according to where they can find their necessary food supply, traveling about in flocks, often considerably beyond the limits of their summer range.

DISTRIBUTION

Range.—Mountainous regions of the Western United States; not regularly migratory:

The range of the pinyon jay extends **north** to central Oregon (Grandview); Montana (probably Missoula, Pompeys Pillars, and Terry); and South Dakota (Rapid City). **East** to western South Dakota (Rapid City and Elk Mountains); Colorado (near Fort Lyon); western Oklahoma (Kenton); eastern New Mexico (Mesa Pajarito and Santa Rosa); and western Texas (Guadalupe Mountains). **South** to southwestern Texas (Guadalupe Mountains); southern New Mexico (San Luis Pass); and northern Baja California (San Pedro Mártir Mountains). **West** to Baja California (San Pedro Mártir Mountains, Vallecitos, and Campo); eastern California (San Bernardino Mountains, Argus Mountains, Inyo Mountains, and White Mountains); western Nevada (Carson); and central Oregon (Bend and Grandview).

Although not a regular migrant, the pinyon jay does considerable wandering, particularly in fall and winter. At such times it has been recorded north to northwestern Montana (Eureka, Fortine, and Columbia Falls); east to eastern Nebraska (Neligh, Norfolk, Lincoln, Harvard, and Red Cloud), eastern Kansas (Lawrence, Baldwin, and Wichita), central Oklahoma (Oklahoma City); and west nearly or quite to the Pacific coast, as California (Los Angeles, Pacific Grove, and Berkeley) and Oregon (Salem and Gaston). It has been recorded in summer without evidence of breeding in south-central Washington (Fort Simcoe); northwestern South Dakota (Short Pine Hills); and northern Nebraska (Holly and Valentine).

Casual records.—A specimen was reported at Eastend, Saskatchewan, on September 16, 1910.

Egg dates.—California: 9 records, April 9 to 21.

Colorado: 13 records, March 23 to May 19.

New Mexico: 34 records, February 16 to June 10; 18 records, March 15 to 28, indicating the height of the season.

Oregon: 4 records, April 10 to 15.

NUCIFRAGA COLUMBIANA (Wilson)

CLARK'S NUTCRACKER

PLATES 47-49

HABITS

Lewis's woodpecker and Clark's nutcracker were named for the two famous explorers who made that historic trip to the sources of the Missouri River, across the Rocky Mountains and down the Columbia River to the Pacific coast, as they were responsible for the discovery of these two unique and interesting birds. Capt. William Clark, who was the first one to mention the nutcracker, referred to it as "a new species of woodpecker"; and Wilson described it as a crow, Clark's crow, *Corvus columbianus*. These impressions are not to be wondered at, for its flight and some of its actions are much like those of woodpeckers, and it resembles the crows in much of its behavior. John T. Zimmer (1911) remarks: "It reminded me of nothing so much as a young Red-headed Woodpecker in that its flight was markedly woodpeckerlike and its grayish body and head and its black wings and tail with white on secondaries gave it, at least superficially, a very close resemblance to the bird mentioned." The first one I saw, while I was crossing the Rocky Mountains in a train, reminded me very much of some large woodpecker bounding across a valley. Its names, both scientific and common, are all well chosen, indicating its feeding habits, its discoverer, and the place of its discovery.

The nutcracker is a mountain bird, ranging from 3,000 feet up to 12,000 or even 13,000 feet, according to latitude and season; its breeding range seems to be mainly between 6,000 and 8,000 feet, or from the lower limit of the coniferous forest up to timber line. It is quite widely distributed in the mountainous regions from southern Alaska and southwestern Alberta to northern Lower California, Arizona, and New Mexico.

Nesting.—Many years elapsed after the discovery of the bird before a nest of Clark's nutcracker was found. This was largely due to the fact that the bird breeds at rather high elevations in the mountains and so early in the season that the ground is covered with deep snow, making traveling very difficult, slow, and limited to small areas. At the time that Major Bendire (1895) wrote his life history of the species he was not aware of the taking of any nests and eggs, except the two taken by Denis Gale in Colorado and those that he took himself near Camp Harney, Oreg., in 1876 and 1878. His account of the finding of his nests is rather interesting:

In March, 1876, I recommenced what looked like an almost fruitless search, in which I had most of the time to tramp through snow from 2 to 4 feet deep; after having examined a great many cavities, mostly in junipers, I was almost ready to give up the task, when I finally examined the pines more closely, and noticed now and then an apparently round ball on the horizontal limbs of some of these trees, which I took to be nests of Fremont's Chickaree, *Sciurus hudsonicus fremonti,* which is very common in this locality. The majority of these supposed squirrels' nests were by no means easily reached, and after trying to dislodge their occupants with sticks, stones, or occasionally with a load of shot, and invariably failing to bring anything to light, I ceased to trouble myself further about them. Being more puzzled than ever, I was about to give up the search for their nests, when, on April 22, after having made more than a dozen fruitless trips, I saw a Clarke's Nutcracker flying quietly and silently out of a large pine about 50 yards ahead of me. This tree had a rather bushy top and was full of limbs almost from the base and was easy to climb. As I could not see readily into the top from below, I climbed the tree. Failing to see any sign of a nest therein, and being completely disgusted, I was preparing to descend when, on looking around, I noticed one of these supposed squirrels' nests placed near the extremity of one of the larger limbs, near the middle of the tree, and 25 feet from the ground; it was well hidden from below, and sitting therein, in plain view from above, I saw not a squirrel, but a veritable Clarke's Nutcracker.

Between April 24 and 30 he found a dozen more nests, all containing young. "In the spring of 1877 I commenced my search for nests on March 15, but failed to see a single bird where I had found them comparatively common during the previous season. Their absence was due in this case to the lack of suitable food. No ripe pine cones were to be found, on the hulled seeds of which the young are at first exclusively fed." The following year, 1878, the birds were back again in their old haunts, and he found his first nest on April 4; it was near the extremity of a small limb of a pine about 40 feet from the ground. "All of the nests found were placed in nearly similar situations, on horizontal limbs of pines, *Pinus ponderosa,* from 15 to 45 feet from the ground, in rather open situations at the outskirts of the heavier forests, and usually on side hills with a southeasterly exposure, at an altitude (estimated) of from 5,000 to 5,500 feet."

He describes an average nest as follows:

The nest proper is placed on a platform of dry twigs, mostly those of the western juniper, *Juniperus occidentalis,* and of the white sage, averaging about three-sixteenth of an inch in thickness, and varying from 8 inches to a foot in length. These twigs, which also help to form the sides of the nest, are deftly matted together and to the smaller twigs of the limb on which the nest is saddled; they are further held together and bound by coarse strips of the inner bark of the juniper tree; these strips are mixed among the twigs and are very suitable for this purpose. The inner nest is a mass of these same bark strips, only much finer, having been well picked into fine fiber; it is quilted together with decayed grasses and pine straw, forming a snug and comfortable structure. No hair or feathers entered into the composition of any of these nests. The outer diameter measures from 11 to 12 inches by about 7 inches in depth; the cup is from 4 to 5 inches wide and 3 inches deep. The quilted inner walls are fully 1½ inches thick; it is quite deep for its size, and the female while incubating is well hidden. Nest building must occasionally begin in the latter part of February, but more frequently in March, and it appears to take these birds some time to complete one of these structures. Both parents assist in this, as well as in incubation, and the male is apparently equally as attentive and helpful as the female. While they are noisy, rollicking birds at all other times, during the season of reproduction they are remarkably silent and secretive, and are rarely seen.

Both of Mr. Gale's Colorado nests were placed in low, scrubby pines, *Pinus ponderosa;* one was only 8 feet from the ground in a tree 20 feet high; and the other was 9 feet up in a 12-foot tree. Several other Colorado nests have been reported by W. C. Bradbury (1917b), who sent H. H. Sheldon to the foothills of the Sangre de Cristo Range, in Saguache County, for two seasons in succession to collect nests and eggs of this species. A number of nests were found, mostly containing young even in March and April. Most of the nests were in pinyon pines at heights ranging from 8 to 16 feet; one was in a juniper at 8 feet and another in a large fir at only 7 feet above ground.

In Mono County, Calif., James B. Dixon (1934) found five occupied nests of Clark's nutcracker on April 9 and 10, 1934; three of these held young and two held eggs. "All of the nests were in juniper trees on steep slopes at the 8000-foot level and contrary to our expectations were located in the coldest spots, where the snow stayed on the ground the longest. It is quite likely that these locations are the freest from the wind which blows so hard at these elevations, and I feel certain the juniper trees are used because of their sturdy build and ability to withstand the wind action. All nest locations seemed to have been selected with protection from the wind in mind, as the nests were either on top of a large limb, or, if supported by a small branch, were surrounded by heavy limbs that gave protection."

J. H. Bowles (1908) found these birds rather plentiful near the west end of Lake Chelan, Wash., "where they seemed to prefer an altitude of a little over 1500 feet. Here on June 13 I located the only nest of the trip, which was disclosed to me by the parent birds carrying food to the young. It was about 150 feet up in a large bull pine, near the top where some disease of the foliage had caused an almost solid cluster four feet in diameter."

J. A. Munro (1919) reports three nests found by him and Maj. Allan Brooks in the Okanagan Valley, British Columbia, on March 9, 1912. "This was in Yellow Pine country; a series of wooded benches overlooking Okanagan Lake." The first two nests were in yellow pines, *Pinus ponderosa,* one 50 feet up and 8 feet out from the trunk, and the other 40 feet above ground; each of these nests held two fresh eggs. The third nest was 25 feet from the ground and 12 feet out from the trunk of a Douglas fir; it contained three partly incubated eggs; one of the birds was seen carrying sheep's wool to the nest.

M. P. Skinner (1916) writes thus of the nesting of the nutcrackers in Wyoming:

About February 1, at Fort Yellowstone, elevation 6300 feet above sea level, the birds are mated and the building of the nest begins, each bird of the pair doing its share. The thick top of a cedar, or other evergreen is selected, with a convenient crotch about twelve feet from the ground. First a rough platform of twigs is built. These twigs are broken from a cedar (western juniper) by a quick, wrenching jerk assisted by the cutting edges of the bill, and carried to the site. Here the material is piled in the crotch till the mass reaches a ball about nine inches in diameter and six inches high. The nest proper is deep and cup shaped, about six inches in diameter, and has walls an inch thick; it is built of cedar or pine needles and the inner lining of grass stems and shredded juniper bark, each strand turned into place by the bird squatting down on it and twisting in it. A few horse hairs and bits of strings are usually included in the lining.

Eggs.—Clark's nutcracker lays, apparently usually, two or three eggs, but often four and occasionally as many as five or even six. They vary in shape from ovate to elliptical-ovate and are only slightly glossy. The ground color is a pale shade of "lichen green," pale grayish green, or very pale, clear green, almost greenish white. They are usually thinly, evenly, and rather sparingly spotted with minute dots, small spots or flecks of pale browns, or shades of olive, gray, or drab. Sometimes the spots are more concentrated about the larger end, and sometimes the smaller end is almost unmarked.

The measurements of 50 eggs average 32.4 by 23.4 millimeters; the eggs showing the four extremes measure **35.0** by 23.4, 34.5 by **25.2,** **29.2** by 23.2, and 33.5 by **21.6** millimeters.

Young.—Both sexes assist in the duties of incubation and care of the young. Bendire (1895) gives the period of incubation, "as nearly as I can judge, about sixteen or seventeen days." This agrees closely with the reported period for the European bird, 17 to 18 days. But M. P. Skinner (1916) says that in Yellowstone Park the eggs "are laid between February 28 and March 3, and the brooding commences immediately. At such a time the brooding bird is subjected to all the vagaries of truly wintry weather." "Often," he says, "she sits through raging snowstorms protected only by the tuft of cedar needles over the nest, and many times has the writer seen the bird actually on the nest with the thermometer below zero. Under such conditions she draws herself down with only her tail feathers and perhaps her bill showing above the rim of the nest. She is very fearless, even submitting to capture rather than leave the nest; when she leaves, she does so quietly, and returns immediately after the intruder is gone. After brooding twenty-two days the young are hatched, naked of course, and with their eyes closed. Four weeks later the young leave the nest and by May 5 are fully feathered and shifting for themselves. Notwithstanding this early start there is no evidence to show that a second brood is raised."

The longer period of incubation and the longer altricial period, as given by Mr. Skinner above, was probably exceptional or due to too much exposure to wintry weather. For Major Bendire (1895) says that the young remain in the nest only about eighteen days; and Mr. Bradbury (1917b) showed the altricial period to be about three weeks.

Mrs. Wheelock (1904) says that the young "were fed on piñon nuts, which were carried to the nest and hulled by the adult while perched just outside on the branch. I could not discover that any other food was brought them. At first this was given by regurgitation, but when the young were a few days old the food was supplied to them direct. As soon as they were ready to leave the nest they were coaxed by short flights to the nut pines, and readily learned to shell the nuts and provide for themselves. Then it would seem a complete change of diet was necessary; for they disappeared from these regions entirely, flocking to a locality where berries, fish, and insects abound. By the middle of June not one was left in the old breeding grounds."

Apparently the young are also fed by regurgitation *after* they have left the nest, for Taylor and Shaw (1927) write: "The nutcracker population becomes most conspicuous when the young have left the nest. The husky youngsters appear quite as large as their parents, and their squawking calls fill the air. They follow the adults with a persistence truly wonderful, awakening the echoes with their stentorian teasing. Presently the parent plunges its bill into the open mouth of the young one, and there ensues what appears to be a struggle for life or death. There is of course no occasion for concern, for this, in nutcracker society, is the orthodox method of feeding the young. The process of regurgitation complete, the participants seem to feel better all around. The respite must seem all too brief to the parent, however, for in a short time the young bird is apparently as hungry and as noisy as ever."

Mr. Dixon (1934) says that "in the event one of the parents left the nest from any cause, the other bird of the pair would immediately assume brooding duties. * * * Actual time records taken on the afternoon of the 9th of April, which was a warm sunshiny afternoon, revealed a change of brooding and feeding duties every thirty minutes on the average."

Plumages.—The young are hatched naked and with eyes closed, but, when perhaps a week or so old, they are partly covered with down (Bradbury, 1917b), the color not mentioned. The juvenal plumage is acquired before the young bird leaves the nest. Charles F. Batchelder (1884) gives the first full description of the juvenal plumage of Clark's nutcracker in more minute detail than seems necessary here. In a general way, the color pattern is much like that of the adult, but the gray portions are browner and the blacks duller. The upper parts are dull brownish gray, darkest on the rump and scapulars. The white eye ring, superciliary stripe, and the other white about the face of the adult are lacking. The general coloring of the under parts is brownish ash, darkest on the breast, most of the feathers tipped with whitish, giving an indistinct barred effect. The wings are much like those of the adult, but the white in the secondaries is more extensive, some of the primaries have a small ashy spot at the tip, the lesser wing coverts are dusky grayish brown, and the other coverts are indistinctly tipped with the same. The tail is like that of the adult, but the black in the white rectrices is more extensive. The black in both wings and tail is duller than in the adult, and the bill and feet are grayer. I have not seen the postjuvenal molt, but I have seen adults molting in July, August, and September.

Food.—Like other members of the crow family, the nutcracker is largely omnivorous, though it seems to show a decided preference for the sweet nuts of the pinyon pine *(Pinus edulis)*, eating the shells as well as the kernels. Grinnell and Storer (1924) report that a female shot on September 22 "held in its throat 72 ripe seeds of the piñon, comprising a volume of about one cubic inch." Another female taken September 25 "held in her distended throat 65 mature seeds of the white-bark pine, and some fragments, all together weighing 10 grams or close to 7 per cent of the weight of the bird, which was 146 grams."

It eats the seeds of several other species of pines and even firs, extracting the seeds from the cones with its crowbarlike beak. Acorns and the berries of the cedar or juniper are included in the diet. Mr. Skinner (1916) says: "Sometimes they will tear the cone to pieces even while the cone is still fast to the branch, often perched at the very tip of a bending branch, or even underneath, clinging in a manner creditable to a chickadee or a nuthatch. More often the cone is detached and carried away to a strong limb where it is held by one foot while the bird strikes strong, downward blows at it with its pickaxlike bill. At times the bird will secure a seed at every second stroke and at the same time tear the cone to shreds."

J. A. Munro tells me that Clark's nutcrackers were noted commonly in a valley in British Columbia on May 17 and 19, 1940. "In one place a flock of twenty-five (plus) rose from a field in which spring wheat was just appearing. It seemed likely they were feeding on the sprouted grain."

Mr. Bradbury (1917b) reports that "the stomachs of the old birds examined always contained masses of pinyon shells, this far exceeding in bulk the mixture of insects and meat of pinyon nuts, about 75 per cent nut food and 25 per cent insects and other matter." One nutcracker was seen feeding on the remains of a deer.

Claude T. Barnes (MS.) thus describes the feeding habits of Clark's nutcracker, as observed at an altitude of 7,050 feet in the Wasatch Mountains of Utah: "Near me an ancient, gnarled limber pine *(Pinus flexilis)* stood on a wind-exposed knoll, raising a broad, open crown on a brown-plated trunk 2 feet in thickness. The ground beneath it was strewn with subcylindrical cones, 5 inches long and 3 inches in diameter at the base. One of the nutcrackers flew to the tree above me, alighting on the outer tip of an upper branch, where a cone was suspended. Standing above the cone and working almost upside down, it pecked with its strong, cylindrical bill at the base of the cone, clinging the while to the branch with its large toes and sharp, much curved claws. Pecking eagerly every two or three seconds, it at times almost tilted over head first, but without releasing the toe holds, fluttered back to balance. Finally, after about 40 pecks and probes at the pencil-thick stalk of the cone, the cone began to fall; whereupon it clasped the cone with its bill and flew to a thick horizontal limb below.

There it took its time in prying out the quarter-inch seeds, two beneath each scale, and swallowed them with apparent satisfaction. * * * Another bird alighted on a small tree and, as I stood only ten feet away, probed out all the seeds of a cone without severing its stalk and without minding my presence."

The nutcracker shares with the jays of the *Perisoreus* group the name of "meat bird" or "camp robber," for, especially in winter when other food is scarce, it comes freely to the camps to pick up whatever scraps of food it can find, and almost anything edible is welcome; at such times it becomes quite bold, frequenting the open-air kitchen and even occasionally entering the tent or cabin. It invades the vicinity of farms and houses, looking for kitchen hand-outs or picking up crumbs of bread or waste grain in the streets. Larger scraps of food are often carried away to be eaten at leisure or hidden for future use.

It sometimes indulges in the bad habit, common to most of the Corvidae, of robbing the nests of the smaller birds and devouring their eggs or small nestlings. J. A. Munro (1919) says that "several nests of Hermit Thrushes, Horned Larks and Pipits, that were under observation, above timber line on Apex Mountain, were destroyed by a pair of Clark's Nutcrackers." And several others have referred to this same habit.

Insects enter largely into the food of this bird during summer and before the pinyon nuts are ripe. Some time is spent on the ground hunting for beetles, ants, grasshoppers, and the destructive black crickets. Some insects are caught on the wing in true flycatcher fashion; from a perch in the dead top of a tree the nutcracker watches for passing insects, darts out and chases them in an erratic course, and returns to its perch. Mrs. Bailey (1928) says that "two were seen near camp on a log, running back and forth chasing sphynx moths that were feeding from the larkspurs bordering the log."

Mrs. Wheelock (1904) writes: "Grasshoppers and the big wingless black crickets he devours in untold numbers, and grows fat on the diet. Butterflies he catches on the wing in flycatcher fashion; grubs he picks from the bark, clinging to the side of the tree trunks and hammering like a woodpecker."

Decker and Bowles (1931) observed the feeding habits of a large flock of over a hundred nutcrackers in the Blue Mountains of Washington:

A good deal of feeding was evidently done high up in the trees, where they spent most of their time, but on frequent occasions the whole flock would come down and feed on the ground. * * * Their food when on the ground consisted principally of the large black ants, beetles and snails. The ants being in enormous numbers were eaten by the thousand, and it was very noticeable to see how careful the birds were to make sure of thoroughly killing them before eating. Every ant was beaten to a pulp before it was swallowed, so there was no chance that the powerful forceps of the insect might injure the throat or stomach of the bird. The beetles received considerably less punishment, but the body of the snails were carefully picked out of the shell before swallowing. We threw them some of our own food, but they did not seem to care especially for it, presumably because they had such an abundance of natural food. The number of black ants these birds eat must be far beyond calculation, as not one of these insects could be found for some time after the birds flew up into the timber.

Victor H. Cahalane (1944) tells an interesting story of a nutcracker's ability to locate food under 8 inches of snow; he flushed one from the ground under a Douglas fir, and found that the bird "had dug a hole three or four inches in diameter at the top, at an angle of perhaps 30 degrees, through the hard-packed snow to the sloping ground. At the bottom of the excavation, frozen to the ground litter, was a Douglas fir cone." The snow was so deep that there was no indication on the surface of the presence of the cone; the bird had done no exploratory digging, but had dug the hole with remarkable accuracy in exactly the right spot.

Behavior.—I always liked the old generic name *Picicorvus,* as it seemed particularly appropriate for a bird that so much resembles the woodpeckers and the crows in behavior and appearance. My first impression of it was that of a large woodpecker, and others, including its discoverer, got the same impression. Its flight, as I remember it, is undulating like that of a woodpecker; and most observers seem to have seen it that way. Dr. Coues (1874), however, says that "the ordinary flight is rapid, straight, and steady, accomplished by regular and vigorous wing-beats; but when flying only from tree to tree, the birds swing themselves in an undulatory course, with the wings alternately spread and nearly closed, much in the manner of the Woodpeckers."

They sometimes soar high in the air like hawks, with wings and tails widely spread, or from some dizzy height make a spectacular dive earthward. Mrs. Wheelock (1904) writes:

It is on the crests of the Sierra Nevada that these birds are found most abundantly. Here they sun themselves on the highest peaks, frolicking noisily in the clear, bracing air. When hungry or thirsty, they dart from their lofty perches and, with wings folded, hurl themselves down the cañon with the speed of a bullet. Just as you are sure they will be dashed to pieces, their wings open with an explosive noise and the headlong fall is checked in a moment. Sometimes the descent is finished as lightly as the fall of a bit of thistle down; sometimes by another series of swift flights; often by one rocket-like plunge. At the foot a mountain brook furnishes food and drink. As the shadows creep up the sides of the cañon, the Nutcrackers follow the receding sunlight to the summit again, mounting by very short flights from tree to tree, in the same way that a jay climbs to the top of a tree by hopping from one branch to another.

Nutcrackers also show their relationship to the jays by their noisy, boisterous habits, their inquisitive curiosity, and their behavior on the ground, where they hop rather awkwardly about foraging for fallen nuts and insects. Their straight forward flight, at times, is also much like that of our blue jay; and their thieving propensities are in keeping with those of the whole corvine tribe. Mr. Munro (1919) says: "Like all corvine birds, they are exceedingly curious and a passing deer or coyote will attract their attention so that the position of game can often be located by their excited cries. They come readily to an imitation of the call of the Pygmy Owl or the Horned Owl and will investigate the caller at close range."

Referring to their methods of securing pine cones, Mr. Skinner (1916) writes: "Being bold, independent free-lances these birds will vary their methods by robbing a pine squirrel of his cone; even going so far as to knock the squirrel from his limb with one blow from their bills at the end of a long, swift swoop. The pine squirrel knows this, too; and it is delicious to see the squirrel, whose own abilities as a robber are not small, glide into some protection and hurl vituperation at his enemies. Nor are the nutcrackers at all backward at 'sassing' back. Many a time the somber, evergreen forests are enlivened by such a squawking match, joined in by all the squirrels and nutcrackers in hearing."

J. Stuart Rowley (1939) says: "Near Virginia Creek on July 4, 1939, I tapped a dead pine stub and was surprised to see several nearly fledged young chickadees 'explode' in my face and fly uncertainly down a ravine. Immediately, two nutcrackers swooped down, concentrating their attack on one individual. One nutcracker seized the fledgling, whereupon it flew to a pine and proceeded to pick off feathers from the tail and wings of the chickadee before tearing it to bits and devouring it."

Voice.—Not much can be said in favor of the voice of Clark's nutcracker. It is generally conceded to be harsh, grating, and unpleasant, especially when heard in volume, as it often is, and when not softened by distance. It is a noisy bird, except when near its nest. Its ordinary guttural, squawking call is variously written as *chaar, char-r-r, chur-r-r, kra-a-a,* or *kar-r-r-r-ack,* each note repeated two or three times.

Mr. Dawson (1923) adds the following variations: "But the Nutcracker's repertory is not exhausted by a single cry. For years I was puzzled by sporadic eruptions of a strange, feline cry, *meack,* or *mearrk,* a piercing and rather frightful sound. The Clark Nutcracker proved at last to be responsible, and he was only at play! The very next morning after the mountain lion scare, we had the versatile birds as musicians. Two of them got out their little toy trumpets, pitched about a fifth apart, and proceeded to give us the Sierran reveille,

      hee hee   hee,   hee hee,
hoo   hoo   hoo hoo   [etc.]. The notes were really quite musical, and the comparison established of children's tin trumpets was irresistible. The effect produced by the two birds sounding in different keys was both pleasant and amusing. * * * The concert lasted for two or three minutes."

Decker and Bowles (1931) say that their incessant cries were heard "during the entire time that they were up in the trees. They would commence a little after daylight and we have never heard such a racket from any other members of the animal kingdom. Besides innumerable calls that must have belonged peculiarly to themselves they also included every known call of the Magpie and the Crow that is uttered by either the young or the old birds of those species. This last may be because both adult and young of the year were present and all screaming at the top of their lungs, making a din that at times grew extremely tiresome and, indeed, almost unbearable. Oddly enough, when on or near the ground they are absolutely silent."

Field marks.—The nutcracker is a fairly large bird, between a small crow and a large woodpecker in size. It has a long, sharp, black bill; its head and body are pale gray; its wings are black, with a large white patch on the secondaries; and its tail is centrally black but largely white laterally. Its behavior and its voice are distinctive, as explained above.

Winter.—"Although the Clark Nutcracker is a characteristic resident of the Hudsonian Zone, it strays both above and below this belt. In summer, after the broods of the year are fledged, some of the birds move down the mountains. * * * And at the same season they sometimes wander up over the rock-strewn ridge crests well above timber line. Some of the nutcrackers which stray to the lower altitudes remain there at least until early winter. * * * But most of the birds remain at the normal high altitudes through the winter months" (Grinnell and Storer, 1924).

In addition to these altitudinal movements the nutcrackers do considerable erratic wandering in winter, appearing unexpectedly at irregular intervals in various small cities and towns, even near the coast, notably in Monterey and Alameda Counties, Calif., where they become familiar dooryard visitors.

Joseph Mailliard (1920) writes of their winter behavior at Carmel, Calif: "The Nutcrackers had discovered that kitchen doors and back yards were good for some free 'hand-outs', and they systematically visited many such. While they fed to some extent on the Monterey pines, apparently more intent upon the tips of young buds than upon the contents of the cones, they picked also a good many scraps and bits of grain or crumbs in the streets, paying no attention to people twenty or thirty feet away, but becoming wary of closer approach. They seemed to have certain hours for being in certain places, and for the first few days of my stay appeared in the street opposite the dining room window while we were at breakfast."

DISTRIBUTION

Range.—Western United States and Canada, and Alaska; casual east to Lake Michigan and the Mississippi Valley; not regularly migratory.

The normal range of Clark's nutcracker extends **north** to central British Columbia (Fort St. James and Moose Pass); Alberta (Jasper House, Banff, and Porcupine Hills); Montana (Glacier National Park, Statesville, and Billings); and northwestern South Dakota (Short Pine Hills). **East** to western South Dakota (Short Pine Hills and Elk Mountains); southeastern Wyoming (Laramie Peak); Colorado (Estes Park, Cheyenne Mountain, and Blanco Peak); and New Mexico (Santa Fe Canyon, Diamond Peak, and San Luis Pass). **South** to southern New Mexico (San Luis Pass); Arizona (Mount Graham and Santa Catalina Mountains); and northern Baja California (San Pedro Mártir Mountains). **West** to Baja California (San Pedro Mártir Mountains and La Grulla); eastern California (Bear Valley, Florence Lake, Yosemite Valley, and Butte Lake); Oregon (Pinehurst and Crater Lake); Washington (Bumping Lake and Lake Chelan); and British Columbia (Alta Lake, Lillooet, and Fort St. James).

Casual records.—Although not a regular migrant, the nutcracker is given to erratic wanderings that sometimes take it considerable distance from its normal range. In Alaska it has been recorded on the southeast coast at Sitka and north to the Kowak River. Other Alaskan records are: Nushagak, November 5, 1885; Takotna, October 1, 1919; Farewell Mountain, September, 1921; Chatanika River, September 1922; and McCarthy, November, 1922. According to Taverner, it also has been collected at Robinson, Yukon. On the coast of British Columbia it has been recorded from Comox, February 18, 1904, and wintering on Graham Island in 1919-20. During the period from October 1919 to April 1920 it appeared in considerable numbers on the coast of southern California as at Pacific Grove, Carmel, and Santa Cruz Island, while one was killed near Hayward on February 16, 1923. One was taken at Coachella, Calif., 44 feet below sea level on September 24, 1935.

There are several records for the Great Plains region east to Manitoba, Margaret, October 1910; Iowa, Boone, September 23, 1894; Wisconsin, Milwaukee, fall of 1875; Illinois, Gross Point, October 9, 1894; Missouri, near Kansas City, about October 28, 1894, and Louisiana, October 12, 1907; and Arkansas, Earl, April 1, 1891.

Egg dates.—British Columbia: 11 records, March 9 to May 25.

California: 32 records, March 7 to April 21; 16 records, March 24 to April 13.

Colorado: 8 records, March 5 to April 16.

Utah: 6 records, March 23 to April 25.

Family PARIDAE: Titmice, Verdins, and Bushtits

PARUS ATRICAPILLUS ATRICAPILLUS Linnaeus

BLACK-CAPPED CHICKADEE

PLATES 50, 51

CONTRIBUTED BY WINSOR MARRETT TYLER

HABITS

The titmice, the family of birds to which the black-capped chickadee belongs, are widely distributed in the two hemispheres and in North America are represented by numerous genera, species, and races from the Atlantic Ocean to the Pacific. Over this vast area, in England, on the continent of Europe, and with us they are well known and very popular birds.

For our black-capped chickadee of the Northeastern United States our regard goes far beyond popularity. The chickadee is perhaps the best-known bird in its range and appears so trustful of man that we look on it with real affection. And no wonder—for chickadees are such cheerful little birds. When we watch a flock of them in winter they remind us of a group of happy, innocent little children playing in the snow. Thinking back to the early days of New England's history, we can imagine that the Pilgrim Fathers, when the chickadees came about the settlement at Plymouth in 1620, watched them as we do now. They were, perhaps, the first friends to welcome the travelers to the New World.

Many writers praise the chickadee. Bradford Torrey (1889) says enthusiastically: "It would be a breach of good manners, an inexcusable ingratitude, to write ever so briefly of the New England winter without noting this [the chickadee], the most engaging and characteristic enlivener of our winter woods; who revels in snow and ice, and is never lacking in abundant measures of faith and cheerfulness, enough not only for himself, but for any chance wayfarer of our own kind." Elsewhere, Torrey (1885) calls the chickadee "the bird of the merry heart."

Spring.—The black-capped chickadee is migratory to some extent, but, as in the case of some other permanent residents, it is often difficult, except at favorable observation points, to determine the time and extent of its northward and southward movements. Taverner and Swales (1908) state: "Our experience with the species at Detroit leads us to believe that it is more migrational than is generally supposed. They are common through the winter, but about the first of April the great bulk of them depart, leaving but a few scattered summer residents behind."

J. Van Tyne (1928) gives a vivid description of a definite migration. He says:

On May 20, 1928, while collecting at the tip of Sand Point (seven miles southwest of Caseville, Michigan), I witnessed a most interesting migration flight of Chickadees *(Penthestes atricapillus)*. Sand Point juts out nearly four miles into Saginaw Bay from the southeast, and apparently forms an important point of departure for many species of birds migrating northward across the bay. The day was clear with but little wind. At 9:30 in the morning I noticed a compact flock of over fifty chickadees flitting rapidly through the brushy growth toward the end of the point. Their strange appearance immediately attracted my attention. They seemed very nervous and tense, with necks outstretched and feathers closely compressed against the body. They made no attempt to feed, but kept moving steadily toward the end of the point. Reaching the last tree, a twelve-foot sapling, the first birds flitted upward to the topmost twigs and there hesitated, lacking the courage to launch forth. But the rest of the flock, following close behind, in a few moments began to crowd upon them. Fairly pushed off the tree-top, the leaders finally launched forth, the rest following in rapid succession. They started upward at an angle of fully forty-five degrees. After climbing perhaps a hundred feet the leaders lost their courage, and, hesitating a moment, they all dropped precipitately back to the shelter of the bushes. But once there they immediately headed for the sapling again and repeated the performance. Finally, after several false starts, they continued out over the lake toward the Charity Islands in the distance.

It was a new experience to me to see chickadees fly by day out across miles of open water.

Courtship.—The chickadee has apparently developed no ritual of courtship other than the pursuit of the female by the male—a common performance of many of the smaller birds. Chickadees are so common and so continually under our observation at close range that if they practiced any marked trait when pairing off, it would certainly have been noticed and described.

Dr. Samuel S. Dickey (MS.) says of the mating of the chickadee: "From what I am able to learn of this process, the birds grow agitated late in March and increase their vivacity during April and early in May. They hurry between aisles of trees and swerve over bypaths, and males dart at and even clasp one another. Then they part, and the more dominant male pursues and chases a female over brush piles and even to the ground. Then up they arise and hurry onward. A few such days of immoderate activity, and their nuptial rites seem completed."

Nesting.—The commonest nesting site of the chickadee is a hole, made by the birds themselves, in a dead stub or branch of a gray birch. From such a tree the decayed wood can easily be removed in dry chips to form a cavity, and the ring of strong bark holds the branch firmly together.

Arthur C. Bent (MS.) says that in Bristol County, Mass., three-quarters of the nests he has found have been in such a location, 4 to 8 feet from the ground. He continues: "Other nests have been in natural cavities in apple trees in orchards, or in other deciduous trees. I believe that chickadees almost always, at least partially, excavate their own nest cavities; I have seen them doing it; they cut through the outer bark of birch stubs with their strong little bills and easily remove the rotten wood from the interior."

Edward H. Forbush (1912) states:

A hole in a decayed birch stump, two or three feet from the ground, a knot-hole in an old apple tree, in a fence-post, or in an elm, forty or fifty feet from the ground, the old deserted home of some Woodpecker, a small milk-can nailed up in a tree, or a nesting-box at some farmhouse window, may be selected by the Chickadee for its home. Commonly it digs out a nest-hole in the decaying stump of a birch or pine. It is unable to penetrate sound wood, as I have seen it repeatedly try to enlarge a small hole in a white pine nesting-box, but it could not start a chip. Often the Chickadee gains an entrance through the hard outer coating of a post or stump into the decaying interior by choosing, as a vantage point, a hole made by some woodpecker in search of a grub. The Chickadee works industriously to deepen and enlarge this cavity, sometimes making a hole nine or more inches deep; and the little bird is wise enough to carry the tell-tale chips away and scatter them far and wide—something the Woodpeckers are less careful about

Sometimes the hole is excavated in the broken top of a leaning stump or tree, and once I found one in the top of an erect white pine stump with no shelter from the storm.

If we come upon a pair of chickadees at work excavating a cavity, we can step up very close to them and watch without interrupting them at all. Both members of the pair work at the same time but visit the nest alternately. Each one digs out a beakful of chips and flies away with it, and no sooner is one gone than the other is back at the nest, excavating. Back and forth they go, working quickly and, except for their faint lisping notes, silently. Mr. Bent (MS.) describes a pair at work. He says: "Both birds took turns at the work, digging out the rotten wood, bringing out a billful each time and scattering it from the nearby trees. Sometimes both birds would be *at* the hole together; one would watch while the other worked, but would not enter until its mate had come out; they were never both *in* the hole at the same time."

Bradford Torrey (1885) comments on such a scene, "the pretty labors of my little architect," thus: "Their demeanor toward each other all this time was beautiful to see; no effusive display of affection, but every appearance of a perfect mutual understanding and contentment. And their treatment of me was no less appropriate and delightful,—a happy combination of freedom and dignified reserve."

The nest proper is placed in the bottom of the cavity and, according to the testimony of Craig S. Thoms (1927) and Dr. Samuel S. Dickey (MS.), is made entirely by the female. The materials of the nest, as listed by Edward H. Forbush (1912) consist "of such warm materials as cottony vegetable fibers, hairs, wool, mosses, feathers and insect cocoons. Every furry denizen of the woods, and some domestic animals, may sometimes contribute hair or fur to the Chickadee's nest."

Aretas A. Saunders (MS.) states that chickadees sometimes add the wool of cinnamon fern to their nest, "the same material commonly used by the ruby-throated hummingbird."

Ora W. Knight (1908) says: "From a week to ten days is required to excavate the hole and three or four additional days to gather together [the materials] * * * which make up the nest proper." He describes a typical nest found at Orono, Maine: "This nest was placed in a cavity eight and a half inches deep near the top of a rotten white birch stub, six and two-thirds feet from the ground. The diameter of the entrance was two and a quarter inches. The nest proper measured two inches in diameter by one inch deep inside."

Eggs.—[AUTHOR'S NOTE: Anywhere from 5 to 10 eggs may be found in the chickadee's nest, but 6 to 8 are the commonest numbers, and as many as 13 have been recorded. These vary from ovate to rounded-ovate, with a tendency toward the latter shape. They have little or no gloss. The ground color is white, and they are more or less evenly marked with small spots or fine dots of light or dark reddish brown; usually these markings are well distributed, but sometimes the larger spots are concentrated about the larger end. The measurements of 50 eggs average 15.2 by 12.2 millimeters; the eggs showing the four extremes measure **16.3** by 12.2, 15.2 by **12.8**, **14.0** by 12.2, and 15.2 by **11.2** millimeters.]

Young.—Dr. Samuel S. Dickey (MS.) writes: "Before the set of eggs is complete, or when they are fresh, the parent, as is the habit of our wild ducks, covers the eggs with the lining of the nest, thus rendering them comparatively safe. I have found that it requires on an average 12 days for the eggs to hatch. When the nestlings are about three days old they agitate their heads, wing stumps, and legs and open their beaks and squeak feebly in anticipation of food. They remain in the nest for approximately 16 days. At this age the nestlings, about to be fledglings, look almost like their parents, but a shagginess or somewhat ill-kempt aspect serves to distinguish them. They are without doubt among the handsomest young birds of our mountain forests." He adds that the male feeds the female during incubation and that both parents feed the young.

Dr. Wilbur K. Butts (1931), in a study made in the State of New York of the dispersal of young banded chickadees, found that as a rule the birds wandered only a short distance, a mile or two, from the nest during the first few months of their lives.

George J. Wallace (1941) concluded, from his study of color-banded chickadees at Lenox, Mass., "that young chickadees, though obviously in company with their parents in late summer, tend to wander away from the more sedentary adults in the fall," and that "the Sanctuary flocks were not made up of family groups in winter."

Plumages.—[AUTHOR'S NOTE: The "pale mouse gray" natal down of the young chickadee is soon replaced by the juvenal plumage, or rather pushed out on the tips of these feathers, and wears away. The juvenal contour plumage closely resembles the spring plumage of the adult, but it is softer, looser, and fluffier; the black of the crown, chin, and throat is much duller; the sides of the head below the eyes are pure white; and the under parts are dull white, washed on the sides and crissum with pale pinkish buff.

About midsummer a partial postjuvenal molt takes place, involving the contour plumage and the wing coverts, but not the rest of the wings or the tail. This produces a first winter plumage, which is practically indistinguishable from the fall plumage of the adult.

Adults have one complete postnuptial molt in July and August, which produces a winter plumage that is more richly colored than the worn and faded plumage seen during spring and summer; the gray of the back and rump is more decidedly buffy; the sides and flanks are deep brownish buff in strong contrast with the white of the abdomen; and the whitish edgings of the larger wing coverts, secondaries, and outer tail feathers are broader.

Wear and fading produces a paler plumage in spring, the buffy tints becoming paler or largely disappearing and some of the white edgings in the wings and tail wearing away.]

Food.—Clarence M. Weed (1898), after a careful investigation of the winter food of the chickadee, states: "The results as a whole show that more than half of the food of the chickadee during the winter months consists of insects, a very large proportion of these being taken in the form of eggs. About five per cent. of the stomach contents consisted of spiders or their eggs. Vegetation of various sorts made up a little less than a quarter of the food, two-thirds of which, however, consisted of buds and bud scales that were believed to have been accidentally introduced along with plant-lice eggs." In his conclusion he says: "The investigations * * * show that the chickadee is one of the best of the farmer's friends, working throughout the winter to subdue the insect enemies of the farm, orchard, and garden."

W. L. McAtee (1926), writing of the chickadee's food throughout the year, says:

About three-tenths of the food of the Chickadee is vegetable, and seven-tenths animal. Mast and wild fruits supply the bulk of the vegetable food. The mast is derived chiefly from coniferous trees, and the favorite wild fruits are the wax-covered berries of bayberry and poison ivy. A good many blueberries also are eaten, but only limited numbers of other wild fruits and seeds.

The important things in the animal food of the Chickadees, in order, are caterpillars and eggs of lepidoptera, spiders, beetles, true bugs of various kinds, and ants, sawflies, and other hymenoptera. The Chickadee certainly consumes a great many spiders (which are moderately useful), but the occurrence seems inseparably connected with the bird's mode of feeding, ever prying as it does, under bark scales and into all sorts of crannies which are the favorite hiding places of spiders. It is just these methods, however, that enables the Chickadee to find so many of the eggs of injurious lepidoptera and plant lice, and scale insects and other minute pests, the consumption of which is so praiseworthy. The good the bird does in consuming these tiny terrors is so great that we must regard as far outweighed, the harm done in feeding upon spiders and parasitic hymenoptera. * * *

Codling moths and their larvae and pupae, the larvae, chrysalids, and adults of the gypsy and browntail moths, birch, willow, and apple plant lice, and pear psylla, and various scale insects are eaten by the Chickadee. Among these scales are one affecting dogwood *(Lecanium corni)*, the black-banded scale *(Eulecanium nigrofasciatum)* which is quite injurious to maples, the scurfy elm scale *(Chionaspis americana)*, and the oyster scale *(Lepidosaphes ulmi)*, which attacks many trees and has been known to kill ashes and poplars in New York.

Among other forest pests attacked by our friend the Chickadee are the flat-headed and round-headed wood borers, leaf beetles, the white pine weevil, nut weevils, bark beetles, tree hoppers, spittle insects, cicadas, leaf hoppers, and sawflies. Other food items of the bird include a variety of beetles, bugs, flies, and grasshoppers, and a few stone flies, dragon flies, daddy-long-legs, millipeds, snails, and small amphibians.

Dr. Dickey (MS.) writes to Mr. Bent: "I have noticed that chickadees like to draw near hunters' cabins at all times of the year, but particularly during the hunting seasons. They arrive within a stone's throw of the shelters, and will inspect and peck at animal hides, fatty substances thrown out from the table, or even entrails of animal carcasses."

Lewis O. Shelley (1926) writes of a curious and evidently unusual habit that he noticed on a warm day in February. He says: "Flying from the piazza, a Chickadee lit in front of a hive. When a bee came out it snapped it up, flew into an elm, and, holding the bee in its foot, picked it to pieces and ate it. I was alarmed for fear the Chickadee would be stung, but it seemed not, for the act was performed again. Neither was it always the same bird that flew down and got a bee, but many different ones."

J. Kenneth Terres (1940) reports seeing a chickadee eating tiny tent caterpillars, too small to be detected in a stomach contents. He says: "On the morning of April 23, 1938, I again observed at close range the destruction of these caterpillars, this time by a Black-capped Chickadee, *Penthestes atricapillus atricapillus,* in a brush-grown field in Broome County, near Nanticoke, New York. When first seen, the chickadee was busily engaged in visiting a number of the newly started nests of the American tent caterpillar located in a nearby wild-apple tree, *Malus pumila.* Using an eight-power binocular at twenty feet, I observed the chickadee closely while it visited three caterpillar nests in succession. It would first tear open the web, then pick up the small worms (on this date about three-eighths of an inch long and a sixteenth of an inch in diameter) and devour them rapidly."

Behavior.—When chickadees visit our feeding shelves what impresses us most is their quickness. They flit in rather slowly to be sure, for so small a bird, and land on the shelf with a thud, often upright, grasping the edge with their strong little claws and then jerking about with such rapidity that the eye can scarcely take in their flashlike movements. When alarmed they disappear as if by magic—we see only the place where they were—an ability that must save them many times from the strike of a bird of prey.

Another chickadee propensity is the assumption of odd attitudes; they often alight up-side-down on the under side of a branch, making, it seems, almost a back somersault as they reach upward to grasp it; and they can hang, back to the ground, steady and secure, from the tip of a swaying branch. Edward H. Forbush (1907) describes thus some of the chickadee's acrobatic tricks:

I once saw a Chickadee attempting to hold a monster caterpillar, which proved too strong for it. The great worm writhed out of the confining grasp and fell to the ground, but the little bird followed, caught it, whipped it over a twig, and swinging underneath, caught each end of the caterpillar with a foot, and so held it fast over the twig by superior weight, and proceeded, while hanging back downward, to dissect its prey. This is one of the most skillful acrobatic feats that a bird can perform—although I have seen a Chickadee drop over backward from a branch, in pursuit of an insect, catch it, and, turning an almost complete somersault in the air, strike right side up again on the leaning trunk of the tree. Indeed, the complete somersault is an every-day accomplishment of this gifted little fowl, and it often swings completely round a branch, like a human acrobat taking the "giant swing." Although the Chickadee ordinarily is no flycatcher, it can easily follow and catch in the air any insect that drops from its clutch.

William Leon Dawson (Dawson and Bowles, 1909), writing of the Oregon chickadee, a subspecies of the black-capped, gives this lively account of its activities: "Chickadee refuses to look down for long upon the world; or, indeed, to look at any one thing from any direction for more than two consecutive twelfths of a second. 'Any old side up without care,' is the label he bears; and so with anything he meets, be it a pine-cone, an alder catkin, or a bug-bearing branchlet, topside, bottomside, inside, outside, all is right side to the nimble Chickadee. * * * Blind-man's buff, hide-and-seek, and tag are merry games enough when played out on one plane, but when staged in three dimensions, with a labyrinth of interlacing branches for hazard, only the blithe bird whose praises we sing could possibly master their intricacies."

There are many instances recorded of the tameness of individual chickadees. The following, by John Woodcock (1913), is a good example:

Although I had fed the Chickadees in winter for several years, none of them were tame enough to feed from the hand until the spring of 1906. A pair were nesting in one of my bird-boxes, and, as I was standing near the nest, one of the birds came toward me. I threw a piece of nut to it, which it picked up and ate. Then I held a piece on my finger-tips, and it came almost without hesitation and carried it off; this was repeated several times. Two days later he would perch on my finger and take a nut from between my teeth, or would sit on a branch and let me touch him while he was eating a nut. * * *

He grew very tame that winter, and would often swing head downward from the peak of my cap, or cling to my lips and peck at my teeth. If I held my hand out with nothing in it, he would always hop to my thumb, and peck the nail two or three times, then hold his head on one side, and look into my eyes, as if to ask me what I meant. * * *

I tamed several more Chickadees that winter; eight out of twelve, as nearly as we could count, were quite tame.

It was rather amusing when I took the 22 rifle to shoot rabbits! After the first shot was fired, I was attended by several Chickadees. They made aiming almost impossible, for every time I raised the rifle, one or two birds would perch on the barrel completely hiding the sights.

Many of us have had somewhat similar experiences.

Harrison F. Lewis (1931) describes an extraordinary experience with a chickadee that he believes was not previously tamed. He writes:

On a chilly day, with drizzling rain, about the year 1915, as I was walking on the outskirts of Wolfville, Nova Scotia, I saw a Black-capped Chickadee *(Penthestes atricapillus atricapillus)* feeding in a leafless alder bush. There was nothing unusual in its appearance, but the fact that it did not seem to heed me in the least when my path led me within a few feet of it attracted my attention. Wondering a little how near the bird I would have to go before it actively evaded me, I paused a moment, then stepped slowly in its direction. When I had advanced to the outer twigs of the bush in which it was busily feeding, it still appeared unaware of my presence, so, while expecting to see it fly away at any moment, I slowly extended my hand toward it. When my fingers were close to it I suddenly closed them upon it and had it securely in my grasp. The Chickadee seemed greatly surprised at this occurrence and struggled violently for a moment in a futile attempt to free itself, but I believe that my own surprise was equal to that of the bird, for I had confidently anticipated its escape rather than its capture.

When I had recovered a little from the first shock of unexpected success, I began to doubt whether the Chickadee could be in good health. "Perhaps," I thought, "it has from some cause lost the ability to fly." I took it into a neighboring house and showed it to one or two other persons, holding it in my hand all the while, then I carried it to the open door and released it. It flew away at once with strong, sustained flight as though in the best of condition.

On the other hand, William H. Longley (MS.) speaks of "a chickadee incubating seven eggs which would bite and buffet our fingers if we put them too close, while the mate fed near by, only occasionally raising its voice expressing what may have been an objection to our presence."

The following quotations refer to the roosting habits of the chickadee. Lynds Jones (1910) says: "On numerous occasions I have started them from their night roost in the thick of a leafy grape vine in midwinter." And Henry D. Minot (1895) recounts the following observation: "February 10th. This afternoon, just before sunset, I noticed two Chickadees, feeding on the ground, and pecking at a bone, to which a remnant of meat was attached. * * * They scarcely left the ground * * * until half-past five, when one flew away over the housetop and disappeared. The other continued to hop about on the ground; and then, without any intimation of his purpose, abruptly flew to the piazza, whether I followed him. He took possession of a Peewee's nest, which stood upon the top of a corner-pillar, adjoining the house, and, having stared at me for a moment, *tucked his head under his wing,* and apparently leaned against the wall. * * * Another retires as regularly at sunset, and sleeps in a hole of a white birch, evidently once a Chickadee's nest, perhaps his own." Eugene P. Odum (MS.) says:

In fall and winter most individuals roosted in dense conifer branches rather than in cavities. However, during the winter, two cavities were discovered where single birds were known to spend the night.

There was a definite tendency for chickadee groups to roost in the same area each night, so that it was possible to station oneself at a known roosting place and observe the birds coming to roost. The flock was usually scattered, individuals seeking places in the dense foliage of different trees. In contrast with the noisy behavior of many species roosting in flocks, chickadees retire with very little calling or ceremony.

As the flocks break up and pairs form in the spring, the winter roosts were abandoned. During early spring movements the pair seems to roost wherever convenient. After the nesting cavity is excavated and the nest material carried in, the female apparently may spend the night in the cavity even before incubation begins. The male roosts outside in some tree nearby. Likewise, during incubation and the feeding of the young the female sleeps in the cavity and the male somewhere outside. After the young are twelve days old, or older, the female may remain outside at night. When the young have left the nest, neither they nor the adult birds were observed to return to the cavity. The first night out the young and adults roosted wherever they happened to be.

If we are near a chickadee when it it flitting about in a tree, making short flights from twig to twig, we hear each time it flies a faint, rustling whir of wings, or sometimes two or more whirs, if the distance be longer. This is the chickadee's method of flight—a delicate, quick flutter, and a pause, then a flutter again. When crossing a wide, open space, the bird flies slowly, undulating in the air a little—each flutter of its wings carries him upward a little way, and during the pause between the flutters he sinks again.

Katharine C. Harding (1932) reports a banded chickadee at least 7½ years old, and Dorothy A. Baldwin (1935) another of the same age. Mr. Wallace (1941) reports one that was 9 years old.

Lester W. Smith (MS.), writing to Mr. Bent, gives an instance of the intelligence of the chickadee. He says: "Among the dozen or more species commonly taken for banding in my Government-type sparrow trap, the black-capped chickadee was the only species with instinctive intelligence to remember its way out. This trap, with its entrance under inward-sloping wires, was successful through the failure of most birds to remember just how and where they came in and the confusion that resulted when escape was found impossible in any general direction, particularly upward. The chickadee, selecting a sunflower seed from among the mixed bait in the trap, went in, not to eat the seed there, but to get it out to where it could be opened on a branch. The little bird at its first visit would walk around the trap until the low entrance was discovered, then dart in, select a seed, and, if nothing disturbed it, head back whence it came and with little investigation find its way out. They rarely became confused as did the juncos, tree sparrows, and purple finches. After the first trip in and out the same individual would fly directly to the entrance and as directly out again after he had grabbed the seed. If I shifted the position of the trap on the same spot, or moved it to a new location, the trail was learned after one trial."

Voice.—The chickadee is a voluble little bird; when two or more are together they are full of conversation, exchanging bright, cheery remarks back and forth. The notes show great variety and extend over a wide range in pitch. Some of the minor ones are very high indeed, closely approaching the insectlike voice of the golden-crowned kinglet and the brown creeper; one, the familiar "phoebe" note, an "elfin whistle" Langille (1884) calls it, is a pure, prolonged tone so low that we can imitate it by whistling; others, lower, but high-pitched, remind us of short words or phrases given in a babylike voice.

The simplest of the notes mentioned above is uttered rather listlessly, thus differing from the kinglet's energetic delivery; it is sibilant but given with a hint of a lisp, suggested by the letters *sth*. It is a faint note, but it may serve to report one bird's whereabouts to another not far away. This note, emphasized and prolonged into *stheep*, is often given in flight, or when a bird is slightly disturbed. It may be doubled. By further emphasis and repetition into a sharp, rapid series, *si-si-si-si*, it serves as a warning or alarm note; we hear this form when a hawk comes near.

Of the "phoebe" whistle, Aretas A. Saunders (MS.) says: "There are two notes of equal length, the second tone lower in pitch than the first. The quality is that of a clear, sweet whistle. The pitch is commonly B-A or A-G, in the highest octave of the piano. Frequently the second note has a slight waver in the middle, as if the bird sang *fee-beyee* instead of *fee-bee*. Rarely a bird drops a tone and a half between the two notes." Not infrequently two birds will whistle the "phoebe" note antiphonally, the second bird picking up the pitch at the end of the first bird's song and then dropping a tone lower, i.e., B-A, and the response A-G, over and over again.

It is a matter for conjecture whether the phoebe note is a true song of the chickadee. It is heard oftenest in spring and early in summer, but we hear it also throughout the winter, sometimes in cold, inclement weather, and it is uttered by both sexes, according to Dr. Jonathan Dwight (1897). Perhaps the deciding point in determining a true song is the manner in which the bird delivers its notes rather than their beauty to our ears. With this in mind, an observation by Bradford Torrey (1885) seems significant. He says:

For several mornings in succession I was greeted on waking by the trisyllabic minor whistle of a chickadee, who piped again and again not far from my window. There could be little doubt about its being the bird that I knew to be excavating a building site in one of our apple-trees; but I was usually not out-of-doors until about five o'clock, by which time the music always came to an end. So one day I rose half an hour earlier than common on purpose to have a look at my little matutinal serenader. My conjecture proved correct. There sat the tit, within a few feet of his apple-branch door, throwing back his head in the truest lyrical fashion, and calling *Hear, hear me,* with only a breathing space between the repetitions of the phrase. He was as plainly *singing,* and as completely absorbed in his work, as any thrasher or hermit thrush could have been. Heretofore I had not realized that these whistled notes were so strictly a song, and as such set apart from all the rest of the chickadee's repertory of sweet sounds; and I was delighted to find my tiny pet recognizing thus unmistakably the difference between prose and poetry.

Francis H. Allen tells me that he has several times heard a chickadee similarly engaged, also early in the morning.

Among the several notes that lend themselves to syllabification is the well-known *chicka, dee-dee.* Aretas A. Saunders (MS.) says of it that it "is more variable than many suppose. While it is most commonly one *chicka* followed by three or four *dees,* it may vary from one to ten *dees,* and there are sometimes two *chickas.* The *chicka* is, as a rule, two tones higher than the *dees,* and the pitch is B on the *chicka* and G on the *dees,* in the next to highest octave on the piano."

Another pretty note may be written *sizzle-ee,* or, when it falls in pitch at the end, *sizzle-oo.* A single bird often gives this phrase over and over, sometimes alternating the two forms, and two birds may make a two-part song of them, singing back and forth. The prettiest note of all, and the most delicate, is a prolonged jingling—as if tiny, silver sleigh-bells were shaking.

Field marks.—The chickadee is a round, fluffy little bird, boldly marked with splashes of gray, black, and white in contrast to the streaks, lines, and pencilings characteristic of many of the smaller birds. The white side of the head, separating the black areas above and below it, shines out brightly and forms a good field mark even in the distance. The short bill and the fur-coat appearance of the plumage distinguish the chickadee from any of the warblers with their slender bills and sleek, elegant stylishness. And the invisible eye, hidden in black feathers, sets the chickadee apart from the kinglets, even when colors are obscured by the dark shadows of evergreens.

Enemies.—The smaller, fast-moving hawks often capture a chickadee, but the little bird is so watchful for danger and so quick in its movements that it sometimes escapes from an attack. Tertius van Dyke (1913) reports a narrow escape of a chickadee (aided by him, to be sure) from the strike of a sparrow hawk.

The northern shrike, too, is the chickadee's enemy, but it is not always successful. Some years ago I (Winsor M. Tyler, 1912) described a case in which a chickadee out-maneuvered a shrike thus:

Jan. 27, 1910. This afternoon (2 P.M.) I watched for five or ten minutes a Shrike attempting to capture a Chickadee. My attention was attracted by the Chickadee's notes, *si-si-si-si, dee-dee-dee,* and I found the bird hiding in an isolated red cedar tree, while the Shrike was doing his best to find him. The Chickadee made no attempt to leave the tree, but kept moving about, chiefly among the inner branches. The Shrike followed his prey as best he could through the network of fine twigs, but often lost sight of it, evidently, and, coming to an outside branch, sat quiet, listening.

When hard pressed, the Chickadee flew out and circled about the tree before diving in among the branches again. After these flights, sometimes he entered the tree low down, and then mounted to the very top by a series of short, rapid hops; sometimes, after flying to the apex of the tree, he passed downward to the lowest branches before flying again. Several times the Shrike hovered in the air, and holding his body motionless and upright, peered into the tree. Finally, although not frightened away, the Shrike gave up the chase.

Chickadee's nests are so carefully hidden away, and the entrance is generally so small, that cowbirds rarely find and enter them. There is, however, an instance of parasitism of unquestionable authority. Fred M. Packard (1936) reports: "On May 25, 1936 a Black-capped Chickadee's nest, containing four Chickadee eggs and two Cowbird's eggs, was found in a nesting box at the Austin Ornithological Research Station at North Eastham, Massachusetts. * * *

"The opening in this box was one and one-half inches in diameter, much larger than the usual entrance to Chickadee nests, and ample to permit the intrusion of Cowbirds."

Dr. Herbert Friedmann (1929) lists another recorded instance from Ravinia, Ill.; an egg was reported to be in a nest of the Carolina chickadee; but the locality would seem to indicate that it was the more northern species.

Harold S. Peters (1936) lists, as external parasites on this chickadee, a louse (*Ricinus* sp.), the larva of a fly *(Ornithoica confluenta),* and a mite *(Analgopsis passerinus).*

Fall.—It is certain that in fall a good many chickadees either migrate or at least wander about extensively. We meet them at this season in localities where they never breed, often in thickly built up sections of large cities. Speaking of the occurrence of chickadees on the Public Garden in Boston, Mass., Horace W. Wright (1909) says: "In the autumn Chickadees are much more in evidence [than in spring], as they quite regularly appear in the Garden and continue their stay into November; and, as already intimated, on two occasions two birds remained through the winter and were seen at intervals up to the end of March. Sometimes small flocks have appeared in October which numbered four, five, or six birds." In September, October, and November I have seen them also in smaller open places in Boston, such as a vacant lot surrounded by several square miles of city blocks.

Dr. Wilbur K. Butts (1931), during an able study of the chickadee by means of marked individuals, attempted to determine the extent of migration of the species at Ithaca, N. Y. Even with the aid of colored bands, the evidence of migration, except in minor degree, seemed not conclusive to him, as his following summaries show. He says:

In considering these evidences of a migratory movement, it should be remembered that even if birds appear to be more numerous during the winter, it is not proved that there really are more individuals present. Many birds are so much more conspicuous in winter than in summer that they may seem to be more abundant. The distributional records show that there is a movement of Chickadees, but it is not proved that there is a distinct north and south migration.

Bird-banding operations at some stations seem to indicate that there is an arrival of Chickadees in the fall and a departure in the spring, but the records have as yet no proof of a distinct north-and-south migratory movement. Published records show only two Chickadee recoveries at points other than the place of banding. These two were recovered at distances of only three and twenty miles. The records do show, however, that there are many permanent resident individuals. The records at most stations do not show whether there are more individuals present in winter than in summer, since at most stations few Chickadees are trapped in the breeding-season. Individuals which are recorded only during the winter months may really be present throughout the year. * * *

The records seem to indicate, also, that there are very few birds passing through Ithaca in the fall. Only four birds were recorded but once. It should be remembered, however, that transient visitants are much less likely to get caught than are the resident individuals. Accordingly, there may have been more individuals passing through than the records seem to indicate. All through the fall many unbanded birds, which may have been transients, were seen.

The evidence shows that there were but few, if any, arrivals from the South in the spring. * * *

Since some of the records in the North indicate a greater abundance of birds in fall and spring, it is possible that there is a migration of birds from the extreme northern part of their range, where we as yet have no records, and this may account for the increase in numbers of the Chickadee in the United States.

Additional evidence of southward migration is furnished by the following note by William Palmer (1885): "This bird has been very abundant here [near Washington, D. C.] during March and April, nineteen specimens having been taken, while many others were seen. Owing probably to the severe winter they were driven south, returning about the middle of March. The first specimens were taken on March 15, and others were taken every week until April 19, when six were shot and many others seen. The weather during April was fine and warm, and the birds were singing and appeared quite at home. But few *P. carolinensis* were seen until the last week in April, showing that they too had been driven much further south."

W. E. Saunders has sent us some notes on the migration of chickadees at Point Pelee from 1909 to 1920, from which it appears that the fall migration there is very irregular. On many days there would be none at all, and then for several days there might be as many as 300 or 400 of these birds. He says: "Usually there are none, but once in a while there is a flight, perhaps (probably) endeavoring to cross the lake; it takes some time to taper off this flight and return to the normal status of none at all. * * * I have always thought this chickadee matter very interesting, and can still remember the first big flight, when, after years of scarcity, all of a sudden chickadees were everywhere; it was fun to watch them down at the last trees, making ineffective little flights up into the air and then settling back into the trees. They had not enough of the migratory instinct to get across. These birds were, doubtless, from stock bred south of the Georgian Bay, and they had never crossed any large body of water."

Mr. Wallace (1941) cites two cases where banded chickadees have been taken at 50 and 200 miles, respectively, southwest of the point of banding; and he says that there are six returns recorded in Washington that might be regarded as long range.

Winter.—Chickadees, collected in small loose flocks, spend the winter roving about the woodland. The birds scatter out a good deal, so much so that they must often lose sight of one another, but they keep continually calling to one another, using their fine, lisping note or the louder *chickadee,* and thus indicating the direction in which the flock is moving. They seldom wander far from the protection of trees and shrubs but occasionally venture out a little way into a field or marsh if there are isolated bushes there in which they can perch and feed. As the flock moves along, each bird examines minutely bark, twigs, and branches, searching for tiny bits of food—spider's eggs, cocoons, or other dormant insect life. The flocks are not large, being seldom composed of more than a dozen birds, but they generally contain too many birds to represent only a single family.

Whenever we go out in the country we meet these cheery little roving flocks—pleasant companions who enliven the dreary, New England winter. Mr. Wallace's (1941) studies indicate that winter flocks "are remarkably constant in individual composition, the same individuals remaining together day after day through the winter, and, as far as survival permits, winter after winter."

DISTRIBUTION

Range.—North America in general, from the limit of trees south to the central United States; not migratory.

The range of the chickadee extends **north** to southern Alaska (near Holy Cross Mission, Knik, and Valdez); southern Yukon (Lake

Marsh); southeastern Mackenzie (probably Fort Simpson and Willow River); northern Alberta (Smith Landing and Fort Chippewyan); northern Saskatchewan (south end of Reindeer Lake); central Manitoba (Eckimamish River); probably northern Ontario (Fort Albany and Moose Factory); southern Quebec (Godbout, Seven Islands, and Natashkwan River); and Newfoundland (Nicholasville and St. John's). The **eastern** limits of the range extend south along the Atlantic coast from Newfoundland (St. John's) to Massachusetts (Nantucket); Long Island (Huntington); and northern New Jersey (Passaic); and in the mountains to western Maryland (Bittinger); southern West Virginia (Cranberry Glades); and western North Carolina (Mount Mitchell). **South** to west-central North Carolina (Mount Mitchell); eastern Tennessee (Mount LeConte); central Illinois (Philo and Rantoul); central Missouri (Marshall and Warrensburg); Kansas (Neosho Falls and Wichita); southern New Mexico (probably Capitan Mountain); Arizona (San Francisco Mountain); and northern California (Callahan). The **western** limits of the range extend northward in Pacific coastal areas from northern California (Callahan) to Alaska (Kodiak Island, Katmai, and near Holy Cross Mission).

While there are records in winter north to the limits of breeding or beyond, as Alaska (St. Michael) and central Quebec (Lake Mistassini) there is sometimes a slight movement southward at this season. Winter occurrences are south to southern North Carolina (Mount Pleasant); southern Indiana (Bloomington and Carlisle); southern Kansas (Independence and Harper); and on the Pacific coast to Eureka, Calif.

Among more than 1,700 return records of chickadees banded in Massachusetts, only two are for points outside of that State. One, banded at Westfield on November 4, 1925, was found dead a short time later (before December 31), at Stratford, Conn. The other, banded at Amherst on October 7, 1932, was caught by a cat at Belvidere, N. J., on December 24, 1932. The files of the Fish and Wildlife Service contain the records of several thousand other banded chickadees, almost all of which were recaptured at the points of banding.

The range as outlined is for the entire species, of which seven subspecies are currently recognized. The typical form *(Parus atricapillus atricapillus)* is found in the Eastern United States, Canada, and Newfoundland, west to Missouri, Illinois, Minnesota, and western Ontario; the long-tailed chickadee *(Parus atricapillus septentrionalis)* is found chiefly in the Rocky Mountain region, breeding from the Kenai Peninsula of Alaska and central Mackenzie south to eastern Oregon and northern New Mexico and ranging eastward to northern Manitoba, western Minnesota, western Iowa, and eastern Kansas; the Oregon chickadee *(Parus atricapillus occidentalis)* is found from southwestern British Columbia south to northwestern California; while the Yukon chickadee *(Parus atricapillus turneri)* occupies the Hudsonian Zone of northern Alaska to the north and west of Cook Inlet. *P. a. bartletti* has been described from Newfoundland; *P. a. practicus* from the central Appalachian region; and *P. a. nevadensis* from northeastern Nevada.

Egg dates.—Alberta: 6 records, May 12 to 23.

Illinois: 17 records, April 20 to June 11; 9 records, May 2 to 16, indicating the height of the season.

Kansas: 16 records, April 10 to June 3; 8 records, April 21 to May 14.

Massachusetts: 27 records, May 7 to July 12; 13 records, May 20 to 29.

Nova Scotia: 5 records, May 21 to June 6.

Oregon: 57 records, April 13 to June 30; 28 records, May 8 to 18.

West Virginia: 13 records, April 22 to May 29.

PARUS ATRICAPILLUS PRACTICUS Oberholser
APPALACHIAN CHICKADEE

This form of black-capped chickadee, which is supposed to range from eastern Ohio to southwestern Pennsylvania and southward through the mountain region to North Carolina, is described by Dr. H. C. Oberholser (1937) as "similar to *Penthestes atricapillus atricapillus,* of Canada, but smaller, particularly the tail; upper parts darker, more grayish, less ochraceous, particularly in winter; wing-coverts and rectrices with narrower white edgings." Dr. Oberholser gives the range as: "Resident and breeds chiefly in the Appalachian Mountains from southwestern North Carolina, north through western Virginia, West Virginia, southwestern Pennsylvania, to central eastern and northeast central Ohio." Presumably the habits do not differ from those of the typical race.

PARUS ATRICAPILLUS BARTLETTI Aldrich and Nutt
NEWFOUNDLAND BLACK-CAPPED CHICKADEE

John W. Aldrich and David C. Nutt (1939) found this chickadee "abundant in all the thickets in eastern Newfoundland." They describe it as "similar to *Penthestes articapillus atricapillus* but darker and more brownish above, and darker buff on flanks and under tail-coverts. White edgings to wing and tail feathers narrower. Bill larger. In color nearer to *P. a. occidentalis* than to any other known race but

larger." It is their opinion "that a very well marked race is represented in Newfoundland with characters most pronounced in the
eastern part of the island and specimens from the western part distinctly
intermediate with *P. a. atricapillus.*"

I saw a few chickadees near Bay of Islands and along Fox Island
River in western Newfoundland but did not collect any specimens.

PARUS ATRICAPILLUS SEPTENTRIONALIS Harris
LONG-TAILED CHICKADEE

HABITS

For nearly a hundred years the long-tailed chickadee has been recognized as a midcontinent race of our familiar little chickadee, ranging
from Alaska to New Mexico in the Canadian and Transition Zones. It
is described by Ridgway (1904) as "similar to *P. a. atricapillus,* but
larger, with wing and tail averaging decidedly longer; coloration paler,
with whitish edgings to greater wing-coverts, secondaries and lateral
rectrices broader, more conspicuous."

P. A. Taverner (1940) has recently made an enlightening study of the
Canadian status of this subspecies, based on a series of 99 specimens in
the National Museum of Canada, including—

17 males and 14 females from southern Ontario, eastward to Nova Scotia;
19 males and 17 females from Manitoba, Saskatchewan and Alberta; 20 males and
12 females from the interior of British Columbia. Specimens from coastal areas
of British Columbia referable to *P. a. occidentalis* were not included. These were
all taken within two hundred miles of the international boundary and can reasonably be assumed to be of local breeding stock. They are from areas separated
from each other by some eight hundred miles and include no specimens between. * * *

In these birds, laid out in comparable seasonal mass groups, a highly critical
eye can detect a slight average color distinction as above postulated [by Ridgway],
between the prairie and the eastern birds, but not enough to be readily detected
and not consistent enough for the recognition of individual specimens. Every
individual in one group can be matched in the other, and no single specimen can
be confidently identified by this character, though winter and early-spring birds,
before the wear and fading of nesting activity, show slightly more color differences than in later season.

His study of the tail measurements shows that "there is a large overlap in extreme measurements: the smallest *atricapillus* is only 0.2 inches
(5.08 mm.) smaller than the smallest *septentrionalis.* The largest *septentrionalis* is only 0.05 inches (1.27 mm.) larger than the largest *atricapillus.*" The wing measurements give "even less definite results. * * *
There is a large overlap in extreme measurements; the smallest *atricapillus* is only 0.1 inch (2.54 mm.) smaller than the smallest *septentrionalis;*
the largest *septentrionalis* is no larger than the largest *atricapillus.*"

A study of Mr. Taverner's paper must convince the reader that *septentrionalis* is one of those millimeter races, based on averages, that many of the individuals of this race cannot be definitely recognized as such from the characters alone, and that his figures "open the question of the desirability of formal recognition of subspecies of which few or no individuals can certainly be referred to their proper race by physical characters without reference to the geographical origin."

In his notes from Montana, Aretas A. Sounders writes: "The only difference I have noted in the field between long-tailed and black-capped chickadees is that of habitat, the long-tailed being chiefly an inhabitant of willow thickets and cottonwood groves, not extending its range up into the evergreen forests of the mountains, as the blackcap does in the Adirondacks. Perhaps this is partly because the evergreen forests are inhabited by mountain chickadees." Other observers seem to agree that throughout its range the favored haunts of this chickadee are the open, sunny, deciduous woods, such as poplar, aspen, willow, and cottonwood groves, and that it seems to avoid the dense coniferous forests.

Nesting.—The nesting habits of this chickadee seem to be similar to those of its eastern relative. An unusually low nest was discovered by Harry S. Swarth (1922) in the Stikine River region "in a tract of rather open woods, mostly of small poplars. It was in a dead poplar stub about three inches in diameter, a mere shell of dead and decayed wood, hardly strong enough to hold the tightly packed and rapidly growing young, who did actually break through the wall at one place. The entrance hole was five inches from the base, the nest itself, flush with the ground. The lining appeared to be entirely of matted moose hair."

E. S. Cameron (1908) says that, in Montana, this chickadee sometimes "nests in small deep holes of high dead pines. On June 15, 1903, a pair of Chickadees were seen to be greatly excited over a strip of rag hung in a pine on Cottonwood Creek, Dawson County. They hovered about it, meditating an attack, but with each breath of wind the flag fluttered, and frightened away the birds which returned when the wind ceased. This strange behavior on their part induced me to investigate, when I found their nest of wool, hair, and grass in a very small hole below the rag. * * * The birds' fears were entirely allayed when I wrapped the offending rag around the branch."

Lee R. Dice (1918) found nests in process of construction in southeastern Washington early in April, "in the decayed wood of orchard or shade trees." He continues:

The process of nest excavation was watched for a short time on April 10, 1914. The nest was being excavated in the rotten heart of a pear tree, and entrance was obtained through the end of a stub about four feet from the ground. The male and female took part equally in the work, and the labors were continued throughout the day. A vigorous pecking could be heard while either bird was at work. The excavated material was carried in the bill a distance of ten yards or more from the nest before being dropped. It was not dropped in the same place each time, but was scattered over a wide area. Usually the birds alighted on some branch before dropping the debris, but sometimes it was dropped while the bird was flying. As soon as one bird left the hole the other entered immediately. Sometimes the bird outside had to wait a short time. Between 12 M. and 1 P.M. the average time each bird spent in the nest hole was thirty seconds and the shortest time four seconds.

Eggs.—The eggs of the long-tailed chickadee are indistinguishable from those of the black-capped chickadee. The measurements of 50 eggs in the United State National Museum average 15.7 by 12.2 millimeters; the eggs showing the four extremes measure 17.5 by 12.7, 15.1 by 12.9, 14.2 by 11.9, and 15.2 by 11.2 millimeters.

Young.—Mr. Swarth (1922) watched the young being fed at the nest referred to above and remarks: "Both parents carried food to the nest assiduously after foraging expeditions that lasted from two to five minutes. In approaching the nest, the old birds came through the trees and bushes until within about eight or ten feet of their destination; then they dropped to the ground and hopped to the entrance [only 5 inches from the ground]. To the casual observer they disappeared at a point some distance from the nest, and it was not until they had been observed for some time that this subterfuge was detected. The staple food that was being brought to the young was a small green caterpillar infesting the poplars at that time; also a white grub, a green katydid, and many small mosquito-like insects."

The plumage changes, food, behavior, and voice of this chickadee are all similar to those of the familiar chickadee of the East and need not be mentioned here. But the following note from Claude T. Barnes is of interest:

"On January 4, 1924, at an altitude of 5,000 feet in the mountains near Salt Lake City, while wading through the deep snow of City Creek Canyon, I was attracted by the thin, oft-repeated *tchip* of half a dozen chickadees that were busy in the upper branches of some birch trees (*B. fontinalis utahensis*). I noticed one take a peck at one of the numerous birch catkins, which, like Kaiser brown caterpillars, were suspended from the branches, and then instantly thereafter work with its bill against a limb, as if trying to get the kernel from a nut. In a second it made another peck followed by another working with its bill against a limb, and so on, hopping from twig to twig and constantly uttering its companionable *tchip*. Rarely did it sing *chick-a-dee-dee-dee*, though three or four times in an hour the familiar notes did come from the flock."

PARUS ATRICAPILLUS NEVADENSIS Linsdale
PALLID BLACK-CAPPED CHICKADEE

Dr. J. M. Linsdale (1938b) gives the above names to a local race of the long-tailed chickadee, which he says is "resident along streams in the Snake River drainage system south of the Snake River, in northeastern Nevada and southern Idaho." He describes it as "similar to *P. a. septentrionalis,* but coloration paler, with whitish edgings to greater wing-coverts, secondaries and lateral rectrices broader, more conspicuous, thus reaching the extreme in these respects for the species, but close to *P. a. turneri* from which it differs in larger size."

PARUS ATRICAPILLUS OCCIDENTALIS Baird
OREGON CHICKADEE

PLATES 52, 53

HABITS

This western form of our familiar blackcap occupies the northwestern coast from extreme southwestern British Columbia to extreme north-western California. It is described by Ridgway (1904) as similar to *P. a. atricapillus* but decidedly smaller (except bill and feet) and colora-tion very much darker; back varying from deep mouse gray or very slight buffy slate-gray in spring and summer to deep hair brown or light olive in fall and winter plumage; sides and flanks (broadly) pale grayish buff in spring and summer, deep brownish buff, wood brown, or isabella color in fall and winter; whitish edgings of innermost greater wing-coverts, secondaries, and exterior rectrices more restricted than in *P. a. atricapillus.*"

Its haunts and habits are similar to those of our familiar eastern chickadee. W. E. Griffee tells me that it "is abundant all through the hardwood timber of western Oregon valleys but is very much less common in the coniferous timber, particularly during the nesting season."

S. F. Rathbun says in his notes that "it is a Transition Zone species and not to be found at any considerable altitude; the highest we have noticed it being 550 feet; it seems to prefer the lowlands," in western Washington.

Nesting.—Mr. Griffee says (MS.): "Probably nine-tenths of the nests are in dead willow, cottonwood, and alder trees and stubs, at least one of these three tree species being found wherever the Oregon chickadee nests. Nesting holes usually start with an irregular opening about 1¼ in diameter and are 8 to 10 inches deep. Often they are within 3 or 4 feet of the ground and rarely over 10 or 12 feet. Cavities, which are about 3 inches in diameter at the bottom, are invariably lined with a layer of green moss, often at least an inch thick. Upon this layer of moss is a thick lining of rodent fur, cow hair, and other hairy material.

"Nests containing incomplete sets practically never are occupied by the birds during the day, a coverlet of fur being drawn over the eggs while the birds are away. Incubating birds sit tightly and try to frighten the intruder by a hiss and flutter of the wings when an inquiring finger is poked into the entrance of the nesting hole."

Eggs.—The Oregon chickadee lays four to ten eggs. Out of 28 sets reported by Mr. Griffee, 18 sets consisted of eight eggs; there were only two of six, four of seven, three of nine, and only one of ten. The eggs are practically indistinguishable from those of the black-capped chickadee. The measurements of 40 eggs average 15.6 by 12.0 millimeters; the eggs showing the four extremes measure **17.0** by 12.0, 16.2 by **12.3, 14.6** by 11.7, and 14.8 by **11.5** millimeters.

The plumage changes, food, behavior, voice, and other habits are apparently similar to those of the closely related eastern race.

Mr. Rathbun observed some of these chickadees eating tent caterpillars, near Seattle, of which he writes in his notes: "It was toward evening, and seeing three of these birds very active in a lilac bush, we stopped to watch them. Near the top of the bush was a tent caterpillars' web of small size; as the worms were crawling toward it, some of them would be seized by the chickadees for food. During a space of less than five minutes the three chickadees captured and ate eight of the worms of large size. When one was caught the bird would beat it about a number of times and then, holding it with one foot on its perch, leisurely tear off pieces with its bill, which it then ate. The birds must have been feeding on the caterpillars for some time before we noticed them, for, on looking over the ground under the bush, many dead and mutilated worms were to be seen strew about."

PARUS ATRICAPILLUS TURNERI Ridgway

YUKON CHICKADEE

HABITS

According to the 1931 Check-list, this chickadee "breeds in the Hudsonian Zone of northern Alaska north and west of Cook Inlet."

Ridgway (1904) describes it as "similar to *P. a. septentrionalis* but slightly smaller, coloration grayer above and more extensively or purely white beneath, and white edgings of greater wing-coverts, secondaries, and outermost rectrices broader, more purely white; in spring and summer plumage the gray of upper parts without perceptible tinge of buff, except on rump and upper tail-coverts, where very faint, and white of sides and flanks very faintly, if at all, tinged with buff; in fall and winter plumage the buffy tinge on sides and flanks very much paler than in *P. a. septentrionalis."*

Under the name long-tailed chickadee, with which this race was included by some of the earlier writers, Dr. E. W. Nelson (1887) writes:

Throughout the wooded region of Alaska, from the moist, heavily-wooded coast in the Sitkan and Kadiak region north throughout the entire Yukon and adjoining country, this bird is a common resident. Specimens were secured both at Cook's Inlet and Kadiak by Dr. Bean. I secured specimens from various places throughout the northern portion of the Territory, at times even along the barren sea-coast, where it only found shelter in the stunted alder or weed patches. Its visits to the coast, however, were mainly in roving parties during spring or fall. A few days of mild weather, at this season, are almost sure to bring some of these familiar birds about the coast settlements, and its familiar *dee-dee-dee* is a welcome sound on the clear frosty mornings which usher in the stinging blasts of winter, or announce the approach of spring. One meets it again while traveling through the silent snow-clad forests of the Yukon, as he tramps wearisomely on, until the mind is unconsciously affected by the lack of animation. At such times, as we move mechanically forward, the shrill, strident note of the Chickadee, as the bird eyes us from its swinging perch on a bush close at hand, breaks the silence and diverts the mind. Frequently the chorus of their Lilliputian cries arise from the bushes all about as the jolly company of harlequins swing and balance their tiny bodies and pass on as though too busily intent upon affairs of importance to stop. After their passage the forest resumes its cheerless silence once more, and the heavy breathing of the icy wind through the tree-tops or the sharp report of the contracting ice in the river are the only accompaniments of the toilsome march.

I have no information on the nesting habits, eggs, food, and other habits of the Yukon chickadee.

PARUS CAROLINENSIS CAROLINENSIS Audubon

CAROLINA CHICKADEE

PLATE 54

CONTRIBUTED BY EDWARD VON SIEBOLD DINGLE

HABITS

The Carolina chickadee, one of the four birds discovered by Audubon in the coastal part of South Carolina, is the low-country representative of the Boreal chickadee *(atricapillus);* yet *carolinensis* by no means confines itself to the Atlantic and Gulf coastal plains. It ascends the Blue Ridge Mountains probably higher than 5,000 feet, according to Brewster (1886), thus falling short by 1,500 or 1,600 feet of attaining the highest point in its range—Mount Mitchell, with an altitude of 6,684 feet. Brewster continues: "Common, and very generally distributed, ranging from the lowlands to at least 5,000 feet, and probably still higher. On the Black Mountains I found it breeding sparingly along the lower edge of the balsam belt, and thus actually mingling with *P. atricapillus.* In one place a male of each species was singing in the same tree, the low plaintive *tswee- dee- tswee - dee* of the *P. carolinensis,* contrasting sharply with the ringing *te - derry* of its more northern cousin. The fact that the two occur here together and that each preserves its characteristic notes and habits, should forever settle all doubts as to their specific distinctness."

Other observers have recorded the birds in summer at altitudes of 3,300, 4,400, and 5,000 feet. But the center of abundance is unquestionably the great swamp areas of the Coastal Plain, where the writer has found it to be one of our commonest birds. Any wooded territory attracts them, it seems, except possibly extensive pine woods. But even small towns and villages often have their chickadees, and the writer has frequently seen it as a backyard resident.

Except during the actual breeding period, chickadees are nearly always seen in small bands—family parties, as it were. Late in summer and in fall they are invariably associated with tufted titmice, yellow-throated and pine warblers, brown-headed nuthatches, and downy woodpeckers; later in the winter their ranks are increased by myrtle warblers and the two kinglets. In such foraging bands, the tufted tits appear as leaders, with the chickadees as next in command.

No bird has endeared itself to us as much as the chickadee; its gentle, confiding ways, soft colors, and saucy air, as well as its readiness to patronize feed trays, render it a universal favorite.

There is evidence that our chickadees, like other members of the titmouse family, remain mated for periods longer than one breeding season; Nice (1933) records a pair of Carolina chickadees in Ohio that were associated for three winters and two summers.

Nesting.—The Carolina chickadee is one of our early breeders; although much depends on whether the season is advanced or late, nest construction in the southern part of its range might begin as early as the first week or 10 days of February or as late as the end of March; in South Carolina excavations generally are begun in March, while in Ohio Wheaton (1882) says: "I have found the nest in this vicinity as early as the 18th of April, ready for the reception of eggs. The female sits very close, and is with difficulty driven from the nest."

The favorite nesting sites are fence posts and decayed stubs of small saplings. And like the chimney swift, the chickadee offers another example of a bird that has partly abandoned primeval nesting conditions in favor of man's more convenient replacements. The birds usually excavate their own burrows and select certain hardwoods or pines that are soft; but often peach and cherry stubs are used, and these have very hard outer bark. Dickey (MS.) lists 17 species of trees, all hardwood, that are used by *carolinensis;* to these the writer could add several pines and the peach. Woodpecker holes are sometimes used, as well as natural cavities in dead or live trees. Dickey writes that he has "known them to build in iron pipes used for clothes lines, pipes to support bridges, small bird houses, etc."

Both sexes excavate, but the female probably does the most in nest construction. The average height of the nest above the ground would be about 5 or 6 feet. Dickey mentions 1 foot as the minimum distance from the ground, while Erichsen's (1919) 22 feet must be considered the maximum.

According to Dickey, two weeks are required on an average to complete a burrow, but obviously this would depend on the depth of the excavation and the softness of the wood. In a peach stub, for example, after the hard shell has been pierced, the going would be easy. On the other hand, oak would be consistently harder to excavate. The same observer gives average measurements of the completed burrows and entrance holes, as follows:

Diameter of aperture ....1⅜ inches
Depth of cavity ....5 inches
Width of cavity at aperture ....2 inches
Width of cavity at nest enlargement ....2⅜ inches

When the hollow has been excavated, nest-building is begun, as Dickey says, "with thick foundation of moss *(Hypnum)*—strips of yellow and brown bark, a few strips of yellow grass and grass culms or panicles, a little thistle down or milkweed pod down, and then such bird feathers as those of sparrows, bluebird or of the parent. The cup is well padded with silvery milkweed or thistledown, animal hair, red hair of the cow, gray fur of the cottontail rabbit and fur also from deer, mice, and other Mammalia."

Erichsen (1919) writes: "Simultaneous with the appearance of the down on the stalk of the cinnamon and royal ferns, which occurs during the middle of March, the chickadee begins nest-building, for this material is used largely by the birds in lining their nests. As far as my observations go, the birds, in gathering the down, always begin at the top of the stalk and work downward. The green moss that collects on the trunks of certain species of hardwoods is also used to a considerable extent, being always placed in the nesting hole first, and upon it the down is deposited."

A constant habit of this bird is to build up one side of the nest higher than the other, thus making a flap, which is used to cover the eggs when the parent is away.

Both sexes take turns in incubating.

Dickey, in describing a nest in a sugar-maple fence post says: "Over a series of at least 15 consecutive seasons a pair of birds bred annually at this habitat."

Eggs.—[AUTHOR'S NOTE: The Carolina chickadee lays five to eight eggs; six seems to be the commonest number. These are practically indistinguishable from the eggs of the black-capped chickadee, with similar variations. They are white, more or less unevenly marked with fine dots, spots or small blotches of shades of reddish brown. Often the spots are concentrated about the larger end. The measurements of 50 eggs average 14.8 by 11.5 millimeters; the eggs showing the four extremes measure 15.9 by 11.8, 15.7 by 12.4, and 12.7 by 10.4 millimeters.]

Young.—According to Dickey, "the period of incubation is exactly 11 days."

The same observer thus describes a newly hatched bird: "Length when spread out 6/8 of an inch. Color a rich pinkish white, say salmon hue. There were mere cilia of gray down on the head, back of wing stumps, lower back or rump. The conspicuous eyeballs slate blue, meas. 3 mm. diameter. Leg length 7/16 of an inch. Wing stumps ⅛ in. long, 1/16 in. wide. Bill ⅛ in. long; ⅛ in. wide at base; a light horn color. * * * It was revealed that they remain inside nests exactly 17 days. They are showy at that stage; well coated with plumage that resembles closely that of the adults, mouse-gray coats and black heads. They are animated enough to snap at and grasp one's fingers."

Plumages.—[AUTHOR'S NOTE: The plumages and molts in this chickadee parallel those of the black-capped chickadee. The young bird leaves the nest in practically full juvenal plumage, but with short wings and tail. In this plumage it is much like the adult, showing the distinctive wing edgings, but the black of the head and throat is duller, and the whole plumage is softer. A partial postjuvenal molt occurs in summer, involving the contour plumage and the wing coverts, but not the rest of the wings or the tail; this produces a first winter plumage, which is practically adult. Adults have a complete postnuptial molt in summer, but apparently no spring molt; the wash of pale pinkish buff on the flanks, and to a lesser extent on the back, characteristic of the fall plumage, disappears by wear and fading before nesting time, producing a grayer bird.]

Food.—Howell (1932) writes: "The food of this species was studied by Beal (1916, pp 24-26), who examined 210 stomachs. Animal matter composed about 72 per cent, and vegetable matter 28 per cent, of the total contents. Nearly half (44 per cent) of the food for the year consisted of moths and caterpillars. Bugs appeared to be next in favor among the insects, including stink bugs, shield bugs, leaf hoppers, tree hoppers, plant lice, and scales. Ants, bees, wasps, beetles, cockroaches, and katydids were consumed in small numbers. Spiders were eaten in considerable numbers, composing more than 10 per cent of the total food. The vegetable food consisted principally of seeds of poison ivy (10 per cent) and of other unidentifiable seeds (12 per cent). A small quantity of blackberries and blueberries was eaten."

Judd (1902) says that "seven Carolina Chickadees *(Parus carolinensis)* were taken during February, April, July, and August. Vegetable matter—mulberry seeds, pine seeds, and ragweed seeds—was present in four stomachs. All the birds had eaten insects. One had eaten 1 bee (Andrenidae), 2 ants, 3 insect eggs, 3 spiders, and 3 caterpillars (measuring worms, Geometridae and hairy Arctiidae, which are usually avoided by birds). One of the stomachs examined contained katydid eggs and two others eggs of the wheel-bug."

The writer has often watched chickadees early in spring feeding on the eggs of certain moths or other insects; these are encased in silky yellow coverings and attached to the under sides of leaves of the live oak. The birds hunt through the trees, inspecting the under sides of the leaves, all the while uttering their soft, conversational notes. The leaves are picked from the twigs and carried to a convenient branch. Then holding the leaf with its feet, the bird tears the silk away and devours the eggs.

Blincoe (1923) records it as feeding late in summer on the seeds of the redbud tree, "swallowing them as fast as they could be removed from the pods."

Brackbill (MS.) describes their manner of eating honeysuckle berries, which consists of holding the berry between the feet, as do blue jays, and hammering with their bills. Further, "examination of some of its discards showed them to be drilled through like beads, and but one seed remained in each. It seemed likely that some pulp and skin had been eaten, also."

Kalter (1932) observed that Carolina chickadees were removing and dissecting flowers of the leafcup *(Polymnia canadensis* L.), and "investigation of the flower heads of this plant showed that most of them were infected with the small striped brown larvae of one of the Noctuid moths. The work of these insects seemed to cause the blossoms to rot and turn brown."

Skinner (1928) says that, in winter, "they are easily attracted to dooryards and about our homes by hanging up bones with bits of meat and gristle attached, to a tree or bush. They will also eat cheese and suet, and pick up bread and doughnut crumbs."

Voice.—There is probably no native bird song more pleasing than the music of this chickadee. It has not the loud, ringing quality of the tufted tit's song, which comes to us from the blossoming dogwoods, half a mile away. Its voice is rather weak and the song a very simple one, but the notes are exquisitely mellow, soft, and satisfying.

Wayne (1910) writes: "The song period begins about the middle of February and the sweet notes are always welcomed as the herald of spring." Dickey (MS.) thus describes the notes: "Usually it is detected as it scolds cats, screech owls, or a human intruder, whereupon it will vent syllables like *dee-dee-dee-dee; chick-ah-dee-dee-dee-dee; sprittle-chick-ah-dee-dee-dee-dee; dee-dee-dee-pee-stick-dee; pee-tee-dee-dee-spee-teetle; spick-spick-ut-uh-dee,* and *phe-bee.*" Sometimes there is a marked liquid flow of notes prior to the *chickadee* series as *sputtle-dee,* but this is hard to put down on paper. Perhaps it is well simply to say that their run of outcries are buzzing exclamations.

"The song has been described by some writers as 'the pumphandle strain,' and that will suggest its nature very well. Indeed this does remind one of the not unpleasant, old-fashioned sounds made by a windlass well. *Spee-deedle-dee-deedle-dee* is what the chickadee seems to say. I have heard it repeatedly on fair days in midwinter; it increases in frequency as March is ushered in; is pronounced everywhere in spring and early in summer (the breeding period); and casually a subdued, shortened song is vented on crisp autumn days, too."

Aretas A. Saunders (MS.) has this to say about the song of *carolinensis:* "In my experience, the best way to identify a Carolina chickadee in the field is by its song. The song is enough different from the black-capped chickadee to name the bird instantly. The other call notes are not so easily distinguished.

"The song consists of two clear whistled notes, but each one is either introduced or followed by a shorter, lower-pitched, sibilant note. That is, instead of the bird singing a simple *fee-bee,* it sings *sūfee-sūbee* or else *feesu beesu.* These sibilant notes are like whispers rather than whistles, and I have known observers not to notice them and to think there was no essential difference in the song of this chickadee and the blackcap. The pitch and pitch interval of the clear, whistled notes are about the same as in the black-capped chickadee, one tone between them, and the notes pitched on B-A or A-G, in the highest octave of the piano. The sibilant notes are on a different pitch than the whistled ones, sometimes higher and sometimes lower. I have comparatively few records of the song of this species, however. Occasionally a bird sings three clear whistled notes, the second and third each lower than the preceding one.

"The season of singing is probably similar to that of the black-capped chickadee, but I have no extended notes except for the spring of 1908, when I was in central Alabama. Then I heard this species sing daily from my arrival early in March till the end of April. In May I heard the song frequently, but not daily, and heard it once or twice early in June. The other notes of this species are similar to those of the blackcap, the *chickadee* call being rendered somewhat faster."

The writer has noticed that the Carolina chickadee, when disturbed on the nest, utters a peculiar note—an explosive little sound like a sneeze. My first experience with this note is described (1922) in part as follows: "Late in the afternoon of March 28, I tapped on the tree; one of the birds was inside and gave a peculiar note, not a hiss such as Mr. Schorger heard, but more like a little sneeze. This was repeated every time I tapped. Several times the bird tapped on the interior of the cavity. Finally it put its head out of the hole and looked calmly at me as I stood about three feet away. I withdrew and it went back into the hollow. No eggs had yet been laid."

Dickey (MS.) describes the sounds as "hisses, serpent-like, and feared by unsuspecting boys." Also, when examining young birds out of the nest, "the parents darted close to my hat, hissed not unlike a black snake, and vented a variety of buzzing and liquid outcries."

The writer has never seen chickadees when they were engaged in hissing, but Pickens (1928) has, and he describes the "defense demonstration" as follows:

One of the most courage-taking sounds that I have encountered in my field studies is the hiss of the copperhead snake *(Agkistrodon mokasen)*. It lacks the animating interest we find in the ringing alarm of the rattlesnakes, and fills one with a kind of nausea. The reptile sounds as if it were inhaling a good part of the surrounding atmosphere and then discharging it in one sudden, explosive puff. There is nothing sedate and leisurely about it. Now the best imitation of that sound that I have heard is the explosive hiss of the brooding Carolina Chickadee *(Penthestes carolinensis).* * * * In preparing for the hiss, the bird, as seen from above, appears to rise slightly on the legs as if to give a freer swing to the movements of the body, while the head is thrown back over the shoulders at a right angle, or even an acute angle. The attitude of the bird is one of tense rigidity. Then, as if with a great effort, the bird nods the head strongly forward. The whole body, with the wings and tail, seems affected. The tail moves, the expanding wings shoot out sideways and strike the surrounding wood inside the cavity, and as the head comes stiffly down the bird emits a strong hiss or puff strikingly like that of the copperhead. The head is brought down quite upon the surface of the lining in front of the bird, and while the noise appears to be produced in part by the stiff rustling of the feathers, and the reverberations within the hollow of the surrounding wood, much, or the greater part of the noise certainly comes from the mouth and throat, and the hiss sometimes dies out in a faint little vocal squeak. All combine to make a fearful noise, and while mimicry is of course an unconscious, or better, an unintentional occurrence, there is no mistaking what the noise is to be taken for.

According to other observers, this snakelike sound is also employed by the northern chickadee *(Penthestes atricapillus)*.

That the hissing note is not confined to North American chickadees is attested to by Jourdain (1929); he states that "the European Titmice produce warning noises in apparently exactly the same manner as the American Chickadee. I have frequently noticed this habit in the case of the British Great Tit *(Parus major newtoni)*, on at least one occasion in the British Coal Tit *(P. ater britannicus)*, and it is also characteristic of the British Blue Tit *(P. caeruleus obscurus)*. Mr. Pickens' description of the movements of the chickadee in producing this explosive hiss applies exactly to those of the Great Tit; but though well known to field-workers, there is little on record in the numerous books on British birds on the subject beyond a few references to 'hissing like a snake,' on the part of the setting Blue Tit."

Enemies.—Like all small birds, our chickadee has to be continually on the watch for small accipitrine hawks, cats, snakes, and small mammals; apparently the screech owl does not prey upon it to any extent, as the little bird tucks itself away at sundown in some hollow, too small for *asio* to enter.

That the cowbird occasionally imposes its domestic duties on the chickadee is shown by Friedmann (1938), who writes of "a nest containing five eggs of the chickadee and two of the cowbird, collected at Piney Point, St. Mary's County, Maryland, April 25, 1934, by E. J. Court, who tells me that he caught the female Cowbird on the nest, about half an hour after daylight."

Peters (1936) includes this bird in his list of avian hosts of external parasites; a Maryland specimen was found infected with lice of the species *Degeeriella vulgata* (Kell.), while mites [*Trombicula irritans* (Riley)], were taken from a South Carolina bird.

DISTRIBUTION

Range.—Southeastern United States; nonmigratory.

The range of the Carolina chickadee extends **north** to southeastern Kansas (Independence); central Missouri (Columbia and St. Louis); Illinois (Carlinville and Ravinia); Indiana (Indianapolis and Anderson); Ohio (Phelps Creek and East Liverpool); Pennsylvania (Washington and Doylestown) ; and central New Jersey (Princeton and Point Pleasant). **East** to the Atlantic coast in New Jersey, Virginia, North Carolina, South Carolina, Georgia, and Florida (Whittier). **South** to southern Florida (Whittier, Fort Myers, St. Petersburg, and Mulat) ; Louisiana (New Orleans, Bayou Sara, and Alexandria) ; and southeastern Texas (Houston and San Antonio). **West** to central Texas (San Antonio, Kerrville, Waco, Fort Worth, and Gainesville) ; Oklahoma (Wichita Mountains, Minco, Tulsa, and Copan) ; and southeastern Kansas (Independence).

The range as outlined is for the entire species, which has been separated into four geographic races. The typical Carolina chickadee *(Parus carolinensis carolinensis)* occupies all the range except Florida, where the Florida chickadee *(P. c. impiger)* is found, and the extreme western portion from northern Oklahoma south to the coast of Texas occupied by the plumbeous chickadee *(P. c. agilis)*. A northern race, *P. c. extimus,* is said to range from New Jersey west to Missouri and south to northern North Carolina and Tennessee.

Casual records.—This species has been recorded as an "accidental visitant" in western New York (Lancaster) ; one was taken at Ecorse, Mich., on July 17, 1899; and a specimen was collected at Keokuk, Iowa, on May 4, 1888.

Egg dates.—Arkansas: 9 records, April 11 to May 15.

Florida: 14 records, March 30 to April 27.

New Jersey: 11 records, April 12 to June 15.

North Carolina: 12 records, April 8 to May 12.

Texas: 45 records, Feb. 16 to May 20; 23 records, March 26 to April 20, indicating the height of the season.

PARUS CAROLINENSIS IMPIGER Bangs

FLORIDA CHICKADEE

HABITS

The Florida chickadee is a small, dark-colored race of the well-known Carolina chickadee. It is fairly common and well distributed over most of the Florida Peninsula, except perhaps the extreme southern part. I have found it almost everywhere that I have been in central Florida, mainly in the live-oak hammocks and around the edges of the cypress swamps.

Arthur H. Howell (1932) records it also in "open pine timber," and says: "A pair noted on the Kissimmee Prairie was occupying a small palmetto thicket, far from any large timber, a very unusual habitat."

Nesting.—Mr. Howell says that the nests "are placed in rotten stubs, usually 10 to 15 feet from the ground." A set of four eggs in my collection was taken by Oscar E. Baynard near Leesburg on March 30, 1910. The nest was about 5 feet up in a dead pine stub, about 5 inches in diameter, that stood on the edge of a small pond surrounded with pine trees. The cavity was about 7 inches deep; the nest was made of dry grass and lined with a few feathers and a considerable quantity of cattle hair and fur from a rabbit.

Frederick V. Hebard writes to me of a nest that was placed in the top of a 4-foot fence post along a road; "the nest was composed chiefly of dried grasses with a webbing of cypress bark strips and hairs, including two of a raccoon and one of a wildcat. The inside of the nest was softened with raccoon and fox-squirrel fur. The coon and wildcat hairs were probably taken from some of those trapped during winter."

Eggs.—Four or five eggs seem to be the usual numbers in the nests of the Florida chickadee, though very few data are available to the writer.

The eggs are practically indistinguishable from those of the Carolina chickadee. The measurements of 39 eggs average 15.1 by 12.1 millimeters; the eggs showing the four extremes measure 18.1 by 14.8, 17.8 by 15.3, 12.9 by 11.5, and 13.8 by 10.7 millimeters.

Except as affected by the difference in environment, the food, behavior, and voice of this chickadee are similar to those of the more northern race, and the plumage changes are apparently the same.

PARUS CAROLINENSIS AGILIS Sennett

PLUMBEOUS CHICKADEE

HABITS

The 1931 Check-list says that this southwestern race of *carolinensis* "breeds in the Lower Austral zone from northern Oklahoma to Refugio and Kendall counties, Texas." Mrs. Margaret Morse Nice (1931) says that it is rare in northwestern Oklahoma, but "resident in central Oklahoma from Tulsa and Hughes to Woodward and Jackson counties." In the region of Austin, Tex., George Finlay Simmons (1925) regards it as a "fairly common permanent and regular resident; appears to be more common during winter because of its preference for civilization at that season." Its haunts seem to be very similar to those of the Carolina chickadee—edges of woods, open woodlands, more or less open country, and, especially in winter, in towns and about houses.

It is slightly larger than specimens of *P. c. carolinensis* from the southern States, paler above and with whiter underparts.

Nesting.—Mr. Simmons (1925) says that, about Austin, the nests are placed anywhere from 1 to 23 feet above ground, commonly about 10 feet, "generally in natural cavities in dead elm, Chinaberry, Spanish oak, live oak, post oak, or blackjack tree or stump; in broken tops of leaning stumps or trees in decayed limbs; in old woodpecker hollows of telegraph pole or fence post; in decaying posts of barbed-wire fences along edges of woodlands; hollow iron hitching posts in town; bird boxes about farm houses and town houses." He says that in central Texas it seldom digs a hole of its own, unless forced to do so because of the scarcity of natural cavities.

He describes the nests as "composed of such warm materials as fine strips of bark (particularly of the cedar), soft green mosses, cowhair, plant fiber, wool, and feathers, with occasionally some rabbit fur, cotton, plant down, straw, bits of string, grass, horsehair, thistle down, and small buds. Lined with soft short cowhair, rabbit hair, plant down, and occasionally soft wool, plant fiber, cotton and feathers. Bottom of cavity filled with a good deal of green moss and occasionally with some cedar bark."

Eggs.—Mr. Simmons (1925) says that the set consists of three to eight eggs, most commonly six. These are apparently indistinguishable from those of other races of the species. The measurements of 40 eggs average 14.9 by 11.7 millimeters; the eggs showing the four extremes measure **16.5** by 12.4, 15.6 by **12.7,** and **13.5** by **10.5** millimeters.

Food.—Mrs. Nice (1931) writes: "In Norman the Chickadees were among the most charming of our feeding shelf guests, announcing their arrival with a cheery *peep,* enjoying everything we had to offer except raisins, but fondest of sunflower seeds and nuts, sometimes taking baths in the water dish in January. * * * These little birds are so friendly, so full of individuality and have so many different notes and pretty ways that they afford a most promising subject for a careful life history study."

Voice.—Mr. Simmons (1925) gives an elaborate account of the voice of the plumbeous chickadee, which would probably apply equally well to that of the other races of *carolinensis.* He says that it is "quite unlike that of the northern Chickadees; a much higher pitched and more hurriedly uttered *chickadee-dee-dee-dee,* characteristic of the Southern species, frequently running into *chick-a-dee-dee-dee-dee-dee-dee-dee; tweesee-dee-dee-dee-dee;* a clearly whistled *pseé-a-dee;* a low plaintive *tswee-dee-tswee-dee,* of four tremulous whistled notes, in sharp contrast to the clear, ringing notes *te-derry,* of the Northern birds; a low *sick-a-dee;* a short *chick-a-da;* a clearer *my watcher key, my watcher key;* a series of *day-day-day* or *dee-dee-dee-dee* notes."

PARUS CAROLINENSIS EXTIMUS (Todd and Sutton)

NORTHERN CAROLINA CHICKADEE

Todd and Sutton (1936) named this northern race, which ranges from New Jersey westward to Missouri and southward to northern North Carolina and Tennessee. They describe it as "similar to *Penthestes carolinensis carolinensis* (Audubon), but averaging larger, sex for sex; pale edgings of wings and tail averaging considerably more conspicuous; sides and flanks brighter reddish brown; and sides of head slightly grayer." The habits presumably are like those of the nominate race.

PARUS SCLATERI EIDOS (Peters)

MEXICAN CHICKADEE

HABITS

The above name now applies to the northern race of a Mexican species, which reaches the northern limits of its distribution in southeastern Arizona and southwestern New Mexico. Several specimens, now in the Brewster collection in the Museum of Comparative Zoology in Cambridge, were collected in the Chiricahua Mountains, Ariz., in March 1881, by Frank Stephens.

Mrs. Bailey (1928) gives the following records for New Mexico: "The Mexican Chickadee was taken in the San Luis Mountains July 19, 1892, and September 29, 1893, at about 7,000 feet (Mearns); and was found rather common among the pines of the Animas Mountains, 7,500-8,000 feet, August 1, 1908 (Goldman). These constitute the only State records of this Mexican species."

The Mexican chickadee *(Parus sclateri sclateri)* of the Mexican highlands is much like our Carolina chickadee in general color pattern, but it is larger, its coloration is darker, and its sides and flanks are olive-gray, with the white on the under parts restricted to a narrow median space, and with the black of the throat spreading fan-shaped over the chest.

The northern race, our bird, is characterized by the describer, James L. Peters (1927), as similar to the above, "but the grayish wash on the sides and flanks paler and grayer, lacking the olivaceous tinge; white area at the lower edge of the black throat and white abdominal median stripe less restricted."

Nesting.—There is a set of six eggs in the Thayer collection in Cambridge collected by E. F. Pope, near Benson, Ariz., on May 16, 1913. The nest of rabbit fur and wool, closely felted together, was in an 8-inch cavity, excavated by the birds, in a dead willow stub about 5 feet from the ground.

A set of six eggs, in the Doe collection in the University of Florida, was also taken by Mr. Pope, on April 4, 1911, in the same locality and similarly located; the nest was "composed of rabbit fur, and plant down, firmly felted together."

Eggs.—The eggs in the Thayer collection are short-ovate and only slightly glossy. They are white and are rather well covered, especially about the larger end, with fine dots and small spots of reddish brown, "hazel," and with a few underlying spots of "pale purple-drab."

The measurements of the above 12 eggs average 14.3 by 11.1 millimeters; the eggs showing the four extremes measure **15.3** by 11.4, 14.0 by **11.5**, **13.5** by 11.0, and 15.3 by **10.6** millimeters.

Plumages.—Young Mexican chickadees, in juvenal plumage, are similar to the adults, but the black of the head is duller, and the plumage is softer, shorter, and less firm. The postjuvenal molt of young birds and the annual postnuptial molt of adults seems to occur mainly in August, beginning late in July and continuing into early September, according to what few molting specimens I have seen. Birds in fall plumage, adults, and young being practically indistinguishable, are slightly more strongly tinged with olive than are the faded spring and summer birds.

Voice.—Dr. Frank M. Chapman (1898) published the following short note on the voice of the type race, as observed by him at Las Vegas, Veracruz: "The call of this Titmouse is a rapid, vigorous, double-noted whistle repeated three times, and not at all like the notes of *Parus atricapillus*. In its conversational 'juggling' notes there is, however, a marked similarity to the corresponding notes of that species."

DISTRIBUTION

Range.—Highlands of Mexico from Veracruz and Oaxaca, north to southern New Mexico and Arizona; nonmigratory.

The range of the Mexican chickadee extends **north** to southern Arizona (Mount Lemmon and the Chiricahua Mountains); and southern New Mexico (San Luis Mountain). **East** to southwestern New Mexico (San Luis Mountain and Animas Mountain); Puebla (Teziutlan); and Veracruz (Orizaba, Los Vigos, and Jalapa). **South** to central Veracruz (Jalapa); Oaxaca (La Parada); and central Guerrero (Omilteme). **West** to eastern Guerrero (Omilteme); Durango (Ciénaga de las Vacas); western Chihuahua (Pinos Altos); and southeastern Arizona (Mount Lemmon).

Two forms of this species are recognized but their respective ranges cannot now be clearly defined. The subspecies found in Arizona, New Mexico, and northern Mexico is *Parus sclateri eidos.*

Casual records.—A specimen of this species has been taken in the Davis Mountains of western Texas.

Egg dates.—Arizona: 3 records, April 4 to May 29.

PARUS GAMBELI GRINNELLI (van Rossem)

GRINNELL'S CHICKADEE

HABITS

This northern race of the white-browed chickadees *(gambeli)* is thus described by Mr. van Rossem (1928): "In relative proportions of wing and tail *Parus gambeli grinnelli* most closely resembles *Parus gambeli gambeli* (Ridgway), from which it differs in smaller size and darker coloration. On interscapular region it is of the identical shade of *Parus atricapillus atricapillus* (Linnaeus)."

Its range includes northern British Columbia, eastern Washington, east-central Oregon, and northern Idaho. Just where it intergrades with the mountain chickadee of the Rocky Mountains and the short-tailed chickadee of northern California does not seem to be definitely known.

The remarks by Dawson and Bowles (1909) on the mountain chickadee, as they found it in eastern Washington, evidently apply to this race. Mr. Dawson found two nests "placed in decayed stumps not above three feet from the ground. One, in a wild cherry stub in northern Okanogan County, contained fresh eggs on the 18th day of May. Their color had been pure white, but they were much soiled thru contact with the miscellaneous stuff which made up the lining of the cavity: moss, cow-hair, rabbits' wool, wild ducks' down, hawks' casts, etc."

At another nest, containing young, "it was an unfailing source of interest to see the busy parents hurrying to and fro and bringing incredible quantities of provisions in the shape of moths' eggs, spiders, wood-boring grubs, and winged creatures of a hundred sorts. Evidently the gardener knew what he was about in sheltering these unpaid assistants. Why, when it comes to horticulture, three pairs of Chickadees are equal to one Scotchman any day."

The measurements of eight eggs average 16.1 by 12.2 millimeters; the eggs showing the four extremes measure **16.3** by **12.3, 14.7** by 12.2, and **16.2** by **12.1** millimeters. They are apparently indistinguishable from those of the mountain chickadee.

PARUS GAMBELI ABBREVIATUS (Grinnell)

SHORT-TAILED CHICKADEE

PLATE 55

HABITS

The 1931 Check-list gives the range of this race of the mountain chickadee as "higher mountains of central and northern California, southern Oregon, and northwestern Nevada south to Mt. Sanhedrin and Mt. Whitney." It thus occupies a range intermediate between the ranges of three other races, but it does not seem to be intermediate in characters and is therefore a good subspecies. Dr. Grinnell (1918) gives its characters as "tone of color on sides, flanks and back the same as in *inyoensis*, though not quite so pale, namely, in fresh plumage, cartridge buff. Tail much shorter than in either *gambeli* or *inyoensis*; and bill averaging smaller than in any of the other three races."

W. E. Griffee writes to me that "this chickadee is common throughout the ponderosa pine and lodgepole pine country of eastern Oregon, particularly in the lodgepole areas."

As to their haunts in the Lassen Peak region, Grinnell, Dixon, and Linsdale (1930) state: "In the higher portions of the section chickadees were observed to frequent lodgepole pines, white firs, hemlocks, yellow pines, and junipers. At lower altitudes, in winter, the birds were observed also in blue oaks and valley oaks. Wherever found, this bird foraged about the ends of branches and over twigs in the outer parts of the foliage of the trees. Apparently the more open portions of the woods, or their marginal portions, were most favorable for the species."

Nesting.—The same authors report several nests found in the above region. Most of the nests were in low stumps in clearings, in open spaces, or on the edges of the woods, either in natural cavities of crevices or in rotting or burnt stubs, so that in most cases the birds had little or no excavating to do. Two nests were in old woodpeckers' holes. Most of the nests were less than 2 meters above the ground; the lowest was only 165 millimeters up; and the highest was nearly 5 meters from the ground. The only lining mentioned was rabbit fur.

Chester Barlow (1901) found a nest under the baseboards of a cabin in which he was camping. The bird had entered through a rough hole in the boards, and "the nest had been built on a joist under the cabin in a space ten inches long and seven and a half inches wide. This had been filled with cow-hair, squirrel fur and hemp picked up from about the dairy, and when the nest was removed it presented a solid mat 2½ inches thick and of the dimensions given. Near the center of the mat a round cavity 2½ inches across and 1½ inches deep held the eight eggs."

J. E. Patterson has sent me a photograph of a nest that was in the under side of the bole of a prostrate pine log, only a few inches above the ground.

Mr. Griffee says in his notes: "Nesting cavities usually are 4 to 8 feet above ground and situated in lodgepole pines when that species is available. Often they will choose a live lodgepole tree, even when punky aspen stubs are available. Nesting pines often have heart rot, which makes the excavating easier than it looks from the outside, but nevertheless the work of excavation must be several times as arduous as that usually performed by the Oregon chickadee. Entrance holes are about 1½ inches in diameter and often go straight back as much as 4 or 5 inches to the heart of the tree before turning down for 8 to 12 inches. The bottom of the cavity, often irregular in shape because of an intruding knot, is lined with more or less shredded bark upon which is piled a lot of rodent fur. As the fur is short and not felted together compactly, the nest will not hold its shape when removed from the cavity.

"In my experience, the short-tailed chickadee always stays on the nest from the time the first egg is laid. Probably this is necessary because of the great abundance of chipmunks, which are small enough to run into a chickadee nesting hole and make a meal of the eggs, if the parent bird were not on the job at all times. The Oregon chickadee, which nests where chipmunks are much less common, and makes a smaller entrance to its nesting cavity, apparently is not so much bothered by chipmunks and so does not hesitate to leave its nest unguarded until incubation begins."

Eggs.—The eggs of the short-tailed chickadee apparently do not differ materially from those of other races of the species. They are mostly plain, pure white, but usually some of the eggs in a set are more or less finely speckled with reddish brown. The numbers I have seen recorded run from five to nine in a set. The measurements of 40 eggs average 16.1 by 12.3 millimeters; the eggs showing the four extremes measure **17.0** by 11.9, 16.5 by **13.2,** and **14.7** by **11.4** millimeters.

Young.—Grinnell, Dixon, and Linsdale (1930) noted one case where two broods were raised in a season in the same nest; probably two broods are often raised. At a nest that they watched the young were fed 11 times within half an hour, between 8.59 and 9.30 A.M., at intervals varying from one to seven minutes, but usually at intervals of three or four minutes. Such food as could be seen consisted of green caterpillars. Both old birds helped in the feeding and in keeping the nest clean, carrying away the excrement sacs.

Mrs. Wheelock (1904) gives the period of incubation as 14 days. She says that the young remained in the nest nearly three weeks, which seems to be an unusually long time for such small birds; and for fully two weeks longer they were begging to be fed. She observed that "the nestlings were fed by regurgitation until four days old, when fresh food was given."

Food.—Strangely enough, nothing specific seems to have been published on the food of any of the races of the mountain chickadee; if there has, I have been unable to find it. But its feeding habits are similar to those of the other chickadees; it has repeatedly been observed examining the twigs, foliage, and crevices in the bark of trees, where it doubtless finds a variety of insect food; and it is fair to assume that its food does not differ very materially from that of other members of the genus. Grinnell and Storer (1924) have seen one thoroughly examining the interior of a rotted-out cavity, where it probably found insect food of some kind.

Behavior.—Grinnell, Dixon, and Linsdale (1930) describe in some detail the intimidation behavior of this species when the nest is invaded:

When a slab of rotten wood was removed the bird lunged, at the same time spreading its wings convulsively, and then gave a prolonged hissing sound—just that order of procedure. The bird repeated this performance nineteen times by count before it suddenly flew from the nest at the close approach and light touch of the observer's hand. The body had been kept closely depressed into the nest cavity. The lunges were rather inane—the bird simply struck out, in one direction and then another. At the moment of the lunge, the black-and-white striping of the head brought her into abrupt and conspicuous view of the observer peering into the cavity—reinforcing the surprise effect of the sounds produced. At times, the hissing sound was produced, the wall of the cavity was struck, and the white of the head moved, all at the same instant. * * * During the winter chickadees regularly made up portions of the companies of birds of several species that foraged together through the day. Some of the individuals that moved to low altitudes in winter joined circulating bands of bush-tits.

Grinnell and Storer (1924) write: "After the nesting season the chickadees and several others of the smaller birds are wont to associate with one another in flocks of varying size. Such a gathering was seen in Yosemite Valley on July 30, 1915. Included in the openly formed yet coherent aggregation were the following species: Mountain Chickadee, Black-throated Gray Warbler, Western Chipping Sparrow, Sierra Creepers, Warbling Vireo, and Cassin Vireo. The birds were foraging through black oaks, incense cedars, and young yellow pines, each kind of bird of course adhering to its own particular niche and own method of getting food."

PARUS GAMBELI BAILEYAE Grinnell

BAILEY'S CHICKADEE

HABITS

This race of the white-browed chickadees occupies the higher mountains of southern California, from Tulare and Monterey Counties to
San Diego County. It was named for that distinguished and popular
ornithologist, Mrs. Florence Merriam Bailey, who has done so much
for western ornithology.

Dr. Joseph Grinnell (1918) gives as his diagnosis of it: "Tone of
coloration on sides, flanks and back distinctly plumbeous—more exactly, on sides and flanks the 'smoke gray' of Ridgway (1912, pl. 46),
and on back near the 'mouse gray' of the same authority (pl. 51). The
tail in this race is short as in *abbreviatus*, but the bill is long and heavy,
averaging thicker through than in any of the other three races."

Like other races of the species, Bailey's chickadee finds its favorite
haunts, during the breeding season at least, in the coniferous forests of
the mountains, from the lower borders of the pines up to 10,000 feet
or higher, or perhaps as far as the evergreen forest extends. In the
San Bernardino Mountains, Dr. Grinnell (1908) found these chickadees in the tamarack pine belt as high as 10,600 feet; and in August
they were common in the pinyons and chaparral, and as far down the
desert slope as Cactus Flat, at 6,000 feet.

Nesting.—J. Stuart Rowley writes to me: "I have found these
chickadees nesting in cavities and woodpecker holes rather abundantly
throughout the higher mountains of southern California. They nest as
high as 35 feet up and down to within a foot of the ground, as a rule,
but on June 9, 1935, while I was collecting on Mount Pinos, Ventura
County, I saw a chickadee go down a squirrel hole underneath a
dead pine stub in a little clearing. Upon investigating, this bird was
seen on a nest on the ground in the excavation under the stub; the
nest contained five eggs ready to hatch. This is the only nest of a
chickadee I have ever found which was actually below the level of the
ground."

In the San Bernardino Mountains, Dr. Grinnell (1908) reports:

A nest found June 17, 1905, near the mouth of Fish Creek, occupied a vertical
slit in a dead black-oak stub. The nest was not more than three feet from the
ground and was made of soft, downy plant fibers, and contained six newly-
hatched young. Another nest was found June 21 on a ridge near Dry lake. This
was twenty feet from the ground in a dead fir stub, and was ensconced behind the
loosened bark. It consisted of fur, apparently from the woodrat and chipmunk,
and contained five eggs in which incubation was well advanced. Another nest
containing seven young was found the same day in a cavity of a pine stub even
with the surface of the ground. A fourth nest in the same locality contained six

small young. In this case the nest was a felted mass of deer hair and woodrat fur, intermingled with a few feathers. It was in a knot-hole of a dead fir sapling, two and one-half feet from the ground. In 1906, at Dry lake, June 15, a set of five slightly incubated eggs of this species was taken from an old sapsucker hole twenty feet above the ground in a dead tamarack pine. The nest was a large mass of reddish deer hair.

There is a set of seven eggs in my collection, taken by Wright M. Pierce near Bear Lake in these mountains on May 19, 1923, from a cavity on the under side of a large, dead, fallen yellow pine.

Eggs.—The eggs of Bailey's chickadee are evidently similar to those of the type race, the mountain chickadee, as described under that form, sometimes pure white, but oftener more or less lightly spotted with fine dots of light browns. I have no record of such large sets as mentioned under the mountain chickadee, but they may occur. The measurements of 40 eggs average 16.2 by 12.7 millimeters; the eggs showing the four extreme measure **17.7** by **13.5,** and **14.6** by **11.6** millimeters.

Voice.—Ralph Hoffmann (1927) gives a slightly different account of the voice of this chickadee from that quoted under the mountain chickadee. He writes:

A visitor from the East misses the familiar Black-capped Chickadee from the rich bird life in the lowlands of southern California. Let him, however, climb a few thousand feet up any of the mountain ranges, among the yellow pines, and there will be, if not his eastern friend, at any rate a close relative. The Mountain Chickadee is so close to the eastern bird, so like in voice, habits and appearance, that it will take a few moments to discover the difference. The bird clings head downward to an outer twig, hammers a seed open on a limb, lisps *tsee-dee-dee* to its fellows and is apparently the same active, cheery mite. The sweet whistled call is more often made up of three (sometimes four) notes than that of the Eastern bird. Sometimes the three notes come down the scale to the tune of 'Three Blind Mice.' At other times the last two are the same pitch *tee-dee-dee.* Occasionally the bird either leaves off the third note or adds a fourth. As the chickadee gleans from twigs, it utters a hoarse *tsick tsick dee dee* or a husky *tsee dee,* and other little gurgling or lisping calls, and a sharp *tsik-a* when startled or excited.

PARUS GAMBELI ATRATUS (Grinnell and Swarth)

SAN PEDRO CHICKADEE

In naming and describing this Lower California subspecies, Grinnell and Swarth (1926) remark:

The race of Mountain Chickadee of the Sierra San Pedro Martir, as compared with related subspecies, exhibits an appreciable darkening of the plumage in the direction not of brown but of slate. This darkening is most apparent on the flight feathers, which are slaty black as compared with the more brownish-hued quill feathers of other races; but it shows also in more leaden-hued flanks and upper parts. This general leaden tone of coloration is quite apparent in fresh plumaged birds, but it is a character that tends to be lost even when the feathers become only slightly worn.

Together with this darkening there is restriction in the area covered by the one conspicuous white marking on this bird, the superciliary stripe, which marking extends forward in fresh plumage to nearly or quite meet its fellow on the forehead. The white on the head of *atratus* is not only less in area occupied, but it is shallower; and birds in breeding plumage, when it is reduced or effaced by wear, come to bear a curious resemblance about the head to *Penthestes atricapillus*.

They gave the range of this form, "so far as known, only the main plateau of the Sierra San Pedro Martir"; and they state that it "adheres closely to the coniferous belt of the Transition and Canadian life-zones." The 1931 Check-list extends its range to include the Sierra Juárez, of northern Lower California.

It probably does not differ materially in its habits from other races of the species.

The measurements of 6 eggs average 16.5 by 12.6 millimeters; the eggs showing the four extremes measure **16.8** by 12.7, 16.5 by **13.0**, and **16.2** by **11.7** millimeters.

PARUS GAMBELI GAMBELI Ridgway

MOUNTAIN CHICKADEE

PLATE 57

HABITS

The species *Parus gambeli* occupies a wide territory in western North America, from British Columbia to Lower California and from the Rocky Mountain region westward to the Pacific coast region, in all suitable mountain ranges. The type race, the subject of this sketch, is confined to the Rocky Mountain region, from Wyoming and Montana southward to Arizona, New Mexico, and central western Texas.

It is well named the mountain chickadee, for, during the breeding season at least and for much of the remainder of the year its favorite haunts are the coniferous forests of the mountains, from 6,000 up to 11,000 feet, but mainly at the higher levels, 8,000 to 10,000 feet. Below the coniferous forests it is largely or wholly replaced by the long-tailed chickadee.

In fall the mountain chickadees, with their young, range up to timberline or even beyond; in fall and during winter they often range down to the foothills and valleys, where they are sometimes seen together with the long-tailed chickadees in the fringes of cottonwoods and willows along the streams.

Mrs. Bailey (1928) says that, in New Mexico, "they may be met with almost anywhere in the forested mountains. In Santa Clara Canyon, where we found them in the oaks, nut pines, and junipers of the south slope, down along the creek, in the turns where the sun came in, they were in the alders and birches together with migrating warblers, vireos, and flycatchers. But they are found in the high, dark, coniferous forests as well, and it is here that their cheery notes are most gratefully heard."

In his paper on the subspecies of the mountain chickadee, Dr. Joseph Grinnell (1918) says: "Among the four subspecies of *Penthestes gambeli* here recognized, color alone is sufficient for distinguishing *P. gambeli gambeli*. The flanks, sides of body and back in this form are pervaded with a distinct tinge of cinnamon—more exactly, the 'pinkish buff' of Ridgway (1912, pl. 29). In addition, this race shows the greatest length of tail, and slenderest bill."

Nesting.—The mountain chickadee does not seem to be at all particular about the choice or location of its nest. It prefers, however, to place its nest in a natural cavity in a tree, or in an old woodpecker hole, and I believe that it does not excavate its own nest cavity if it can find one already made for it. Its nest has been found at heights ranging from 2 to 80 feet above the ground, the extreme heights being very rare; apparently very few nests are more than 15 feet from the ground, and many are less than 6 feet up. J. K. Jensen (1923) says that, in New Mexico, he often finds this chickadee nesting in bird boxes, and he has "found the nests in cavities in pine stumps, in quaking aspens and under rocks." Later (1925) he writes:

"May 15, 1925, I made a trip ten miles southeast of Santa Fe intending to examine a number of bird boxes. One of the boxes contained a set of six eggs of the Mountain Chicadee *(Penthestes gambeli gambeli)* and three eggs of the Gray Titmouse *(Baeolophus inornatus griseus)* with the Chicadee incubating. I took out the six eggs of the Chicadee and left those of the Titmouse. May 22, the Chicadee was incubating four Titmouse eggs, all of which hatched. June 8, I again visited the box and found the Chicadee busy feeding four young Titmice."

This chickadee is a very close sitter, reluctant to leave its nest, and it has developed to a very high degree the intimidation reaction to the approach of an intruder, a habit shared by other species of chickadee. This consists of a loud hissing noise and a rapid fluttering of the wings, when the nest is invaded; it might be enough to frighten away some smaller enemy, but to man it serves only to illustrate the devotion of the brave little bird to its eggs or young; the heartless egg collector seldom rewards the devoted mother for her bravery.

The nest is made of soft mosses and the fur and hair of mammals, being warmly lined with finer material of the same kind, much like the nests of other chickadees.

Eggs.—The mountain chickadee seems to lay very large sets. Sets of less than six are probably incomplete; and from that the numbers run up to 12, sets of 9 being common. There are 16 sets of eggs of this chickadee in the collection of the University of Colorado Museum; in this series there are three sets of 7, seven sets of 9, three sets of 10, two sets of 11, and one set of 12 eggs.

The eggs are normally ovate in shape, with variations toward rounded-ovate. The shell is smooth and practically lacking in gloss. The ground color is pure, dead white, with much variation in the extent of the markings. In most sets a small or a large proportion of the eggs are entirely unmarked; and in some sets all the eggs are spotless pure white. Some eggs are faintly marked all over the entire surface with fine dots of pale or reddish brown. In others these small spots are more or less concentrated about the larger end. An occasional set is brilliantly marked with a ring of bright reddish brown or orange-rufous spots or small blotches.

The measurements of 40 eggs average 15.6 by 12.3 millimeters; the eggs showing the four extremes measure **16.5 by 12.9,** and **13.8 by 11.5** millimeters.

Young.—I have no information on the period of incubation for this race.

Dean Amadon has sent me the following note on a nestful of young mountain chickadees that he observed at 9,800 feet on the east slope of Snowy Range, Medicine Bow Mountains, Wyo., about 20 feet from the ground in a natural crevice of a large lodgepole pine: "Both parents were feeding the young very frequently at the time, which was 7 P.M., just before dark. Once, when the two parents appeared with food at the same time, they crouched on limbs near the nest, opening and quivering their wings, as do young birds when begging for food. The young were noisy enough to be heard for about 30 feet from the nest tree; I was told that they had left the nest and were not seen two days later, on July 8. Nest was in rather dense pine stand, but only 40 feet from the clearing made by a main road."

Claude T. Barnes tells me that five tiny fledgings, in a nest that he examined, "hissed in the manner of a snake" when he reflected light into the nest. The nest was in a quaking aspen, which, "twisted by winds and snow, had cracked, making a cozy hole about 4 inches deep and 3 inches in diameter."

Plumages.—In the juvenal plumage, young birds are much like adults, but the white stripe over the eye is more imperfect, less distinct, and grayer; the black of the crown and throat is duller; and the edgings of the greater wing coverts and tertials are tinged with pale buff. The body plumage is softer and fluffier, less compact. The annual molt seems to occur mainly in August, though I have seen very few molting birds. Fall birds have the white superciliary stripe broader and more distinct, and the gray portions of the plumage slightly more buffy than in spring birds; in spring birds the superciliary stripe is reduced to a broken series of white streaks.

Voice.—Aretas A. Saunders sends me the following note on the song of this chickadee, as heard by him in Montana: "The song of the mountain chickadee is similar in quality to that of the black-capped chickadee, a sweet, clear whistle, but it usually consists of three notes of equal length, each lower in pitch than the preceding, so that it is like *fee-bee-bay*. The bird occasionally responds to an imitation and comes to the observer, but not so readily as the black-cap does. The song is rather infrequently used."

Claude T. Barnes says in his notes: "Two chickadees, which from the white superciliary stripe I took to be *gambeli,* flitted about the maple trees uttering every second or so a single note *chip.* Once in a while one would issue a burring sound like *trrrrrrrrrp.* Overcome with curiosity at my immobility, they edged their way by little flits and various maneuvers behind tree trunks until they were within six feet of me, when, evidently being satisfied, they as casually worked their way from me."

Edward R. Warren (1916) writes: "They seemed to say 'chick-a-dee-a-dee-a-dee', not 'chick-a-dee-dee' as the Black-caps do. And the tone was also different, but I cannot describe it."

Field marks.—Although similar in general appearance and behavior to our more familiar black-capped chickadees, this species can be easily recognized by the usually conspicuous white stripe that runs from the bill over the eye and into the black cap. Its song is different, as described above, and its ordinary chickadee note is somewhat hoarser and more deliberate. Also, it is oftener found in the coniferous forests than is *atricapillus.*

DISTRIBUTION

Range.—Mountains of the Western United States, Canada, and northern Baja California; nonmigratory.

The range of the mountain chickadee extends **north** to northern British Columbia (Atlin and Nine Mile Mountain); Alberta (Smoky Valley and Banff); and central Montana (Gold Run and Fort Custer).

East to eastern Montana (Fort Custer and Red Lodge); eastern Wyoming (Wheatland and Laramie); Colorado (Golden, Pikes Peak, and Fort Garland); New Mexico (Willis and Capitan Mountain); and western Texas (Davis Mountains). **South** to southwestern Texas (Davis Mountains); southwestern New Mexico (Pinos Altos Mountain); southern Arizona (Santa Catalina Mountains); and northern Baja California (San Pedro Mártir Mountains). **West** to Baja California (San Pedro Mártir Mountains and Sierra Juárez); central California (Barley Flats, Big Trees, Marysville, and Weed); Oregon (Pinehurst and Fort Klamath); Washington (Mount Rainier and Spokane); and British Columbia (Summerland and Atlin).

The range as outlined is for the entire species, which has been separated into at least six subspecies. The typical mountain chickadee *(Parus g. gambeli)* is found in the Rocky Mountain region from Wyoming and Montana south to western Texas, New Mexico, and Arizona; Grinnell's chickadee *(P. g. grinnelli)* occupies the northern parts of the range from British Columbia south to central Oregon and northern Idaho; the short-tailed chickadee *(P. g. abbreviatus)* is found from southern Oregon south to central California and northwestern Nevada; Bailey's chickadee *(P. g. baileyae)* is found in the higher mountains of southern California; the San Pedro Chickadee *(P. g. atratus)* is found in the mountains of northern Baja California; and the Inyo chickadee *(P. g. inyoensis)* occupies the mountain areas of eastern California chiefly in Mono and Inyo Counties.

Egg dates.—California: 82 records, May 4 to July 11; 42 records, May 22 to June 13, indicating the height of the season.

Colorado: 33 records, April 5 to June 23; 17 records, June 3 to 11.

Oregon: 15 records, May 20 to June 12; 8 records, May 30 to June 6.

PARUS GAMBELI INYOENSIS (Grinnell)

INYO CHICKADEE

This is a pale race of the white-browed, or mountain, chickadees found in the more arid mountain regions of eastern California. Dr. Grinnell (1918) describes it as "the palest colored race of the four; sides, flanks and back, in unworn plumage, pervaded with pale buff— the 'cartridge buff' of Ridgway (1912, pl. 30). Wear or fading, or both, removes most of this buff tone, so that the resulting effect, in spring and summer birds, is of an ashy tone of coloration, distinctly lighter than in any of the other three subspecies, in same stage. It seems probable that there is a paler tone to the underlying plumage parts and that this becomes revealed by loss of the superficial pigment-bearing portions through the gradual progress of feather abrasion. *Inyoensis* shows nearly as long a tail as does *gambeli*. Its bill is somewhat smaller."

He gives as its range "the higher mountains of eastern California lying east and southeast of Owens Valley, from the vicinity of the Mono Craters and the White Mountains, in Mono County, south to the Panamint Mountains, in Inyo County."

I cannot find anything in print about the haunts, nesting, eggs, food, or other habits of this subspecies, which are probably similar to those of mountain chickadees elsewhere.

The measurements of 9 eggs average 16.8 by 12.8 millimeters; the eggs showing the four extremes measure **17.1** by 12.9, 17.0 by **13.0, 15.7** by 12.8, and 17.0 by **12.5** millimeters.

PARUS CINCTUS ALASCENSIS (Prazák)

ALASKA CHICKADEE

HABITS

The Alaska chickadee is our representative of an Old World species, widely distributed in northern Siberia and Russia, that has crossed Bering Strait and become established, as an American subspecies, in the Hudsonian Zone of northern Alaska and northwestern Mackenzie. Closely related races occur in eastern Siberia *(P. c. obtectus)* and in northern Europe *(P. c. cinctus)*. The range of our subspecies extends, so far as known, from the Kowak River and St. Michael, Alaska, on the west to the Anderson River, Mackenzie, on the east. Dr. E. W. Nelson (1887) wrote: "Its range does not appear to extend to the south along the Upper Yukon, as a considerable series of Titmice, brought me from that region by the fur traders, does not contain a single example. From the vicinity of Nulato, thence down the Yukon, and to the north and northeast, this form appears to be as abundant as the Hudsonian Titmouse, whose range it shares in this region."

Olaus J. Murie (1928) writes of the haunts of this chickadee in the Old Crow River district, Yukon, as follows:

The habitat of the Alaska Chickadee was spruce and willow woods covering the valley of the Old Crow River, about a mile in width, but much more in some places. The edge of this valley rises abruptly about 75 feet above the stream to the level of the flat tundra, which stretches away for many miles on either side. The tundra is covered with numerous ponds and lakes, and except for a few small groups of spruces here and there it is practically treeless, although near the mouth of the river, where it enters the general wooded area of the Porcupine River district, there is a more general distribution of forest. The wooded valley of the Old Crow and some of its tributaries, therefore, carry the spruce woods in the form of long sinuous belts through a tundra region. These narrow belts of forest are the home of the Alaska Chickadee, apparently to the exclusion of other Chickadees. *Penthestes hudsonicus* was not collected on the Old Crow nor was it identified among those observed. * * *

A consideration of the circumstances under which the Alaska Chickadee has been observed tends to show that it prefers the edges of forest tracts or regions where spruce forest is broken up, as contrasted with more extensive, continuous forest areas. * * * Grinnell says, 'The Alaskan Chickadee was never seen in company with the other species and was an inhabitant of the spruce tracts along the base of the mountains rather than in the river bottoms.' Other records show that *alascensis* was found even in the willows beyond the spruce woods.

The range of this bird borders the northern tree limit, where habitat conditions mentioned here may be found.

Joseph S. Dixon (1938) found it "well distributed in the aspen and spruce forests," in Mount McKinley National Park. "People living in the region told us that these chicadees disappear in the spring and are rarely seen all summer, but that in the fall they again gather about the cabins to be fed."

Our Alaskan bird differs from the Siberian race in having a smaller bill and slightly darker coloration. It resembles the Hudsonian chickadee in a general way, but the sides of the neck are white instead of gray, and the coloration is paler throughout.

Nesting.—The first nest reported of the Alaska chickadee was that found by Roderick MacFarlane (1891) in the Anderson River region in northern Mackenzie, of which he states: "On 1st June, 1864, a nest of this species, containing seven eggs, was found near Fort Anderson, in a hole in a dry spruce stump, at a height of about 6 feet from the ground. It was composed of a moderate quantity of hare or rabbit fur, intermixed with a sprinkling of dried moss. The female parent was snared on the nest, but the male was not seen. The contents of the eggs were tolerably fresh."

There is a set of six eggs in the Thayer collection, taken by Bishop J. O. Stringer near the Peel River on June 30, 1898; the nest was in a hole in a tree stump. The European race is said to make its nest in natural cavities in trees, or in old woodpecker holes. Dr. E. W. Nelson (1887) says that it "has been known to eject the rightful owner of a cavity and seize upon the site for its own nest. Its eggs are usually placed upon a mass of hairs of the lemming and hare, combined with fine moss or vegetable down."

Eggs.—Dr. Nelson (1887) says that "it lays from seven to nine eggs, which are grayish-white, with reddish-violet and reddish-brown spots often collected at the larger end. The eggs are broad in proportion to **their length."**

The eggs in the Thayer collection are ovate, with hardly any gloss. The ground color is pure white and is finely speckled, more thickly around the larger end, with small dots of reddish brown.

The measurements of 31 eggs average 16.3 by 12.1 millimeters; the eggs showing the four extremes measure **17.3** by 12.7, 16.5 by **13.2,** and **15.0** by **11.0** millimeters.

Voice.—Mr. Murie (1928) says: "The characteristic call of this Chickadee generally consisted of two notes, which I described variously in my field notes as 'dee-deer,' 'chee-ee,' or 'pee-vee,' with emphasis on each syllable and a tone and accent which seemed to imply a peevish or complaining state of mind. Nelson describes it as '*pistéé-téé,* uttered in a hissing tone.' I imagined this call differed from that of the Hudsonian Chickadee, but in the absence of the latter for comparison at the time I cannot rely too greatly on this impression."

Field marks.—The Alaska chickadee most closely resembles the Hudsonian chickadee, but it is slightly larger, its general coloration is paler, its crown is grayer, less brownish, the sides of its neck are white instead of gray, and it lacks the rich-brown flanks of *hudsonicus,* the latter being paler and grayer. It can be distinguished from any race of the black-capped chickadee by the gray instead of the black crown and by the absence of white edgings on the tail feathers.

So little is known about our Alaskan subspecies that it seems worthwhile to include the following notes on the European race, kindly contributed by B. W. Tucker: "In the forest belt of Arctic Europe the Lapptit *(Parus cinctus cinctus),* the racial representative of the Alaska chickadee in that region, is one of the characteristic small birds. It has been described as more characteristic of the pine forest than of the birch and alder, but in Finnish and Norwegian Lapland I found it chiefly among birch or mixed birch and pine on the borders of the pine forest proper. In habits this species does not differ from the black-capped tits. In the latter part of June I found it chiefly in family parties of parents and fledged young, the latter in one or two cases still being fed by the old birds, but I also found a pair still feeding young in the nest on June 21. The nest in question was in an old woodpecker hole at about eye level in a dead pine trunk on the edge of a swamp, woodpecker holes being, it is stated, the favorite nesting site of this tit. Whether it will also excavate its own hole, like some of its allies, seems not to be recorded.

"The common note of the Lapptit may be represented as a low-pitched, grating *tchaa-tchaa* or *tzaa-tzaa,* in which the *a* sound must be given a pronounced nasal quality, as in *twang,* but more exaggerated. It is often quite indistinguishable to my ear from the typical note of the willowtit, but at other times it is obviously softer and less harsh, and easily distinguished. Sometimes it is repeated rapidly in a way that would doubtless recall the chickadee to an American observer but that to a Britisher recalls the marshtit, for in Europe oddly enough it is the marshtit that has (among others) a rapid *chicka-bee-bee-bee* note and not the willowtit, although the latter is more closely allied to the chickadee and indeed is currently regarded as a form of the same species. The call note is a hoarse little *tzeep* or *tzip* and *tzee-ip* and the same note, or duplications of it, commonly precedes the *tzaa—tzeet-tzaa-tzaa, tzitzi-tzaa-tzaa,* etc. It has also a very shrill keening note, again like the willowtit. The song also recalls the song of the willowtit, as opposed to the marshtit, in being distinctly warbling, and I have described it in my notes as a sustained, low, liquid, ripply or bubbling warble."

DISTRIBUTION

Range.—Arctic regions of northern Asia and northwestern North America; nonmigratory.

In northern Asia the range of this chickadee extends eastward from the Yenisei River to the coast of Bering Sea and south to Lake Baikal. In North America the species is found **north** to northern Alaska (Kowak River and Altana River); northern Yukon (Old Crow); and northern Mackenzie (Richards Island and Fort Anderson). **East** to northwestern Mackenzie (Fort Anderson). **South** to northwestern Mackenzie (Fort Anderson); and central Alaska (Twelvemile Creek, probably Dishna River, and St. Michael). **West** to western Alaska (St. Michael and Kowak River).

Three races of this species are recognized, the typical form of northern Europe *(P. c. cinctus);* the form of eastern Siberia, Kamchatka, around the shores of the Okhotsk Sea *(P. c. obtectus);* and the form found in North America known as *P. c. alascensis.*

Egg dates.—Alaska: 2 records, June 10 and 24.

Arctic Canada: 1 record, June 30.

Mackenzie: 1 record, June 1.

Siberia: 1 record, June 1.

PARUS HUDSONICUS HUDSONICUS Forster

HUDSONIAN CHICKADEE

HABITS

The type race of these brown-capped chickadees was described and named many years ago from specimens taken on the west coast of **Hudson Bay,** at Severn River. For nearly 100 years it remained the only recognized form of the species. The characters by which other recognized subspecies have been separated will be mentioned under the respective races. This type race has the widest breeding range of the three races recognized in the 1931 Check-list, from northwestern Alaska to Hudson Bay and south to central Manitoba and central Ontario, whence it wanders farther south in winter.

Dr. Joseph Grinnell (1900a) says of its status in the Kotzebue Sound region of Alaska: "At our winter camp on the Kowak this species was common up to the middle of September. After that date and up to the first of April, but one or two at a time were seen and then only at long intervals. Early in September groups of four to seven were noted nearly every day in the spruces around our cabin. * * * Those chickadees observed during the winter were all in the dense willow thickets along Hunt River. They were there quieter and, by nature of their retreat, hard to find. It may have been that at the advent of cold weather all the chickadees left the spruces and betook themselves to the shelter of the willow-brush; but I am rather inclined to believe that there was a partial migration to the southward. By the first of May the chickadees were back again roving through the woods in pairs."

Lee R. Dice (1920) says that in the interior of Alaska "they occur in willows and alders and in white spruce and paper birch forest."

I cannot find that the Hudsonian chickadee differs materially in any of its habits from the better-known Acadian chickadee, to which the reader is referred.

Nesting.—Dr. Samuel S. Dickey (MS.) tells of a nest probably of this race that he found while exploring along the Moose River, west of James Bay. It was in "a singular glade, along the moist north terrace of Moose Island. It was hedged in with black spruce trees of moderate size and a natural fence of speckled alder and mountainholly shrubs *(Nemopanthus mucronata)*. The green moss carpet that lay beyond was seen to contain in its midst several short, weathered, barkless spruce stubs, and in the side of one of them a dark hole loomed up. It was about the size of a 25-cent piece. With a stout pocket ax I parted the hard resistant wood opposite the entrance. Thus I could view the interior. It contained a compactly woven nest, in whose cup lay a clutch of six somewhat incubated eggs. The nest was better composed than are those of the eastern chickadees. Its foundation was a mass of dead and also green cedar moss *(Hypnum cristatum)*. Upon this was the soft bed of fur of the snowfoot hare. The eggs were nicely adjusted in the cup in the center of this mass. Further examination of this nest showed that at its base was decayed moss and hair of former nests of other years. Evidently the much worn cavity and entrance had served this pair for many seasons.

"The eggs were light creamy white in ground color and were sprinkled nearly all over the entire shell surface with light reddish brown. They resembled more closely the run of eggs of the winter wren (*Nannus hiemalis*) than of specimens from other eastern chickadees."

The stub was 30 inches high, and the entrance to the nest was 18 inches above the bog surface; the cavity was 6 inches deep.

A. D. Henderson has sent me two photographs of nesting stumps of this chickadee. One was an old tamarack stump in a muskeg; the entrance hole was only 5½ inches above the moss, and the nest was practically at ground level. The other was a stub in an old cutting, 13 inches high, and the bird was seen to enter through a hole in the ax cut.

Eggs.—The eggs are apparently similar to those of the Acadian chickadee. The measurements of 22 eggs average 15.3 by 12.3 millimeters; the eggs showing the four extremes measure **16.3** by 12.2, 15.4 by **12.7**, 14.9 by 12.4, and 15.2 by **11.7** millimeters.

Fall.—Frank L. Farley writes to me from Camrose, Alberta: "This species is partially migratory in this latitude. It is a common summer resident of the spruce forests along the foothills of the Rockies and in much of the territory north of the Saskatchewan River. Numbers of them, at the approach of winter, wander south and east to scattered areas of spruce, which persist on the northern exposures of the larger rivers in central Alberta. Here they remain until early in March. It has been noted that these visitors always feed in the higher branches of the spruces, while the black-capped chickadees hunt the lower parts of the trees. During very severe cold spells, when the thermometer shows 50° below zero, I frequently find dead blackcaps in the woods, no doubt as a result of intense cold or inability to secure food, but I have never known the Hudsonian to succumb at such times. They are, no doubt, the hardier of the two species."

Numerous reports of periodic southward migrations of Hudsonian chickadees have been recorded. The latest one comes to me in a letter from O. E. Devitt, of Toronto, who writes: "In October 1937 a rather remarkable influx of these northern titmice occurred in this area. They were first noted at Nancy Lake, near King, on October 29, by R. D. Ussher; several turned up at Ashbridges Bay, Toronto, on October 31, and on this date I saw several near Bradford and collected a male. This specimen was later submitted to P. A. Taverner, of Ottawa, with the view that it might be the Acadian race. However, it was identified by him as the '*nigrescens*' type of the brown-headed chickadee (*Penthestes hudsonicus hudsonicus*). Individuals remained about Nancy Lake well on into the winter."

DISTRIBUTION

Range.—Northern North America; not regularly migratory.

The range of the Hudsonian chickadee extends **north** to northern Alaska (Hunt River, Tanana, White Mountains, and Bern Creek); Yukon (Robinson Camp); Mackenzie (Fort Goodhope, Lake Hardisty, and Fort Rae); northern Manitoba (Lake Du Brochet and Mosquito Point); northern Quebec (Chimo); and Labrador (Okkak). **East** to Labrador (Okkak, Davis Inlet, and Cartwright); Newfoundland (Paynes Cove, St. John's, and Makinsons Grove); Nova Scotia (Dartmouth and Seal Island); New Hampshire (Washington); and southeastern New York (Big Indian Valley). **South** to New York (Big Indian Valley); possibly Massachusetts (see text); possibly Pennsylvania (see text); southern Ontario (Brule Lake and Sayma Lake); northern Michigan (Blaney); northern Wisconsin (Mamie Lake); northern Minnesota (Sandy Lake and Deer River); northern Montana (south fork of Teton River); and southern British Columbia (Schoonover Mountain). **West** to British Columbia (Schoonover Mountain, Salmon River, and Ninemile Mountain); and Alaska (Houkan, Fort Kenai, Nushagak, Nulato, Kowak River, and Hunt River). The species has been reported in the mountains of eastern Pennsylvania (Laanna and Pocono Lake) during the summer, and it may be found to breed there.

While not regularly migratory there appears to be an occasional fall movement that extends the range in winter south to southern Michigan (Lansing); northern Illinois (Waukegan Flats); southern Minnesota (Fairmont); and northern North Dakota (Upsilon Lake); Massachusetts; and New York.

As outlined the range is for the entire species, of which four subspecies are recognized. The typical Hudsonian chickadee *(P. h. hudsonicus)* is found from northern Alaska and Mackenzie southwestward to Ontario; the Columbian chickadee *(P. h. columbianus)* is found from Montana and Alberta west through British Columbia and Alaska to the Kenai Peninsula; and the Acadian chickadee *(P. h. littoralis)* occupies that part of the range that, in general, is east of Michigan and Ontario.

The Cascade brown-headed chickadee *(P. h. cascadensis)* is apparently confined to the northern Cascade Mountains in Washington.

Egg dates.—Alaska: 1 record, June 9.

Alberta: 4 records, May 17 to June 13.

Labrador: 5 records, June 8 to 12.

Manitoba: 5 records, June 2 to 15.

Nova Scotia: 26 records, May 16 to June 9; 14 records; May 29 to June 5.

PARUS HUDSONICUS COLUMBIANUS Rhoads
COLUMBIAN CHICKADEE

This extreme western race of the Hudsonian chickadee ranges from the Kenai Peninsula in Alaska, through the Rocky Mountain region of western Canada, southward to southern British Columbia, Alberta, and northern Montana, but the limits of its range are not accurately outlined.

According to Ridgway (1904), it is "similar to *P. h. hudsonicus,* but slightly darker and less brown above, especially the pileum and hind neck; chin and throat more decidedly black; bill relatively larger."

Nothing seems to have been published on any of the habits of this race, which are probably similar to those of the other races. The type specimen was taken at Field, British Columbia, in the heart of the Canadian Rockies, at an elevation of about 5,000 feet.

PARUS HUDSONICUS LITTORALIS Bryant
ACADIAN CHICKADEE

PLATES 58, 59

HABITS

The eastern race of the Hudsonian chickadee breeds in the spruce and balsam forests, north to the limits of these forests, east of Hudson Bay, in Ungava, Labrador, Quebec, Newfoundland, and Nova Scotia. The 1931 Check-list says that it breeds south to Maine, "the mountains of northern Vermont and central New Hampshire, and the Adirondacks of New York." But there is some evidence to indicate that it may breed, at least occasionally, in some of the mountains farther south. It has been seen in June in Plymouth County, Mass., and at least twice during that month in the Pocono Mountains in Pennsylvania.

As to the Massachusetts evidence, Dr. Arthur P. Chadbourne (1896) wrote many years ago: "While walking through some dense old-growth pine woods *(Pinus strobus* and *P. rigida)* * * * I was greeted by the snarl *chee-dè-e-e-ɛ-e-ah* of a Hudson Bay Titmouse. . . . The woods in which I saw the Chickadee were only a few rods from a large cedar swamp, said to be a couple of miles wide, which is seldom visited except by lumbermen in winter; and in many portions the original growth of huge white cedars *(Cupressus thyoides)* and hemlock *(Abies canadensis)* has never been cut. In this old timber one seems to be in northern Maine or New Hampshire, instead of in Massachusetts;—the subdued half twilight of the damp cool forest, with its rocks and fallen trees, covered with a rich carpet of green moss and ferns might well tempt this and other northern birds to make it their summer home."

The Pennsylvania evidence is equally suggestive.   J. Fletcher Street (1918) found a pair of these chickadees at Pocono Lake on June 17, 1917.   "The location was at the edge of a sphagnum swamp amid a dense grove of dwarf spruces.   When discovered the birds evidenced considerable excitement and came and scolded within three feet of me."

Regarding the other Pennsylvania record, Thomas D. Burleigh (1918) writes:

In company with Richard C. Harlow, Richard F. Miller and Albert D. McGrew, I spent three weeks in the field in the spring of 1917 about La Anna, Pike County, Pa., and June 3, while searching a large sphagnum bog for a nest of the elusive Northern Water-Thrush, two brown-capped chickadees were seen. * * * They showed a preference for a certain part of the bog that we had been floundering through but although several suspicious looking holes were found, we could detect no signs of their nesting.   I returned to the spot the next day, and had no difficulty in finding the birds again.   This time I spent two hours trailing them but with no success other than leaving with the conviction that they were mated and if not as yet nesting here, would undoubtedly do so.   Not satisfied, however, all of us returned the following day and made another attempt but with no more luck though we again found them at the same place. * * * The situation in which they were found was typical of that much farther north, being indeed a northern muskeg in every sense of the word, with lichen covered tamarack, deep beds of sphagnum moss and scattered pools."

The localities in which these birds were found in Pennsylvania would match almost exactly the favorite haunts of the species in the north woods of Canada, coniferous forests of spruces, firs, cedars, pines, and larches, especially in the vicinity of peat bogs and muskegs, sometimes mixed with a few small paper birches and mountain-ashes, with a ground cover of sphagnum moss, Labrador tea, and other northern plants.

The only place where I have ever found the Acadian chickadee really common was on Seal Island, Nova Scotia.   This island is situated about 15 miles off the southwest coast and is about 4 miles long; it is divided in the middle by a low stretch of sandy beach and marsh, from which the land rises at both ends and is heavily wooded with a dense forest of small spruces and firs, sometimes growing so closely together as to be nearly impenetrable.   These forests, nourished by the fog banks that envelop that coast almost constantly, were rich and luxuriant, thickly carpeted with soft mosses and ferns, the fallen trees, as well as the trunks and branches of the standing trees, supporting a flourishing growth of various mosses and lichens.   In these dense and shady retreats, where the trees were often dripping with moisture, were the favorite haunts of Bicknell's thrushes and Acadian chickadees, the two most conspicuous birds.   It was the first week in July and the young of both species were on the wing. We saw the chickadees feeding their young, and, through the kindness of John Crowell, the owner of the island, we were allowed to collect a few specimens.

Courtship.—At Averill, Vt., on May 24, 1915, Frederic H. Kennard watched a pair of these chickadees in what seemed to be courtship behavior. His notes state that "the male was very thoughtful and attentive to his little mate, whom he fed frequently. She kept up an almost continuous twittering call, and would flutter her little wings and ruffle her feathers on his near approach. We saw him give her several green cankerworms and many other small bugs; and I saw them copulate once."

Nesting.—W. J. Brown writes to me that the Acadian chickadee is fairly well distributed in the County of Matane on the south shore of the Gulf of St. Lawrence, where he has found some ten nests. The lowest nest was one foot from the ground in an old spruce stub in a bog, and the highest elevation was about 8 feet. Out of the ten nests under observation, two had the entrance hole on the top of the stub, facing all kinds of weather, while the others had the hole at the side of a dead stub. All the nests were made of rabbit hair, with some feathers, well put together.

Philipp and Bowdish (1919) record three nests found in New Brunswick as follows: "On June 5, 1917, a nest was found, nearly or quite completed, in a natural cavity in a cedar stump about two feet from the ground. On June 16 the bird was sitting hard on five eggs, and was persuaded to come out only with great difficulty. As she laid no more, this was apparently her full laying. On June 24 a nest containing seven quite small young was found in a knot hole in a small live spruce. On June 13, 1918, another nest with young was found in a cavity in the top of a dead and rotten stub, about ten feet from the ground. This nest was very near the site of the 1917 nest with young, very possibly belonging to the same pair of birds."

Mrs. Eleanor R. Pettingill (1937), with her husband, Dr. Olin S. Pettingill, Jr., had an interesting experience with a family of Acadian chickadees on Grand Manan, New Brunswick. The nest was about 10 feet from the ground in a hole, made perhaps by a hairy woodpecker, in a tall dead spruce stump. The nest contained seven young that were "fitted compactly together on a comfortable bedding of birch-bark shreds and bits of moss. The base of the nest was about 6 inches below the entrance."

Aaron C. Bagg (1939) discovered a nest containing young in Somerset County, Maine. He writes:

For some moments the two chickadees flew about nearby, uttering faint scolding notes of alarm and continued to gather insects from the ends of spruce boughs. It was not until I had seated myself and remained quiet that one of the birds flew to a clump of spruce trees, a lone birch and a cedar. Presently the chickadee dropped down low behind the cedar for an instant. Suspecting it might be the nest, I approached and discovered the nesting cavity hidden in the cedar. Fifteen inches above ground the cedar, not over six inches in diameter, divided into two upright branches at an easy angle. Where they joined on the inside, a space fully ten inches in length, the secret lay—an ideal cavity with a narrow opening. Within the nest were young birds well feathered, nearly ready to leave. When I revisited the nest on July 5, they had departed. Search of the vicinity failed to discover any of the fledglings or of their parents. The nesting cavity contained finely shredded cedar peelings and bits of decayed cedar. But the bulk of the nest consisted of moss and deer hair.

Several other nests have been described by other observers, all in very similar situations and made of similar materials. The cavity is often filled with dry or green mosses, bits of lichens, fine strips of the inner bark of cedars and the down from ferns; but the main part of the nest, where the eggs are hatched, is a well-felted bag composed of deer hair and the soft fur of rabbits or other small mammals, making a warm bed for the young.

Eggs.—The Acadian chickadee has been known to lay four to nine eggs to a set, but apparently the two extremes are uncommon. The eggs vary in shape from ovate to short-ovate, and some are rather pointed. They have very little or no gloss. They are white and more or less sparingly and unevenly sprinkled with fine dots or small spots of reddish brown, "hazel"; occasionally the spots are concentrated in a ring about the larger end.

The measurements of 40 eggs average 16.0 by 12.2 millimeters; the eggs showing the four extremes measure 17.4 by 11.6, 15.8 by 12.9, 14.7 by 12.7, and 15.3 by 11.4 millimeters.

Young.—We have no definite information on the period of incubation, which is probably about the same as that of other chickadees, or about two weeks; other young chickadees usually remain in the nest 14 to 18 days.

The Pettingills (1937) watched their family of Acadian chickadees on Grand Manan for an entire day, from sunrise to sunset, each taking turns, with the following results:

Both sexes shared equally in the feeding. Frequently they arrived together and one stood by awaiting its turn to feed while its mate entered the hole with food. On several occasions one bird passed the food to its mate returning from the nest who immediately re-entered the nest to carry it below. * * *

During the course of the day the birds fed their young 362 times. The first feeding took place not long after break of day, at 4:46 A.M. One bird was already on the nest, where it had spent the night, when its mate arrived with food. The brooding bird left the nest and permitted its mate to enter and feed. They left the vicinity of the nest together. Before 6 A.M. they fed their young 34 times. During the day the average feedings were 24 times per hour. The last feeding took place at 7:40 P.M. when the bird, returning with food, entered and remained in the nest to brood.

Plumages.—I have seen no very young nestlings of this species, but Mrs. Pettingill (1937) says that the young, when "approximately three days of age," were naked except for a few wisps of down on their developing tracts, and that their eyes were only partially opened.

The juvenal body plumage is acquired before the young leave the nest, but the wings and tail are not fully grown until later. The color pattern is much like that of the adult, but the colors are duller and paler, and the plumage is softer and less compact. Dr. Dwight (1900) describes it as "above, brownish mouse-gray, the pileum pinkish drab-gray. Wings and tail dull slate-gray whitish edged, the coverts edged with pale wood-brown. Below, including suborbital region and auriculars dingy white and washed on the sides and crissum with pale cinnamon, the chin and throat dull black."

He says that the first winter plumage is "acquired by a partial post-juvenal moult, beginning early in August in eastern Canada, which involves the body plumage and wing coverts, but not the rest of the wings nor the tail, young and old becoming practically indistinguishable." In the first winter plumage the pileum is "darker and the back browner, contrasting but slightly with the cap; the flanks, sides and crissum rich Mars-brown; the black on the throat deeper and the white of the sides of the head and lower parts clearer."

Adults have a complete postnuptial molt in July and August, replacing the decidedly worn, ragged, and somewhat faded plumage of the breeding season with the fresh and more richly colored autumn plumage.

Food.—These and other chickadees are among the best conservators of the forest trees that we have. They are very active at all seasons, inspecting the trunks, branches, and twigs of the trees in search of the minute insects that are so injurious and are too small to be noticed by the woodpeckers and other large birds; nothing escapes the chickadees' keen little eyes. The food of all the species is very similar in character, varying only with the latitude of the habitat of each and the kind of trees it frequents. The Acadian chickadee is essentially a bird of the northern coniferous forests and is especially useful in protecting the spruces, firs, and pines in these woods. All summer it may be seen carefully inspecting the branches of these trees even to the outermost twigs, often hanging head-downward to search among the bases of the needles, looking for caterpillars, moths, beetles, and other insects with which to feed its large and growing family. And in the winter it continues the good work by prying into crevices in the bark and searching the branches and twigs for hibernating insects, pupae, or egg clusters; many of the insects thus destroyed are among the worst enemies of the conifers.

Lucien M. Turner says in his Ungava notes: "It is not uncommon to find their beaks covered with gum from the spruce and larch. I accounted for it by supposing that during the summer months, when the gum exudes plentifully and is so soft, many insects adhere to it, and when winter comes the birds search for just such places to obtain the insects."

While visiting New England in fall and winter, and probably at other seasons, this chickadee apparently feeds to some extent on various seeds; it comes occasionally to feeding stations and seems to be fond of fatty foods.

Horace W. Wright (1917) received a letter from Richard M. Marble, of Woodstock, Vt., in which he states that an Acadian chickadee fed at a feeding station from November well into January. Mr. Wright saw some of these birds feeding frequently "upon stalks of golden rod and aster"; and he quotes William Brewster as having seen them "pecking at gray birch seed-cones."

He (Wright, 1914) saw them "picking seeds on the ground" and "picking at the green undeveloped berries of a red cedar."

Behavior.—Judging from my limited experience with them on their breeding grounds, I should say that Acadian chickadees are just as tame and confiding as our little blackcaps; we had no trouble in watching them at short range, and they fed their nearly full-grown young within a few feet of us. They are less active than our familiar chickadees, more deliberate in their movements, and less noisy. They were traveling about in small family parties. Some observers have referred to it as shy, but others have found it almost tame enough to be caught by hand. Oliver L. Austin, Jr. (1932), found that, in Labrador, "while occupied with their nesting duties, they are very quiet and unobtrusive." He continues:

They keep out of sight and are seldom noticed, even by one looking for them. I had to hunt, and hunt persistently, to get three specimens at Sandwich Bay, July 16, 1928. But the moment the young leave the nest, usually during the last two weeks in July, the Chickadees alter their demeanor, and the woods suddenly seem overrun with them. Small bands of young and old together, probably parents with their broods, meander inquisitively through the spruce and larch forests amid the clouds of mosquitoes and black flies. Their tameness, in contrast to their previous reticence, is surprising. Their buzzing statements of facts greet you on all sides. They seem to follow you about. Hunting industriously for aphids, they explore every nook and cranny of the larch branch within two feet of your face. Perched upside down, rightside up, in all manner of conceivable and inconceivable attitudes, they regard you perkily with their bright beady eyes, burst suddenly into their commonplace *atser-day-day-day,* and flutter past your ear to the next promising branch. At this season you could almost catch them with a butterfly net.

W. J. Brown tells me of an usually tame Acadian chickadee that he saw. An old trapper had a nest of this bird about 15 feet from his back door in a natural cavity in a balsam fir. He had tacked a small branch directly under the entrance hole for a perch. When he tapped the side of the hole the chickadee came out and alighted on his hand to be fed. The man said that he had fed the bird in this way several times a day for the past ten days and that it would perch on any finger held out to it.

Acadian chickadees may usually flock by themselves, especially in their summer haunts, but they are often seen to join the merry little winter parties of black-capped chickadees, kinglets, nuthatches, and small woodpeckers that roam through the woods at that season. William Brewster (1938) watched such a gathering at Lake Umbagog in January. He says of the behavior of the Acadians: "For a time they kept high up in the tops of some tall balsams working among the cones, apparently extracting and eating the seeds. The Nuthatch was with them here for several minutes, but the Black-cap chickadees remained lower down. The Hudsonians differed from the Black-caps as follows: —they were much less noisy (often passing minutes at a time in absolute silence); they seldom hung head downward; they hopped and flitted among the branches more actively and ceaselessly, spending less time in one place; their shorter tails were less in evidence; they flirted their wings much more with a more nervous, tremulous motion very like that of Kinglets."

Mr. Wright (1917) noted that "Golden-crowned Kinglets have proved to be the closest companions of these Northern Chickadees on many occasions. Indeed, they seem to be their natural associates. Black-capped Chickadees are rather their incidental companions, with whom they occasionally come in touch, but do not habitually move."

Voice.—My impression of the ordinary note of the Acadian chickadee was that it resembled the corresponding note of the black-capped chickadee, but it was easily recognizable as not so sharp, clear, or lively as that of our more familiar bird; it was fainter, more lisping, hoarser, and more nasal; I wrote it in my notes as *chicka-deer-deer,* or *chicka-deer.* But I like Mr. Austin's rendering, *atser-day-day-day,* as quoted above.

William Brewster (1938) describes it as "a sharp *che-day, day,* very different from any note of the common Chickadee. * * * The ordinary chirp is much louder and more petulant. Another note frequently heard is a sharp *chip, chee-chee, chee* sometimes preluded by a sharp *che-chit* or *chee-chit-chit.*" Again (1906) he writes it "a nasal, drawling *tchick, chee-dày, dày.*"

A number of observers have heard this chickadee give a sweet, warbling song of three or four notes, and considerable has appeared in print about it. Other equally keen observers have failed to hear it. Aretas A. Saunders writes to me: "I am inclined to think that the brown-capped chickadees have no notes that correspond to the 'phoebe' note of the black-capped chickadee. I have seen quite a bit of the Acadian chickadees in the Adirondack Mountains, and Hudsonian chickadees in Montana, but have never heard a song of any kind from them."

Mr. Brewster (1906), after mentioning some published accounts of the song, writes: "I have never heard anything of the kind from the Hudsonian Chickadee, although I am reasonably familiar with that species, having had abundant opportunities for studying its notes and habits in the forests of northern New England where I have met with it on many different occasions and during every month of the year excepting April."

If two such experienced and keen observers have failed to hear the song, it must be of rather rare occurrence.

Townsend and Allen (1907) add to the discussion and quote Dr. Townsend's description of a song, heard at Cape Breton Island in August 1905: "It was a low, bubbling, warbling song, which I vainly attempted to describe in my notes. It began with a *pset* or *tsee;* followed by a sweet but short warble."

Francis H. Allen (1910) gives a number of references to published accounts of the song, and then adds his own observation, made on Mount Moosilauke, N. H., September 29, 1909: "I observed it for some time at close range and heard it sing again and again. The song was a short one but took two or more forms, one of which I set down at the time as bearing some slight resemblance to the syllables *wissipawiddlee,* though this rendering conveys no clear impression of its warbling quality. The final syllable was sometimes trilled and sometimes pure. It seemed to me that the song corresponded exactly to the *phoebe* song of the Black-capped Chickadee *(P. atricapillus),* but it was also strangely suggestive of the song of Bicknell's Thrush!"

Field marks.—The Acadian chickadee bears a general resemblance to our common chickadee, but it can be easily recognized by its brown instead of black cap and by the rich chestnut-brown of the sides and flanks; the black throat patch seems somewhat duller or less conspicuous. Its behavior and its voice are different, especially the latter, which is easily recognized.

Enemies.—Chickadees have to be on the watch for all the mammals and birds that prey on small birds, such as small hawks and squirrels, but their eggs and young are usually safe from many nest robbers in their snug retreats. Mrs. Pettingill (1937) writes: "Once during the day a red squirrel appeared in a near-by birch sapling. The Chickadees, on discovering him, were beside themselves with fright. * * * Both birds began to call noisily and to jump from one branch to another, keeping at a safe distance from the animal, but threatening him by making fake darts in his direction. The fact that one of them feigned injury came as a surprise. Perching at first upright on a branch, it suddenly but slowly toppled over backward until it clung upside down on the branch, wings fluttering helplessly. It did not drop to the ground but let go and alighted near-by. This time, on another branch, it seemed to fall over sidewise, nevertheless holding its legs stiff and not letting go."

In the nest examined by Mr. Bagg (1939) were found an adult *Protocalliphora,* a blood-sucking fly, and several pupae probably of this species, as well as the larvae of some less harmful species of insects. These fly maggots may be, at times, sufficiently numerous to kill not only the very young birds but those nearly fully fledged.

Fall.—Most of these northern chickadees remain on or near their breeding grounds all winter, but on certain seasons, at very irregular intervals, there has been a well-marked southward migration into New England and perhaps farther south. Two of the largest incursions were in 1913 and 1916, of which Horace W. Wright (1914 and 1917) has written full accounts. In November 1904 he saw only four birds, but in four different localities. But in the 1913 flight he had the records of over 40 birds, seen by himself and others, in 15 different towns or distinct localities. The 1916 flight seems to have been even heavier and more widely extended, reaching as far south as Staten Island, N. Y. The birds began to appear in October and some remained almost through January, the crest of the wave coming between November 9 and December 10. He referred the birds of the 1916 flight to the supposed Labrador race, *nigricans,* which is now regarded as the first winter plumage of the species; as all the birds taken seemed to be of this race, it suggests that these southward flights may be made up mainly of immature birds.

Some of these birds evidently spent the winter of 1916 and 1917 in

New England, or not much farther south. Mr. Wright (1918) gives another full account of the return flight in the spring of 1917; the latest record seems to be of a bird seen at West Roxbury, Mass., on May 18.

PARUS HUDSONICUS CASCADENSIS A. H. Miller

CASCADE BROWN-HEADED CHICKADEE

Dr. Alden H. Miller (1943b), who is responsible for the above names, says that this race "differs from *Parus hudsonicus columbianus* by darker, much sootier pileum (near Chaetura Drab in fresh plumage rather than Bister or Sepia); interscapular area somewhat darker and gray of sides of neck darker and more extensive proximally in auricular area; chestnut of sides darker (Prout's Brown rather than Cinnamon Brown). Size much as in *columbianus,* although measurements of the small sample available suggest somewhat greater average dimensions for the Cascade population * * *.

"The race *cascadensis* is even darker and duller on the pileum than is *P. h. littoralis* of the eastern seaboard and it is larger and notably longer-billed as are *P. h. hudsonicus* and *P. h. columbianus.* The contrast in coloration between *cascadensis* and *columbianus* is almost as great as that between *P. h. hudsonicus* and *Parus cinctus alascensis,* which, although evidently closely related, breed side by side in the Kowak Valley of Alaska.

"Range. *Cascadensis* is known at present only from the northern Cascade Mountains in the vicinity of Monument 83 on the United States-Canadian boundary in northwestern Okanogan County, Washington."

PARUS RUFESCENS RUFESCENS Townsend

CHESTNUT-BACKED CHICKADEE

HABITS

The visitor from the Eastern States is accustomed to seeing roving bands of chickadees, kinglets, creepers, and other small birds trooping through the winter woods and is not surprised when he finds just such jolly companies of friendly little feathered mites foraging through the dark coniferous forests of the humid Northwest coast. The kinglets and the creepers are so much like their eastern representatives that he does not recognize the difference, as he sees them in life; they are just familiar friends, kinglets and creepers. But the chickadees are different; they do not fit into memory's picture of our New England woods; their caps are not so black as those of eastern birds, and the rich chestnut of their backs and sides is strikingly new. We get the thrill of a new bird, seen for the first time. But, as we watch them we see that they are still chickadees, with all their manners, activities, and cheery notes, just old familiar friends in more richly colored garments, but just as sociable, friendly, and intriguing; they win our affection at once.

The chestnut-backed chickadee, of which there are three recognized subspecies, occupies a long narrow range from the Sitka district of southern Alaska southward along the humid coast belt to a little south of Monterey Bay in California. Dr. Joseph Grinnell (1904), in his interesting paper on the origin and distribution of this species, says that its range is very "narrow, only within the confines of Oregon and Washington exceeding one hundred miles and elsewhere usually much less, save for one or two isolated interior colonies." The type race, the subject of this sketch, occupies the greater part of this range, from Alaska southward to Marin County, Calif., where it intergrades with the Nicasio chickadee *(P. r. neglectus)*. The interior colonies referred to are in suitable coniferous forest environments in Idaho and western Montana, west of the Continental Divide. As to the possible origin of the chestnut-backed chickadee, Dr. Grinnell (1904) calls attention to a certain degree of resemblance, as to appearance and habitat, between it and the Hudsonian chickadee; and he suggests that the two species may have been derived from common ancestry, the dark, rich coloring of *rufescens* having evolved under the influence of the humid coast belt in which it lives; though no intermediates between the two are now known to exist, these may have been eliminated by the existence of insurmountable natural barriers and thus the two have become separate species. It is an interesting theory!

The favorite haunts of the chestnut-backed chickadees are the heavy, dark forests of firs, spruces and pines, dense cedar, tamarack, and hemlock woods, and, in California, the redwood forests. In the woods about Seattle and Kirkland, Wash., where we found this and the Oregon chickadees in 1911, there was then some of the primeval forest of lofty firs still left, but much of it had been lumbered and overgrown with second-growth firs of two or three species, with a considerable mixture of hemlock and a very handsome species of cedar; there was also some deciduous growth consisting of large alders and maples, with flowering dogwood in full bloom and a wild currant with pink blossoms.

Dr. Samuel S. Dickey writes to me that on the coast of southern Alaska "it is a bird that rather inclines to remain much of the time in the forests of gigantic evergreens, the Alaska cedars, Sitka spruces, western hemlocks and the firs. But not infrequently it will rove into glades and bogs, and it often comes down to the edges of the sea beaches, and only a stone's throw from cabins and totem poles of the native Indians."

Nesting.—The only nest of this species that I have seen was shown to me by D. E. Brown near South Tacoma, on May 14, 1911, while we were hunting through some of his favorite collecting grounds, smooth, level land with a fine growth of firs and cedars scattered about; the local species of firs were most abundant and were growing to perfection in the open stand, where they were well branched down to the ground. The chickadee's nest was 5 feet from the ground in the trunk of a large dead pine, in a cavity evidently excavated by the birds in the rotten wood behind the bark; since it contained one fresh egg, it was not closely examined.

The nests of the chestnut-backed chickadee have been placed at widely varying heights, according to various observers, but all seem to agree that most of the nests are less than 10 feet above ground. Thomas D. Burleigh (1930) writes of the nests that he has found about Tacoma: "The usual situation was in a fir stub, varying in height from a foot and a half to nine feet from the ground, although one nest was twelve feet from the ground in a knot hole in the trunk of a large dead oak at the edge of a stretch of open prairie, while another was five feet from the ground in a cavity in the thick bark of a large Douglas fir in a short stretch of open woods."

J. H. Bowles (1909) has published an excellent account of this chickadee, in which he says:

At the approach of the nesting season the Chestnut-backs retire to the most arid section of the country to be found, the more exposed it is to the sun the better, and it is only in such locations that one may ever expect to find them during the breeding season. The nesting site is chosen about the middle of April, most often in the dead stub of some giant fir or oak. On one occasion only have I found the nest near water, this being in a small willow on the edge of a swamp.

The birds almost invariably dig their own hole, but I once found a nest in the winter burrow of a Harris Woodpecker. One peculiarity about them, which greatly increases the difficulty of finding their nests, is that they almost never start the hole for themselves. Instead they select some place where a fragment of the wood or bark has been split away, or else they will often take the oval hole made by the larva of one of our largest beetles. These holes are not altered at the entrance in any way and, as the dead trees are full of them, it is extremely difficult to locate the one containing the nest.

He says that these chickadees sometimes nest in loose colonies; in one locality he "found no less than seven occupied nests inside a very small area, some not more than fifty yards apart." The lowest nest he found was only "two feet up in a tiny fir stub," but he says that "it is nothing unusual to find them fifty feet up in the giant fir stubs, remnants of long past forest fires." Dawson (1923) says that he has "found nests as high as eighty feet in a fir stub; and in two instances in a dead tree wholly surrounded by water."

As to the nest itself, Mr. Bowles (1909) writes: "The cavity is usually about seven inches in depth, seldom any more, tho occasionally much less. Almost any soft substance to be found in the vicinity is used to make up the nesting material, but there is always a substantial foundation of green moss. Cotton waste from factories, hair of cows, squirrels, rabbits and goats, and small feathers are most often used, one very beautiful nest in my collection being composed almost entirely of feathers from the Kennicott Screech Owl (*Otus asio kennicottii*). No matter how large the bottom of the cavity may be, it is always packed tight, and I have sometimes removed a nest that would easily fill both hands."

Mr. Burleigh's (1930) nests contained similar materials, including horsehair, feathers of a Steller's jay, rope fiber, and white fur of a cat.

Eggs.—The chestnut-backed chickadee most commonly lays a set of six or seven eggs, sometimes only five, frequently eight, and sometimes as many as nine. Mr. Bowles (1909) says that "the eggs vary greatly in both shape and size, some being shaped like a quail's egg, others like a murre's egg." But most of the eggs that I have seen are ovate or short-ovate, though some are slightly pointed.

The ground color is pure white, and they are sparingly sprinkled with reddish brown or light red dots, sometimes with "sayal brown" or "snuff brown"; the markings are sometimes concentrated about the larger end of the egg, but oftener they are irregularly distributed; some eggs are evenly sprinkled with fine dots, and some are nearly or quite immaculate. The measurements of 40 eggs average 15.3 by 12.0 millimeters; the eggs showing the four extremes measure **16.2** by 12.2, 14.9 by **12.6, 14.4** by 12.0, and 15.9 by **11.3** millimeters.

Young.—The exact period of incubation does not seem to have been determined for this species. Dawson (1923) and Bowles (1909) both state that incubation begins when the first egg is laid, as the sizes of the embryos in a set of eggs vary considerably. Perhaps the bird does not incubate all through the laying period, but she covers the eggs when she leaves the nest, which keeps them warm, and furthermore, the nest is usually fully exposed to the heat of the sun, which helps the progress of embryo development.

The chickadees are very brave in the defense of their nest and resort to the common chickadee habit of hissing and fluttering their wings in a startling manner when an intruder looks into the nest.

Undoubtedly both parents assist in the care and feeding of the young, though we have no definite data on the subject. They seem to lose all fear of the human intruder in their anxiety to defend their young, and Dawson (1923) says: "Not infrequently, if the young are kindly treated, the parent bird will venture upon the hand or shoulder to pursue its necessary offices."

Plumages.—I have seen no very young chickadees of this species. In the juvenal plumage, which is mainly acquired by the time the young bird leaves the nest, except that the wings and tail are not fully grown, the color pattern is the same as in the adult; the chestnut of the back, however, is duller, and that of the sides and flanks is duller and paler; the plumage is also softer and less blended. A partial post-juvenal molt, involving the contour plumage and the lesser wing coverts, but not the rest of the wings or the tail, takes place in August; by the first week in September the young bird has acquired a first winter plumage that is practically indistinguishable from that of the adult. Adults have one complete postnuptial molt in August and September.

The birds show some signs of wear and fading before summer, but assume the darker and more richly colored plumage in fall.

Food.—Prof. F. E. L. Beal (1907) studied the contents of 57 stomachs of the chestnut-backed chickadees, taken in every month except March, April, and May, and found the food to consist of nearly 65 percent of animal matter and 35 of vegetable. Caterpillars constituted 18 percent of the animal food, taken in nearly every month, even in December and January; the largest amount, 53 percent, was taken in August. Hemiptera, leafhoppers, treehoppers, and olive and other scales were the most important item of food, amounting to about 25 percent. "Wasps were eaten to the extent of 13 percent of the food, but no ants were found. Beetles amount to less than 2 percent of the food, but nearly all are noxious; weevils appeared in one stomach." No trace of flies or grasshoppers was found. Spiders amounted to 7 percent for the year but amounted to nearly 16 percent in August.

"The vegetable portion of the food consists of fruit pulp 8 percent, seeds nearly 20 percent, and miscellaneous matter 7 percent. Fruit pulp was found only in a few stomachs taken in fall and winter and was probably waste fruit. The seeds eaten were mostly those of coniferous trees."

Dr. Dickey (MS.) writes: "When they have wearied themselves in taking food from the trunks of trees, they will pass to huge crumbling logs, peck at green moss pads for certain items, and flit into and glean the tangled branches of such shrubs as salal, elderberry, dwarf blueberry, depauperized paper birch, alders, dogwood, etc. By mid-March, hordes of small winged gnats arise from the newly unfolding buds of the elderberry bushes, and such items are cherished by the tits. I noticed that, after the birds had fairly bounced the 'bugs' out of such cover, then they would arise and, craning their heads back and forth and thrusting forward their bodies, cleanse the air of this kind of provender. They lower themselves to boulders and rocks, and cleanse boats on dry docks of hiding insects, nor will they hesitate to peck at 'barnacle scales'."

Behavior.—The chestnut-backed chickadees are almost exact counterparts of our familiar eastern chickadees in traits of character and behavior—fearless, sociable, and full of friendly curiosity. Taylor and Shaw (1927) describe their behavior very well, as follows:

Establish yourself in an inconspicuous position in the general vicinity of a flock and make a squeaking sound with the lips on the back of the hand. Presently a tiny chestnut back will appear in the needle clumps not far away, calling excitedly and seeking the author of the unusual call note. If you remain quiet and continue the squeaking you may soon find that the trees about you are alive with chestnut-backs, buzzing about and hurling imprecations at you as if you were a new kind of owl. Incidentally in the group of excited birds you may catch sight of a number of other species.

Chestnut-backed chickadees are sociable bodies, indeed, even gregarious. Seen characteristically in flocks of 4 or 6 to 20 of their own kind, they approve, if they do not actively cultivate, loose association with several others, including western golden-crowned kinglets, red-breasted nuthatches, Shufeldt juncos, mountain chickadees, and lutescent and Townsend warblers. Of these the kinglets are most often found in their company.

Dr. Dickey, in Alaska, found them associated in March with winter wrens, brown creepers, red-breasted nuthatches, sooty song sparrows, and Oregon juncos. "They are quick to react to noises in their haunts; they come hurrying out of plant cover, a gleam in their beady eyes; and they dangle upon dried seed heads and flower clusters of several kinds of shrubs or undergrowth of this wild region. They will swing up and down on bending branches, vent squeaks and low chirps, varied with buzzing 'dizzes.' "

Voice.—The chestnut-backed chickadee seems to have a great variety of notes, perhaps of conversational value, most of which bear little resemblance to the *chick-a-dee* note of our blackcap, though Dawson (1923) says that this becomes *kissadee*, and Taylor and Shaw (1927) say that "when scolding us at short range they interjected numerous fine *tseek a dee dees* into their conversation."

Taylor and Shaw found some of the notes to be very similar to "certain calls of the western golden-crowned kinglet, so as to be distinguishable with difficulty. Several much-used expressions were caught and set down in our field books as follows: *Toot seet seet see!* repeated several times; this was sometimes varied to *tsweet tsweet tsweet!* One note is like *whist,* uttered once or repeated as many as three times. This triple note sometimes sounds like *twit twit twit.* The principal call note heard during one exceedingly foggy, rainy day early in September was *tseet tseet!* or *tseet tseet tsew,* a very kingletlike utterance. This was quite certainly a flock location call, for the birds continued it when we were no longer in view."

Both Bowles (1909) and Dawson (1923) refer to a song resembling that of the chipping sparrow. The latter says of it: "When the emotion of springtime is no longer controllable, the minikin swain mounts a fir limb and raps out a series of notes as monotonous as those of a Chipping Sparrow. The trial is shorter and the movement less rapid, so that the half dozen notes of a uniform character have more individual distinctness than, say, in the case of the Sparrow: *Chick chick chick chick chick chick.* Another performer may give each note a double character, so that the whole may sound like the snipping of a barber's shears: *Chulip chulip chulip chulip chulip.*"

Field marks.—While traveling through the treetops in loose flocks, as is its customary habit, this chickadee might easily be mistaken for one of the other western chickadees. But at lower levels and at short range its chestnut back and sides are quite conspicuous. In juvenal plumage the young of this and the Hudsonian chickadee look somewhat alike, but the former is considerably darker. The chestnut-backed chickadee frequents, as a rule, more heavily timbered and drier country than the Oregon chickadee, but there are exceptions to this rule. Its voice is also different from that of the others.

Enemies.—In addition to the ordinary enemies of all small birds, there is, according to Mr. Bowles (1909), another unusual enemy, which "is no other than the common black-and-yellow bumble bee. This insect has a veritable mania for living in holes in trees, and a chickadee nest appears to be the acme of its desires. It seems to like the nesting material and prefers the nest before the eggs are laid, but will often drive the bird away from an incomplete set, pulling up most of the nesting and leaving the eggs underneath."

Fall.—S. F. Rathbun (MS.) says: "In October there appears to be a movement to the lowlands of those individuals that have spent the summer in the higher altitudes; and one who may happen to be in the mountains at that period will see, each day all through the month, numbers of these active little birds trooping by, this continuing well into November. It is common throughout the winter in the sections adjacent to tidewater, and also breeds in such localities, for the species can be found at all seasons in all parts of the region [around Seattle], although there is little doubt that the majority of the birds spend the summer in the more elevated districts, judged from the numbers that come from such in fall."

Winter.—The chestnut-backed chickadee is a permanent resident throughout its range. Alfred M. Bailey (1927) says that "they are to be noted the year around, throughout the whole of southeastern Alaska," where "they are probably the most numerous of the winter birds. * * * These cheery little creatures are often the only signs of bird life to be seen in the winter woods, and their quiet, comrade-like call, as they drift from one tree to another, can often be heard when the birds are obscured by the falling snow."

DISTRIBUTION

Range.—Pacific coast region from Prince William Sound, Alaska, south to southern California and east to western Montana; not migratory. The range of this species extends **north** to Alaska (Resurrection Bay, Glacier, and Skagway); and northern British Columbia (Flood Glacier). **East** to British Columbia (Flood Glacier, Blackwater Lake, and Kootenay); western Montana (probably Glacier National Park, Clark Fork, and Fort Sherman); northeastern Oregon (Blue Mountains and Powder River Mountains); and California (Mount Shasta, McCloud, Hayward, and Cambria). **South** to southern California (Cambria). The **western** limits extend northward along the coast from southern California (Cambria) to the Kenai Peninsula, Alaska (Resurrection Bay).

As outlined the range is for the entire species, of which three subspecies are recognized. The typical form *(Parus rufescens rufescens)*, occupies all of the range south to central California; the Nicasio chickadee *(P. r. neglectus)* is confined to coastal areas of Marin County, Calif.; and Barlow's chickadee *(P. r. barlowi)* is found on the coast of California in the general area between San Francisco Bay and Monterey Bay.

Egg dates.—California: 58 records, March 12 to June 7; 30 records, April 23 to May 16, indicating the height of the season.

Washington: 30 records, April 7 to June 10; 16 records, May 9 to 24.

PARUS RUFESCENS NEGLECTUS Ridgway

NICASIO CHICKADEE

In a short and narrow coastal region of middle California, the humid Transition Zone of Marin County, there lives a subspecies of the chestnut-backed chickadees that seems to be strictly intermediate in coloration between the other two races, one north and one south of it. Mr. Ridgway (1904) describes it as "similar to *P. r. rufescens* but with much less of chestnut on sides and flanks, which exteriorly are pale gray, the chestnut also paler and duller." Dr. Joseph Grinnell (1904), in discussing this group, remarks that "this southward paling of the lateral feather tracts seems to be parallel to the relative decrease in the humidity of the regions occupied." And this paling is carried still farther, with the practical disappearance of the chestnut sides in the next race to the southward, *barlowi*. The Nicasio chickadee seems to occur in its purity only in Marin County; it intergrades with *P. r. rufescens* in Sonoma County to the northward; and San Francisco Bay and the Golden Gate seem to form an effectual barrier between it and *P. r. barlowi* to the southward.

In support of his theory that the two species *Parus hudsonicus* and *P. rufescens* were derived from common ancestry, Dr. Grinnell (1904) calls attention to the fact that "the young of *barlowi* has the sides paler rusty than *neglectus*, *neglectus* slightly paler than *rufescens*, but *rufescens* has the sides slightly more rusty than *hudsonicus*, a sequence which accords well with the present theories of origin." His hypothesis is based, of course, on the generally accepted theory that immature individuals more closely resemble the ancestral form than the adults do.

We have no reason to think that the habits of the Nicasio chickadee are very different from those of the chestnut-backed chickadees elsewhere. The eggs are characteristic of the species, as described under the preceding form. The measurements of 39 eggs average 15.7 by 12.3 millimeters; the eggs showing the four extremes measure **17.6** by **12.2**, 16.5 by **13.0, 14.3** by 12.0, and 15.2 by **11.9** millimeters.

PARUS RUFESCENS BARLOWI Grinnell
BARLOW'S CHICKADEE

PLATE 56

HABITS

Farther south along the coastal area of middle California, from San Francisco Bay to a little south of Monterey Bay, we have this paler, less rufous-sided race of the chestnut-backed chickadees. In his original description of this race, which he named in honor of Chester Barlow, Dr. Grinnell (1900b) characterizes it as "similar to *P. rufescens neglectus*, but the sides pure smoked gray without a trace of rusty."

The San Francisco Bay region seems to form the northern barrier to the distribution of this race farther northward, and climatic conditions apparently prevent its extension farther south, a certain degree of humidity being required by this species. Dr. Grinnell (1904) remarks that "even the Santa Cruz District with its gray-sided *barlowi* has very much greater rainfall and cloudiness than regions immediately to the southward and interiorly. Too abrupt aridification with accompanying floral changes apparently forms the present barrier to further distribution in these directions."

One would hardly expect Barlow's chickadee to differ materially in its habits from the other two races of the species, and this seems to be the case, though differences in environment may cause it to select unorthodox nesting sites, as the following quotations will show. Dudley S. De Groot wrote to Milton S. Ray (1916) of three nests found in Golden Gate Park: "A nest found April 7, 1916, which contained six badly incubated eggs lying in a thick bed of rabbit fur, was located eight feet up in a hole in the side of a log cabin. Another was in a small cavity fifteen feet up in a eucalyptus and contained young almost ready to fly. The third nest was remarkable for its situation, being placed in a pipe leaning against an out-building. The nest was about one and a half feet down the pipe, which was only three inches in diameter, and contained, in very cramped quarters, young birds about half grown."

Joseph Mailliard (1931) located a nest in the same park that was in "the large mandibular foramen on the inside of the left mandible of the huge Sulphur-bottom Whale skeleton under the shed!"

The eggs of Barlow's chickadee are practically indistinguishable from those of the chestnut-backed chickadee. The measurements of 40 eggs average 15.5 by 11.9 millimeters; the eggs showing the four extremes measure 17.3 by 12.2, 15.2 by 12.7, and 14.2 by 11.2 millimeters.

PARUS BICOLOR Linnaeus

TUFTED TITMOUSE

PLATES 60, 61

HABITS

As we travel southward, weary of the rigors of the northern winter and anxious to meet spring halfway, one of the first southern birds to greet us from the still leafless woods is this cheery little tomtit. Long before we reach the land of blooming jessamine, hibiscus, and oleanders, we may hear its loud, whistling *peto, peto,* welcoming us to his home in the southland.

To the novice, these notes may at first seem to bear a resemblance to the three-syllabled notes of the Carolina wren or to the spirited whistles of the cardinal, which one hears so frequently all through the southern States. I have sometimes been puzzled, when hearing them for the first time each season, but one soon learns to recognize them, for all three are quite distinct. Being a rather noisy bird and rather somberly colored, the tufted titmouse is generally heard before it is seen, though it is not particularly shy. We may look for it with best success in the deciduous woods.

In Pennsylvania, near the northern limit of its range, according to Dr. Samuel S. Dickey (MS.), it is oftener found in beech timber, where its coloration harmonizes and where it finds congenial openings in trunks and branches for nesting purposes. "But it will inhabit a variety of cover in modern times, and is found among oaks, tulip poplars, yellow locusts, sycamore, and thickets of mixed saplings, as well as orchards."

In the Middle West and southward it frequents the river-bottom forests in spring and summer but wanders about at other seasons in more open areas, among the shade trees and about houses in the villages. In the Gulf States and Florida it is often found in the live oaks, which shade the streets of villages and towns, but its favorite haunts are the mixed hammocks and the smaller cypress swamps that are scattered through the flat pine woods; it is seldom seen in these flat woods outside of these little cypress heads.

Courtship.—After wandering about all through fall and winter in small flocks by themselves, or mixed with other species, they begin their courtship activities early in spring and prepare to separate into pairs. Dr. Dickey (MS.) says that "males pursue and chase females among branches and often end up in brush piles. They will glide between avenues of trees along the courses of streams. Sometimes pairs thus continue up the forest-clothed flanks of slopes and cliffs. Males are combative, and are seen clasping and falling among vegetation. They will end up on the forest floor, there to part company and to continue their advances to the female of their choice."

Nesting.—As I have never been fortunate enough to find a nest of the tufted titmouse, I must rely on the observations of others. Dr. Dickey (MS.) writes to me, of the nesting habits in Pennsylvania and West Virginia, as follows:

"Naturally tufted tits breed inside cavities, mostly in trees. They utilize both live and dead growth, often cavities that are surrounded with hard resistant wood. A number like to breed in knot holes, slits left where lightning has struck, and long weathered squirrels' holes.

Not a few pairs inhabit the abandoned holes of downy, hairy, red-bellied, red-headed, and pileated woodpeckers and the cavities of the flicker. I have three records of their breeding in crude wooden bird boxes in town yards; and one instance was a hollow metal pipe, 4 inches in diameter, beside the yard of a farmhouse."

He thinks this bird shows a preference for beech trees for nesting sites, but he has found nests in white, red, and blue oaks, tupelos, syca-mores, pines, hemlocks, apple trees, mulberries, water and sugar maples, yellow locusts, white ashes, and chestnuts. "It will come out from cover and nest in fence posts along roads and borders of fields, and occasionally it selects some stub or post in an open pasture. The run of pairs prefer the woods in which to breed, but not a few enter and nest in orchards, groves, parks, and even shade trees along community streets. Nests are found at both low and high elevations; they range from 3 feet up to as many as 85 or 90 feet. They will continue to use the identical cavity for years, if unmolested.

"Nest building begins late in April, although birds are seen to carry odd leaves and trash into holes even as early as late in March. They begin by carrying in strips of bark and dead deciduous leaves; those of white oak and maple are common. Then they add sprays of green moss and dry grass, and round out the interior with pads of hair from cattle, rabbit, deer mouse, and others, and bits of rags, strings, or cloth."

In South Carolina, according to Wayne (1910), "this species deposits its eggs in natural cavities of trees or in the deserted holes of the smaller woodpeckers and does not appear to excavate a hole for itself. It seems to have a preference for hollows in chinquapin and dogwood trees, and the hole ranges from four to forty-five feet above the ground. While nest-building, the birds carry large quantities of material at every trip and one generally accompanies the other to and from the site. The nest is composed of wool, cotton, hair, leaves, fibrous bark and snake skins, the last material being indispensable to this species, as it is to the Crested Flycatcher. * * * The birds are the closest of sitters and have to be removed from the nest before it can be examined. Only one brood is raised and these follow the parents for many months."

Mr. Wayne (1910) also tells of an unsuccessful attempt of a pair of these titmice to nest in a tuft of Spanish moss. The nest was built in a very large mass of the hanging moss (*Tillandsia usneoides*) and five eggs were laid, but a violent rainstorm occurred and the nest and eggs were blown out. The bird was undismayed, rebuilt the nest again in the same bunch of moss, and laid three eggs; but she was disappointed again, for another storm came up before the set was completed, and the nest and eggs were found on the ground the following day.

M. G. Vaiden, of Rosedale, Miss., reports an unusually lofty nest of the tufted titmouse that was located in a cavity of a black locust near there some 97 feet above the ground. Another nest was found in a gatepost about 8 feet up; it was composed of horsehair and a few feathers.

The nesting cavities occupied by this titmouse vary greatly in size and shape, which means that in some cases a large quantity of material has to be collected to fill up extra space. C. S. Brimley (1888) mentions a nest that he found in a hollow in a small dogwood, near Raleigh, N. C. "The opening of the hollow was about two feet from the ground, and the hollow reached to the earth, but for half the distance three sides of it were gone. So the birds had piled up moss, leaves, etc., from the ground right up into the hole and then lined the nest at the top with white cat fur and a few pieces of snakeskin, the eggs being at least eighteen inches from the bottom of the nest."

The material used to fill up large cavities consists largely of leaves, preferably those that are damp enough to pack well without crumbling, but not too wet; these are picked up near a brook, or in some damp place in the woods. Lucy H. Upton (1916) made an interesting observation on the use of damp leaves by a titmouse that she was watching: "Having chosen a damp brown oak leaf from the ground, it flew with it into a bare tree, and, holding the leaf with its claw firmly against the branch, it drew itself to its full height, raised its head like a Woodpecker, and with all the might of its tiny frame gave a forcible blow to the leaf with its bill. This process was kept up nearly half an hour. * * * At last its purpose seemed to be accomplished. It rested, and lifted the leaf by the petiole. We then saw that the hammering had made it into a firm brown ball nearly as large as an oak gall."

The bird flew away with the leaf and probably placed it in its nest.

Freda L. Hood (1916) adds by way of explanation, that these birds "line their nests with a pulpy substance not unlike a sponge. They carry a large number of these damp leaf-balls to their nest-hole and there pull them into shreds. * * * The Titmouse uses this sort of lining for its nest only when they build in damp weather. They do not seem to be able to use dry leaves in this manner."

R. B. McLaughlin (1888) adds that about moist places "she gets a supply of green moss and mixes in a modicum of dirt. After she has accumulated the desired amount of such materials, we will find her at the bed of the flying squirrel (*Pteromys volucella*), or some other mammal which collects the thin inner bark of trees. and she does not hesitate to appropriate as much as she needs. Then she is off for the farmer's barn, and any bunch of cornsilks about his granary is used.

Again she is over where he curried his horse or butchered his pig, in quest of hair."

But the titmouse is not content with picking up stray hairs, or even bunches of fur from dead animals, and often becomes bold enough to collect this needed nesting material from living mammals, including human beings.

J. Harris Reed (1897) noticed a tufted titmouse that was apparently trying to drive a red squirrel away. "The squirrel was lying flat on the upper side of a large sloping limb, and the Titmouse would approach cautiously from behind and catch at its tail. It was not long before I noticed that the bird had collected quite a mouthful of the hairs, with which it flew off to a hole near by where it was deposited."

Ward Reed (1927) writes: "While walking through the woods looking for Crows' nests about the first of May, I came upon an unusual sight. On a branch of a tree a few feet from the ground sat a Woodchuck *(Marmota monax)*, while bobbing up and down above it a tufted Titmouse *(Baeolophus bicolor)* was engaged in plucking hair from its back. On a near-by twig the bird's mate was perched, with its mouth already full of hair, and in a few minutes they flew away together."

Mrs. Vitae Kite (1925) generously allowed an aggressive titmouse to help himself to some of her silvery locks. "Without the least warning he lit squarely on top of my head, giving me such a start that it was with great difficulty I controlled myself and sat still. At first I thought he was trying to frighten me away but soon changed my mind, when he began working and pulling at my hair with all his might. Now my hair has been very white for many years, but I still have plenty of it, and was more than willing to divide with this little bird, so I steadied myself and 'held fast' while that energetic 'Tom' had the time of his life gathering 'wool' to line his nest, for that was what I now felt sure he was doing. He didn't seem to have much luck with the coils on top, so he worked around over my ear, where there were short loose hairs, and I could hear and feel him snip-snip as he severed them—not one by one, but in bunches, it seemed to me."

Even man is not immune to such attacks, for E. Irwin Smith (1924) had a similar experience. He was seated on a stump on the edge of some woods, with his hat off, when he noticed a titmouse flitting about his head. "It flew back into the bushes, only to return and flutter above my head as before. Yet the third time it came back, but this time, instead of flying away again, it lit on my head, and, in a very diligent manner, began to pick the hairs therefrom. The pricking of its sharp little toes on my scalp and the vigor of the hair-pulling was a trifle too much for my self-control, and I instinctively moved my head. Away it flew, but only for a moment, and then it was back at work, harder than before."

Other somewhat similar cases have been reported, but the above will suffice to illustrate this strange habit.

Eggs.—Four to eight eggs may be found in the nest of the tufted titmouse, but oftener there are either five or six. The eggs vary from ovate (commonly) to elongate-ovate, and there is practically no gloss. The ground color is usually pure white, but often creamy white, or rarely pale "cream color." They are generally more or less evenly speckled all over the entire surface with very small spots or fine dots; often these markings are thickest at the larger end, where they are sometimes concentrated into a wreath; rarely this concentration is at the small end. The markings are in various browns, "hazel," "cinnamon-rufous," "vinaceous-rufous," "burnt sienna," or "chestnut"; some eggs have a few underlying shell markings of "lilac-gray" or "drab-gray." The measurements of 50 eggs average 18.4 by 14.1 millimeters; the eggs showing the four extremes measure **20.7** by 13.5, 18.5 by **14.7, 16.8** by 14.0, and 17.0 by **12.7** millimeters.

Young.—Dr. Dickey (MS.) tells me that in several nests that he watched the period of incubation proved to be "exactly 12 days"; and he says that young remain in the cavity 15 or 16 days. A young bird, 4 days old, had a light salmon-pink body, with eyes only partly open, and was naked except for "feather tufts of dusky grey down" on the top of the head, at the base of the skull and in the middle of the back. When 6 days old, the "body had blue-gray down and rows of conspicuous slate-blue pin-feather shafts"; the eyes were now open. Two days later, "the gray down was falling away from head and sides; the back mouse-gray; the flanks under back of wings tinged with light brown; pin-feather scabbards of wings not entirely unsheathed, but fast disintegrating." When ten days old, the young were well feathered and closely resembled the adults, but they remained in the nest five days more.

Mrs. Margaret Morse Nice (1931) writes: "In the Wichita Reserve June 6, 1926, we discovered we had fastened our tent to a black jack in a cavity of which five fully feathered titmice were housed; happily the parents accepted the situation with equanimity. I watched the nest from 2 to 4 P.M. the first day, from 10:40 to 12:10 the next. Despite the hot weather mother Tit brooded 3 and 8 minutes the first day, 8 and 15 the next, father in the meantime giving the food he brought to her. Both birds kept their crests depressed, both often twitched their wings—the female more than her mate—and both used a great variety of notes. During the first two hours 18 meals were given, during the last hour and a half, seven."

I cannot find it definitely so stated, but apparently incubating and brooding devolve mainly, if not wholly, on the female. Both sexes help to feed the young for some time after they leave the nest, and *both* young and old travel about together in a family party during summer, until they all join the mixed parties of their own and other species that roam the woods during fall and winter.

Plumages.—The young nestling is mainly naked, except for a scanty covering of dusky, or bluish gray, down on the head and back. The development of the first plumage is outlined above. In full juvenal plumage, when it leaves the nest, the young is much like the adult, but all its colors are duller and less distinct, and the plumage is softer, less compact. The crest is not fully developed; the forehead is dusky rather than black and not so clearly defined against the gray of the crest and the white of the sides of the head; and the sides and flanks are tinged with pinkish buff, instead of the richer brown of the adult. A partial postjuvenal molt takes place in August, or later, which involves the contour plumage and the wing coverts, but not the rest of the wings or the tail. In this first winter plumage, young and old are practically indistinguishable; the crest is distinct, the forehead is black, the sides of the head are more decidedly white, and the sides and flanks have become a deep "russet" or "Mars brown." There is apparently no spring molt. Adults have one complete annual, postnuptial molt in August. Adults in fresh fall plumage are more or less tinged with olive-brownish on the back and with pale buffy brownish on the chest. The sexes are practically alike in all plumages, but the colors of the adult female are usually somewhat duller than those of the male.

Food.—The 186 stomachs of the tufted titmouse examined by Professor Beal (Beal, McAtee, and Kalmbach, 1916), were irregularly distributed throughout the year, and were considered by him too few "to afford more than an approximation of the bird's economic worth." However, the results show that, so far as his investigation goes, the bird is beneficial and has no bad food habits to offset the good it does.

The food consisted of 66.57 percent animal matter and 33.43 percent vegetable. He says that the food "includes one item, caterpillars, which form more than half the animal food, and two items, caterpillars and wasps, which are more than half of the whole food." Beetles make up 7.06 percent, of which only one-tenth of 1 percent are useful species; the cotton-boll weevil was found in four stomachs. Ants are eaten occasionally, and other hymenopterous food, bees, wasps, and sawfly larvae, amounted to 12.5 percent. Other items include stink bugs, treehoppers, scales, only one fly, eggs of katydids, egg cases of cockroaches, spiders (found in 40 stomachs examined in May, 12.67 percent, only a trace in June, and in 3 stomachs in July, 16.33 percent, evidently a makeshift food), and a few snails. Caterpillars are the largest item, 38.31 percent of the whole food for the year. No grasshoppers or crickets were found.

Of the vegetable food, corn was discovered in one stomach, evidently taken on trial. Fruit was eaten to a moderate extent (5.15 per cent), mostly in midsummer, and included raspberries, blackberries, and strawberries, which might have been of cultivated varieties, but probably were not. The wild fruits were such as grow by the wayside and in swamps, as elderberries, hackberries, blueberries, huckleberries, and mulberries. Seeds of various kinds, as sumac—including poison ivy—bayberry, or wax myrtle, aggregate 4.07 per cent. It is difficult to draw the line between broken seeds and mast in stomachs of the tufted tit, but, together considered as mast, these form more than two-thirds of the vegetable food. While largely composed of acorns, there is no doubt that chinquapins and beechnuts and many smaller seeds enter into its composition. As thus defined, mast amounts to 23.4 per cent of the whole food, comprising 95 per cent of that eaten in November, 50.42 per cent in January, and 55.97 per cent in February; in fact, it is the principal vegetable food eaten from August to February. That such small birds should crush such hard nuts as acorns and chinquapins is surprising, but the broken fragments found in the stomachs well demonstrate their ability.

M. P. Skinner (1928) writes:

During the winter at least [in North Carolina], the favorite food of Titmice is the acorns supplied by the innumerable shrub oaks, post oaks and turkey oaks. From January to March, I found them hunting acorns, occasionally on the ground, but generally in the trees themselves. Quite often they knocked the acorn from its twig and then flew down to the ground after it. Titmice do not open their bills wide enough to admit the whole acorn, but they sometimes pick it up by its stem, or more often they simply spear the nut with their sharp, closed bill and fly up to a limb with it that way. Once on a suitable limb, the acorn is firmly held between the bird's two feet and strong downward blows are rained upon it. This hammering is rapid and very effective, so that it does not take long to scale off the shell, and then the soft interior meat is eaten in small pieces. * * * At times they spring out after insects flying by them, and sometimes they tear the tent nests of caterpillars to pieces. On February 11, 1927, near the Mid Pines Club, a Titmouse picked up an oak apple an inch or more in diameter, carried it in its bill to the crotch of a tree and there dug through its half inch of tough material to feed on the hundred or so small white grubs in the center.

Dr. H. C. Oberholser (1896) saw one hammering away at a half-punctured cocoon of a Polyphemus moth. B. J. Blincoe (1923) saw one feeding on cultivated Concord grapes. Dr. Dickey (MS.) has seen it catch flying insects, including a small butterfly or moth; the wings of Lepidoptera are torn off and only the soft body is eaten. He includes wild cherries and service-berries in the food. Others have added dogwood berries and those of the Virginia-creeper. In winter, titmice will come readily to feeding stations to eat suet and bread and doughnut crumbs; a dish of water will also attract them. Mabel Gillespie (1930) says that the berries of the Japanese honeysuckle, alder seeds, and the seeds of tuliptree pods are favorite foods.

A. L. Pickens writes to me: "They are among the friendliest of all our southern birds, exceeding the Carolina chickadee by far. I have taught several individuals to take food from my hand. The great drawback is that they are so thrifty that they empty a food box and store all the surplus food before the more backward chickadees, wrens, and nuthatches arrive. For the last two forms I was able to overcome this difficulty by using a block instead of a box for the food. In this block I bored holes with a small auger. Then at the bottom of the holes I placed bits of nuts, and the wrens and nuthatches, with their longer beaks, were able to reach deeper than the titmice and so retrieve the food."

Behavior.—This lively little titmouse is one of the most popular of the southern birds, with its active, vivacious manners, as it flits about in the foliage of the trees, often hanging head downward from some terminal cluster of leaves, or clings to the trunks and branches, searching in the crevices of the bark for its insect food. It attracts attention and endears itself to us with its tame, confiding manners, as it is not at all shy, but comes freely into our orchards and gardens, even close to houses, and partakes of our hospitality at our feeding stations; it appears utterly fearless of human presence. As Edmund W. Arthur (Todd, 1940) says:

We should probably ascribe to him without hesitation the first place in our hearts. He presents many claims to the rank of first nobleman of the forest realm. His presence is genial and pleasing, his plumage attractive, his alertness conspicuous; and his habits are good. * * *

Each pair of tufted titmice has a domain of its own during mating season. Over this the birds exercise a jealous sway, at least in so far as errant titmice are concerned. Enter upon this domain and without too much fuss begin to whistle the titmouse challenge. Directly you will excite vigorous replies from the lord of the manor. If you persist—and you probably will—he will approach to within a few feet of you. If you carry in your hand a hat or a sizable piece of dark cloth or a box, his lordship seems to think you have another bird in captivity. He will shake himself as if with rage, or in defiance, and drop, scolding, almost within arm's length, where as long as you continue to answer him, he will remain to scold and protest.

At other times, too, these inquisitive birds show their curiosity by reacting to the sound of human voices. Dr. Dickey tells me that they are "seen to react to the voices and noises made by road workers, drillers, and farmers. They hurry forward from shelter in twos or threes. Even when a visitor calls at the door of a house and starts to talk, then the titmouse arrives, evidently curious at a stranger in its habitat. I sometimes hesitate to wonder if such birds do not discriminate between the natives and strangers, for they have a sagacity that is hard to fathom."

Mabel Gillespie (1930) gives them credit for great intelligence and individuality about her banding traps. She says that they quickly learned how to find their way out of the traps, "with no time lost in searching about for the entrance." They repeatedly entered the trap, picked up some food, and went out again immediately, time after time. She says that she could not confuse them, as she often did other birds, by running toward the trap; for "approaching danger seems only to stimulate their keenness and composure, for they most containedly and successfully seek the exit at the first hint of hazard." She thinks that this ability to find the exit so quickly is due to accuracy of memory, and relates the following incident to illustrate it: "During one night there was a fall of very soft snow, with a succeeding drop in temperature. The traps were all removed but one lest they should become frozen in the ice crust. After the freezing the outline of each trap was clearly visible in the crust. A Titmouse was seen to fly to the ground at the spot directly in front of the outlined mark of an entrance funnel. This showed that the bird clearly remembered the location of the funnel. Then, however, just as it was about to run forward, it appeared to realize that the trap was not there. The food was directly in front of the bird with no intervening obstruction. Yet the bird hesitated, looked about, and observed that another trap was in its accustomed place. It flew to this trap and entered for food."

The above incident does illustrate accuracy of memory, but it also indicates suspicion of food under unusual conditions, or a sense of security in taking food from traps, based on past experience.

The tufted titmouse is quick and active in all its movements, flitting upward among the branches or gliding down between them, but it seldom indulges in long flights; its short flights from tree to tree, or across an open space, are undulating, irregular, much like that of a chickadee, it seems to me; but Dr. Dickey (MS.) calls it "rather precise; the short, rounded wings and well spread tail, with vibrating vanes, press the atmosphere." It reminds him somewhat of the flight of the blue jay.

Voice.—The notes of the tufted titmouse are many and varied, mostly loud and generally pleasing; it is a noisy bird. Aretas A. Saunders has sent me the following excellent notes on the subject: "The loud, whistled call of the tufted titmouse, commonly translated as *peto, peto,* is in about the same status as song as the *phoebe* whistle of the chickadee. That is, it is used by both sexes and, apparently, at almost any season of the year. Also, like the chickadee, the birds respond to an imitation and come to the imitator very readily.

"The song is loud, clear, and lower-pitched than the chickadee's *phoebe*. It is also quite variable; I have a number of records and no two of them are alike. The song consists of a two-note phrase, repeated over and over three to eleven times, according to my records. The two-note phrase is more frequently with the first note high and the second low. The interval between may be one, one and one-half, or two tones. The pitch of the notes varies in different songs, or different individuals, from A″ to A‴, that is, between the highest two A's on the piano. The majority of songs in my records are between E‴ and C‴.

"Sometimes the two-note phrase sounds like *peto,* at other times like *wheedle* or *taydle*. When the pitch goes up, instead of down, the phrase is commonly written *daytee;* the same pitches and pitch intervals are common but it often sounds like *toolee,* and sometimes the first note is short, and it is like *tleet* or *tlit*. I have recorded all these variations in the field, writing down what each particular song sounded like to me at the time it was heard.

"An occasional phrase slurs down, like *teeoh,* and there are rarely phrases of three notes, such as *wheedleoh,* or of one note, *whee,* each repeated a number of times. Sometimes a song begins or ends with notes unlike the rest, as *tidi, waytee, waytee, waytee,* etc., or *wheedle, wheedle, wheedle, whee, whee*."

Dr. Dickey (MS.) mentions a number of slightly different interpretations of some of the above notes, and adds some that are quite different, such as *piper-tee, piper-tee, piper-tee; ah-peer, ah-peer; chee-chu, chee-chep*; and *wheep-did-er-ee,* ending with *purty-purty-purty*.

Nuttall (1832) devotes considerable space to the voice of the tufted titmouse, and aptly remarks that "though his voice, on paper, may appear to present only a list of quaint articulations, * * * yet the delicacy, energy, pathos, and variety of his simple song, like many other things in nature, are far beyond the feeble power of description." He mentions a very lively and agreeable call of *'whip-tŏm-kĭlly-kĭlly*; and then, "in a lower, hoarser, harsh voice, and in a peevish tone, exactly like that of the Jay and the Chickadee, *dāy-dāy-dāy-dáy,* and *day-dăy-dăy-day-dáit;* sometimes this loud note changed into one which became low and querulous. On some of these occasions he also called *'tshica dee-dee*. The jarring call would then change occasionally into *kai-tee-did did-dit-did*."

Several other observers have noted the resemblance of some of these notes to the notes of the Carolina chickadee. The single whistled call sounds like the whistle of a man calling his dog. It can readily be seen from a study of the above interpretations how easy is it for a novice to confuse the voice of the titmouse with that of the Carolina wren, the chickadee, or even the cardinal. All observers agree that the titmouse is a loud and persistent singer for nearly all the year; it is a joy to hear it tuning up in January, when so many other birds are silent. The song increases in frequency and intensity when the nuptial season approaches in February; early in spring its oft-repeated *peto* note is given so constantly that it may become monotonous and even tiresome. No wonder that the bird is locally known as the "Peter bird."

Field marks.—The tufted titmouse may be recognized as a small gray bird, less than English sparrow size, with a prominent, blackish crest, and chestnut-brown flanks. The colors are duller in the female than in the male, but otherwise they are much alike. Mr. Skinner (1928) suggests that "its big black eyes show a strong contrast to its trim gray plumage. * * * When the crest lies back on the crown, its long feathers stick out behind so that it is noticeable then as well as when erect."

Enemies.—Titmice are doubtless subject to attack by the ordinary enemies of all small birds, cats, hawks, owls, and snakes, but published records are not plentiful. The enterprising cowbird finds and enters the nesting cavity to deposit its unwelcome egg occasionally. Dr. Friedmann (1929) records four cases, and probably others have occurred since, but sometimes the entrance hole is too small for the parasite to enter.

Harold S. Peters (1936) lists, as external parasites on the tufted titmouse, two lice *(Myrsidea incerta* and *Philopterus* sp.), a mite *(Trombicula irritans),* and a tick *(Hoemaphysalis leporis-palustris).*

Fall and winter.—Mabel Gillespie (1930), referring to the vicinity of Glenolden, Pa., writes: "During the late spring, summer, and early fall, Titmice tend to disappear. This disappearance indicates a period of retirement during nesting and the subsequent annual molt. At this season the birds are in the secluded depths of woods and are unaccustomedly silent. In the fall they appear in small groups, which, as far as they can be counted, vary from two to at least six. Presumably there is more or less wandering at this time, but the tendency apparently is to choose a favorable location in which to spend the winter, and then to remain within a rather limited area. * * * In winter small groups suggesting family units occupy very definite and limited areas, never overlapping."

This last statement hardly agrees with the observations of several others; for instance, Dr. Dickey (MS.), referring to Pennsylvania and West Virginia, says: "Particularly in autumn and winter, tufted tits **are rovers.** They tend to assemble with such birds as Carolina chickadees, cardinals, various sparrows, several local woodpeckers, Carolina wrens, goldfinches, tree sparrows, and juncos. Bands of such species enter patches of weeds, flit along the courses of streams, cross country roads and highways, and peer forth from cover at farm yards. I was interested, during my many trips among these birds in fall and winter, to learn that often individuals roost, or spend dark drab days, inside orifices of woodpeckers and in natural cavities of posts and stubs. Not long ago I came upon a tit; it was drowsy and almost could be taken in the hand. Whether the species invariably roosts in such manner at night, I do not know, but I have read of campers routing the birds from holes in stubs."

Several other observers have reported winter wanderings of titmice in association with such other species of winter gleaners as are named above. Mr. Skinner (1928) says that, in North Carolina in winter, "sometimes these Titmice seem to join with Chickadees, Juncos or White-throated Sparrows. With Fox Sparrows, Field Sparrows, Blue Jays, Cardinals and Myrtle Warblers, their association is probably only accidental and very temporary."

Tufted titmice are practically permanent residents in even the more northern portions of their range, being regularly found in winter as far north as New Jersey, Pennsylvania, Indiana, and Illinois. Though largely woodland birds at all seasons, gleaning their food from the trunks and branches of trees, or rustling among the leaves on the ground, they are more inclined in winter to roam about in the open, or visit the neighborhood of houses, along with the chickadees and blue jays, to pick up scraps of refuse, or visit the well-stocked feeding stations. On the feeding shelf the tit seems to be the dominant character; only the blue jay refuses to make way for him.

DISTRIBUTION

Range.—Eastern United States; not migratory. The range of the tufted titmouse extents **north** to central Iowa (Ogden and Independence); southern Wisconsin (Maxomanie and probably Racine); southern Michigan (Grand Rapids and Detroit); probably southern Ontario (London and Toronto); New York (Hamburg, Potter, and Goshen); and northern New Jersey (Mahwah and Englewood). The **eastern** limits extend south along the coast from New Jersey (Englewood), to southern Florida (probably Royal Palm Hammock). **South** to southern Florida (probably Royal Palm Hammock, Fort Myers, Tallahassee, and Choctawatchee Bay); Louisiana (New Orleans); and southeastern Texas

(Houston and Giddings). **West** to eastern Texas (Giddings, Corsicana, Decatur, and Gainesville); Oklahoma (Norman, Arnett, and Copan); eastern Kansas (Wichita, Junction City, and Manhattan); and central Iowa (Ogden).

During summer the species has been recorded from eastern Nebraska (Red Cloud, Lincoln, and Neligh), from northern New York (Long Lake), and from Maine (Orono), while during fall or winter it has been recorded from southeastern South Dakota (Vermillion), Minnesota (Fosston, Hutchinson, Minneapolis, and Northfield), Connecticut (New Haven and Norwalk), and eastern Long Island (Easthampton).

Egg dates.—Florida: 5 records, May 2 to 9.

Maryland: 5 records, April 26 to May 26.

Oklahoma: 6 records, April 16 to May 27.

South Carolina: 7 records, April 18 to 30.

Texas: 9 records, March 26 to April 20.

PARUS ATRICRISTATUS ATRICRISTATUS Cassin

BLACK-CRESTED TITMOUSE

HABITS

The above name was originally applied to the species as a whole, which is to be found over much of eastern Texas and eastern Mexico, but the species has been split into two subspecies, and the subject of this sketch is now restricted in its distribution to the southern portion of the range, from the Rio Grande River southward through eastern Mexico to Coahuila, San Luis Potosi, and northern Veracruz.

George B. Sennett (1878 and 1879) seems to have been the first to make us acquainted with the black-crested titmice of the Rio Grande Valley. He says in his first paper: "These lively and sweet singers were everywhere abundant, especially in old lagoon-beds, now largely grown up with the mesquite and lignum vitae." We found these sprightly and attractive little birds to be common everywhere around Brownsville, in the heavy timber along the resacas, in the open groves of deciduous trees, and in the shade trees in the town. Their behavior and their delightful songs were strikingly like those of the familiar tufted titmouse.

Nesting.—Mr. Sennett (1879) was also the first to find the nest of this bird. His first nest, found in 1877, contained young, but on April 20, 1878, his assistant brought him a nest containing five young and an addled egg, probably the first one ever collected. He says of this nest:

The nest was situated some six feet from the ground, in a hollow limb of a half-dead willow, which was leaning on some brush, and was discovered by the bird's flying into its opening. It lay some ten or twelve inches from the opening, and is composed chiefly of wool intermixed with strips of soft inner bark, and now and then bits of snake-skins; the whole being much firmer and thicker than is usual with nests that are built in hollow stubs. All other nests found with young were situated higher, with one exception; the distance varying from four to twelve feet from the ground. I found them to occupy usually the abandoned holes of the Texas Woodpecker, *Picus scalaris;* but split forks of trees were sometimes put in use. They prefer living trees to dead ones, and in every case in my experience the opening had to be enlarged, sometimes with difficulty, before examination of the nest could be made. The localities mostly selected for nesting are groves or open timber free from undergrowth, whether in old lagoon beds, which receive the overflow from the river, or on the driest knolls. They do not avoid human habitations, as two nests were found on the ranch in ebony-trees, near buildings much frequented. The parents guard their treasures well, and are much disturbed when the nest is invaded; though not until they see that their nest is actually being handled do they give any cry of alarm, or other intimation of uneasiness than their near presence.

Others have referred to this bird's use of pieces of cast-off snake-skin in its nests, which seems to be a common practice with the species.

Eggs.—The black-crested titmouse probably deposits a set of five or six eggs ordinarily, perhaps sometimes as many even as seven, or as few as four.

Of the single egg, referred to above, Mr. Sennett (1879) says: "The ground color is clear dead white; distributed unevenly over the whole surface, and not very sparingly, are flecks and blotches of fawn-color of various shades, the sides having rather more than either end." A set of four fresh eggs he describes as follows: "The ground color is pinkish-white. The spots of reddish-brown are small and few in number, and scattered over the greater part of the egg, but at the larger end they are large and numerous, covering nearly the whole end, though in no case forming a ring."

The measurements of 50 eggs average 17.0 by 13.5 millimeters; the eggs showing the four extremes measure **18.9** by 13.5, 16.2 by **14.4, 15.0** by 13.4, and 15.1 by **12.7** millimeters.

Young.—The period of incubation does not seem to be known. Mr. Sennett (1879) found that both sexes share the duties of incubation, for he took a male on the nest, and noticed that other males had bare and wrinkled bellies. Apparently two broods are reared in a season.

Plumages.—According to Mr. Sennett (1879), the "young just from the egg are nude, with the exception of a few long, dark, downy feathers on the back, nape, and over the eyes."

Ridgway (1904) describes the young in juvenal plumage as "essentially like adults, but black of crown and crest much duller (the feathers **often narrowly tipped with grayish**), less sharply defined laterally and posteriorly against the gray, and anteriorly invading the forehead almost (sometimes quite) to base of culmen; throat and chest pale gray; color of sides and flanks much paler (cinnamon-buff instead of rufous-cinnamon); back sometimes suffused with sooty or blackish."

What few molting birds I have seen indicate that the postnuptial molts of adults, and probably the postjuvenal molt, take place mainly in August; September birds seem to be in fresh plumage.

Food.—Little seems to be known about the food of this titmouse, and practically nothing has been published on it. Austin Paul Smith (1910) writes: "Having but one true Titmouse, the Black-crested *(Baeolophus atricristatus),* we especially appreciate him, though he is omnipresent, even into the heart of the city [Brownsville]. They inspect any object of size, that may arouse suspicion of harboring caterpillars or other insects. They are very fond of the caterpillar of the butterfly *(Libythea bachmanni)* which so persistently attacks our hackberry trees, as to have surely defoliated them this summer, but for the combined efforts of the Titmouse and Sennett Oriole."

Voice.—Mr. Smith (1910) says further: "The Black-crested Tit is rarely silent, the usual notes being a continuation of sounds like 'pete-chee-chee-chee'; more rarely 'peter-peter.' By April the young have appeared on the scene."

DISTRIBUTION

Range.—South-central Texas and northeastern Mexico; not migratory. The range of the black-crested titmouse extends **north** to central Texas (Chisos Mountains, San Angelo, and Lomita). **East** to central Texas (Lomita, Austin, Corpus Christi, and Brownsville); and Veracruz ("highlands"). **South** to Veracruz ("highlands"); and San Luis Potosí (Tamazunchale and Valles). **West** to San Luis Potosí (Valles); western Tamaulipas (Victoria and Montemorelos); western Nuevo León (Monterey and Lampazos); and southwestern Texas (Chisos Mountains).

The range outlined is for the entire species, which has been separated into two subspecies, the typical black-crested titmouse *(Parus atricristatus atricristatus),* occupying the southern portion of the range, north to the Rio Grande Valley, while Sennett's titmouse *(P. a. sennetti),* is found in Texas.

Egg dates.—Texas: 68 records, March 7 to May 26; 34 records, April 3 to May 2, indicating the height of the season.

PARUS ATRICRISTATUS SENNETTI (Ridgway)
SENNETT'S TITMOUSE

HABITS

More than 50 years elapsed after the species was discovered before this northern race of the black-crested titmouse was named and described, as distinct subspecifically from the race found in the Rio Grande Valley and in eastern Mexico. Ridgway (1904) describes it as "similar to *B. a. atricristatus,* but decidedly larger; upper parts much clearer gray, with little, if any, olive tinge; adult female with crest feathers more often and more extensively tipped with gray, and both sexes with forehead more often tinged with brown or rusty, sometimes deeply so." He goes on to say in a footnote that "any pronounced rusty tinge to the color of the forehead indicates, in the writer's opinion, admixture of *B. bicolor* blood." And he suggests that the two subspecies described by Sennett, from Bee County, Tex., are merely hybrids between the two species, *bicolor* and *atricristatus.*

John Cassin (1852), who first described the species and published the first colored plate of it, says that it was discovered in Texas by John Woodhouse Audubon, and it was described by Cassin in the Proceedings of the Academy of Natural Sciences of Philadelphia, vol. 5, p. 103, 1850. The only information he had regarding its habits came from Dr. Samuel W. Woodhouse, from whose journal he quoted as follows:

While our party was encamped on the Rio Salado in Texas, near San Antonio, in March, 1851, I observed this handsome little chickadee for the first time. It was busily engaged in capturing insects among the trees on the bank of the stream, and like the other species of its family, was incessantly in motion and very noisy. At our camp at Quihi, on the eighth of May, I again found it very abundant among the oaks. The young males, which were then fully grown, much resembled the adult females, both wanting the black crest which characterizes the male. Afterwards I noticed this species, occurring sparingly, along our route, as far as the head waters of the San Francisco river in New Mexico. [The birds seen in New Mexico were probably the gray titmouse, *P. inornatus ridgwayi.*]

I observed it almost entirely in trees bordering streams of water. * * * It occurred in small parties, appeared to be very sociable and lively in its habits, and in general appearance and in nearly all its notes which I heard, it so very much resembled the common crested chickadee of the Northern States as scarcely to be recognized as a distinct species at a short distance.

The distribution now given for this subspecies in the 1931 Check-List is the "Lower Austral Zone of central Texas, from Tom Green and Concho counties east to the Brazos River, and from Young County south to Nueces and Bee counties."

George Finlay Simmons (1925) lists as its haunts, in the Austin region, "woodlands along creeks; scrub oaks and cedars on hillsides; oak groves or mottes in open or semi-open country; city shade trees; ravines, gullies, and canyons among the hills; oak-clad crests of hills; telegraph poles along railroad rights-of-way."

Mrs. Bailey (1902) says that "Mr. Bailey found the black-crest one of the most abundant birds of the Upper Sonoran zone, flying about conspicuously among the junipers, nut pines, and scrub oaks."

Nesting.—Like others of the genus, Sennett's titmouse builds its nest in natural cavities in trees, stumps, or posts, in old woodpecker holes, or in bird boxes. Mr. Simmons (1925) says that the nest may be placed anywhere from 3 to 22 feet above ground, averaging about 10 feet; he says the commonest locations are "hollows in live oaks, next commonest in fence posts and bird boxes, then boxed fence posts, elms, hachberries, cedars, telephone poles and post oaks." He found one nest on the top of an old mockingbird's nest in a well-sheltered tree.

He says that the nest consists of "a mass of rubbish composed of cowhair, rabbit fur, green lichens, cedar bark, green moss, cotton, feathers, and oak blossoms; and occasionally small bits of inner bark and bark fiber, grass, wool, soft down, leaves, rootlets, a stick or two, hemp string, tissue paper, onion skins, and felting materials." It is "lined with soft short cowhair, rabbit fur, opossum hair, wool, cotton, and occasionally horsehair, soft down, feathers, and snake skin." The bottom of the cavity is filled with green moss to a depth of from 2 or 3 inches in one case to as much as 36 inches. He and others have reported that pieces of cast-off snakeskin are usually found in the nests; these and the tissue paper and onion skin may be used to give resiliency or ventilation to the nest; this is evidently a characteristic habit of the species, as it is with the crested flycatcher, under which it is more fully discussed.

Eggs.—The eggs of Sennett's titmouse vary in number from four to seven; of 25 sets, recorded by Mr. Simmons (1925), 5 were sets of seven, 12 of six, 5 of five, and 3 of four eggs each. The eggs are apparently indistinguishable from those of the other black-crested titmouse, under which they are more fully described. Albert J. Kirn writes to me that he once found a nest containing twelve eggs; and once he found a nest with two eggs of the titmouse and two of the dwarf cowbird, which was afterwards deserted. The measurements of 40 eggs average 17.9 by 14.1 millimeters; the eggs showing the four extremes measure **19.3** by 14.3, 18.0 by **15.0**, **16.1** by 14.0, and 18.4 by **12.6** millimeters.

Food.—Although no careful analysis of the food of this titmouse seems to have been made, it is probably similar to that of the closely related tufted titmouse, about two-thirds insects in various forms and about one-third vegetable matter. It seems to be very fond of caterpillars and searches the limbs and trunks of trees for hidden insects, their larvae and their eggs; scales and spiders are probably eaten to some extent, as well as berries, wild fruits, and some soft-shelled nuts and acorns. H. P. Attwater (1892) says that "the favorite food of the Black-crested Titmouse during winter is the pecan nut; they hold them on the horizontal limbs, or place them in the cracks of the bark, and break them open by knocking with their bills, like Woodpeckers."

Behavior.—The behavior of both races of the black-crested titmouse is so much like that of the more familiar tufted titmouse that one would hardly recognize the difference between the two species at a little distance. They are both friendly, confiding, cheery, busy, active, and noisy little birds that attract our attention and our interest. Their elevated crests give them a jaunty appearance that is very pleasing. During the breeding season they are usually seen in pairs; after the young are on the wing, rather early in the season, they may be seen traveling about in family parties, and in the winter in loose companies, when they are often seen about human habitations and in towns. Mr. Simmons (1925) says that they become attached to one locality; "one pair nested for six consecutive years in a hollow of an old persimmon tree."

Illustrating the confiding nature of this bird, Henry Nehrling (1893) relates the following experience that came to him while he was watching a white-eyed vireo's nest:

While I stood there in perfect silence, with my gaze steadfastly fixed on the pretty lichen decorated domicile, a Black-crested Titmouse came very close to me. It first perched on a small bush, then it flew to the ground, and finally, growing very bold, clung fast to my trousers. When I moved, it flew back to the bush, contemplating me curiously, but finally, convinced of my good will, it returned and clambered up and down my back, pecking me, and again vigorously thrusting its bill into my clothes. It was obviously looking for insects, especially wood-ticks which in such places creep over ones clothes in great numbers. Whenever I made a noticeable movement, it fled a little distance, but invariably returned. Finally when I went and seated myself on a prostrate tree, it followed me again. It became so bold, that it not only climbed up and down my back, but fearlessly crept over to my shoulders and arms, and even onto the hat. This Titmouse stayed near me as long as I remained in that part of the woods, and pursued me a short distance, screaming loudly *Wait-wait-wait-wait*, while I continued my way through the forest. Never before have I seen such boldness and confidence exhibited by a wild bird.

Voice.—Mr. Simmons (1925) says that the song is "quite similar to that of the Eastern Tufted Titmouse," consisting of "a cheery, abbreviated *Peté, Peté, Peté, Peté;* a series of monotonous whistles, *hew, hew, hew, hew, hew, hew, hew, hew;* a whistled *tseee ep;* a rasping, scolding *eck-eck.*"

Enemies.—Mr. Simmons (1925) lists this bird as a local victim of the cowbird.

Field marks.—The long crest, usually erect, will mark this species as one of the crested titmice. The jet-black crest of the male and the dark crest of the female, much darker than that of the other crested titmice, will mark the bird as a black-crested titmouse; the subspecies are hardly recognizable in the field. This species has a white forehead, often tinged with brownish in the northern subspecies, whereas the tufted titmouse has a black forehead. The females and young have more or less admixture of gray in the crown and crest.

PARUS INORNATUS SEQUESTRATUS (Grinnell and Swarth)

OREGON TITMOUSE

This northern race of the plain tits seems to occupy a very narrow range in Jackson County, Oreg., and Siskiyou County, Calif., between the Coast and Cascade Ranges. The describers, Grinnell and Swarth (1926), state:

[It] differs from *Baeolophus inornatus inornatus,* to which it is nearest geographically, in slightly smaller size and in grayer, more leaden color throughout, with but a trace of the brownish tinge that shows clearly on the upper parts of *inornatus;* lower surface less purely white, more suffused with gray. Similar to *B. i. griseus,* but smaller, with especially shorter tail, and darker in color, much less ashy in tone. Similar to *B. i. affabilis* but bill much smaller, and coloration not quite so deeply leaden, especially as to wing and tail feathers. * * *

Careful examination of the Museum's series of all the races here concerned has convinced us that both the Lower California and Oregon forms are deserving of names. It is a curious fact that, though the intervening forms are different from either, these two subspecies, so far apart geographically, should be strikingly alike in the matter of their relatively dark, brown-less coloration. The outstanding difference between them lies in the bill. The Oregon bird is smaller billed even than typical *inornatus;* the San Pedro Martir race is large billed, like *murinus.*

The Oregon titmouse does not seem to differ materially in any of its habits from its more southern relatives.

The measurements of eight eggs average 17.8 by 13.2 millimeters; the eggs showing the four extremes measure **18.1** by **12.4,** and **17.5** by **14.0** millimeters.

PARUS INORNATUS INORNATUS Gambel

PLAIN TITMOUSE

PLATE 62

HABITS

This is, indeed, a *plain* titmouse, without a trace of contrasting colors in its somber dress; *inornatus,* unadorned, is also a good name for it. But it is a charming bird, nevertheless, with its jaunty crest, like a miniature jay, its sprightly manners, and its melodious voice. Its gray coat blends well with the trunks and branches of the oaks among which it forages. It is the western counterpart of our familiar eastern tufted titmouse, which it resembles in appearance, behavior, and voice and for which it might easily be mistaken, unless clearly seen.

The species, of which there are at least nine subspecies, occupies a wide range in western North America, from the Rocky Mountain region to the Pacific coast, and from Oregon to Lower California. The type race is now restricted to northern and central California.

The favored haunts of the plain titmouse are the oak-clad, sunny slopes of the foothills, where the foliage of the evergreen oaks provides shelter and a good food supply all the year around; and here it is practically resident at all seasons. It seldom seems to range above 3,500 feet.

In the Lassen Peak region, according to Grinnell, Dixon, and Linsdale (1930), "a considerable variety of larger plant species furnished situations favorable for some major activity of this bird species. The trees and shrubs that were definitely recorded as foraged through by the plain titmouse are the following: willow, cottonwood, sycamore, valley oak, live oak, black oak, blue oak, golden oak, several species of orchard trees, digger pine, and buckbrush. We were left with the impression that the blue oak is the tree within the Lassen section used for feeding place, nesting, and shelter by the largest number of these birds."

Courtship.—Dr. John B. Price (1936), who studied for six seasons the family relations of the plain titmouse at Stanford University, Calif., makes no mention of a courtship performance, nor can I find it mentioned elsewhere. But his studies reveal a tendency to remain mated for more than one year; he says:

"A titmouse usually keeps the same mate from year to year and there was only one known case of 'divorce'. Of a total of 14 pairs recaptured, 11 were mated together for at least two years and only 3 were not. No sex difference was found in the retention of territory from year to year. If a bird lost its mate the survivor, whether male or female, remained in the nesting territory and secured a new mate. In one case the new mate was known to be a juvenile of the year before. * * * An interesting fact is that there was only one case of 'divorce' where a bird took a new mate while its former mate was still known to be living. In all other cases where a titmouse took a new mate the former mate was never recaptured anywhere and quite probably was dead, especially as in several cases it was known to be several years old." One male "was banded as an adult in 1928 and was recaptured nesting in the same box in 1934 when it must have been at least seven years old." It was absent in 1935.

Nesting.—The birds that Dr. Price studied all nested in bird boxes; evidently they prefer to nest in boxes where these are available. Sixty-four adults were caught in the boxes and banded; there were 33 cases of adults renesting in the same box (a bird nesting in the same box for three years would be two cases of renesting); there were 17 cases of adults nesting in boxes from 43 to 90 yards away, which were practically in the same territory; there was only one case of nesting in a box 200 yards away, and none at a greater distance. "If the changing of nest-boxes were really a change of nesting territory one would expect that the former territory would be taken over by another pair of titmouses nesting in the first box. With the exception of the female that moved 200 yards, this never took place. The first box was always either empty or used by bluebirds or chickadees. Often a bird would alternate between two boxes from year to year."

The plain titmouse normally nests in holes in trees, either old woodpecker holes or natural cavities in the trunks or limbs of trees, often partially excavated by the birds in soft or rotten wood. Dawson (1923) says it is a mistake to think that this bird cannot excavate its own nest, and says:

"Two of the nests I have found (and *not* rifled) were excavated in the heart wood of live limbs of the blue oak *(Quercus douglasi)*, not less than ten inches in diameter. * * * I once traced a Plain Titmouse to a hole about twenty feet up in one of those cliffs of mingled gravel and 'dobe' which line the banks of the San Jacinto River. * * * We found a neat, round aperture in the earth, which must have been barely large enough to admit the bird, being, in fact, so snug that it showed two separate 'scores' for the feet. This opened rapidly into an ample chamber with extensive inner recesses—a monument of toil. The nest proper, a great bed of rabbit-fur, was placed about one foot from the entrance."

Grinnell and Storer (1924), referring to the Yosemite region, say that "old woodpecker holes are used when available, but many, perhaps a majority, of nests are placed in naturally rotted-out cavities." The height from the ground varies according to the location of these cavities; one of their nests was only 33 inches and another 10½ feet from the ground; I have seen one that was 32 feet aloft, which is probably unusual. One of their nests was only 17 inches from a western bluebird's nest; it was "in a natural cavity of rather large size. The bottom held a mass of fine dry grasses, perhaps 4 inches in depth, and on top of this was a heavy felted lining of cow hair and rabbit fur. The top of this mat was 5½ inches * * * below the margin of the entrance." A nest in the A. D. DuBois collection was made of moss, grass, weed stems and fibers, and was lined with a few feathers and rabbit fur.

Eggs.—Six to eight eggs seem to be the commonest numbers laid by the plain titmouse, with seven the prevailing number. Among 22 sets in the Museum of Vertebrate Zoology, 14 were sets of seven. In 62 complete sets of eggs recorded by Dr. Price (1936), the numbers ranged from three to nine; there were six eggs in 12 nests, seven eggs in 17 nests, eight eggs in 14 nests, and nine eggs in 7 nests. Ernest Adams (1898) records a set of 12 eggs, which might have been the product of two females.

The eggs are mostly ovate, sometimes elongated to elliptical-ovate, and have practically no gloss. They are pure white and often entirely unmarked, but usually some of the eggs in a set, and sometimes all of them, are faintly marked with minute dots of very pale reddish brown. These pale markings are sometimes evenly distributed over the entire egg and sometimes very sparingly scattered. Whole sets are sometimes pure, unmarked white.

The measurements of 40 eggs average 17.4 by 13.4 millimeters; the eggs showing the four extremes measure **19.3** by 13.3, 18.4 by **14.2**, and **16.3** by **12.7** millimeters.

Young.—Mrs. Wheelock (1904) gives the period of incubation as 14 days, which matches the figures given for other titmice. Dr. Price (1936) was convinced, by his examination of "brood patches," that only the female incubates; "only females ever were captured incubating the eggs."

Mr. Adams (1898) says that the female is a very close sitter and has to be removed by hand, clinging tenaciously to the nest material and often bringing some of it out with her; even when thus forcibly removed, she returns to the nest immediately; he had to put one in his pocket to keep her out of the nest while he was removing the eggs.

Both parents assist in feeding the young, and the large broods keep them quite busy. Mrs. Wheelock (1904) says: "My theory that most young birds are fed by regurgitation *at first* was confirmed in this case by the fact that, although I was within twelve feet of the nest whenever either bird entered it during that first day, not once was any food visible in the beak of either. After the fourth day the worms and insects carried were frequently projecting on each side of the small beak, but up to that time there had been none seen, though a careful watch was kept with both opera glasses and naked eyes." Apparently, the young birds that she watched were about ready to leave the nest on the sixteenth day.

Evidently the young are driven away from the home territory as soon

as they are able to care for themselves. Dr. Price (1936) banded 145 young birds, but only two "were recaptured nesting the following year, and both were more than a quarter of a mile distant from the box where hatched."

Plumages.—The juvenal plumage, even with a decided crest, is mainly assumed before the young bird leaves the nest. This is much like the adult plumage, but it is softer and less compact; the wing coverts are somewhat paler and indistinctly buffy at the tips. Most of the molts occur in August, but I have seen a bird in full juvenal plumage as late as August 29 and an adult molting as early as July 14. The postjuvenal molt includes the contour plumage and the wing coverts, but not the flight feathers. The postnuptial molt of adults is complete.

Food.—Prof. F. E. L. Beal (1907) examined 76 stomachs of the plain titmouse and says that "unlike most of the titmice, the plain tit eats less animal than vegetable food, the proportion being 43 percent of animal and 57 of vegetable." Bugs (Hemiptera) seem to be the favorite, 12 percent; the injurious black olive scale forms nearly 5 of this 12 percent; in the month of August, 34 percent of the contents of nine stomachs consisted of these scales, and one stomach was filled with them; other hemipterous food included leafhoppers, jumping plant-lice, and tree-hoppers. Caterpillars amounted to nearly 11 percent. Beetles formed nearly 7 percent, all harmful species, including the genus *Balaninus*, weevils with long snouts; these "insects, by means of this long snout, bore into nuts and acorns, wherein they deposit eggs, which hatch grubs that eat the nut. The tit finds these beetles while foraging upon the oaks. One stomach contained the remains of 13 of them, another 11, a third 8, and a fourth 7, while others contained fewer." Some of these were probably found while the bird was feeding upon the acorns. Hymenoptera, ants and wasps, made up about 6 percent of the food, and other insects, daddy-long-legs and grasshoppers, amounted to a little more than 5 percent; one stomach contained the remains of 13 grasshoppers. A few spiders were eaten, less than 1 percent.

In the vegetable food, fruit amounted to nearly 32 percent, three times as much as eaten by the linnet, and this appeared to be of the larger cultivated varieties; no seeds of wild berries were found. "Cherries were identified in a number of stomachs, and the pulp of the larger fruits was abundant. * * * Oats were found in a number of stomachs and constituted nearly 30 percent of the contents of two stomachs taken in January. * * *

"Leaf galls, seeds of poison oak, weed seeds, unidentifiable matter and rubbish make up the remainder, 24 percent, of the vegetable food."

Practically all the insects eaten by this titmouse are harmful, the scales exceedingly so; it is therefore very beneficial in protecting the trees of forest and orchard; it is not sufficiently abundant to do any very serious damage to cultivated fruits.

Behavior.—With all its somber colors the plain titmouse is a most attractive little bird, always cheery, active, and friendly as it forages among the oaks for its insect prey, pecking and prying into every crack and crevice, with its crest erected like a jaunty little jay and greeting us with its varied notes. Mr. Adams (1898) gives his first impression of it as follows:

After searching the tree to which my attention had been called for some time, my curious gaze rested upon a little gray bird which, with crest erected and with his whole frame seemingly alert, was pecking furiously at the bark of the oak, evidently in search of food. Now and then a single sharp note came to my ears, and occasionally one slightly prolonged and possessing a greater degree of authority. At times he seemed to be angry, and then his notes came faster and harsher, but when a fat insect fell to his lot, he at once became pacified, his notes were subdued, his crest lowered, and the once miniature Jay had become peaceful *Parus inornatus* once more. * * *

This Titmouse is not very sociable and never gathers into large flocks—in fact I have rarely seen more than three together at any time of the year. Like many others of the feathered tribe, he has an inherent hatred towards owls. I remember finding a nest of the California Screech Owl in a hollow trunk of an oak and on the outside a cavity containing the nest of a titmouse. The thin partition separating the two sitters was not such as to prevent the scratching of the owl being distinctly audible to the other. The female would often appear at the entrance of her home greatly agitated. Sometimes she would mount the rim of the trunk and peer down into the darkness as if to ascertain the cause of such a commotion. The male when he visited his mate, would perhaps at her request, fly repeatedly at the poor owl.

Mrs. Ruth Wheeler tells me that these titmice "are the most inquisitive of the smaller birds. Whenever any disturbance is caused in the woods, they are the first birds to arrive and raise an outcry. Whenever we set the camera up near a nest or a feeding station, a titmouse usually is soon on the scene carefully investigating."

Voice.—Ralph Hoffmann (1927) writes: "Spring comes to the brown hillsides of California as soon as the first rains break the long autumn drought; the cuckoo-flower and ferns push up through the mould, the gooseberries blossom and the Plain Tit begins his lively if monotonous refrain from the live oaks. Different birds have various forms of this spring song, *witt-y, witt-y, witt-y* or *ti-wee, ti-wee, ti-wee*. It is always a high clear whistle with a marked accent, and a persistence that shows the relationship of the bird to the Tufted Tit, with its *pée-to*, in the river bottoms of the Middle West. The rest of the year, when the

Plain Tit is hunting leisurely through the oaks, his commonest note is a scratchy *tsick-a-dee-dee* or *tsick-a-dear,* which has to make up to a California bird lover in the lowlands for the absence of the Black-capped Chickadee."

Mrs. Bailey (1902) says: "There is an indefinable charm about the slow, clearly enunciated *tu-whit, to-whit, tu-whit,* that echoes through the oaks, telling of the presence of the plain titmouse." Mrs. Wheelock (1904) says that "his common note of *tsee-day-day* is not unlike that of the mountain chickadee, and occasionally he indulges in a whistled *peto, peto* that reminds one of his pretty Eastern cousin."

W. L. Dawson (1923) adds a few more similar notes, and a *ssic-rap sssicrap,* and one that "sounded like *di di di tipoong, di di di tipoong,* the *di* notes very wooden and prosaic, the concluding member suddenly and richly musical."

Field marks.—A plump, chunky bird, plainly clad in somber grayish brown above and plain gray below, with a prominent crest, usually erected, could be no other than a plain titmouse. Its favorite haunts are among oaks, its behavior suggests the chickadees, and its notes though varied are characteristic of the family.

Enemies.—Titmice, like all other small birds, have to be constantly on the alert to avoid all the well-known predatory birds and mammals, but this species has an important enemy in the California jay. Dr. Price (1936) says: "Jays are often seen about nesting boxes containing young titmouses and sometimes perch on the box and peer inside. When the young birds leave the nest the jays often dive at them and kill them."

DISTRIBUTION

Range.—Western United States and northwestern Mexico; not migratory.

The range of the plain titmouse extends **north** to southern Oregon (Gold Hill and probably Blitzen Canyon); Utah (Boulder); and Colorado (Douglas Springs and Canyon City). **East** to central Colorado (Canyon City); extreme western Oklahoma (Kenton); New Mexico (Santa Fe, Corona, and Capitan Mountain); and western Texas (Guadalupe Mountain). **South** to Texas (Guadalupe Mountain); southern New Mexico (Silver City); and northern Baja California (Valladares). **West** to western Baja California (Valladares and Las Cruces); California (Twin Oaks, Pasadena, Watsonville, and Red Bluff); and western Oregon (Ashland, Medford, and Gold Hill).

At outlined the range is for the entire species, of which nine geographic races are recognized. The typical subspecies *(Parus inornatus*

inornatus) is found in California from Shasta and Mendocino Counties south to Kern and San Luis Obispo Counties; the Oregon titmouse *(P. i. sequestratus)* is found between the Coast and Cascade Ranges from southwestern Oregon south through Siskiyou County, Calif.; the San Diego titmouse *(P. i. transpositus)* is found in southwestern California from Santa Barbara County to San Diego County; the San Pedro titmouse *(P. i. murinus)* occupies the northern part of Baja California south to about latitude 30°; the ashy titmouse *(P. i. cineraceus)* is found in the Cape region of Baja California; while the gray titmouse *(P. i. ridgwayi)* occupies the Rocky Mountain and Great Basin portion of the range. The lead-colored plain titmouse *(P. i. plumbescens)* is found in southern Arizona and New Mexico; the Kern Basin plain titmouse *(P. i. kernensis)* in Kern County, Calif.; and the Warner Valley titmouse *(P. i. zaleptus)* in southern Oregon.

Egg dates.—California: 101 records, March 20 to July 16; 50 records, April 4 to 29, indicating the height of the season.

New Mexico: 19 records, May 3 to 28.

Mexico: 28 records, April 22 to May 31; 14 records, April 30 to May 14.

PARUS INORNATUS TRANSPOSITUS (Grinnell)

SAN DIEGO TITMOUSE

The San Diego titmouse occupies an area in southwestern California extending from Santa Barbara County to San Diego County.

The above scientific name was suggested by Dr. Grinnell (1928) to replace the name *murinus*, which was formerly applied to the plain tits of the above region; the latter name is now restricted to the birds of northwestern Lower California, the San Pedro titmouse. He gives as its characters, "as compared with *B. i. inornatus,* slightly larger and grayer, bill much heavier. As compared with *B. i. murinus,* browner, decidedly less leaden gray in cast of coloration; bill and feet less blackish, rectrices and remiges brownish rather than plumbeous."

The habits of this titmouse are apparently similar to those of the plain titmouse found farther north, which need not be repeated here.

The measuresments of 40 eggs average 17.4 by 13.3 millimeters; the eggs showing the four extremes measure 19.0 by 14.0, and 13.0 by 9.5 millimeters; the latter egg might be considered a runt; the next smallest egg measures 16.0 by 12.0 millimeters.

PARUS INORNATUS MURINUS (Ridgway)

SAN PEDRO TITMOUSE

The San Pedro titmouse was originally described, as a new subspecies, by Dr. Joseph Grinnell (Grinnell and Swarth, 1926) under the subspecific name *affabilis,* as the race confined to the Sierra San Pedro Mártir. This region was included in the range of Mr. Ridgway's *murinus,* which he (1904) understood to extend from Los Angeles and San Bernardino Counties to the San Pedro Mártir Mountains.

Dr. Grinnell (1928), in a later paper, explains why Ridgway's name should apply to the San Pedro Mártir bird, why the name *affabilis* should be discarded, and why it was necessary for him to give a new name to the San Diego bird, which is now called *transpositus,* an appropriate name under the circumstances!

Dr. Grinnell (Grinnell and Swarth, 1926) says that the San Pedro titmouse "is the darkest, most leaden colored, of any of the subspecies of *Baeolophus inornatus,* showing no trace of the brown tinge that is apparent strongly in *inornatus* and somewhat less so in *murinus*" [=*transpositus*].

As its habitat seems to be in the live-oak association, we have no reason to think that its habits are any different from those of other races of the species.

The eggs are similar to those of the plain titmouse. The measurements of 40 eggs average 18.0 by 13.7 millimeters; the eggs showing the four extremes measure **19.8** by 14.5, 19.3 by **14.6, 16.2** by 13.1, and 16.5 by **12.2** millimeters.

PARUS INORNATUS CINERACEUS (Ridgway)

ASHY TITMOUSE

Far removed from its nearest allies, with none of the species in the wide intervening area, the ashy titmouse lives in a very restricted region in the mountains of the Cape region, near the southern tip of Lower California. William Brewster (1902) says that "it is a bird of the pine forests which cover portions of the summit and upper slopes of the high mountains near the southern extremity of the Peninsula. Here, according to Mr. Belding, it is 'common from 3,000 feet altitude upward.' On the Sierra de la Laguna Mr. Frazar found it quite as numerous in December as in May and June. None of the specimens killed at the latter season showed any indications of being about to breed, and the eggs, like those of many other birds which inhabit these mountains, are probably not laid much before midsummer."

Ridgway (1904) characterizes it as "similar in coloration of upper parts to *B. i. griseus* [= *ridgwayi*], but under parts much paler, and size slightly smaller."

Nothing seems to have been published on its habits.

PARUS INORNATUS RIDGWAYI Richmond

GRAY TITMOUSE

HABITS

This eastern race of the plain titmice has the widest range of the nine subspecies. The 1931 Check-list says that it "breeds in the Upper Austral Zone of the mountains from northeastern California, Nevada, southern Idaho, Utah, southwestern Wyoming, and Colorado to southeastern California, southern Arizona, southeastern New Mexico, and central western Texas."

Dr. Joseph Grinnell (1923), writing of the status of the gray titmouse in California, was "almost tempted to propose full specific status" for it, and says: "The Gray Titmouse is a very distinct form, separated sharply from the Plain Titmouse geographically as well as on the basis of phylogenetic characters. No intergradation between these two titmouses is known to take place. The Gray Titmouse in California is a rare bird. It has been found to exist only in small numbers and at a few widely scattered points. The general territory in which it occurs lies east of the Sierran divide, in the arid Great Basin faunal division. The life-zone occupied is the Upper Sonoran, and the association the piñon-juniper."

From the above and following quotations. it will be seen that the haunts of the gray titmouse are quite different from those of the plain titmouse, which seems to prefer mainly the various oak associations.

Ridgway (1877) says: "In the pine forests of the eastern slope of the Sierra Nevada, especially in their lower portion, and among the cedar and piñon groves on the desert ranges immediately adjacent to the eastward, the Gray Titmouse was a rather common species; but it did not seem to be abundant anywhere." According to Baird, Brewer, and Ridgway (1874), "Dr. Woodhouse met with this species in the San Francisco Mountains, near the Little Colorado River, New Mexico. He found it very abundant, feeding among the tall pines in company with the *Sitta pygmaea, S. aculeata,* and *Parus montanus.*" Mrs. Bailey (1928) writes of its haunts in New Mexico: "The attractive Gray Titmouse with its prettily crested head and soft Quaker-gray plumage is intimately associated with pleasant camps in the low, sun-filled junipers and nut pines of the mesas, the low desert ranges, and the foothills of the Rocky Mountains." She says that it ranges over the whole mountainous part of the State, and records it at various elevations from 4,600 to 8,000 feet. Although this titmouse is said to occur in the Chiricahua Mountains, in southern Arizona, we did not see it there, or anywhere else in that region; it is apparently a rare bird in Arizona.

Nesting.—The nesting habits of the gray titmouse seem to be similar to those of its California relatives. It nests in woodpecker holes and natural cavities in trees and stumps, or in bird boxes. W. E. Griffee writes to me: "My experience with this titmouse was limited to the spring of 1934 when, while living at Albuquerque, N. Mex., I put up a string of nesting boxes in the junipers and pinyon pines of the Sandia Mountain foothills, 20 miles east of Albuquerque. Evidently the birds were very common, because early in the season I procured four sets of eggs from the 25 or 30 (out of 50) boxes, which wood-chopping Spanish Americans had failed to find and knock down. Bottoms of the boxes were covered with a mixture of grass, bark strips, and dirt. A heavy lining of rabbit fur, rodent hair, etc., was well cupped to receive the eggs. Incubating birds sat tightly, almost like chickadees, and were difficult to flush, even after the tops of the boxes were removed."

Eggs.—What eggs of the gray titmouse I have seen are indistinguishable from those of the plain titmouse, as they might be expected to be. Mrs. Bailey (1928) calls them "plain white"; probably many eggs and even full sets of eggs are entirely unmarked. The measurements of 40 eggs average 17.5 by 13.7 millimeters; the eggs showing the four extremes measure **19.1** by 14.0, 18.4 by **14.2, 16.0** by 14.0, and 18.4 by **12.9** millimeters.

Behavior.—Henshaw (1875) writes: "Its habits much resemble those of its eastern congener *(L. bicolor).* It spends much of its time on the ground, searching for insects, and quite likely the piñon nuts and acorns may, during the fall and winter, form a part of its food, though I have never seen them pay any attention to these. It has much curiosity, and, though somewhat timid, will occasionally remain within easy distance of an intruding person; keeping a careful watch upon his motions, uttering its harsh, scolding notes, expressive alike of anger and fear. It has, in the early summer, a short, disconnected song, which, however, is often sweet and pleasing. I have never seen more than three or four together, even in the fall; but, in every company of the other Titmice, Warblers, or Bluebirds, a few of this species is always found."

Voice.—Although the notes of the gray titmouse, as well as all its habits, are similar to those of the species elsewhere, I am tempted to quote the following attractive account from the pen of Mrs. Florence Merriam Bailey (1928):

[In the] low sunny groves the wayfarer hears many of its small notes, delightfully homelike and conversational in tone, including its rapid *wheed-leah, wheed-leah, wheed-leah,* repeated three or four times in quick succession, and its chickadee-like *tsche-de-dee, tu-we-twee-twee,* sometimes used to preface its loud clear *pe-to* calls. But its most conversational notes are best heard at the nest, where you may perhaps listen to a variety of small talk, such as the infantile, lisping notes of the hungry, brooding bird coaxing her mate to feed her; the tender note of her mate calling her to come to the door for the food he has brought; pretty conjugal notes of greeting and farewell; the chattering scold and cries of anger, anxiety, and terror, heard when enemies threaten; sharp notes of warning to the young, and wails of grief when harm has come to the nestlings. Such notes, given emphasis by vivacious, eloquent movements and gestures, interpret the thoughts and feelings of these intense little feathered folk, almost as clearly as elaborate conversations do the emotions of less demonstrative human beings.

PARUS INORNATUS PLUMBESCENS (Grinnell)

LEAD-COLORED PLAIN TITMOUSE

In naming this subspecies, Dr. Joseph Grinnell (1934) says of it: "As compared with *Baeolophus inornatus griseus,* from the eastern part of the Great Basin region, north of the Colorado River: similar in general features, but bill smaller, especially shorter; tail shorter; coloration darker, more leaden hued, this tone most pronounced dorsally but pervading the lower parts also. Color of back, close to Deep Mouse Gray (of Ridgway, 1912, pl. LI)."

He gives as its range "New Mexico (at least southwestern) and parts of Arizona south of the Colorado and Little Colorado rivers."

PARUS INORNATUS KERNENSIS (Grinnell and Behle)

KERN BASIN PLAIN TITMOUSE

Dr. Joseph Grinnell and William H. Behle (1937) have named this local race of the plain titmouse, giving the following diagnosis: "Compared as to coloration with *B. i. inornatus,* dorsum grayer, less brownish, and flanks and underparts generally slightly less buffy, clearer whitish; compared with *B. i. transpositus,* less olivaceous dorsally, and paler gray below; less clearly gray dorsally, but paler below, than in *zaleptus.* In size characters, closest to *inornatus;* bill decidedly shorter, less massive, than in *zaleptus,* and less massive even than in *transpositus.*"

They give the range as "drainage basin of Kern River, within southeastern rim of San Joaquin Valley, in Kern County, and extreme southern Tulare County, California."

PARUS INORNATUS ZALEPTUS (Oberholser)

WARNER VALLEY TITMOUSE

The above name has been applied by Dr. H. C. Oberholser (1932) to the birds of this species, living in the Warner Valley region of southern Oregon. He describes them as "similar to *Baeolophus inornatus griseus,* but much more grayish above with practically none of the brownish tinge so evident in the latter race; also paler above; and somewhat lighter, more clearly grayish below, with little or no buffy wash." He says that this race is "very different from *Baeolophus inornatus sequestratus* Grinnell in its larger size and much paler, more grayish coloration." The habits, so far as known, are not distinctive.

PARUS WOLLWEBERI ANNEXUS Cassin

BRIDLED TITMOUSE

HABITS

The oddly marked bridled titmouse, with its sharply pointed crest, its black-and-white-striped head, and its vivacious and friendly manners, is, to my mind, the prettiest and the most attractive of all the crested tits. Only those of us who have traveled in Arizona or New Mexico, or farther south into the highlands of Mexico, have been privileged to see it, for it is a Mexican species that finds its northern limits in a rather restricted area in southwestern New Mexico and southern Arizona. We found it rather common in the oak-clad foothills of the Huachuca Mountains, where the striking color pattern of its pretty head gave it an air of distinction and always attracted our admiration. We found it most commonly from the base of the mountains, about 5,000 feet, up to about 6,000 feet. Harry S. Swarth (1904) writes: "This, one of the characteristic birds of the mountains of Southern Arizona, is found in the greatest abundance everywhere in the oak regions of the Huachucas, breeding occasionally up to 7,000 feet, but most abundant below 6,000 feet. On one occasion, late in the summer, I saw a Bridled Titmouse in a flock of Lead-colored Bush Tits on the divide of the mountains at about 8,500 feet, but it is very unusual to see the species at such an altitude."

Mrs. Bailey (1928) says of its haunts in New Mexico: "Small flocks of about half a dozen each, probably families, were eagerly met with among the blue oaks, junipers, and nut pines of San Francisco Canyon, where they were associated with Lead-colored Bush-Tits and Gray Titmice. Other small flocks of the prettily marked Bridled were later discovered in sycamores in the open valley, at the junction of White

Water Creek and San Francisco River; but they are more characteristically birds of the oak country."

Nesting.—We did not succeed in finding a nest of the bridled titmouse. The natural cavities in which they habitually nest are too
numerous to be thoroughly explored, and it is almost impossible to
make the sitting bird leave its nest by rapping on the tree. Once a bit
of white in the bottom of a deep, dark cavity, which looked like a sitting
bird, induced us to chop it out, but there was no nest in the hole, much
less a bird!

My companion, Frank Willard, sent me his notes on two nests found
in that locality in 1899. One, found on May 17, was in a natural cavity
in an oak tree well up on the side of the mountain at about 6,800 feet;
the entrance was through a small knot hole in live wood, 12 feet from
the ground; the nest was made of grass and lichens. The other, found
on May 20, was in a natural cavity, 15 feet up, in a dead oak by the
roadside, well down the canyon, between 5,000 and 5,500 feet elevation.

W. E. D. Scott (1886) was, I believe, the first to publish an account
of the nesting of this species, of which he writes: "On the two occasions
that I have discovered the species breeding the nests were located in
natural cavities in the live-oaks, close to my house. The first of these
was found on May 9, 1884. I took the female as she was leaving the
nest, which was in a cavity, formed by decay, in an oak stump. The
opening of this hole was about three and a half feet from the ground;
its diameter was about three inches inside, and it was some eighteen
inches deep. The entrance was a small knot-hole where a branch had
been broken off, and was only large enough to admit the parent birds.
The hollow was lined with cottonwood down, the fronds of some small
rock-ferns, and some bits of cotton-waste."

Of the other nest, found May 8, 1885, he says: "The small entrance
was some six feet from the ground, and the cavity was a foot deep, and
two and a half inches in diameter. It was lined on the bottom and well
up on the sides with a mat composed of cottonwood down, shreds of
decayed grasses, some hair from a rabbit, and many fragments of
cotton-waste, gathered by the birds from refuse waste that had been
used to clean the machinery of a mill hard by."

There are four nests in the Thayer collection, all taken in southern
Arizona and all in natural cavities in oaks, at various heights from 4 to
28 feet above ground; all these nests are lined, more or less profusely,
with soft, cottony substances that may have been the downy coverings
of leaf buds, blossoms, or catkins, or possibly the cocoons of insects
or spiders; one nest was made almost entirely of this material; the
foundations of these nests consist of strips of coarse weed stems, fine

grass stems and soft weed leaves, more or less mixed with the cottony substances. Herbert Brandt has sent me his notes on a nest that "was made entirely of silvery, curly leaves, an inch or two in length, fashioned into a shallow basket form, but so loosely held together that the nest was removed, intact, with difficulty."

Eggs.—Five to seven eggs make up the usual set for the bridled titmouse. The eggs are ovate and have little or no gloss. They are white and quite immaculate. The measurements of 50 eggs average 16.1 by 12.6 millimeters; the eggs showing the four extremes measure **18.0** by **14.0**, **13.7** by 12:3, and 14.0 by **11.5** millimeters.

Plumages.—I have seen no very young birds of this species. In the juvenal plumage the young bridled titmouse looks very much like the adult, but the "bridling" marks on the head are less sharply defined, the throat patch is mainly grayish, only the chin being clear black, and the whole plumage is softer and less compact. This plumage is worn until about the middle of July, when a molt of the contour feathers, but not the flight feathers, produces a first winter plumage, in which adults and young are alike.

Adults have a complete postnuptial molt beginning in July. In fresh fall plumage the back and rump are more olivaceous than in the more grayish worn plumage of the spring and summer. The sexes are alike in all plumages.

Food.—Nothing seems to have been recorded on the food of the bridled tit, but it probably does not differ materially from that of other titmice of this genus, all of which live in similar habitats and spend much of their time foraging in crevices in the bark, on the trunks and branches of the oaks, hunting for insects, their larvae and their eggs. Their food habits are probably mostly, if not wholly, beneficial.

Behavior.—After the young are strong on the wing, late in August, these active, sprightly little titmice may be seen flitting through the oak woods in family parties, the young learning to forage for themselves. Later in the fall they form larger groups. Henshaw (1875) considered that this species differs somewhat in its flocking habits from some of the other titmice. He writes:

Instead of being found in small companies or as stragglers on the skirts of the large flocks of other species, it habitually moves about in flocks, composed often of twenty-five, and even more, of its own species; its exclusiveness in this particular being quite noticeable, though once or twice I have seen a few on intimate terms of companionship with the other Chickadees. It pays especial attention to the oaks, in which trees they move about slowly from limb to limb, scrutinizing each crevice and fold of bark which is likely to serve as a hiding place for insects. They are thus very thorough in their search, but have less of the rapidity of movement and nervous energy which characterize other members of this group. They are less noisy, too; their notes, though Chickadee-like, being weaker and fainter, and not infrequently one may, when watching one or two of these birds, find himself surrounded by a large number, which have silently closed in around while he was wholly unconscious of their presence. They are strictly arboreal, sharing only to a slight degree the terrestrial habits which are common to the other Titmice, especially of this genus.

Mrs. Bailey (1928) says: "One that I watched hopped up the branches of a tree quite in the manner of a jay climbing his tree ladder."

Field marks.—The bridled titmouse is well marked and need not be mistaken from anything else. Its chickadeelike behavior and voice are obvious. Its pointed black crest distinguishes it from the white-cheeked black-capped chickadees; and the peculiar color pattern of its black and white face distinguishes it from the other crested titmice, and from most other birds as well.

DISTRIBUTION

Range.—Mexico, southern Arizona, and New Mexico; not migratory. The range of the bridled titmouse extends **north** to central Arizona (Prescott and probably Fort Verde); southern New Mexico (Alma and Pinos Altos); and Tamaulipas (Galindo). **East** to Tamaulipas (Galindo); Puebla (Chachapa); and Oaxaca (La Parada). **South** to Oaxaca (La Parada) and Morelos (Cuernavaca). **West** to Morelos (Cuernavaca); Durango (Arroya del Buey and Ciénaga de la Vacas); western Chihuahua (Mina Abundancia); Sonora (Oposura, La Chumata, and San Rafael); and central Arizona (Baboquivari Mountains, Santa Catalina Mountains, probably Sacaton, and Prescott).

This species has been separated into two subspecies, the typical form *(Parus wollweberi wollweberi)* being confined to southern Mexico. The race found in the United States and in Sonora and Chihuahua is known as *P. w. annexus.*

Egg dates.—Arizona: 21 records, April 19 to May 26; 11 records, April 25 to May 17, indicating the height of the season.

Texas: 6 records, March 13 to May 19.

AURIPARUS FLAVICEPS ORNATUS (Lawrence)

EASTERN VERDIN

PLATES 63, 64

HABITS

This little olive-gray bird, with a yellow head and chestnut shoulders, is one of the characteristic birds of the southwestern desert regions. I made its acquaintance in a dry wash in southeastern Arizona, where the hard stony soil supported a scanty growth of low mesquites, hackberries, hawthorns, catclaws, and other little thorny shrubs, with a few scattering chollas.   Here it was living in company with cactus wrens, crissal thrashers, and Palmer's thrashers.   Elsewhere in that region we found it on the mesquite plains, on the greasewood and cholla flats, and on the low hillsides dotted with picturesque giant cactus.   It and the other desert birds seem to make a living in the harsh and cruel desert, far from any water, where the soil is baked hard and dry and every living plant is armed with forbidding thorns; even the "horned toad," which is really a lizard, carries a crown of thorns on its head and lesser spikes on its body, to protect it.   But the verdin is equal to the occasion and builds its own armored castles, protected by a mass of thorny twigs, in which to rear its young, and to which it can retire at night.

Dr. Joseph Grinnell (1914) says of the haunts of the verdin in the lower valley of the Colorado River: "The only essential condition for the presence of this species appeared to be stiff-twigged thorny bushes or trees of some sort.   This requisite was met with in a variety of situations, as in the screwbeans of the first bottom, mesquites of the second bottom, and catclaw, ironwood, palo verde and daleas of the desert washes. * * *   While the birds were often seen in willows, arrowweed, and even low shrubs of *Atriplex* and sandburr, these were always within a limited radius of nests.   As far as observation went, these birds do not need to visit water; some were met with as much as three miles away from the river up desert washes."

Near Brownsville, Tex., we found the verdin only in the high, dry, thorny chaparral, and not in the dense and heavy timber along the resacas.   George Finlay Simmons (1925) says that, in the Austin region, "bushy mesquite valleys" are preferred to open deserts.

Nesting.—The remarkable nests of the verdin are much in evidence throughout its habitat, especially large for so tiny a bird, wonderfully well made, and surprisingly conspicuous, for they are usually placed at or near the end of a low limb, without a shred of leafage to hide them. The number of nests that one sees would seem to indicate that the bird is more abundant than it really is.   But it is well known that the verdin builds roosting nests, or winter nests, as well as breeding nests.   They are most industrious little nest builders.   Furthermore, the nests are so firmly made that old nests probably persist for more than one year, perhaps several years.   Most of the nests that we saw in Arizona were in mesquites, hackberries, or catclaws, but others have found them in palo verdes, stunted live oaks, *Zizyphus* bushes, chollas, and a variety of other thorny trees and shrubs.

Of the numerous descriptions of the marvelous nest of the verdin that have been published, I have selected Herbert Brandt's (1940) as best worth quoting:

This structure is a globular or oval interlacement of thorned twigs that may measure up to eight inches in outside diameter, although usually its axis is shorter in one direction than the other. The thorny sticks are so placed that on the bristling exterior their free ends project upward, quill-like, and protect the nest. Each twiglet measures from two to five inches, and the free end may protrude an inch or more, in orderly fashion, so at first glance the twig ends give the appearance of a strange little hedgehog sitting in a convenient crotch with quills ever ready.

The number of twigs employed in this abode depends largely on its size and varies from a few hundred to more than two thousand; in fact I have a nest, collected in Arizona, that by careful estimate has over two thousand twigs! Thus the amount of labor required to interweave and lodge each individual freshly broken twiglet is indeed prodigious and is perhaps unequalled in the nest-building of our small birds. * * *

One of the most interesting features of the Verdin's dwelling is its strength; so strongly is it built that it is really difficult to tear one apart, and even then it is necessary to wear heavy gloves because the multitude of thorns will effectively repel one's bare hands. The nest is so woven around the tough branching limbs that to remove it intact is impossible without cutting off the support; in consequence the winds of the sandstorms cannot harm it or its feathered tenants. In fact, this is the most perfectly defended and anchored fortress that any desert bird has devised, and the reason so many are seen is because they endure year after year in various disrepair despite the angry elements that sweep across the sandy plains, often for months on end.

These nests are found as low as two feet from the ground and up to a height of nearly twenty feet, yet they are usually placed in the lower half of the tree in which they are built, or in the upper part of bushes, but always well out toward the end of a branch. Here the structure is woven about a pronged fork, with its entrance almost invariably opening outward and downward. The stick quills usually bristle backward away from this threshold, pointing along the limb, and act as a barbed repellent to any crawling invader trying to reach it from behind. It is through this small hidden orifice, itself well thorned, that the bird enters, and when it does it must pass over a raised threshold that is built up so high that one can scarcely touch the contents of the interior without pressing down the elevated doorstep. The whole fortified little stronghold seems designed to keep safe its semi-concealed entrance.

After suggesting that the male probably builds several nests, among which his mate may select her choice, he continues:

Once she has decided on her apartment she proceeds to line it snugly. The first step is to cover the many rough, protruding rafter ends with a padding of leaves and grass fibers, and finally to line cozily the whole cavity with an abundance of feathers, large and small. This room, before the feather finish, measures up to four inches across and between two and three inches high, but after the lining is placed, loose feathers more or less fill the entire chamber. Then the walls are smooth and silky to the touch, and so carefully padded that seldom does a thorn jab through. In one instance I found the nest of a Verdin only four feet below that of a Pallid Horned Owl containing two large young. The Verdin's tiny stick-built suite was crammed full of fluffy reddish feathers seized from the tiger of the air that dwelt in the master apartment above.

Mr. Simmons (1925) describes the nest much as above and then adds that "woolly and sticky fibers and weeds bind the nest together, and the whole structure is well interwoven with cobwebs and spider webbing, with some plant down and thistle down used to block up crevices"; the inside is "first lined with spider webs, forming a very thin coat which covers the cup-bottom; then lined with a coat of small grass leaves and bits of dead leaves; then an inner coat of plant down, thistle down, spider webbing, bits of silk-like cocoons, and a number of tiny soft bird feathers, on which the eggs rest."

Mary Beal (1931), of Barstow, Calif., has seen a verdin's nest in an almond tree and in a tamarisk. The pair that built the nest in the almond tree had first built their usual nest in a mesquite in her back yard, but a pair of English sparrows took possession of it. The verdins started to build in a Chinese elm, but the sparrows tore the nest to pieces; and "every location the Verdin considered was made impossible by the annoying tactics of the sparrows. Finally, the harassed Verdin chose the branch of an almond tree in the midst of the adjoining orchard. The tree was blossoming when the nest was built, the lovely fragrant flowers almost concealing it. By the time the petals had fallen the leaves were out, keeping the nest securely hidden from casual sight. Here, in this unusual setting, the little family was reared in peace."

M. French Gilman (1902) says that, on the Colorado desert of southern California, "most frequently the nests are found in mesquite trees and the smoke tree or *Dalea spinosa*, Daley's thorn tree. But any spiny shrub will answer, as I have found nests in the screw-bean, cholla cactus, desert willow, tree-sage, catsclaw, Eriodictyon, and last month I found one in a grapevine growing up in a cottonwood tree. The nests will average about five feet from the ground though I have found them as low as 2½ feet and as high as ten or twelve feet."

Dr. Walter P. Taylor has sent me the following notes on nest-**building**:

The male seemed to be much busier than his mate; his yellow head and brilliant red epaulettes were prominent. With his mouth full of nesting material he took a look at me. Deciding I was quite safe he went to the nest, climbing into it from the bottom, and remaining for several moments, undoubtedly arranging his material. He reappeared, looked me over again, then resumed his search for nest material on the ground beneath a nearby catsclaw. As he proceeded, he picked up many small twigs and dropped them again, being, apparently, very uncertain as to just what kind of a stick he wanted, and very wasteful of his energy. One twig in particular he handled for some time, then dropped it, apparently unintentionally, but seemed immediately to forget all about it, and went right on looking.

One that he watched on October 31, 1919, was apparently gathering material to line a winter nest: "One seen this morning picking cotton from a wad I had affixed to a hackberry to mark the site of a mouse trap. It pecked down on the cottony wad several times and until its bill held as much as possible. Then it flew to another branch of the hackberry and rearranged the cotton in its bill. Then it flew by 'flirty' flights to a mesquite not far away, then made off for a long flight over the mesa to a point where I could not follow it. This bird, or another, was back in a few moments for another load of cotton."

Mr. Sennett (1878) had several experiences indicating that the verdin will desert its nest, if the eggs are handled before the set is completed.

James B. Dixon tells me that the verdin apparently raises two broods in favorable seasons on the Colorado desert, as he has found nests with eggs in late February and early March and then found them with fresh eggs again during the first week in May of the same year.

Eggs.—The cozy nest of the verdin may contain anywhere from three to six eggs, but normally four; sets of five are not especially rare; Mr. Gilman (1902) states that sets of four and five are "about evenly divided as to frequency," though most others agree that four is by far the commonest number. The delicate little eggs are dainty and beautiful. The ground color is light bluish green, greenish blue, or bluish or greenish white. This is usually rather sparingly and irregularly marked with fine dots, small spots, or rarely with very small blotches of reddish brown. Usually the markings are much scattered, but sometimes they are concentrated in a ring around the large end. The measurements of 50 eggs average 15.3 by 11.0 millimeters; the eggs showing the four extremes measure **16.3** by 11.2, 15.2 by **12.2, 14.0** by 11.2, and 14.7 by **10.2** millimeters.

Young.—Whether both sexes incubate does not seem to be definitely known, but, as the male is known to build his own roosting nest, perhaps incubation is performed entirely by the female. The incubation period of most of the Paridae is about 14 days, but Mrs. Wheelock (1904) says:

In ten days after the last bluish white egg was laid, there were three infinitesimal bits of naked bird life, huddled tightly together in the middle of the feather-lined hollow. A slit carefully cut at this time and fastened shut after each observation enabled me to keep an exact record of the development of the brood. Although I could not watch the mother feeding the young, I am positive it was done by regurgitation, for she would eat as unconcernedly as if merely occupied with her own dinner, and fly at once with apparently empty mouth into the nest, emerging shortly to repeat the performance. During the first five days the male was not seen to go into the nest, but sang right merrily near by. After that time the young began to make themselves heard in hungry cries, and he began to carry food to them, which we could see in his bill. This food consisted almost exclusively of small green worms, and eggs and larvae of insects. The young Verdins remained in the nest quite three weeks, and long after their debut they returned to the nursery every night to sleep.

The young are fed by their parents for some time after they leave the nest and are well able to fly. They sit around in the bushes waiting to be fed, and uttering notes much like those of the adults but shorter and weaker, while their parents are foraging for food.

Plumages.—I have seen no very young verdins, but apparently they are hatched naked, and they probably acquire the juvenal contour plumage before they leave the nest. Young birds in juvenal plumage have no yellow on the head and no chestnut on the lesser wing coverts. The entire upper parts, crown, back, rump, and lesser wing coverts are uniform grayish brown, "hair brown"; the under parts are very pale brownish gray, nearly white posteriorly; the wings, except the lesser coverts, and the tail are as in the adult. The postnuptial molt of the adult begins before the end of July and may last well through September; I have seen molting adults as early as July 28 and as late as September 29. Probably the postjuvenal molt comes at about the same time.

Food.—Very little has been published on the food of the verdin, which seems to consist of insects and their larvae and eggs, and of wild fruits and berries. Dr. Taylor tells me that he has seen one eating the berries of the hackberry tree. Mrs. Bailey (1928) writes: "These little desert birds would seem almost independent of water, nests having been found at least ten miles from any known water. The question is whether by means of their insect food and berries they are made largely independent of other liquid."

Behavior.—In its movements and all its activities the verdin shows its relationship to the chickadees, as it flits about actively in the bushes searching for its food, clinging tenaciously to branches swayed by the wind, or hanging head downward from the tips of twigs, hunting every little crevice and the under sides of the foliage. During the nesting season it is shy and retiring; perhaps we should say sly, rather than shy, for it slips away unobserved from its nest and keeps out of sight in the nearest thicket; it is difficult to surprise one on its nest, and one is seldom seen near the nest unless it has young, when the little bird becomes quite bold, flitting about in the vicinity, chippering and scolding. But in winter its behavior is somewhat different, according to Mr. Brandt (1940), who says that "like the chickadee, the Verdin becomes more sociable, responding readily to my summons, and is one of the first birds to approach, showing little fear as it 'cheeps' in a thin voice and moves quickly about in anxious inquisitiveness."

Voice.—The verdin has a remarkable voice for so small a bird, one that would do credit to an oriole or a thrush. Dr. Taylor says in his notes: "The song of the verdin I put down as *tswee, tswee, tswee, tsweet!* All the calls are whistled. Another call is *tsee, tsoo, tsoor!* and *tsee, tsoo, tsooy!*" He writes the common short call note as *tsit, tsit, tsit, tsit;* the notes are run together more or less and are repeated very rapidly.

Mary Beal (1931) says: "The Verdin's song of three clear notes all on one key has rather a plaintive, resonant quality. One summer when I lay ill for all the weeks of June in a sleeping porch in the midst of many trees, those liquid bird-notes came to me across the orchard at intervals throughout the day. The vital quality compelled attention and puzzled me all that season, for I did not place the singer until the next spring when I saw a tiny Verdin in the act of sending forth those rich, full notes. The depth and carrying power of the tones are amazing in such a small bird—so different from its quick, sharp call-note."

Mrs. Wheelock (1904) writes: "The usual note of the adult Verdins is a chickadee-like 'tsee-tu-tu' uttered while hunting, chickadee fashion, among the terminal buds and under the leaves for their insect food, and this the nestlings mimic in two syllables as soon as they leave the nest,— 'tsee-tee, tsee-tee.' It is a cry of hunger, and never fails to bring the parent with food."

Enemies.—Dr. Herbert Friedmann (1929) records the verdin as "a poorly known victim of the Dwarf Cowbird." He found a cowbird's egg in one nest, near Brownsville, Tex., and Roy W. Quillen wrote to him that he has found "eggs of the Cowbird in a few nests of the Verdin." In all cases the entrances to the nests were much enlarged, and perhaps deserted.

Field marks.—The verdin is a tiny bird, about the size of a bushtit and much like it in general appearance and behavior, but it has a shorter tail and other distinctive markings; it is brownish gray above and paler below; in the male the head is largely quite bright yellow and in the female somewhat duller yellow; the lesser coverts on the bend of the wing are bright, reddish chestnut, but these do not show conspicuously unless the wings are partially open. Young birds in their first plumage in summer have no yellow on the head and no chestnut on the wings. The verdin's voice is quite distinctive.

Winter.—The verdin is a permanent resident in at least the warmer portions of our southwestern deserts; it does not have to migrate, for **it finds sufficient insect food** in one form or another and lives well on various wild fruits and berries. But it builds most interesting winter homes for its protection and warmth. Mrs. Bailey (1928) says that "in southern Arizona, out of fifteen Verdin nests that I found in one small tract, ten showed signs of winter occupation and nine were found to contain roosting birds, the small occupants being flushed at intervals from 4.28 P.M. until after sunset, at various dates from December 9, 1920, to March 13, 1921. Two of the little birds seen going to their nests went half an hour or more before sunset, when it was light enough to be seen by Sharp-shinned Hawks and any other too observant neighbor."

She noticed that these nests were usually under thick thorny trees or bushes, which gave the bird some additional protection as it went to its nest.

Mr. Gilman (1902) writes: "Last December I found two female winter nests and later saw several of both sexes. One of them in a mesquite tree was ten or twelve feet from the ground and measured more than eight inches long by seven wide and seven deep. Lining was about one and one-quarter inches thick and composed of feathers— quail, chicken and others. * * * The nests of male and female differ a little, the former being less elaborate, smaller, with not so much lining in it. The female *winter* nest differs but little from the *breeding* nest and I am inclined to believe in some cases is used as such."

W. L. Dawson (1923) says that "two male lodges in the M. C. O. are each only three inches in length over all, with openings at either end, and about two and a half inches of clear space inside—not room enough to turn around in, but just sufficient protection from the pounce of an Elf Owl. * * * Verdins are not gregarious, like Bush-tits; but also they are never solitary, for they roam the desert in pairs, or, in small family groups, or in loose association. It is here that the remarkable penetrative, or carrying power, of the *silp* note serves the Verdin in good stead, for it allows mated birds to hunt, say, a hundred yards apart, without actually losing each other."

DISTRIBUTION

Range.—Southwestern United States and northern Mexico; not regularly migratory.

The range of the verdin extends **north** to southern California (Victorville, Barstow, and Death Valley); southern Nevada (Corn Creek and Bunkerville); southwestern Utah (St. George); southern New Mexico (San Antonio, Tularosa Flats, and Carlsbad); and southwestern Texas (Monahans, Castle Mountains, Kerrville, and Sequin).

East to central Texas (Sequin, Corpus Christi, and Brownsville); and eastern Tamaulipas (Matamoros). **South** to northern Tamaulipas (Matamoros); southern Coahuila (Saltillo and Saral); probably Durango (Durango); and southern Baja California (San Jose del Cabo and Cape San Lucas). **West** to Baja California (Cape San Lucas, Magdalena Bay, San Ignacio, and San Felipe); and California (San Diego, Palm Springs, and Victorville).

Three subspecies of the verdin are now recognized by the American Ornithologists' Union (1944) committee on nomenclature. Modern research, for which the references are given in their nineteenth supplement to the Check-list, indicates that the Cape verdin *(Auriparus flaviceps flaviceps)* should stand as the type race of the species, as Sundevall's type apparently came from the southern half of Baja California. The eastern verdin *(A. f. ornatus)* occupies southern Texas, southern New Mexico, and southern Arizona; and Grinnell's verdin *(A. f. acaciarum)* ranges from southern Nevada and southwestern Utah to northern Baja California.

Egg dates.—Arizona: 40 records, March 4 to June 18; 20 records, April 17 to May 18, indicating the height of the season.

California: 56 records, March 6 to June 8; 28 records, March 28 to April 11.

Mexico: 26 records, March 22 to June 25; 14 records, April 7 to May 8.

Texas: 30 records, April 1 to June 5; 16 records, April 18 to May 6.

AURIPARUS FLAVICEPS ACACIARUM Grinnell

GRINNELL'S VERDIN

Dr. Joseph Grinnell (1931) made an exhaustive study of the nomenclature of this species and an extensive search for Sundevall's type, to which the reader is referred for the somewhat confusing details and a history of the nomenclatorial changes. In the course of his study he concluded that the verdins of southeastern California needed a new name, for which he proposed the subspecific name *acaciarum*. He describes this race as "similar to *Auriparus flaviceps flaviceps* (Sundevall), but with yellow of fore parts somewhat less intense and extensive; body color averaging a trifle browner, especially on dorsum; tail and wing, more notably the former, averaging a little longer; bill apparently averaging smaller. Similar to *A. f. ornatus* (Lawrence), but paler and a little smaller."

The new Check-list will give the range of this form as from southern Nevada and southwestern Utah to northern Lower California.

AURIPARUS FLAVICEPS FLAVICEPS (Sundevall)

CAPE VERDIN

HABITS

Many years ago Baird (1864) called attention to the principal characters distinguishing the verdins of the southern part of Lower California from those of Arizona and Texas, but he did not propose a new name for the race. The most concise description is given by Ridgway (1904), who says that it is similar to the northern race "but decidedly smaller (except bill), with yellow of head averaging brighter, and the forehead more frequently (?) tinged with orange-rufous; young, howeven, distinctly different in coloration from that of" the northern race, "the upper parts being olive, strongly tinged with olive-green, and the under tail-coverts (sometimes most of under parts) tinged with olive-yellow."

According to the 1931 Check-list, the Cape verdin occupies the "Lower Austral Zone in the southern part of Lower California, south of about lat. 30°, and in southwestern Sonora." William Brewster (1902) says that Frazar found it "abundant everywhere in the Cape Region except on the Sierra de la Laguna, where none were met with. It was breeding at La Paz in March, at Triunfo in April, and apparently at San José del Cabo in *November,* for on the third of that month Mr. Frazar found two nests about half completed on which the birds were busily at work." These November birds were probably building winter nests.

Griffing Bancroft (1930) writes of its haunts in central Lower California:

This is the most widely distributed and, I believe, the most abundant of the local birds. It occurs in every association of the region under discussion, excepting only the littoral sand dunes. On the lava-strewn mesas, where vegetation is barely able to maintain a foot-hold and where animal life seems almost impossible, isolated pairs of these fascinating little workers are to be found regularly and in surprising numbers. In the irrigated river beds they seek open spots where here and there a stray mesquite or a bit of cholla has been permitted to remain. Brush-covered mountain slopes, plains dotted with cholla, cardón, and tree yucca, dry river beds supporting dense mesquite and palo verde, and cañons where the palo blanco grows are equally their home sites.

Nesting.—The same observer writes: "Fifty percent of the nests are in cholla, forty percent in mesquite, and the other ten percent scattering. The latter include anything from elephant trees to matilija poppies." He continues:

When nest building begins both birds work industriously. They find an arrangement of cholla stems in which it is possible to construct a suitable circle about five inches in diameter. They build one of fine weed twigs or of grass stems which often have leaves still attached. These inch-long bits are fastened to the cactus with a layer of plant down, the bird standing within the rim and tucking in the material with a most business-like air. The next step carries the outside super-structure backward from the ring to supporting arms of cholla. The frame is of the same material as the original circle. The builders continue to work from the inside and soon the frame becomes a shell. That, in turn, is added to and padded until a thickness of perhaps half an inch is reached.

The result is a flexible nest. In marked contrast to those of Arizona and particularly the Vizcaíno Desert and Sonora it is hardly ever protected on the outside with reinforcement in the shape of thorns or larger twigs. * * *

The opening to the hollow globe is completed last. It is left just large enough to permit the entrance and egress of the parents and it is so placed as to face away from the plant on which the nest is built. It is almost level with the bottom (only once did I observe a hole squarely in the center), and it is often somewhat concealed with an overhang of building material. The interior design permits of the low entrance being safely used. The tunnel runs upward. At the interior end the wall of the nest drops abruptly or even outwardly. So the eggs lie directly below the entrance. It is interesting to note that this is not true of the nests of any other race of verdin.

The nests are lined with feathers and are located in the trees much as are those of the Arizona verdin. In fact, J. Stuart Rowley, who saw many nests in the same general region, tells me that the nesting habits are the same as in Arizona.

There are five nests of the Cape verdin in the Thayer collection in Cambridge. One of these, taken at Comondu on April 24, was 4 feet up in a mesquite and is quite unique, unlike any verdin's nest I have ever seen. It is made almost wholly of white woolly or cottony substances, reinforced with a few short pieces of fine twigs and weed stems, mixed with a few dry leaves and feathers; there are no long or thorny twigs even on the exterior, and the whole nest is conspicuously white.

The other four nests, collected at Purissima in May and June, are quite unlike the above and very different from any nests of the Arizona verdin in the almost complete absence of thorny or large twigs. They are all very much alike and are, or rather were, more or less globular structures, made up of great compact masses of short fine twigs, weed and grass stems, flower clusters, fine straws, insect cocoons, and various kinds of plant material; they contain considerable plant down in the lining, but very little of the white woolly material and only a few feathers; only one contains a few small thorny twigs on the outside. There is nothing in any of these nests that suggests the bristling, thorny fortresses of the Arizona birds.

Eggs.—Mr. Rowley tells me that his sets of the Cape verdin consist of three eggs, except for one set of five, taken April 27, 1933, near San Ignacio. Mr. Bancroft (1930) writes:

The number laid is definitely not more than three: I have seen but one set of four out of a hundred examined. Clutches of three outnumber two in a ratio of approximately four to three. Incubated singles comprise about ten per cent of the total.

Eggs of the Cape Verdin run through a wide range of sizes, shapes and colors. Many are half again as large as the average and many are fifty per cent smaller. There is the elongated type, one end almost a hemisphere and the other a cone-shaped point. On the other hand it is not rare to find them as perfectly elliptical as the typical humming bird egg. The ground color is green, the markings gray—facts established for us by an oculist with the proper instruments. The shade of green varies until almost blue is reached.

The great variation in size, as indicated above, is not shown in the series of measurements I have collected, and the 15 eggs of this race I have examined are practically indistinguishable from those of the Arizona verdin. The measurements of 40 eggs average 15.3 by 11.2 millimeters; the eggs showing the four extremes measure **16.7** by 11.0, 15.2 by **12.2**, 14.2 by 11.1, and 14.7 by **10.2** millimeters.

In all other respects, plumage changes, food, general behavior, and voice, the Cape verdin does not seem to differ from the Arizona bird.

PSALTRIPARUS MINIMUS MINIMUS (Townsend)

COAST BUSHTIT

PLATES 65-68

HABITS

Almost anywhere in California, except in the desert regions and in higher elevations in the mountains, and at almost any season of the year, except during the breeding season, one may observe jolly bands or loose flocks of these tiny, gray, long-tailed birds drifting through the live oaks and shrubby thickets, uttering high-pitched, twittering notes to keep in touch with each other as they feed. These charming little birds are so widely distributed and so universally abundant in southern California that they form one of the most interesting features in the landscape, a lively bit of bird life that one never tires of watching.

Authorities have differed somewhat as to the distribution of the California forms of the bushtit, but I believe it is now understood that this type race occupies a narrow coastal strip from extreme southwestern British Columbia southward to the Mexican border, and perhaps beyond, with a wider range in San Diego County, where it is especially abundant. Harry S. Swarth (1914), who has made an exhaustive study of the subject, seems to have shown that only one race, *P. m. minimus*, occurs in southern California, where formerly the inland race, *P. m. californicus*, was supposed to occupy a range in the interior.

"*P. minimus minimus* as compared with *P. m. californicus,* is darker colored throughout, birds seasonably comparable being contrasted, the under parts are heavily suffused with dusky, and the flanks are more distinctly vinaceous. These differences are quite as apparent in the juvenal plumage as in the adult, sometimes more so." (Swarth, 1914.)

Courtship.—Alice Baldwin Addicott (1938) has published a very full and detailed account of the behavior of the bushtit in the breeding season, from which I shall quote freely. As to the beginning of the mating season, she says:

As early as January and February flocks of bush-tits which have remained intact during the fall and winter start dividing into smaller and smaller groups. At the same time a few pairs may be found which have separated from the flocks and which have wandered off in search of nesting territory.

* * * At the beginning of the mating season, courtship may be observed in pairs which have separated from the flocks as well as within the small flocks which are existent during this part of the year. Courting consists chiefly of excited location notes, trills and sexual posturing. * * *

Territorial ownership appears to be poorly developed. When a stray bird enters the territory of a nesting pair, the latter may respond by chasing the intruder for a few seconds, giving utterance to excited alarm notes and trills, until the intruder leaves. However, in many instances a stray bird is ignored, and it may even be allowed to forage with the mated pair.

She goes on to show that toleration of stray birds in the nesting territory may often go so far as to permit a third bird to forage in the territory for an hour at a time and even to take part in the nesting activities, suggesting that the gregarious habits of the species may carry over into the breeding season.

Nesting.—The curious and beautiful nests of the bushtits are works of art and marvels of avian architecture. In marked contrast to the thorny castles of the verdins, the long, gourd-shaped, hanging pockets of the bushtits are made of the softest materials and are often prettily decorated or camouflaged. Some nests are more or less concealed among the foliage or among hanging bunches of beardlike lichens, but most of them are suspended in plain sight, where one might easily be overlooked as a stray wisp of lichens or an accumulation of plant debris.

The nests are hung in a variety of trees, saplings, or bushes at various heights, although a large majority are not over 15 feet above the ground.

Mrs. Addicott (1938) remarked that most of the nests she studied at Palo Alto, Calif., were in oaks, but this does not seem to be always true elsewhere. Grinnell, Dixon, and Linsdale (1936) have given the data for 38 nests on Point Lobos Reserve, Calif.; the kinds of trees occupied and the heights of the nests from the ground were as follows: 16

nests were in *Ceanothus* bushes, both living and dead, a 4½ to 11 feet; 12 were in pines at from 6 to 50 feet, only 4 of which were above 15 feet; 4 were in live oaks at from 7 to 20 feet; 3 were in sage bushes at from 4 to 4½ feet; 2 were in cypresses at from 7 to 10 feet; and 1 was in a *Baccharis* bush at 6 feet. In the northern portion of the range, many nests are found in conifers, spruces, firs, and hemlocks, often suspended from the ends of limbs 15 to 25 feet from the ground and in plain sight; but I found two nests near Seattle that were in "spirea" bushes 9 to 10 feet up. Nests have been found in eucalyptus and pepper trees, in willow and alder saplings, in *Kuntzia* and hazel bushes, and probably in a variety of other trees and shrubs.

Mrs. Addicott (1938) gives a full description of the bushtit's nest and how it is built by the busy pair of little architects; her paper is well illustrated with sketches and photographs. I quote from it in part:

The nest built by the Coast Bush-tit is an intricate, pendant structure, hung in a concealing clump of leaves of an overhanging branch, and it is built of materials which blend with its surroundings, such as mosses, lichens, oak leaves and spider web. The entrance consists of a hole, usually placed on one side near the top, either above or below the supporting twigs. Above the entrance is the hood which is carefully woven around several twigs and which covers the top of the nest. Below the entrance is the neck which is the passage to the bowl, where the eggs are laid. The nest is entered horizontally, but the passage bends immediately and is vertical in the neck. At the bottom the passage flares to make the bowl. The neck is the slenderest, and usually the thinnest, part. The widest portion is the bowl, and here the walls are much thicker and are heavily lined. These features combined with the thick floor of the bowl, make the latter a warm place for the development of eggs and young. * * *

The first step in the construction of any bush-tit nest is the building of the rim. This is a delicate circle of nest material bound together with spider web and supported between the forks of a twig or between two adjacent twigs to which it is firmly fastened. This circular rim is almost always horizontal, or nearly so, and in no case which has been observed does this hole ever remain as the final entrance of the nest. It is rather the rim of a preliminary open bag.

After the rim has been built, nest construction may proceed in either of two ways. In the first and most prevalent method, a small bag, perhaps an inch in depth, is hung from the rim, loosely woven and very thin and delicate. In building, the birds cling to the edge of the rim and hang head down into the bag, adding materials to strengthen and thicken it. After this first tiny sac has been made strong it is stretched and extended from within and then the thin places which result are filled in with more nest material. As the nest becomes longer, the birds enter it head first to carry on the work. After a bird disappears inside, the nest shakes violently and bulges out in place after place as the new material is added in the thin sections. The shaking apparently serves to stretch the structure. Most of the work is done from the inside, but some of the thick parts of the bowl are added to from the outside, the birds clinging to the sides as they work. The nest is now a long pendant bag, open at the top.

When the hood and final entrance of the nest are built, material is added to the back and sides of the original rim. Material is brought over the top until the original hole has been roofed over and the entrance thus shifted to one side of the top. Rarely one finds a nest with the entrance hole only partly roofed over.

As the nest nears completion, a lining of spider web, down, or feathers is made for the passageway and the bowl. The bowl is thickened and filled in, and the walls above the floor of the nest for at least an inch are made quite thick. Material is added to the nest from time to time until the eggs have hatched, probably because the nest, being pendant, needs continual repairing.

In the second method, mentioned above, "a long extremely loose bag is constructed of strands of material hung from the rim"; this may be 5 to 8 inches long, instead of one an inch deep; otherwise the construction proceeds as above. Of the materials used in nest-building, she says: "Bush-tit nests are built of materials found commonly in the breeding territory. Lichens, mosses, grasses and the staminate flowers of the live oak, *Quercus agrifolia*, are constituents of almost all nests. These are woven together with hundreds of strands of spider web. Other materials found less commonly are filaree fruits, bark fibers, various plant downs, fir or spruce needles, oak leaves, acacia blossoms, blossoms of other plants such as broom and the pappi of composite flowers, feathers, bits of paper and string, and insect cocoons."

I have a series of six nests before me as I write, no two of which are alike; they vary greatly in size, shape, and bulk, as well as in the color of the material used. They vary in length from 7 to 10 inches and in the width of the bowl from barely 3 to over 4 inches, though these figures give only a faint idea of the variations in bulk. Nests 12 inches long are fairly common, some are as short as 6 inches, and Dr. B. W. Evermann (1881) measured one that was 21 inches in length. He also mentions nests hung in bunches of mistletoe, and says that others have been "found in sage and greasewood bushes, and one in a bunch of cactus."

The materials in the nests before me are quite varied and produce very different color effects, probably of concealing value. One, collected near Seattle, is constructed largely of very fine twigs and fine rootlets, firmly interwoven into a short bag, 7 inches long; it is decorated with bright green lichens or mosses and lined with soft down feathers or fur; it was in a "spirea" bush. Another similar nest is 10 inches long, and is interwoven with the twigs of a fir; it is made almost wholly of bright, yellowish-green mosses. Three California nests are still different; in one the dull dark-brown tone of the lichens and mosses is relieved by only a few bits of gray lichen; another is almost covered with soft, whitish, curly leaves or weed tops and cocoons or nests of spiders or insects, giving it a very pale tone; and in the third nest, the light, reddish brown blossoms, of which the nest is almost entirely made, produce that tone of color prettily offset by a circle of light buff plant down around the entrance.

According to Mrs. Addicott (1938) the time devoted to nest-building varies greatly and is subject to some interruption on account of bad weather or other disturbance. Nests started in February were actually worked on for 51 and 49 days in two cases, and for from 42 to 34 days in three other cases.

A nest started in April was completed in 13 days. "Second nests are built more quickly than first nests. Pairs disturbed during building, egg-laying or incubation frequently desert and build second nests, usually with new mates. * * * When one bird deserts a nest another mate is found to take its place. If both desert, they separate and seek new mates. * * * It might be suggested that the gregarious habits of the species during the rest of the year and the constant presence of unmated birds in small flocks during the nesting season are of significance here."

Leslie L. Haskin tells me that "after the nest is completed the birds seem to leave it for a time before actual occupancy begins. One nest that I observed closely was thus left unused for nearly thirty days, after which eggs were laid, and a thrifty brood brought forth. I am inclined to believe that during this time the parents use the nest for a nightly roost. I do know that, *after* the eggs are laid, both the male and female bird spend the night in the nest, a very cozy and comfortable bedroom. The same pair of birds appears to return to the same site, or one near by, year after year. When an occupied nest is found, it is not at all uncommon to find last year's nest close by. On one occasion I found a nest containing young, built upon a horizontal fir limb about 12 feet from the ground. At intervals of about 2 feet apart along the same limb, three other nests, or rather their remains, left from previous years, were hung, each one a little older and more bedraggled than the preceding one. It was evident that for at least four years this branch had been used by the same pair of birds."

W. E. Griffee writes to me of a nest that was used for three sets of eggs in one season: "The only occupied nest I have found in recent years was one which, on May 26, 1935, held young 4 or 5 days old. After this brood had flown, I took a set of 6 eggs, incubated 2 to 4 days, from the same nest on June 19th. As the nest was soiled with excrement from the first brood, I did not collect it, so the birds completed a third set of 6 eggs in it on about July 3rd or 4th." This was near Portland, Oreg., where he thinks bushtits are not so common as they were 30 years ago. From this and other reports, it seems that the bushtits often raise two broods in a season.

Carroll Dewilton Scott tells, in the notes sent to me, of the difficulties encountered by a pair of bushtits in their attempts to build a nest in a eucalyptus tree: "I could see the birds alight time after time in a cluster of limbs, about 12 feet from the ground, and often see bits of material fall to the ground. The bewildered birds would hop all about the spot, trying to find the stuff they had brought. But the eucalyptus limbs and leaves were so smooth that nothing would stick to them. All this day and the next they worked without making even the beginning of a nest. Twice they changed the nesting site without effect. Their ancestral experience had been with native plants, thick-leaved, usually prickly or rough. Finally, on March ninth, their persistence and patience were rewarded. They got a little circle of moss and fibres caught on the limbs and soon completed the nest."

S. F. Rathbun writes to me: "The bushtit [in western Washington] seems to favor localities that are more or less open, where the sunlight breaks among a growth that overruns old logged-off sections; this growth is in the nature of the hazel, small alders, young dogwoods, and more particularly the beautiful ocean-spray, or 'spirea' (*Holodiscus discolor*). The little glades that will be found among such spots are especially liked by this bird, and the partiality shown by it for such as have a growth of *H. discolor* is very marked. In fact, it may be said that, wherever this occurs to any extent, one can expect to find the bushtit, and usually does. The ocean-spray is the hallmark of the bushtit.

"I have found the nests in many locations, but the birds have a fondness for building in the 'spirea'; and I think the reason for this is that, as the bush always has numbers of the dry flower clusters of the preceding year in evidence, the nest is better protected from view, as in a way it resembles a panicle of the dry blooms. I have several times found a nest attached to one of these dry flower clusters, becoming a part of it, as it were, and in this way the concealment became more effective." (See pl. 67.)

Eggs.—The commonest numbers of eggs in the nests of the bushtit run from 5 to 7, seldom fewer or more, but as many as 12, 13, and even 15 have been found; probably these large numbers are the products of two females, as 3 birds have been seen visiting a nest. W. C. Hanna has a set in his collection containing an egg of the dwarf cowbird; it would be interesting to know how it was deposited. The eggs are mostly ovate in shape and have little or no gloss. They are pure white and unmarked. The measurements of 50 eggs average 13.7 by 10.1 millimeters; the eggs showing the four extremes measure 14.7 by 10.2, 13.8 by 11.2, and 12.2 by 9.1 millimeters.

Young.—Authorities seem to agree that the incubation period for the bushtit is 12 days, the young hatching on the 13th. Based on her study of two nests, Mrs. Addicott (1938) makes the following statements:

Incubation apparently is started on the day the last egg is laid, or on the day before. The burden of incubation seems to be shared equally by the male and female. * * * On cold days the eggs are incubated almost constantly, the parents alternating about every ten minutes on the nest. On warm days much shorter periods are spent on the eggs, and the parents forage together to some extent. Both birds spend the nights in the nest.

The young are naked at hatching and down does not develop. The eyes are closed and remain closed at least through the seventh day.

The young are fed solid undigested food from the first day on, and lepidopterous larvae are carried into the nest a few hours after hatching.

During early stages the young are brooded and fed as much by one bird as the other. Toward the end of the period, one parent does about two-thirds of the foraging and feeding. The mate spends most of the time moving about in the nest tree. The young apparently leave the nest on the fourteenth or fifteenth day. They become independent of the parents within eight days after leaving the nest, but have been seen to feed themselves on the day of nest-leaving.

She says that while the young are being fed in the nest "feeding occurs from eight to twelve times an hour" and that the "fecal sacs were carried as far as fifteen feet from the nest." Of the behavior at the time of nest-leaving, she writes:

When the nestlings are ready to fly, the slightest disturbance sends them out of the nest. As one juvenile starts to leave, the impulse apparently spreads rapidly to the others. So quickly do they pop out of the nest, that one has the feeling that the nest has suddenly exploded. There is an incessant medley of juvenal trills.

The juveniles fly a little awkwardly and to a lower level when they leave the nest. They scatter in all directions, often alighting in the grass, uttering the trill all the while. The parents immediately become excited, uttering a rapid succession of alarm notes as they dash from one young bird to another in an evident effort to protect them and to get them together. This is quite a task, for the juveniles fly as far as twenty-five yards from the nest tree.

The parents spend from fifteen minutes to half an hour gathering the scattered family in low bushes or in a small tree. * * * Half an hour after nest-leaving the young start following the parents in their search for food, begging with trills as they go. * * *

As soon as the juveniles are able to fly well enough to follow the parents, the family moves about in typical flock formation, the parents doing all of the foraging for the young. This takes place at least the day after nest-leaving, and frequently only a few hours afterward on the same day.

* * * Feeding by the parents is continued from eight to fourteen days. One family was fed till the eighth day after leaving, when two of the juveniles were seen foraging for themselves, clumsily hanging upside down in search of food. It seems likely that the shortness of the juvenal tail makes this process difficult, since it has been observed that it is impossible for an adult without a tail to do this. * * *

Begging and feeding were observed in another family up to the fourteenth day after nest-leaving. Feeding of the entire brood lasted nine days, after which one or two individuals were fed at long intervals.

Plumages.—The young bushtit is hatched naked, with eyes closed; the eyes are not open until the eighth day or later, according to Mrs. Addicott (1938). Mrs. Wheelock (1904) says that on the sixth day the young were "covered with a hairlike grayish white down." This is very scanty and is worn for only a short time, as the juvenal plumage soon pushes it out, and the young bird is practically fully fledged when it leaves the nest and is able to fly for a short distance. The juvenal plumage is much like the faded and worn plumage of the spring and summer adult; Mr. Swarth (1914) says that this plumage, "with shorter and fluffier feathers, and with more extensively light-colored bases, gives a general effect that is rather mottled and uneven." This plumage is worn for about 2 months, or until toward the end of July, when the postjuvenal molt begins and continues through September. This is an incomplete molt, which does not involve the flight feathers; it produces a first winter plumage which is practically indistinguishable from that of the adult, dense and lustrous and dark colored. Adults have one complete postnuptial molt each year at about the same time, but beginning a little earlier in July and continuing a little later, or into October. Mr. Swarth (1914) says that the molt is "of quite long duration, about three months for adults, and a little less for the post-juvenal molt." Considerable wear and fading take place during the winter and spring, so that before the annual molt takes place the birds become quite dull in coloration, worn, and shabby.

Food.—Prof. F. E. L. Beal (1907) examined 353 stomachs of bushtits from California. He says:

The first analysis of the food of the year gives nearly 81 percent animal matter, composed entirely of insects and spiders, to 19 percent of vegetable. * * * The largest item in the insect portion of the bird's food consists of bugs (Hemiptera), which amount to over 44 percent of the whole. * * * Moreover, the particular families of Hemiptera so extensively eaten by the bush tit are the two that are most destructive to the interests of horticulture—namely, the plant-lice (Aphididae), and bark-lice, or scales (Coccidae). The last amounts to nearly 19 percent of the year's food, and are eaten in every month. * * * The large black olive scale *(Saissetia oleae)* was identified in 44 stomachs, but other species were also found. * * * While the San Jose scale was not positively determined [probably because its distinctive characters are too minute to be recognized in a mass of semi-digested food], another species of the same genus, the greedy scale *(Aspidiotus rapax)*, was found in 4 stomachs, and scales not specifically identified were found in 113. Of a total of 353 stomachs, 158 held scales; several were entirely filled with them, and in quite a number upwards of 90 percent of their contents consisted of these insects. * * * The remaining portion of the hemipterous food of the bush tit, over 31 percent, is made up of plant-lice, tree-hoppers

(Membracidae), leaf-hoppers (Jassidae), some jumping plant-lice (Psyllidae), and a considerable number of false chinch bugs *(Nysius angustatus)*, with a few lace-bugs (Tingitidae).

Other insect food includes beetles, "somewhat over 10 percent," caterpillars, 16 percent, wasps and ants, 1½ percent, and the remainder of animal food, "about 8 percent," consists mainly of spiders. Of the vegetable food, less than 1 percent consists of fruit, pulp and skins, and the remainder is composed of a few seeds, granules of poison-oak, leaf galls, and rubbish, much of which is probably taken accidentally.

His analysis of the stomach contents of eight nestlings is of interest: "The animal matter comprised, approximately: Beetles 2, wasps 2, bugs 8, caterpillars and pupae 80, and spiders 7 percent. The point of greatest interest, however, lies in the fact that every one of these stomachs contained pupae of the coddling moth, distributed as follows: Two stomachs contained 2 each, two contained 3 each, one contained 4, one 7, one 9, and one 11, making 41 in all, or an average of over 5 to each."

From the above it will be seen that the bushtit is one of the most useful of the birds of California; it does practically no harm, except to eat a few beneficial ladybugs, in such small numbers as to be negligible, and it destroys immense numbers of the most harmful insects, most of which are so minute that only the microscopic eyes of these little birds could find them; and the larger birds would not bother to eat them. The great expansion of the fruit-growing industry in California has enabled these scale insects to increase enormously, and the scale-eating birds have not kept pace with them, so that artificial means must be employed to keep them under control.

Mr. Scott writes to me: "On January 20, 1917, I was surprised to find a flock of 24 bushtits feeding on ripe olives, both on the trees and on the ground. The birds were evidently hungry, as half a dozen flew to the ground within a few feet of me to peck the meat of the fallen olives, just as the house finches were doing."

Behavior.—Bushtits are sociable, friendly little birds, not only toward their human neighbors but among themselves, possessing many of the lovable and trustful traits of their relatives the chickadees. They show no fear of man and carry on their vocations with confident indifference to his near presence. But they are even fonder of their own society. Except during the short season when the pairs are occupied with their nesting activities, they travel about all through the rest of the year in loose flocks of varying sizes; small bunches of family parties join later into larger groups. I have often enjoyed watching a cloud of these little gray, long-tailed birds drifting through the trees and shrubbery on the edge of an arroyo in Pasadena.

The flock is somewhat scattered, and one cannot tell at first how many birds there are in the company, but they keep in touch with one another as they feed, with gentle twittering notes. They seem to be in constant motion as they travel along, hurriedly crossing the open spaces between the bushes, a few at a time, then more and more, all traveling in the same general direction; when 20 or 30 have crossed, and we think that all have gone, there are always a few stragglers hurrying along to catch up with the procession.

These flocks may consist of anywhere from half a dozen to 20 or 30 birds. Dawson (1923) has counted over 70, but usually there are less than 20. And, in winter, they may be accompanied by other small birds that forage with them, such as kinglets, wrens, and chickadees; these birds do not, I believe, follow along in the procession of bushtits.

Dr. Joseph Grinnell (1903) writes of the behavior of these flocks:

During such slowly moving excursions, each individual is rapidly gleaning through the foliage, assuming all possible attitudes in its search for tiny insects among leaves and twigs. * * * At times, especially towards evening, the flocks become more restless and move along from bush to bush and tree to tree much more rapidly than when feeding, the birds straggling hurriedly after each other in irregular succession. During these hurried cross-country excursions, the simple location-notes are pronounced louder and are interlarded at frequent intervals with a shrill quavering note. The faster the band travels, the louder and more oft-repeated becomes these all-important location-notes; for the greater becomes the danger of individuals becoming separated from the main flock. Bush-tits are usually hidden from each other in dense foliage. They have no directive color-marks; therefore, being gregarious birds, the great value of their location-notes becomes apparent.

Should a bush-tit lag far behind as to be beyond hearing of his fellows, he may suddenly come to a realization of his loneliness; he at once becomes greatly perturbed, flitting to the tallest available perch, and uttering the last mentioned note reinforced into a regular cry for his companions. This is usually heard by the distant band and several similar answering cries inform the laggard of the direction the flock has taken. Off he goes in zigzag precipitation and joins his fellows with evident relief.

He describes a peculiar behavior, also noted by others, that is evidently effective as a protective device:

A flock of bush-tits will be foraging as usual, with the ordinary uncertain medley of location-notes, when suddenly one or two birds utter several of the sharp alarm notes and then begin a shrill quavering piping. This is taken up by the whole flock, until there is a continuous monotonous chorus. At the same time every member of the scattered company strikes a stationary attitude in just the position it was when the alarm was first sounded, and this attitude is maintained until the danger is past. In nearly every case the danger is in the shape of a hawk, more especially of the smaller species such as the sharp-shinned or sparrow hawks. No matter how close the hawk approaches, the shrill chorus continues and even intensifies until the enemy has passed. The remarkable thing about this united cry, is that it is absolutely impossible to locate any single one of the birds by it. The chorus forms an indefinably confusing, all-pervading sound, which I know from personal experience to be most elusive. It may be compared in this respect to the sound of the cicada. * * * It seems reasonable to infer that this monotonous chorus of uncertain direction, at the same time as it sounds a general alarm, serves to conceal the individual birds, all of which at the same time maintain a statuesque, motionless attitude. * * * Scarcely any attention is ever paid by the bush-tits to large hawks, such as buteos, or to other large birds such as turkey vultures, pigeons, or jays. The bush-tits seem to be able to easily identify their real enemies at surprisingly long range.

Dr. Robert C. Miller (1921) has published an interesting paper on the flock behavior of the coast bushtit that is well worth reading; I quote from his summary as follows:

The flock moves from place to place by what may be termed the spread of impulse. An individual bird, moved no doubt by the hunger instinct, takes temporary leadership, and is followed to a new location by the others. There are no regularly assigned leaders, though probably the most venturesome birds assume the leadership most often. * * *

The method of flock movement makes evident the extreme improbability of there being any definite forage routes. The direction taken by the flock at any time is a matter of caprice, or the circumstances of the moment. Due to their dislike for crossing open spaces, however, the birds are likely to frequent areas where the vegetation is continuous and will generally avoid those where it is discontinuous, so that an impression of regularity in their forage movements may thus secondarily be given.

In all their movements the bushtits remind one strongly of the chickadees. Their flight is weak and apparently uncertain, though well adapted to their mode of life; they flit about rapidly and accurately in the foliage or bare twigs of trees and bushes; they seldom make long flights, and when they have to cross open spaces between covers they do so hurriedly and with a weak undulating flight. As gymnasts they are fully equal to the chickadees, and perhaps even more expert, for their longer tails give them even better control of movement and balance in foraging for their food about the tips of slender swaying twigs. Mrs. Bailey (1902) states it very well as follows: "Flitting from branch to branch they fly up to light upside down on the underside of a bough, and then without taking the trouble to turn right side up drop backwards to catch upside down on the tip of another twig, where they bend double over the terminal buds looking for food."

Carroll Dewilton Scott has sent me the following notes on rather unusual behavior of bushtits: "Only once in fifty years' association with brownie bushtits have I seen any show of belligerency. A flock of about 15 were hopping about in a small tree near the house and flying into a bed of flowers nearby. Suddenly two of the tiny birds were rolling about on the ground, fighting in dead earnest. Then they flew back to the tree, only to bounce around a few minutes later in a wheelbarrow that was standing under the tree. Now the most dramatic thing happened. The whole flock became excited and fluttered about the contestants, calling their shrill alarm notes and showing the greatest solicitation as long as the mites were fighting. In a few minutes all was as peaceful and merry as ever."

Of an unusual roosting habit, he says: "It seems strange to see these lowly midgets roosting in tall, open-branched eucalyptus trees, especially in winter. Several times in Balboa Park, San Diego, I had seen a flock fly into eucalyptus trees after sundown to roost. So, on December 27, we went to the locality about sunset. Between sunset and dark, a flock of 50 flew out of a brushy canyon to a dead acacia that stood between the canyon and a grove of eucalyptus, thence from one eucalyptus to another, some going a hundred feet or more before settling down for the night. Another instance of similar roosting in tall, swaying eucalyptus occurred on my place at Pacific Beach in 1935. All through November and December, even in gusty, rainy weather, a flock roosted every night in several tall eucalyptus near the house."

Illustrating the flocking habits of bushtits, Robert S. Woods writes to me: "No better demonstration of the intensely gregarious disposition of this species can be found than that which is afforded by a visit of the flock to a bird bath. Though there may be ample room around the rim of the basin, all will crowd together on one side, while newcomers and those from the outskirts of the group continually alight on the backs of the others, trying to force their way into the middle of the line. Even when only three are present, one of those on the outside may fly up and try to wedge itself between the other two. Apparently they feel safe or content only when flanked on both sides by others of their kind."

Mr. Rathbun, in his notes, tells of an incident that illustrates the sociability of bushtits; he found a nest toward dusk and says: "I lightly touched the branch to which the nest was attached, and, to my surprise, four tits flew from the nest; as this was too early in the season for any young of the year, I thought I would see what the nest contained and found six eggs that were well incubated." Apparently, two of the birds were the owners of the nest and the other two had taken up quarters in the nest for the night.

Voice.—Some of the notes of the bushtit are mentioned above in the quotations from Dr. Grinnell's (1903) excellent paper on this subject. He recognizes five distinct notes, or variations or combinations of notes, each of which seemed to him to signify some particular state of mind of the birds:

1. The simple location-note uttered while the birds are feeding and undisturbed, "*tsit, tsit; tsit; tsit.*"

2. "From one to five of the simple notes uttered somewhat more loudly and followed by a rather shrill quavering note of longer duration," uttered when traveling more rapidly and not feeding, "*tsit, tsit, tsit, sre-e-e-e; tsit, sre-e-e-e.*"

3. The same as the last "but pronounced with much more volume and emphasis," uttered by lone individuals when separated from the flock, *tsit, tsit, sre-e-e-e.*

4. Similar to the first note but greatly intensified, the alarm note, uttered when the nest is disturbed or an enemy discovered, "*tsit; tsit, tsit; tsit.*"

5. "A shrill quavering trill, of the same quality as described under number 2 above, but without the preceding simple notes, and chanted continuously in a monotone by all members of a flock for as long as two minutes." This is the chorus, confusing note uttered in the presence or approach of an avian enemy, *sre-e-e-e-e,* etc.

Field marks.—There is no bird in California quite like the bushtit. It is a tiny, brownish-gray bird with a long tail and with no prominent recognition marks. The subspecies may be recognized by their coloration, as explained under each. A loose flock of feathered mites drifting through the tree tops or bushes could be safely recorded as bushtits about as far off as they could be seen.

Winter.—Bushtits are permanent residents throughout nearly all their range. Their behavior in winter has been referred to above.

DISTRIBUTION

Range.—Western United States, southern British Columbia, Mexico, and Guatemala; nonmigratory.

The range of the bushtit extends **north** to southwestern British Columbia (Point Grey and Chilliwack); northeastern Oregon (Canyon City Mountain); probably southwestern Wyoming (Green River); and central Colorado (Colorado Springs). **East** to central Colorado (Colorado Springs, Turkey Creek, Beulah, and Trinidad); western Oklahoma (Kenton); eastern New Mexico (Santa Rosa and Carlsbad); western Texas (Mount Ord and Castroville); Veracruz (Las Vigas and Mirador); and Guatemala (San Mateo and Tecpam). **South** to Guatemala (Tecpam and Huehuetenango); Chiapas (San Cristóbal); Oaxaca (Tehuantepec and La Parada); Michoacán (Patzcuaro); Jalisco (Haciendo el Molino and La Laguna); Tepic (Santa Teresa); and southern Baja California (Miraflores). **West** to Baja California (Miraflores, Victoria Mountains, San Fernando, and

El Rosario); California (San Diego, Santa Barbara, Palo Alto, East Park, and Yreka); western Oregon (Mosquito Ranger Station, Corvallis, and Portland); Washington (Olympia, Seattle, and Bellingham); and southwestern British Columbia (Boundary Bay and Point Grey).

The range as outlined is for the entire species, which has been separated into several geographic races, some of which are found only in Mexico and Guatemala. Those occurring in the United States, Canada, and northern Mexico are the coast bushtit *(Psaltriparus minimus minimus)*, which ranges along the Pacific coast from southwestern British Columbia south through California to the Mexican border; the California bushtit *(P. m. californicus)*, found in the interior from south-central Oregon, south to Kern County, Calif.; the black-tailed bushtit *(P. m. melanurus)* of northern Baja California; Grinda's bushtit *(P. m. grindae)*, which occurs in the mountains of the Cape district of Baja California; the lead-colored bushtit *(P. m. plumbeus)*, ranging in the Rocky Mountain region from eastern Oregon and western Wyoming south to Texas and northern Sonora; and Lloy'd bushtit *(P. m. lloydi[1])*, which is found from southern New Mexico and western Texas south into northern Mexico.

Egg dates.—Arizona: 33 records, April 8 to June 12; 17 records, April 17 to May 15, indicating the height of the season.

California: 124 records, February 26 to July 15; 62 records, April 1 to May 7.

Mexico: 7 records, April 9 to June 3.

New Mexico: 5 records, April 20 to June 3.

Texas: 3 records, April 11 to June 21.

Washington: 11 records, April 13 to July 3.

PSALTRIPARUS MINIMUS CALIFORNICUS Ridgway
CALIFORNIA BUSHTIT

This interior race of the bushtits is described by Ridgway (1904) as "similar to *P. m. minimus* but decidedly paler, the pileum light broccoli brown in spring and summer, the back, etc., olive-gray instead of deep smoke gray. (The autumnal and winter plumage very similar to the spring and summer plumage of *P. m. minimus.)*" Swarth (1914) says of the distinguishing characters of *californicus:* "As compared with *P. minimus minimus*, of clear gray and white tones of color, rather than of the brownish hue of that subspecies. Typical *californicus* is often almost pure white beneath, noticeably so in the juvenal plumage. Sides and flanks slightly or not at all tinged with vinaceous."

---

[1] Recognized as a subspecies of *Psaltriparus melanotis* after this account was written.

The range of the California bushtit, as now understood, includes the interior valley regions, chiefly in the Lower and Upper Sonoran Zones, exclusive of desert regions, from Jackson County, Oreg., to Kern County, Calif. Mountain ranges and intervening deserts both seem to serve as effective barriers at certain places between the two subspecies.

I cannot find anything in the literature and have received no contributed notes to indicate that any of the habits of this inland race differ materially from those of the coast form. As it lives in a somewhat less humid environment, it probably selects some different species of trees and shrubs for its nesting sites, such as madrones, junipers, and sage on occasions, though it evidently shows a preference for oaks. It apparently builds its nests of similar materials and eats similar kinds of food. In short, practically all that has been written about the coast bushtit would apply equally well to the present form. Their eggs are indistinguishable. The measurements of 40 eggs of *californicus* average 14.1 by 10.5 millimeters; the eggs showing the four extremes measure 15.3 by 10.0, 13.0 by 11.1, 12.8 by 10.1, and 14.0 by 9.9 millimeters.

Enemies.—Dr. Herbert Friedmann (1929) reports the finding of an egg of the dwarf cowbird in a nest of the California bushtit, but, as it was found in Riverside County, it probably refers to the coast bushtit. This seems to be the only known time that the species has been imposed upon. It is a mystery how the cowbird can enter such nests as those of the bushtit and the verdin without so enlarging the entrance as to cause the owners to desert.

In this connection, the reader is referred to what E. C. Stuart Baker has to say about similar difficulties encountered by the Himalayan cuckoo, in his contribution to my Bulletin 176, page 87.

PSALTRIPARUS MINIMUS MELANURUS Grinnell and Swarth

BLACK-TAILED BUSHTIT

In describing and naming this Baja California subspecies, Grinnell and Swarth (1926) say that it is "as compared with *Psaltriparus minimus minimus,* of southern California, of darker, more plumbeous general coloration. In *minimus* there is a brownish tinge to the plumage, above and below, that is lacking in *melanurus.* This difference is most outstandingly apparent on wings and tail; *melanurus* is, comparatively speaking, a 'black-tailed' bushtit. * * * The color differences distinguishing *melanurus* and *minimus* are readily apparent in fresh plumaged birds. Bush-tits change rapidly in appearance with wear and fading of the feathers, and worn-plumaged individuals of these two subspecies no longer exhibit the differences that are so easily seen in early fall.

Melanurus in its dark slaty color is readily distinguished at any season from the paler subspecies, *P. m. californicus;* and it is, of course, as readily told from the still paler species, *P. plumbeus."*

This is one of several new subspecies that have been described from the Sierra San Pedro Mártir region. Its known range seems to extend from the United States boundary southward to about latitude 30°, and from the crest of the above mountains to the sea. It occupies chiefly the Upper Sonoran Life Zone but extends locally into the Transition and Lower Sonoran Zones.

It probably does not differ materially in its habits from other adjacent races of the species. J. Stuart Rowley writes to me: "Near Socorro, Lower California, a nest of this bird was found on April 18, 1933. The nest was placed in a low bush near our camp and contained five fresh eggs. The nest and habits of these birds were similar to those of the California races."

The measurements of 40 eggs average 13.7 by 10.3 millimeters; the eggs showing the four extremes measure **14.8** by 10.7, 14.0 by **11.1,** **12.8** by 10.1, and 13.4 by **9.7** millimeters.

PSALTRIPARUS MINIMUS GRINDAE Ridgway
GRINDA'S BUSHTIT

HABITS

Grinda's bushtit was long regarded as a distinct species, and Mr. Ridgway (1904) treated it as such in his Birds of North and Middle America. It was named by Lyman Belding, in his manuscript sent to Mr. Ridgway, in honor of his friend Don Francisco C. Grinda, of La Paz. Ridgway's (1883) original description of it is as follows:

Entire pileum uniform light brown, or isabella-color (exactly as in some specimens of *P. minimus);* side of head similar, but paler, and gradually fading into white on chin and throat; remaining lower parts very pale smoky-gray, with a faint lilac tinge (exactly as in *P. minimus*). Upper parts light plumbeous-gray, in very marked and abrupt contrast with the brown of the nape. * * *

This pretty new species, while combining, to a certain degree, the characters of *P. minimum* and *P. plumbeus,* is yet apparently quite distinct from both. In the brown head and color of the under parts it agrees exactly with the forr er. but the resemblance ends there. From the latter it differs in much whiter throat and decidedly clearer, more bluish, shade of the upper parts, in both of which respects there is a close resemblance to *P. melanotis.* The bill is very slender, like that of *P. plumbeus.*

The range of Grinda's bushtit seems to be limited to the Cape region of Lower California, chiefly in the Upper Sonoran Zone and in the mountains. William Brewster (1902) says that Mr. Frazar found it *"occurring almost as numerously about San José del Rancho as on the

Sierra de la Laguna. It is a sedentary species, of which each individual bird probably spends its entire life within a very limited area, for Mr. Frazar noticed no marked seasonal variations in the number of its representatives at any of the localities which he visited."

Nesting.—Mr. Brewster (1902) writes: "A nest found on May 24 in the top of a small pine about eight feet above the ground, on the Sierra de la Laguna, is similar in shape to the nests of *P. m. californicus* and *P. plumbeus.* It is nine inches long, with a diameter varying from two to two and one half inches. The entrance hole is in one side near the top. The walls are composed of small, dry leaves, fern-down, catkins, spiders' cocoons, yellowish usnea and grayish lichens, all these materials being felted into a thick, tenacious fabric of a generally mixed brown and grayish color. There were no eggs, the nest being not quite finished when taken."

There is a set of three eggs in the Doe collection, University of Florida, that was collected by J. Stuart Rowley near Miraflores on May 3, 1933. The nest was in an "oak-like tree," about 6 feet from the ground, and was made of mosses, fine downy fibers, and feathers.

Eggs.—The eggs of Grinda's bush-tit are indistinguishable from those of other bushtits. The measurements of eight eggs average 13.5 by 10.1 millimeters; the eggs showing the four extremes measure 14.2 by 9.9, 11.5 by 10.1, and 12.2 by 10.4 millimeters.

Plumages.—Mr. Brewster (1902) describes the juvenal plumage as "differing from the adult in being ashier beneath, with a decided purplish tinge on the sides; the back paler bluish, the crown light purplish brown; the outer tail feathers with their outer webs ashy white to the shaft; the secondaries and wing coverts edged and tipped with grayish or rusty white."

He says that a bird in "first winter plumage," taken November 28, is "similar to the young just described, but with the crown deep purplish brown; the back darker or more slaty than in the adult; the wings and tail more bluish; the outer tail feathers with exceedingly narrow light margins on their outer webs." It seems to me that this bird, which I have seen, is an adult in fresh autumn plumage, as it does not have the juvenal tail, described above; the postjuvenal molt does not involve the flight feathers, and, if this were a first winter bird, it should still have the juvenal tail.

I have also examined the molting specimen he refers to, taken July 28, which is also evidently an adult. Adults apparently have a complete post nuptial molt, beginning in July, and they are darker and more richly colored after this molt.

In all its other habits, food, behavior, and voice, it probably does not differ from other races of the species, if due allowance be made for the difference in environment.

LEAD-COLORED BUSHTIT

HABITS

This bushtit, which was long considered to be a distinct species, is now regarded as the easternmost representative of a western species. It has a wide distribution in the general region of the Rocky Mountains, according to the 1931 Check-list, "from eastern Oregon and western Wyoming south to northern Sonora and western Texas, and from eastern California to central Colorado." Its range in California seems to be a very narrow one, which Swarth (1914) defines as limited to the "desert region of the southeastern portion of the State, in Mono, Inyo, and northern San Bernardino counties. A discontinuous range, being confined to the Upper Sonoran zone of the various desert mountain chains and the east slope of the Sierra Nevada, these tracts being separated by vast expanses of Lower Sonoran, uninhabited by the species."

The lead-colored bushtit evidently intergrades with the California bushtit at the western border of its range in eastern California and with Lloyd's bushtit at the southeastern border of its range in western Texas. Mr. Swarth (1914) treated the lead-colored bushtit as a full species and seemed loath to regard it as a subspecies, though he discussed the subject quite fully and evidently felt that the three forms are very closely related. Grinnell, Dixon, and Linsdale (1930) discuss in considerable detail the status of certain specimens collected in the Lassen Peak region, and seem to favor the subspecies theory.

Referring to the Tobaye region of Nevada, Dr. Jean M. Linsdale (1938b) reports this bushtit as "a common bird of the lower part of the mountains, between 6000 and 8000 feet; noted once as high as 8700 feet. * * *

"Prominent among the kinds of plants frequented by bushtits were the following: tall sage brush, piñon, mountain mahogany, birch, willow, dogwood, limber pine, and aspen. The species occurred over the ridges and along the streams; the greatest number being found on the floor of cañons near their mouths at the base of the mountains."

We found the lead-colored bushtit common and well distributed in all the mountain ranges of southern Arizona but entirely absent from the low, arid regions of the southwestern part of the State. In the Huachucas, it ranged well up toward the summits but was most abundant between 4,000 and 7,000 feet in the mouths of canyons, in the foothills, and on the lower mountain slopes, especially where the large trees were scattered, leaving large, open, sunny areas, overgrown with scrub oaks, small madrones, scattered junipers, and a variety of other shrubbery.

The lead-colored bushtit is intermediate in its characters between the California bushtit on the west and Lloyd's on the southeast, being a plainly colored bird without the distinctive head markings of either of the other two forms.

Nesting.—The nesting habits of the lead-colored bushtit are similar to those of the species elsewhere, though some different species of trees and shrubs are utilized as nesting sites and the nests are made of such material as is locally available. Of the six nests recorded in our Arizona notes, one was at the end of a pendant limb of a live oak, three were in oak scrub, one was in a small juniper, and one in a madrona bush; the juniper nest was only 6 feet from the ground and the nest in the live oak was 9 feet up, the others being at intermediate heights. They were all at altitudes ranging from 4,000 to 7,000 feet. The madrona nest, now before me, is a beautiful structure; it was made largely of green mosses, whitish lichens, and buff plant down, mixed with a few small dry leaves, fine plant fibers, etc., all firmly matted together; it was profusely lined with small, gray, downy feathers, and what looks like mouse fur.

Bendire (1887) says that the nests are not always "strictly pensile, but are woven into and supported by small twigs and branches of the oak bushes (*Quercus undulata?*) in which they are built. Several nests were placed in bunches of a species of mistletoe (probably *Phoredendron flavescens*), and in these cases the nests are supported and placed directly in the forks of this plant. * * * The nests are outwardly composed of the dried, curled-up leaves of the white sage, plant-down of a pinkish tint, spider webs, small bits of mosses and lichens, and are thickly lined inside with soft, small feathers. * * * The nests are placed in about equal proportions in low oak bushes, from 5 to 7 feet from the ground, generally well concealed by the foliage, or in bunches of mistletoe in oak or mesquite trees, from 15 to 20 feet high."

Swarth (1904) says that the earlier nests are all in the lower foothill regions, "but later in the season [in the Huachucas] they nest abundantly in the higher altitudes, sometimes high up in the pine trees. I saw one nest at the very top of a tall pine, but the tree was growing on a steep hill side, and the nest was about on a level with the trail from which I saw it."

Mrs. Bailey (1928) says that in the Chisos Mountains, Tex., the nests are made in nut pines, 12 to 15 feet from the ground. And, in New Mexico, nests were built in junipers and cottonwoods; one nest that she saw was made mostly of sheep wool with small woolly leaves and oak tassels interwoven. Mrs. Wheelock (1904) reports blackberry vines as favorite nesting sites in eastern California.

Eggs.—The lead-colored bushtit usually lays five or six eggs, which are just like other bushtits' eggs, ovate in shape and pure white in color.

The measurements of 48 eggs average 13.4 by 10.2 millimeters; the eggs showing the four extremes measure **14.5** by 10.2, 14.2 by **10.7**, **12.5** by 10.2, and 13.0 by **9.9** millimeters.

Young.—Two broods are often, perhaps usually, raised in a season and often at very short intervals. A. J. van Rossem (1936) flushed from a nest what he estimated was a brood of seven or eight young just before dusk. The next morning, four young, just able to fly, left the nest as he approached; and, much to his surprise, he found in the nest five perfectly fresh eggs. The birds that flew from the nest the previous evening may have been adult birds that had gone into the nest to roost, but the interesting point is that the eggs were laid before the young left the nest; thus the first brood had helped to start incubation on the second brood!

Plumages.—The sequence of plumages and molts in the lead-colored bushtit is the same as in the other races of the species, but the juvenal plumage exhibits some very interesting variations. Mr. Swarth (1914) writes of this plumage:

Practically like adult. The gray of the upper parts is duller, less of a blue-gray, the brown cheeks are not so sharply contrasted against the rest of the head, and there is a faintly indicated black line over the auriculars and on the nape. Unlike the adults, the young of *plumbeus* exhibit considerable diversity in markings, and while the above described specimen represents the plumage perhaps most frequently encountered, there is a large proportion of birds with more or less extensive black markings on the head. In a few specimens there is no trace of head markings, a number have them faintly indicated, as in the specimen described above, and in others the patterns vary from a narrow line extending backward from the eye, to nearly as extensive a black marking as in the adult male of *P. m. lloydi* (See Swarth, 1913, pp. 399-401).

It was these black markings on the heads of young *plumbeus* that led to the erroneous extension of the range of *lloydi* into southern Arizona and New Mexico.

The postjuvenal molt of *plumbeus* begins during the last week of July, or earlier. I have seen molting adults from August 8 to September 13.

I can find nothing reported on the food, behavior, or voice of this bushtit that differs in any respect from what has been written about the California races, but shall quote the following passage from the facile pen of Dr. Elliott Coues (1878), which so aptly describes the flocking habit of the species: "They are extremely sociable—the gregarious instinct common to the Titmice reaches its highest development in their case, and flocks of forty or fifty—some say even of a hundred—may be seen after the breeding season has passed, made up of numerous families, which, soon after leaving the nest, meet kindred spirits, and enter into intimate friendly relations. Often, in rambling through the shrubbery, I have been suddenly surrounded by a troop of the busy birds, perhaps unnoticed till the curious chirping they keep up attracted my attention; they seemed to *pervade* the bushes. If I stood still, they came close around me, as fearless as if I were a stump, ignoring me altogether."

PSALTRIPARUS MELANOTIS LLOYDI Sennett

LLOYD'S BUSHTIT

HABITS

Lloyd's bushtit was originally described by George B. Sennett (1888) as a full species; Ridgway (1904) treated it as a subspecies of *Psaltriparus melanotis,* the black-eared bushtit of Mexico. In his original description Sennett stated that in the adult male the sides of the head are "glossy black, which extends backward on each side, meeting and forming a collar on lower back of neck." The adult female has the "ear-patches clear glossy brown instead of black. * * * This species is distinct from *P. melanotis,* Black-eared Bush-Tit, by reason of total absence of both brown on back and rufous on underparts. It is easily distinguished from *P. plumbeus* by the collar, and by the black instead of ashy brown on sides of head. Aside from the head markings it is more like *P. plumbeus* in color than *P. melanotis,* but it has a much whiter throat and a larger bill."

The range of Lloyd's bushtit, as given in the 1931 Check-list, includes the "mountains of the southeastern desert region, mainly in the Upper Austral Zone, from southern New Mexico and central western Texas (mountains between Pecos River and Rio Grande) south into Sonora and Chihuahua."

Ridgway (1904) extended the range into southern Arizona, but Swarth (1913) has shown that this was an error. The Arizona specimens of the supposed *lloydi* race, originally named *P. santaritae,* are all birds in juvenal plumage. Mr. Swarth (1913) has demonstrated that the young of *plumbeus,* in juvenal plumage, have more or less black markings on the head of the male, even in specimens taken as far away from the supposed range of *lloydi* as Nevada. This matter is fully discussed in his paper, to which the reader is referred.

The status of Lloyd's bushtit, as at least of casual occurrence, in southern New Mexico, seems to be established, for Mrs. Bailey (1928) says that "a full plumaged adult male was taken in the San Luis Mountains, July 19, 1892 (Mearns), and is now in the collection of the United States National Museum." There may be other specimens of females and young, which are not so readily distinguished from *plumbeus*.

Van Tyne and Sutton (1937) do not agree with the concept that *lloydi* is a subspecies of *minimus;* they treat it as subspecies of *melanotis,* and evidently regard *minimus* and *melanotis* as specifically distinct. They say:

All of the forty-three Brewster County [Texas] specimens we have examined are without exception clearly *melanotis* or *minimus*. In this region *melanotis* is confined to higher altitudes more definitely than is *minimus* but the breeding ranges of the two forms overlap widely and their relations seem to be those of two distinct species. The only bit of contrary evidence we noted was the fact that on May 5, 1932, Dr. Peet collected at Boot Spring an adult male *melanotis* in company with a breeding female *minimus*. The two birds seemed to be traveling together, and no other Bush-tits were seen about at the time. The whole problem of the relation between these two forms is a fascinating one which calls for more study in this critical region.

Nesting.—Mr. Sennett (1888) reports a nest taken in Presidio County, Tex., on June 21, 1887, by William Lloyd, for whom he named the species. It was found in Limpia Canyon, at an altitude of 6,200 feet, and was fastened to twigs of a cedar seven feet from the ground. "The cedar tree was twenty-five feet high, situated on a divide between two ravines."

There is a set of five eggs in the Doe collection in Florida, said to be of this subspecies; it was taken by Capt. R. W. Barrell on April 20, 1890, in Grant County, N. Mex., which is in the southwestern corner of the State and within the *possible* breeding range of this race; the nest was placed 5 feet up in a live oak. There is a set of seven eggs in my collection, taken by E. F. Pope in the Comanche Mountains, Tex., on April 11, 1912; the nest "was suspended on the end of a drooping branch of a small pine, 12 feet from the ground."

Eggs.—The eggs of Lloyd's bushtit are apparently just like those of other bushtits. The measurements of 23 eggs average 13.8 by 10.5 millimeters; the eggs showing the four extremes measure **14.5** by 10.6, **14.0** by **11.4**, **13.2** by **10.0** millimeters.

Plumages.—The following remarks by Van Tyne and Sutton (1937), based on specimens from Brewster County, Tex., are interesting as showing some of the characters of *lloydi* and some of its variations in plumage:

The best characters for distinguishing this species from *Psaltriparus minimus* are: black on the sides of the nape and on the auricular region, gray crown contrasting strongly with the more olivaceous back, throat usually much whiter than the rest of the under parts, strong vinaceous wash on sides and flanks. In addition the "face" is not Drab or Light Drab as in *plumbeus*, but is Hair Brown (in males this brown is usually suffused or mottled with black in varying degrees). A dozen of our Brewster County males can be arranged in a series to show perfect gradation from a face that is nearly pure Hair Brown to one that is practically black. Many of the intermediates show a curious mottling of black on a brown background. We were at first inclined to consider that the more brown-faced specimens were immature, but three juvenal-plumaged males collected by A. C. Lloyd on May 25, 1933, are among the most black-faced of the whole series. In these juvenile birds the vinaceous wash on the flanks is almost imperceptible. The black-faced juveniles make it obvious that the description by Ridgway, which characterizes the young males of *lloydi* as having only black ear coverts, is inadequate. Apparently Sennett was correct in stating that the young are 'similar to adults.' Of the three adult females examined, one (June 5) has a Hair Brown face, well-marked auricular region, and a strong vinaceous wash below; the other two, more worn females (July 21), are separable from *plumbeus* only by grayer faces and by blackish auricular feathers.

In all other respects, food, behavior, and voice, this bushtit does not differ materially from the other bushtits.

LITERATURE CITED

Abbott, Clinton Gilbert.
    1929. Watching long-crested jays. Condor, vol. 31, pp. 124-125.

Adams, Ernest.
    1898. Notes on the plain titmouse. Osprey, vol. 2, pp. 81-82.

Addicott, Alice Baldwin.
    1938. Behavior of the bush-tit in the breeding season. Condor, vol. 40, pp. 49-63.

Aiken, Charles Edward Howard, and Warren, Edward Royal.
    1914. Birds of El Paso County, Colorado. Colorado College Publ., gen. ser., No. 74 (sci. ser., vol. 12, No. 13, pt. 2), pp. 497-603.

Aldous, Clarence Moroni.
    1937. Eastern crows nesting on or near the ground. Auk, vol. 54, pp. 393-394.

Aldous, Shaler Eugene.
    1942. The white-necked raven in relation to agriculture. U. S. Fish and Wildlife Res. Rep. 5, 56 pp.

Aldrich, John Warren, and Nutt, David C.
    1939. Birds of eastern Newfoundland. Sci. Publ. Cleveland Mus. Nat. Hist., vol. 4, pp. 13-42.

Alexander, Frank McDaniel.
    1930. Notes on the birds of south central Kansas. Wilson Bull., vol. 42, pp. 241-244.

Alexander, Horace Gundry.
    1927. A list of the birds of Latium, Italy, between June 1911 and February 1916. Compiled from the notes and letters of the late C. J. Alexander. Ibis, ser. 12, vol. 3, pp. 659-691.

Allen, Francis Henry.
    1910. Warbling song of the Hudsonian chickadee. Auk, vol. 27, pp. 86-87.
    1919. The aesthetic sense in birds as illustrated by the crow. Auk, vol. 36, pp. 112-113.

American Ornithologists' Union.
    1931. Check-list of North American birds, ed. 4.
    1944. Nineteenth supplement to the American Ornithologists' Union check-list of North American birds. Auk, vol. 61, pp. 441-464.
    1945. Twentieth supplement to the American Ornithologists' Union check-list of North American birds. Auk, vol. 62, pp. 436-449.

Anderson, Rudolph Martin.
    1907. The birds of Iowa. Proc. Davenport Acad. Sci., vol. 40, pp. 125-417.

Anthony, Alfred Webster.
    1889. New birds from Lower California, Mexico. Proc. California Acad. Sci., ser. 2, vol. 2, pp. 73-82.
    1893. Birds of San Pedro Martir, Lower California. Zoe, vol. 4, pp. 228-247.

Attwater, Henry Philemon.
    1892. List of birds observed in the vicinity of San Antonio, Baxter County, Texas. Auk, vol. 9, pp. 337-345.

Audubon, John James.
    1842. The birds of America, vol. 4.

AUSTIN, OLIVER LUTHER, JR.
1932. The birds of Newfoundland Labrador. Mem. Nuttall Orn. Club, No. 7.
AVERILL, A. B.
1895. Nest of the magpie. Oregon Nat., vol. 2, p. 136.
BAGG, AARON CLARK.
1939. Nesting of the Acadian chickadee in Maine. Auk, vol. 56, p. 190.
BAILEY, ALFRED MARSHALL.
1927. Notes on the birds of southeastern Alaska. Auk, vol. 44, pp. 351-367.
BAILEY, FLORENCE MERRIAM.
1902. Handbook of birds of the Western United States.
1904. Additional notes on the birds of Upper Pecos. Auk, vol. 21, pp. 349-363.
1928. Birds of New Mexico.
BAILEY, HAROLD HARRIS.
1913. The birds of Virginia.
1925. The birds of Florida.
BAILEY, VERNON.
1903. The white-necked raven. Condor, vol. 5, pp. 87-89.
BAIRD, SPENCER FULLERTON.
1864. Review of American birds, in the museum of the Smithsonian Institution,
pt. 1: North and Middle America. Smithsonian Misc. Coll., No. 181.
BAIRD, S. F.; BREWER, THOMAS MAYO; and RIDGWAY, ROBERT.
1874. A history of North American birds: Land birds, vols. 1 and 2.
BAKER, ROLLIN HAROLD.
1940. Crow depredation on heron nesting colonies. Wilson Bull., vol. 52, pp.
124-125.
BALDWIN, DOROTHY A.
1935. Black-capped chickadee age-records. Bird-Banding, vol. 6, p. 69.
BALL, STANLEY CRITTENDEN.
1938. Summer birds of the Forillon, Gaspé County, Quebec. Can. Field Nat.,
vol. 52, pp. 95-103.
BANCROFT, GRIFFING.
1930. The breeding birds of central Lower California. Condor, vol. 32, pp.
20-49.
BARLOW, CHESTER.
1901. Some characteristics of the mountain chickadee. Condor, vol. 3, pp.
111-114.
BARROWS, WALTER BRADFORD.
1912. Michigan bird life. Spec. Bull. Dept. Zool. and Physiol. Michigan Agr.
College, 822 pp.
BATCHELDER, CHARLES FOSTER.
1884. Description of the first plumage of Clark's crow. Auk, vol. 1, pp. 16-17.
BEAL, FOSTER ELLENBOROUGH LASCELLES.
1897. The blue jay and its food. U. S. Dept. Agr. Yearbook for 1896, pp.
197-206.
1907. Birds of California in relation to the fruit industry, pt. 1. Biol. Surv.
Bull. 30.
1910. Birds of California in relation to the fruit industry, pt. 2. Biol Surv.
Bull. 34.

BEAL, F. E. L.; McATEE, WALDO LEE; and KALMBACH, EDWIN RICHARD.
  1916. Common birds of southeastern United States in relation to agriculture.
    Farmers' Bull. 755.
BEAL, MARY.
  1931. Verdin ways. Bird-Lore, vol. 33, pp. 399-400.
BECK, ROLLO HOWARD.
  1895. Notes on the blue-fronted jay. Nidologist, vol. 2, p. 158.
BEEBE, CHARLES WILLIAM.
  1905. Two bird-lovers in Mexico.
BENDIRE, CHARLES EMIL.
  1887. Notes on a collection of birds' nests and eggs from southern Arizona
    Territory. Proc. U. S. Nat. Mus., vol. 10, pp. 551-558.
  1895. Life histories of North American birds. U. S. Nat. Mus. Spec. Bull. 3.
BERGTOLD, WILLIAM HENRY.
  1917. A study of the incubation periods of birds.
BERRY, SAMUEL STILLMAN.
  1922. Magpies versus livestock: An unfortunate new chapter in avian de-
    predations. Condor, vol. 24, pp. 13-17.
BLINCOE, BENEDICT JOSEPH.
  1923. Random notes on the feeding habits of some Kentucky birds. Wilson
    Bull., vol. 35, pp. 63-71.
BOLLES, FRANK.
  1896. At the north of Bearcamp water.
BOWLES, JOHN HOOPER.
  1900. The Northwest crow. Condor, vol. 2, pp. 84-85.
  1908. A few summer birds of Lake Chelan, Washington. Condor, vol. 10,
    pp. 191-193.
  1909. Notes on *Parus rufescens* in western Washington. Condor, vol. 12, pp.
    65-70.
BOWLES, J. H., and DECKER, FRANK RUSSELL.
  1930. The ravens of the State of Washington. Condor, vol. 32, pp. 192-201.
  1931. The ferruginous rough-leg in Washington. Murrelet, vol. 12, pp. 65-70.
BRADBURY, WILLIAM CHASE.
  1917a. Notes on the black-crowned night heron near Denver. Condor, vol.
    19, pp. 142-143.
  1917b. Notes on the nesting habits of the Clarke nutcracker in Colorado.
    Condor, vol. 19, pp. 149-155.
  1918. Nesting of the Rocky Mountain jay. Condor, vol. 20, pp. 197-208.
BRADSHAW, FRED.
  1930. Unusual nesting sites. Can. Field Nat., vol. 44, pp. 149-150.
BRALY, JOHN CLAUDE.
  1931. Nesting of the piñon jay in Oregon. Condor, vol. 33, p. 29.
BRANDT, HERBERT WILLIAM.
  1940. Texas bird adventures.
BRAUND, FRANK WILLIAM, and McCULLAGH, ERNEST PERRY.
  1940. The birds of Anticosti Island, Quebec. Wilson Bull., vol. 52, pp. 96-123.
BREWSTER, WILLIAM.
  1883. Crows fishing. Bull. Nuttall Orn. Club, vol. 8, p. 59.

1886. An ornithological reconnaissance in western North Carolina. Auk, vol. 3, pp. 173-179.

1902. Birds of the Cape region of Lower California. Bull. Mus. Comp. Zool., vol. 41, No. 1, pp. 1-241.

1906. The birds of the Cambridge region of Massachusetts. Mem. Nutall Orn. Club, No. 4.

1937. The birds of the Lake Umbagog region of Maine, pt. 3. Bull. Mus. Comp. Zool., vol. 66, pt. 3, pp. 408-521.

1938. The birds of the Lake Umbagog region of Maine, pt. 4. Bull. Mus. Comp. Zool., vol. 66, pt. 4, pp. 525-620.

BRIMLEY, CLEMENT SAMUEL.

1888. Nesting of the tufted tit in 1888. Ornithologist and Oologist, vol. 13, p. 142.

BROCK, SYDNEY E.

1910. Incubation and fledging periods in birds. Zoologist, vol. 14, pp. 117-118.

BROOKS, ALLAN.

1931. The relationships of the American magpie. Auk, vol. 48, pp. 271-272.

BROOKS, WINTHROP SPRAGUE.

1920. A new jay from Anticosti Island. Proc. New England Zool. Club, vol. 7, pp. 49-50.

BROUN, MAURICE.

1941. Migration of blue jays. Auk, vol. 58, pp. 262-263.

BROWN, D. E.

1930. Nesting habits of the Steller jay *(Cyanocitta stelleri stelleri)* in western Washington. Murrelet, vol. 11, pp. 68-69.

BROWN, NATHAN CLIFFORD.

1879. A list of birds observed at Coosada, central Alabama. Bull. Nuttall Orn. Club, vol. 4, pp. 7-13.

BRYANT, WALTER (PIERC)E.

1888. Unusual nesting sites, 2. Proc. California Acad. Sci., ser. 2, vol. 1, pp. 7-10.

1889. Descriptions of the nests and eggs of some Lower California birds, with a description of the young plumage of *Geothlypis beldingi*. Proc. California Acad. Sci., ser. 2, vol. 2, pp. 20-24.

BURKITT, J. P.

1936. Young rooks, their survival and habits. British Birds, vol. 29, pp. 334-338.

BURLEIGH, THOMAS DEARBORN.

1918. The Hudsonian chickadee *(Penthestes hudsonicus* subsp. ?) in northeastern Pennsylvania in June. Auk, vol. 35, p. 230.

1930. Notes on the bird life of northeastern Washington. Auk, vol. 47, pp. 48-63.

BUTTS, WILBUR KINGSLEY.

1931. A study of the chickadee and white-breasted nuthatch by means of marked individuals. Bird-Banding, vol. 2, pp. 1-26.

CAHALANE, VICTOR HARRISON.

1944. A nutcracker's search for buried food. Auk, vol. 61, p. 643.

CAMERON, EWEN SOMERLED.
    1907. The birds of Custer and Dawson Counties, Montana. Auk, vol. 24, pp. 389-406.
    1908. The birds of Custer and Dawson Counties, Montana. Auk, vol. 25, pp. 39-56.

CASSIN, JOHN.
    1852. Illustrations of the birds of California, Texas, Oregon, British and Russian America.

CHADBOURNE, ARTHUR PATTERSON.
    1896. The Hudsonian chickadee *(Parus hudsonicus)*, red-breasted nuthatch *(Sitta canadensis)*, and golden-crowned kinglet *(Regulus satraps)* in Plymouth County, Mass., in summer. Auk, vol. 13, p. 346.

CHAMBERLAIN, MONTAGUE.
    1884. Prehensile feet of the crow *(Corvus frugivorus)*. Auk, vol. 1, p. 92.

CHAPMAN, FRANK MICHLER.
    1898. Notes on birds observed at Jalapa and Las Vigas, Vera Cruz, Mexico Bull. Amer. Mus. Nat. Hist., vol. 10, pp. 15-43.

COBURN, CHARLES ARTHUR.
    1914. The behavior of the crow, *Corvus americanus,* Aud. Journ. Anim. Behavior, vol. 4, pp. 185-201.

COLLINGE, WALTER EDWARD.
    1924. The food of some British wild birds.

COTTAM, CLARENCE, and KNAPPEN, PHOEBE.
    1939. Food of some uncommon North American birds. Auk, vol. 56, pp. 138-169.

COUES, ELLIOTT.
    1871. The long-crested jay. Amer. Nat., vol. 5, pp. 770-775.
    1874. Birds of the Northwest.
    1878. Birds of the Colorado Valley, pt. 1. U. S. Geol. Surv. Terr. Misc. Publ. 11.

CRAM, ELOISE BLAINE.
    1927. Bird parasites of the nematode suborders Strongylata, Ascaridata, and Spirurata. U. S. Nat. Mus. Bull. 140.

CRIDDLE, NORMAN.
    1923. The American magpie in Manitoba. Can. Field Nat., vol. 37, pp. 25-26.
    1927. A tale of four crows. Can. Field Nat., vol. 41, pp. 179-183.

CROOK, COMPTON.
    1936. A winter food supply for the crow. Auk, vol. 53, pp. 337-338.

CRUICKSHANK, ALLAN DUDLEY.
    1939. The behavior of some Corvidae. Bird-Lore, vol. 41, pp. 78-81.

CURRIER, EDMONDE SAMUEL.
    1904. Summer birds of the Leech Lake region, Minnesota. Auk, vol. 21, pp. 29-44.

CUSHING, JOHN ELDREDGE, JR.
    1941. Winter behavior of ravens at Tomales Bay, California. Condor, vol. 43, pp. 103-107.

DALES, MARIE.
    1925. A maurading blue jay. Wilson Bull., vol. 37, p. 224.

DARCUS, S. J.
  1930. Notes on birds of the northern part of the Queen Charlotte Islands in 1927. Can. Field Nat., vol. 44, pp. 45-49.
DAWSON, WILLIAM LEON.
  1923. The birds of California, vol. 1.
DAWSON, W. L., and BOWLES, JOHN HOOPER.
  1909. The birds of Washington, vol. 1.
DECKER, FRANK RUSSELL, and BOWLES, JOHN HOOPER.
  1931. Summer birds of the Blue Mountains, Washington. Murrelet, vol. 12, pp. 12-14.
DICE, LEE RAYMOND.
  1917. Habits of the magpie in southeastern Washington. Condor, vol. 19, pp. 121-124.
  1918. The birds of Walla Walla and Columbia Counties, southeastern Washington. Auk, vol. 35, pp. 40-51, 148-161.
  1920. Notes on some birds of interior Alaska. Condor, vol. 22, pp. 176-185.
DICKEY, DONALD RYDER, and VAN ROSSEM, ADRIAAN JOSEPH.
  1938. The birds of El Salvador. Publ. Field Mus. Nat. Hist., zool. ser., vol. 23, pp. 1-635.
DICKS, FORD.
  1938. Activities of Steller's jays in a filbert orchard. Murrelet, vol. 19, p. 17.
DILLE, FREDERICK MONROE.
  1888. Nesting of the black-billed magpie. Ornithologist and Oologist, vol. 13, p. 23.
DINGLE, EDWARD VON SIEBOLD.
  1922. Peculiar note of Carolina chickadee. Auk, vol. 39, pp. 572-573.
DIXON, JAMES BENJAMIN.
  1934. Nesting of the Clark nutcracker in California. Condor, vol. 36, pp. 229-234.
DIXON, JOSEPH SCATTERGOOD.
  1938. Birds and mammals of Mount McKinley National Park, Alaska. Nat. Park Serv. Faunal Ser., No. 3.
DOOLITTLE, EDWARD ARTHUR.
  1919. A strange blue jay flight. Auk, vol. 36, p. 572.
DuBOIS, ALEXANDER DAWES.
  1918. An albino magpie. Condor, vol. 20, p. 189.
DWIGHT, JONATHAN, JR.
  1897. The whistled call of *Parus atricapillus* common to both sexes. Auk, vol. 14, p. 99.
  1900. The sequence of plumages and moults of the passerine birds of New York. Ann. New York Acad. Sci., vol. 13, pp. 73-360.
EATON, ELON HOWARD.
  1903. An epidemic of roup in the Canandaigua crow roost. Auk, vol. 20, pp. 57-59.
EIFRIG, CHARLES WILLIAM GUSTAVE.
  1905. A one-legged crow. Auk, vol. 22, pp. 312-313.
ELLICOTT, GRACE.
  1906. Note on food of blue jay. Guide to Nature, vol. 1, p. 168.
ELTON, CHARLES, and BUCKLAND, FRANK.
  1928. The gape-worm (*Syngamus trachea* Montagu) in rooks (*Corvus frugilegus* L.). Parasitology, vol. 20, pp. 448-450.

ERICHSEN, WALTER JEFFERSON.
    1919. Some summer birds of Liberty County, Georgia.
ERICKSON, MARY MARILLA.
    1937. A jay shoot in California. Condor, vol. 39, pp. 111-115.
ERRINGTON, PAUL LESTER.
    1935. Food habits of mid-west foxes. Journ. Mamm., vol. 16, pp. 192-200.
EVERMANN, BARTON WARREN.
    1881. Least titmouse. Ornithologist and Oologist, vol. 6, p. 19.
FARGO, WILLIAM GILBERT.
    1927. Feeding station habit of fish crow. Auk, vol. 44, pp. 566-567.
FARLEY, FRANK LeGRANGE.
    1925. Changes in the status of certain animals and birds during the past
        fifty years in central Alberta. Can. Field Nat., vol. 20, pp. 200-202.
    1932. Birds of the Battle River region.
FERRY, JOHN FARWELL.
    1910. Birds observed in Saskatchewan during the summer of 1909. Auk,
        vol. 27, pp. 185-204.
FINLEY, WILLIAM LOVELL.
    1907. Two studies in blue. Condor, vol. 9, pp. 121-127.
FISHER, WALTER KENRICK.
    1902. A trip to Mono Lake, ornithological and otherwise. Condor, vol. 4,
        pp. 1-11.
FLEMING, JAMES HENRY, and LLOYD, HOYES.
    1920. Ontario bird notes. Auk, vol. 37, pp. 429-439.
FORBUSH, EDWARD HOWE.
    1907. Useful birds and their protection.
    1912. The chickadee. Bird-Lore, vol. 14, pp. 372-375.
    1927. Birds of Massachusetts and other New England States, vol. 2.
FOX, HERBERT.
    1923. Disease in captive wild mammals and birds.
FRAZAR, MARSTON ABBOTT.
    1887. An ornithologist's summer in Labrador. Ornithologist and Oologist,
        vol. 12, pp. 33-35.
FRIEDMANN, HERBERT.
    1925. Notes on the birds observed in the lower Rio Grande Valley of Texas
        during May, 1924. Auk, vol. 42, pp. 537-554.
    1929. The cowbirds: A study in the biology of social parasitism.
    1938. Additional hosts of the parasitic cowbirds. Auk, vol. 55, pp. 41-50.
GARDNER, LEON LLOYD.
    1926. Experiments in the economic control of the western crow (Corvus
        brachyrhynchos hesperis). Auk, vol. 43, pp. 447-461.
GEIST, OTTO WILLIAM.
    1936. Notes on a fight between Alaska jays and a weasel. Condor, vol. 38,
        pp. 174-175.
GIBSON, LANGDON.
    1922. Bird notes from north Greenland. Auk, vol. 39, pp. 350-363.
GILL, GEOFFREY.
    1920. Blue jay vs. mouse. Bird-Lore, vol. 22, p. 161.
GILLESPIE, MABEL.
    1930. Behavior and local distribution of tufted titmice in winter and spring.
        Bird-Banding, vol. 1, pp. 113-127.

GILMAN, MARSHALL FRENCH.
 1902. Notes on the verdin. Condor, vol. 4, pp. 88-89.
 1907. Magpies on the La Plata. Condor, vol. 9, pp. 9-12.
 1908. Birds on the Navajo Reservation in New Mexico. Condor, vol. 10, pp. 146-152.
GOODHUE, ISABEL.
 1919. The song of the blue jay. Auk, vol. 36, pp. 111-112.
GOSS, NATHANIEL STICKNEY.
 1885. *Cyanocitta stelleri frontalis* nesting in holes in trees. Auk, vol. 2, p. 217.
 1891. History of the birds of Kansas.
GREEN, MORRIS MILLER.
 1889. Biological Survey field notes, Canaveral, Fla., April 1889.
GRINNELL, JOSEPH.
 1900a. Birds of the Kotzebue Sound region, Alaska. Pacific Coast Avifauna, No. 1.
 1900b. New races of birds from the Pacific coast. Condor, vol. 2, pp. 127-128.
 1901. The long-tailed jay. Auk, vol. 18, p. 188.
 1903. Call notes of the bush-tit. Condor, vol. 5, pp. 85-87.
 1904. The origin and distribution of the chestnut-backed chickadee. Auk, vol. 21, pp. 364-382.
 1908. The biota of the San Bernardino Mountains. Univ. California Publ. Zool., vol. 5, pp. 1-170.
 1914. An account of the mammals and birds of the lower Colorado Valley. Univ. California Publ. Zool., vol. 12, pp. 51-294.
 1918. The subspecies of the mountain chickadee. Univ. California Publ. Zool., vol. 17, pp. 505-515.
 1923. The present state of our knowledge of the gray titmouse in California. Condor, vol. 25, pp. 135-137.
 1928. Notes on the systematics of west American birds, 2. Condor, vol. 30, pp. 153-156.
 1931. The type locality of the verdin. Condor, vol. 33, pp. 163-168.
 1934. The New Mexico race of plain titmouse. Condor, vol. 36, pp. 251-252.
GRINNELL, JOSEPH, and BEHLE, WILLIAM HARROUN.
 1937. A new race of titmouse, from the Kern Basin of California. Condor, vol. 39, pp. 225-226.
GRINNELL, JOSEPH; DIXON, JOSEPH; and LINSDALE, JEAN MYRON.
 1930. Vertebrate natural history of a section of northern California through the Lassen Peak region. Univ. California Publ. Zool., vol. 35.
 1936. Vertebrate animals of Point Lobos Reserve, 1934-35.
GRINNELL, JOSEPH, and STORER, TRACY IRWIN.
 1924. Animal life in the Yosemite.
GRINNELL, JOSEPH, and SWARTH, HARRY SCHELWALD.
 1926. New species of birds (*Penthestes, Baeolophus, Psaltriparus, Chamaea*) from the Pacific coast of North America. Univ. California Publ. Zool., vol. 30, pp. 163-175.
GUTHRIE, JOSEPH EDWARD.
 1932. Snakes versus birds; birds versus snakes. Wilson Bull., vol. 44, pp. 88-113.
HAGERUP, ANDREAS THOMSEN.
 1891. The birds of Greenland.

HARDING, KATHERINE CLARK.
  1932. Age record of black-capped chickadee 93789. Bird-Banding, vol. 3, p. 118.
HARLOW, RICHARD CRESSON.
  1922. The breeding habits of the northern raven in Pennsylvania. Auk, vol. 39, pp. 399-410.
HENSHAW, HENRY WETHERBEE.
  1875. Report upon ornithological collections made in portions of Nevada, Utah, California, Colorado, New Mexico and Arizona during the years 1872, 1873, and 1874. (Wheeler Survey.)
HICKS, LAWRENCE EMERSON, and DAMBACH, CHARLES A.
  1935. Sex ratios and weights in wintering crows. Bird-Banding, vol. 6, pp. 65-66.
  1936. A statistical survey of the winter bird life of southeastern Ohio— Muskingum County. Wilson Bull., vol. 48, pp. 273-275.
HOFFMANN, RALPH.
  1904. A guide to the birds of New England and eastern New York.
  1927. Birds of the Pacific States.
HOOD, FREDA L.
  1916. Nesting habits of the tufted titmouse. Bird-Lore, vol. 18, p. 178.
HORNING, JAMES E.
  1923. Crows building in low willows—Alberta. Auk, vol. 40, pp. 327-328.
HOWELL, ALFRED BRAZIER.
  1917. Birds of the Islands off the coast of southern California. Pacific Coast Avifauna, No. 12.
HOWELL, ARTHUR HOLMES.
  1913. Descriptions of two new birds from Alabama. Proc. Biol. Soc. Washington, vol. 26, pp. 199-202.
  1932. Florida bird life.
HUEY, LAURENCE MARKHAM.
  1942. Two new wrens and a new jay from Lower California, Mexico. Trans. San Diego Soc. Nat. Hist., vol. 9, pp. 427-434.
IMLER, RALPH H.
  1939. Comparison of the food of the white-necked ravens and crows in Oklahoma. Wilson Bull., vol. 51, pp. 121-122.
ISELY, DWIGHT.
  1912. A list of the birds of Sedgwick County, Kansas. Auk, vol. 29, pp. 25-44.
JACOBS, JOSEPH WARREN.
  1935. Red type of crow eggs. Auk, vol. 52, pp. 189-190.
JELLISON, WILLIAM LIVINGSTON, and PHILIP, CORNELIUS BECKER.
  1933. Faunae of nests of the magpie and crow in western Montana. Can. Ent., vol. 65, pp. 26-31.
JENSEN, JENS KNUDSEN.
  1923. Notes on the nesting birds of northern Santa Fe County, New Mexico. Auk, vol. 40, pp. 452-469.
  1925. Mountain chickadee with an adopted family. Auk, vol. 42, p. 593.
JONES, LYNDS.
  1910. The birds of Cedar Point and vicinity. Wilson Bull., vol. 22, pp. 172-182.
JOURDAIN, FRANCIS CHARLES ROBERT.
  1929. Protective mimicry of the chickadee. Auk, vol. 46, p. 123.

1938. The handbook of British birds, vol. 1. (By H. F. Witherby, F. C. R. Jourdain, Norman F. Ticehurst, and Bernard W. Tucker.)

JUDD, SYLVESTER DWIGHT.
1902. Birds of a Maryland farm. Biol. Surv. Bull. 17.

JUNG, CLARENCE SCHRAM.
1930. Notes on birds of the delta region of the Peace and Athabasca Rivers. Auk, vol. 47, pp. 533-541.

KAEDING, HENRY BARROILHET.
1897. Notes on the yellow-billed magpie. Oologist, vol. 14, pp. 15-17.

KALMBACH, EDWIN RICHARD.
1920. The crow in its relation to agriculture. Farmers' Bull. 1102.
1927. The magpie in relation to agriculture. U. S. Dept. Agr. Techn. Bull. 24.
1939. The crow in its relation to agriculture. Farmers' Bull. 1102, rev. ed.

KALMBACH, E. R., and ALDOUS, SHALER EUGENE.
1940. Winter banding of Oklahoma crows. Wilson Bull., vol. 52, pp. 198-206.

KALTER, LOUIS BIESER.
1932. Birds attracted to small-flowered leaf cup. Auk, vol. 49, p. 365.

KELSO, JOHN EDWARD HENRY.
1926. Birds of the Arrow Lakes, West Kootenay District, British Columbia. Ibis, ser. 12, vol. 2, pp. 689-723.

KENNARD, FREDERIC HEDGE.
1898. Habits of the blue jay. Auk, vol. 15, p. 269.

KITE, VITAE.
1925. A tufted titmouse goes wool-gathering. Bird-Lore, vol. 27, pp. 180-181.

KNEELAND, SAMUEL.
1883. Prehensile feet of the crow. Science, vol. 2, pp. 265-266.

KNIGHT, ORA WILLIS.
1908. The birds of Maine.

KNIGHT, WILBUR CLINTON.
1902. The birds of Wyoming.

KRAMER, GUSTAV.
1930. Stimme von Raben- und Nebelkrähe. Orn. Monatsb., Jahrg. 38, p. 146.

KUMLIEN, LUDWIG.
1879. Contributions to the natural history of Arctic America. U. S. Nat. Mus. Bull. 15.

KUMLIEN, LUDWIG, and HOLLISTER, NED.
1903. The birds of Wisconsin. Bull. Wisconsin Nat. Hist. Soc., new ser., vol. 3, Nos. 1-3.

LACEY, HOWARD.
1903. Notes on the Texan jay. Condor, vol. 5, pp. 151-153.

LANGILLE, JAMES HIBBERT.
1884. Our birds in their haunts.

LEMMON, ISABELLA McC.
1904. A blue jay household. Bird-Lore, vol. 6, pp. 89, 90.

LEWIS, HARRISON FLINT.
1931. Unsuspecting chickadees. Can. Field Nat., vol. 45, pp. 30-31.

LILFORD, Lord.
1860. See Powys, Hon. Thomas Lyttleton.

LINCOLN, FREDERICK CHARLES.
  1927. Returns from banded birds, 1923 to 1926. U. S. Dept. Agr. Techn. Bull. 26.
  1939. The migration of American birds.
LINSDALE, JEAN MYRON.
  1937. The natural history of magpies. Pacific Coast Avifauna, No. 25.
  1938a. Environmental responses of the vertebrates in the Great Basin. Amer. Midl. Nat., vol. 19, pp. 1-206.
  1938b. Geographic variation in some birds in Nevada. Condor, vol. 40, pp. 36-38.
LORENZ, KONRAD Z.
  1937. The companion in the bird's world. Auk, vol. 54, pp. 245-273.
LOW, SETH.
  1934. Notes on Cape Cod crow movements. Bird-Banding, vol. 5, pp. 192-193.
MACFARLANE, RODERICK ROSS.
  1891. Notes on and list of birds and eggs collected in Arctic America, 1861-1866. Proc. U. S. Nat. Mus., vol. 14, pp. 413-446.
MACMILLAN, DONALD BAXTER.
  1918. Four years in the White North.
MACOUN, JOHN, and MACOUN, JAMES MELVILLE.
  1909. Catalogue of Canadian birds, ed. 2.
MACPHERSON, ARTHUR HOLTE.
  1929. A list of the birds of inner London. British Birds, vol. 22, pp. 222-244
MAILLIARD, JOHN WARD.
  1912. Concerning nesting sites of the California jay. Condor, vol. 14, p. 42.
MAILLIARD, JOSEPH.
  1900. California jay again. Condor, vol. 2, pp. 58-59.
  1904. California jays and cats. Condor, vol. 6, pp. 94-95.
  1908. Cooper hawks attacking crows. Condor, vol. 10, p. 129.
  1920. Notes on nutcrackers in Monterey County, California. Condor, vol. 22, pp. 160-161.
  1927. The birds and mammals of Modoc County, California. Proc. California Acad. Sci., ser. 4, vol. 16, pp. 261-359.
  1931. Unique nesting site of Santa Cruz chickadees. Condor, vol. 33, pp. 220-221.
MAYAUD, NOEL.
  1933. Notes et remarques sur quelques corvides. La pie *Pica pica* (L.). Alauda, ser. 3, vol. 5, pp. 362-382.
MAYNARD, CHARLES JOHNSON.
  1896. The birds of eastern North America.
MCATEE, WALDO LEE.
  1914. Birds transporting food supplies. Auk, vol. 31, pp. 404-405.
  1926. The relation of birds to woodlots in New York State. Roosevelt Wildlife Bull., vol. 4, pp. 7-152.
MCLAUGHLIN, RICHARD BURTON.
  1888. Nesting of the tufted titmouse. Ornithologist and Oologist, vol. 13, pp. 61-63.
MCMANUS, REID.
  1935. Feedings habits of the raven in winter. Auk, vol. 52, p. 89.
MEISE, WILHELM.
  1928. Der Verbreitung der Aaskrähe (Formenkreis *Corvus corvus* L.). Journ. für Orn., 1928, pp. 1-203.

MERRIAM, CLINTON HART.
  1885. A remarkable migration of Canada jays. Auk, vol. 2, p. 107.
MERRILL, JAMES CUSHING.
  1876. Notes on Texan birds. Bull. Nuttall Orn. Club, vol. 1, pp. 88-89.
  1878. Notes on the ornithology of southern Texas, being a list of birds observed in the vicinity of Fort Brown, Texas, from February, 1876, to June, 1878. Proc. U. S. Nat. Mus., vol. 1, pp. 118-173.
MILLER, ALDEN HOLMES.
  1933. The Canada jays of northern Idaho. Trans. San Diego Soc. Nat. Hist., vol. 7, pp. 289-297.
  1943a. A new race of Canada jay from coastal British Columbia. Condor, vol. 45, pp. 117-118.
  1943b. A new race of brown-headed chickadee from northern Washington. Occ. Pap. Mus. Zool. Louisiana State Univ., no. 14, pp. 261-263.
MILLER, ROBERT CUNNINGHAM.
  1921. The flock behavior of the coast bush-tit. Condor, vol. 23, pp. 121-127.
MINOT, HENRY DAVIS.
  1895. The land and game-birds of New England, ed. 2.
MITCHELL, CHARLES A., and DUTHIE, R. C.
  1929. Tuberculosis in crows. Amer. Rev. Tuberculosis, vol. 19, No. 2.
MITCHELL, HORACE HEDLEY.
  1915. Crows nesting on the ground. Auk, vol. 32, p. 229.
MOORE, WILLIAM HENRY.
  1904. The Canada jay. Ottawa Nat., vol. 18, pp. 142-144.
MORTIMER, D. [=B. (BENJAMIN)].
  1890. Notes on habits of a few birds of Orange County, Florida. Auk, vol. 7, pp. 337-343.
MOUSLEY, HENRY.
  1916. Five years personal notes and observations on the birds of Hatley, Stanstead County, Quebec—1911-1915. Auk, vol. 33, pp. 57-73.
  1924. Further notes and observations on the birds of Hatley, Stanstead County, Quebec. Auk, vol. 41, pp. 572-589.
MUNRO, JAMES ALEXANDER.
  1919. Notes on some birds of the Okanagan Valley, British Columbia. Auk, vol. 36, pp. 64-74.
  1929. Notes on the food habits of certain Raptores in British Columbia and Alberta. Condor, vol. 31, pp. 112-116.
  1935. Black-headed jay mimicking loon. Condor, vol. 37, p. 170.
MURIE, OLAUS JOHAN.
  1928. Notes on the Alaska chickadee. Auk, vol. 45, pp. 441-444.
NEHRLING, HENRY.
  1893. Our native birds of song and beauty, vol. 1.
NELSON, ARNOLD LARS.
  1934. Some early summer food preferences of the American raven in southeastern Oregon. Condor, vol. 36, pp. 10-15.
NELSON, EDWARD WILLIAM.
  1887. Report upon natural history collections made in Alaska between the years 1877 and 1881. U. S. Signal Service Arctic Ser., No. 3.
NETHERSOLE-THOMPSON, CAROLINE and DESMOND.
  1940. Display of the hooded crow. British Birds, vol. 34, p. 135.

NETHERSOLE-THOMPSON, DESMOND, and MUSSELWHITE, DONALD WOODWARD.
1940. Male rook incubating. British Birds, vol. 34, p. 44.

NICE, MARGARET MORSE.
1931. The birds of Oklahoma, rev. ed.
1933. Robins and Carolina chickadees remating. Bird-Banding, vol. 4, p. 157.

NICHOLSON, DONALD JOHN.
1936. Observations on the Florida blue jay. Wilson Bull., vol. 48, pp. 26-33.

NICHOLSON, EDWARD MAX, and KOCH, LUDWIG.
1936. Songs of wild birds.

NIETHAMMER, GÜNTHER.
1937. Handbuch der deutschen Vogelkunde, vol. 1.

NUTTALL, THOMAS.
1832. A manual of ornithology of the United States and of Canada. Land birds.

OBERHOLSER, HARRY CHURCH.
1896. A preliminary list of the birds of Wayne County, Ohio. Bull. Ohio Agr. Exp. Stat., vol. 1, No. 4, pp. 243-353.
1897. Critical notes on the genus *Auriparus*. Auk, vol. 14, pp. 390-394.
1914. Four new birds from Newfoundland. Proc. Biol. Soc. Washington, vol. 27, pp. 43-54.
1917. The status of *Aphelocoma cyanotis* and its allies. Condor, vol. 19, pp. 94-95.
1918. The common ravens of North America. Ohio Journ. Sci., vol. 18, pp. 213-225.
1920. Washington region. Bird-Lore, vol. 22, p. 106.
1921. The geographic races of *Cyanocitta cristata*. Auk, vol. 38, pp. 83-89.
1932. Descriptions of new birds from Oregon, chiefly from the Warner Valley region. Sci. Publ. Cleveland Mus. Nat. Hist., vol. 4, no. 1, 12 pp.
1937. Description of a new chickadee from the Eastern United States. Proc. Biol. Soc. Washington, vol. 50, pp. 219-220.

OLDHAM, CHARLES.
1930. The shell-smashing habit of gulls. Ibis, ser. 12, vol. 6, pp. 239-243.

OSGOOD, WILFRED HUDSON.
1901. Natural history of the Queen Charlotte Islands, British Columbia. North Amer. Fauna, No. 21, pp. 7-38.

PACKARD, FRED MALLERY.
1936. A black-capped chickadee victimized by the eastern cowbird. Bird-Banding, vol. 7, pp. 129-130.

PALMER, WILLIAM.
1885. Abundance of *Parus atricapillus* near Washington. Auk, vol. 2, p. 304.
1890. Notes on the birds observed during the cruise of the United States Fish Commission schooner *Grampus* in the summer of 1887. Proc. U. S. Nat. Mus., vol. 13, pp. 249-265.

PATCH, CLYDE LOUIS.
1922. A biological reconnaissance on Graham Island of the Queen Charlotte group. Can. Field Nat., vol. 36, pp. 133-136.

PEARSE, THEED.
1938. A remarkable influx of ravens into the Comox district, Vancouver Island, B. C. Murrelet, vol. 19, pp. 11-13.

Peters, Harold Seymore.
1936. A list of external parasites from birds of the eastern part of the United States. Bird-Banding, vol. 7, pp. 9-27.

Peters, James Lee.
1920. A new jay from Alberta. Proc. New England Zool. Club, vol. 7, pp. 51-52.
1927. Descriptions of new birds. Proc. New England Zool. Club, vol. 9, pp. 111-113.

Pettingill, Eleanor Rice.
1937. Grand Manan's Acadian chickadees. Bird-Lore, vol. 39, pp. 277-282.

Philipp, Philip Bernard, and Bowdish, Beecher Scoville.
1919. Further notes on New Brunswick birds. Auk, vol. 36, pp. 36-45.

Philipson, W. Raymond.
1933. The rook roosts of South Northumberland and the boundaries between their feeding territories. British Birds, vol. 27, pp. 66-71.

Pickens, Andrew Lee.
1928. Auditory protective mimicry of the chickadee. Auk, vol. 45, pp. 302-304.

Pierce, Fred John.
1922. A crow that nearly "looped the loop." Wilson Bull., vol. 34, p. 115.

Piers, Harry.
1898. Remarkable ornithological occurrence in Nova Scotia. Auk, vol. 15, pp. 195-196.

Potter, Laurence Bedford.
1927. Freak nesting site of a magpie. Condor, vol. 29, p. 249.
1932. Unusual nesting sites. Can. Field Nat., vol. 46, p. 49.

Powys, Thomas Lyttleton (afterward Lord Lilford).
1860. Notes on birds observed in the Ionian Islands, and the Provinces of Albania proper, Epirus, Acarnania, and Montenegro. Ibis, vol. 2, pp. 1-10, 133-140, 228-239.

Price, John Basyl.
1936. The family relations of the plain titmouse. Condor, vol. 38, pp. 23-28.

Pycraft, William Plane.
1918. Some neglected aspects in the study of young birds. Trans. Norfolk and Norwich Nat. Soc., vol. 10, pp. 408-416.

Rathbun, Samuel Frederick.
1911. Notes on birds of Seattle, Washington. Auk, vol. 28, pp. 492-494.
1916. The Lake Crescent region, Olympic Mountains, Washington, with notes regarding its avifauna. Auk, vol. 33, pp. 357-370.

Ray, Milton Smith.
1916. More summer birds of San Francisco County. Condor, vol. 18, pp. 222-227.

Reed, James Harris.
1897. A novel idea of a tufted titmouse. Auk, vol. 14, p. 325.

Reed, Ward.
1927. A note on a habit of the tufted titmouse. Wilson Bull., vol. 39, p. 107.

Rhoads, Samuel Nicholson.
1894. A reprint of the North American Zoology, by George Ord, pp. 290-361.

RIDGWAY, ROBERT.
1877. United States geological exploration of the fortieth parallel, pt. 3: Ornithology.
1882. Descriptions of several new races of American birds. Proc. U. S. Nat. Mus., vol. 5, pp. 9-15.
1883. Descriptions of some new birds from Lower California, collected by Mr. L. Belding. Proc. U. S. Nat. Mus., vol. 6, pp. 154-156.
1904. The birds of North and Middle America. U. S. Nat. Mus. Bull. 50, pt. 3.
1912. Color standards and color nomenclature.
ROBERTS, THOMAS SADLER.
1936. The birds of Minnesota, ed. 2, vol. 2.
ROCKWELL, ROBERT BLANCHARD.
1907. The Woodhouse jay in western Colorado. Condor, vol. 9, pp. 81-84.
1909. The use of magpies' nests by other birds. Condor, vol. 11, pp. 90-92.
1910. An albino magpie. Condor, vol. 12, p. 45.
ROCKWELL, R. B., and WETMORE, ALEXANDER.
1914. A list of birds from the vicinity of Golden, Colorado. Auk, vol. 31, pp. 309-333.
ROWLEY, JOHN STUART.
1935. Notes on some birds of Lower California, Mexico. Condor, vol. 37, pp. 163-168.
1939. Breeding birds of Mono County, California. Condor, vol. 41, pp. 247-254.
RUBEY, WILLIAM WALDEN.
1933. Flight maneuvers of the raven. Bird-Lore, vol. 35, pp. 143-145.
RUSSELL, RICHARD JOEL.
1931. Dry climates of the United States. Univ. California Publ. Geol., vol. 4, pp. 1-41.
SAGE, JOHN HALL; BISHOP, LOUIS BENNETT; and BLISS, WALTER PARKS.
1913. The birds of Connecticut.
SAUNDERS, ARETAS ANDREWS.
1914. The birds of Teton and northern Lewis and Clark Counties, Montana. Condor, vol. 16, pp. 124-144.
1921. A distributional list of the birds of Montana. Pacific Coast Avifauna, No. 14.
SCHALOW, HERMAN.
1904. Die Vogel der Arktis. Fauna Arctica (Römer and Schaudinn), pp. 82-288.
SCOTT, WILLIAM EARL DODGE.
1886. On the breeding habits of some Arizona birds. Auk, vol. 3, pp. 81-86.
SCOTT, WILLIAM LOUIS.
1884. The winter Passeres and Picariae of Ottawa. Auk, vol. 1, pp. 156-161.
SELOUS, EDMUND.
1901. Bird-watching.
1927. Realities of bird life.
SENNETT, GEORGE BURRITT.
1878. Notes on the ornithology of the lower Rio Grande of Texas, from observations made during the season of 1877. Bull. U. S. Geol. and Geogr. Surv. Terr., vol. 4, pp. 1-66.

1879. Further notes on the ornithology of the lower Rio Grande of Texas, from observations made during the spring of 1878. Bull. U. S. Geol. and Geogr. Surv. Terr., vol. 5, pp. 371-440.

1888. Descriptions of a new species and two new subspecies of birds from Texas. Auk. vol. 5, pp. 43-46.

SETON, ERNEST THOMPSON.

1890. The birds of Manitoba. Proc. U. S. Nat. Mus., vol. 13, pp. 457-643.

SHELLEY, LEWIS ORMAN.

1926. Chickadees eating bees. Bird-Lore, vol. 28, pp. 337-338.

SILLOWAY, PERLEY MILTON.

1903. Birds of Fergus County, Montana. Fergus County Free High School Bull. 1.

SIMMONS, GEORGE FINLAY.

1925. Birds of the Austin region.

SIMPSON, ALAN H.

1926. Hooded crows killing a lamb. British Birds, vol. 19, p. 283.

SIMPSON, GEORGE BUCHANAN.

1925. Oregon jays. Can. Field Nat., vol. 39, pp. 29-30.

SKINNER, MILTON PHILO.

1916. The nutcrackers of Yellowstone Park. Condor, vol. 18, pp. 62-64.

1921. Notes on the Rocky Mountain jay in the Yellowstone National Park. Condor, vol. 23, pp. 147-151.

1928. A guide to the winter birds of the North Carolina sandhills.

SMITH, AUSTIN PAUL.

1910. Miscellaneous notes from the lower Rio Grande. Condor, vol. 12, pp. 93-103.

1916. Additions to the avifauna of Kerr Co., Texas. Auk, vol. 33, pp. 187-193.

SMITH, E. IRWIN.

1924. As a bird saw it. Bird-Lore, vol. 26, p. 177.

SMITH, WILBUR FRANKLIN.

1905. Blue jays at home. Bird-Lore, vol. 7, pp. 268-271.

SOPER, JOSEPH DEWEY.

1928. A faunal investigation of southern Baffin Island. Nat. Mus. Canada Bull. 53.

STOCKARD, CHARLES RUPERT.

1905. Nesting habits of birds in Mississippi. Auk, vol. 22, pp. 146-158.

STONE, WITMER.

1899. A search for the Reedy Island crow roost. Bird-Lore, vol. 1, pp. 177-180.

1926. Changed habits of blue jay at Philadelphia. Auk, vol. 43, p. 239.

1937. Bird studies at Old Cape May.

STREET, JOHN FLETCHER.

1918. Hudsonian chickadee on the Pocono Mountain, Pa. Auk, vol. 35, pp. 230-231.

SUTTON, GEORGE MIKSCH.

1929. Can the Cooper's hawk kill a crow? Auk, vol. 46, p. 235.

1932. The birds of Southampton Island. Exploration of Southampton Island, Hudson Bay, pt. 2: Zoology, sect. 2.

1934. Notes on the birds of the western panhandle of Oklahoma. Ann. Carnegie Mus., vol. 24, pp. 1-50.

1935. A new blue jay from the western border of the Great Basin. Auk, vol. 52, pp. 176-177.

SVIHLA, ARTHUR.

1933. An abnormally colored black-billed magpie. Murrelet, vol. 14, pp. 44-45.

SWAINSON, WILLIAM, and RICHARDSON, JOHN.

1831. Fauna Boreali-Americana, vol. 2: Birds.

SWANN, HARRY KIRKE.

1913. A dictionary of English and folk-names of British birds.

SWARTH, HARRY SCHELWALD.

1904. Birds of the Huachuca Mountains, Arizona. Pacific Coast Avifauna, No. 4.

1911. Birds and mammals of the 1909 Alexander Alaska Expedition. Univ. California Publ. Zool., vol. 7, pp. 9-172.

1912. Report on a collection of birds and mammals from Vancouver Island. Univ. California Publ. Zool., vol. 10, pp. 1-124.

1913. The status of Lloyd's bush-tit as a bird in Arizona. Auk, vol. 30, pp. 399-401.

1914. The California forms of the genus *Psaltriparus*. Auk, vol. 31, pp. 499-526.

1918. The Pacific coast jays of the genus *Aphelocoma*. Univ. California Publ. Zool., vol. 17, pp. 405-422.

1922. Birds and mammals of the Stikine River region of northern British Columbia and southeastern Alaska. Univ. California Publ. Zool., vol. 24, pp. 125-314.

1926. Report on a collection of birds and mammals from the Atlin region, northern British Columbia. Univ. California Publ. Zool., vol. 30, pp. 51-162.

TAVERNER, PERCY ALGERNON.

1919. The birds of the Red Deer River, Alberta. Auk, vol. 36, pp. 248-265.

1926. Birds of western Canada. Victoria Mem. Mus. Bull. 41.

1940. Canadian status of the long-tailed chickadee. Auk, vol. 57, pp. 536-541.

TAVERNER, P. A., and SWALES, BRADSHAW HALL.

1907. The birds of Point Pelee. Wilson Bull., vol. 19, pp. 133-153.

1908. The birds of Point Pelee. Wilson Bull., vol. 20, pp. 107-129.

TAYLOR, WALTER PENN.

1912. Field notes on amphibians, reptiles and birds of northern Humboldt County, Nevada, with a discussion of some of the faunal features of the region. Univ. California Publ. Zool., vol. 7, pp. 319-436.

TAYLOR, W. P., and SHAW, WILLIAM THOMAS.

1927. Mammals and birds of Mount Rainier National Park.

TERRES, JOHN KENNETH.

1940. Birds eating tent caterpillars. Auk, vol. 57, p. 422.

THAYER, HARRIET CARPENTER.

1901. Our blue jay neighbors. Bird-Lore, vol. 3, pp. 50-53.

THOMS, CRAIG SHARPE.

1927. Winning the chickadee. Bird-Lore, vol. 29, pp. 191-192.

TODD, WALTER EDMOND CLYDE.
  1928. A new blue jay from southern Florida. Auk, vol. 45, pp. 364-365.
  1940. Birds of western Pennsylvania.
TODD, W. E. C., and SUTTON, G. M.
  1936. Taxonomic remarks on the Carolina chickadee, *Penthestes carolinensis.*
      Proc. Biol. Soc. Washington, vol. 49, pp. 69-70.
TORREY, BRADFORD.
  1885. Birds in the bush.
  1889. A rambler's lease.
  1904. Nature's invitation.
TOWNSEND, CHARLES WENDELL.
  1905. The birds of Essex County, Massachusetts. Mem. Nuttall Orn. Club,
      No. 3.
  1918. A winter crow roost. Auk, vol. 35, pp. 405-416.
  1923. The voice and courtship of the crow. Bull. Essex County Orn. Club,
      vol. 5, pp. 4-8.
  1927. Notes on the courtship of the lesser scaup, everglade kite, crow, and
      boat-tailed and great-tailed grackles. Auk, vol. 44, pp. 549-554.
TOWNSEND, C. W., and ALLEN, GLOVER MORRILL.
  1907. Birds of Labrador. Proc. Boston Soc. Nat. Hist., vol. 33, pp. 277-428.
TUCKER, BERNARD WILLIAM.
  1938. The handbook of British birds, vol. 1. (By H. F. Witherby, F. C. R.
      Jourdain, Norman F. Ticehurst, and B. W. Tucker.)
TYLER, JOHN GRIPPER.
  1913. Some birds of the Fresno district, California. Pacific Coast Avifauna,
      No. 9.
TYLER, WINSOR MARRETT.
  1912. The shrike in action. Bird-Lore, vol. 14, pp. 351-352.
  1920. An odd note of the blue jay. Bird-Lore, vol. 22, pp. 160-161.
TYRRELL, WILLIAM BRYANT.
  1934. Bird notes from Whitefish Point, Michigan. Auk, vol. 51, pp. 21-26.
UPTON, LUCY H.
  1916. Interesting performance of a tufted titmouse. Bird-Lore, vol. 18, pp.
      125-126.
USSHER, RICHARD JOHN, and WARREN, ROBERT.
  1900. The birds of Ireland.
VAN DYKE, TERTIUS.
  1913. A narrow escape. Bird-Lore, vol. 15, pp. 112-113.
VAN ROSSEM, ADRIAAN JOSEPH.
  1927. Eye shine in birds, with notes on the feeding habits of some goatsuckers.
      Condor, vol. 29, pp. 25-28.
  1928. A northern race of the mountain chickadee. Auk, vol. 45, pp. 104-105.
  1931. Descriptions of new birds from the mountains of southern Nevada.
      Trans. San Diego Soc. Nat. Hist., vol. 6, pp. 325-332.
  1936. A note on the nesting of the bush-tit. Condor, vol. 38, p. 170.
VAN TYNE, JOSSELYN.
  1928. A diurnal local migration of the black-capped chickadee. Wilson Bull.,
      vol. 40, p. 252.
  1929. Notes on some birds of the Chisos Mountains of Texas. Auk, vol. 46,
      pp. 204-206.

VAN TYNE, JOSSELYN, and SUTTON, GEORGE MIKSCH.
    1937. The birds of Brewster County, Texas. Mus. Zool. Univ. Michigan Misc.
        Publ. 37.
WALLACE, GEORGE JOHN.
    1941. Winter studies of color-banded chickadees. Bird-Banding, vol. 12, pp.
        49-67.
WARNE, FRANK L.
    1926. "Crows is crows." Bird-Lore, vol. 28, pp. 110-116.
WARREN, EDWARD ROYAL.
    1912. The magpies of Culebra Creek. Bird-Lore, vol. 14, pp. 329-333.
    1916. Notes on the birds of the Elk Mountain region, Gunnison County,
        Colorado. Auk, vol. 33, pp. 292-317.
WARREN, OSCAR BIRD.
    1899. A chapter in the life of the Canada jay. Auk, vol. 16, pp. 12-19.
WAYNE, ARTHUR TREZEVANT.
    1910. Birds of South Carolina. Contr. Charleston Mus., No. 1.
WEED, CLARENCE MOORES.
    1898. The winter food of chickadees. New Hampshire Coll. Agr. Exp. Stat.
        Bull. 54.
WHEATON, JOHN MAYNARD.
    1882. Report on the birds of Ohio. Geol. Surv. Ohio, vol. 4: Zoology and
        Botany, pp. 187-628.
WHEELER, HARRY EDGAR.
    1922. The fish crow in Arkansas. Wilson Bull., vol. 34, pp. 239-240.
WHEELOCK, IRENE GROSVENOR.
    1904. Birds of California.
WHITE, STEWART EDWARD.
    1893. Birds observed on Mackinac Island, Michigan, during the summers of
        1889, 1890, and 1891. Auk, vol. 10, pp. 221-230.
WIDMANN, OTTO.
    1880. Notes on birds of St. Louis, Mo. Bull. Nuttall Orn. Club, vol. 5, pp.
        191-192.
    1907. A preliminary catalog of the birds of Missouri.
WILLARD, FRANCIS COTTLE.
    1912. Migration of the white-necked raven. Condor, vol. 14, pp. 107-108.
WILLETT, GEORGE.
    1941. Variation in North American ravens. Auk, vol. 58, pp. 246-249.
WILSON, ALEXANDER, and BONAPARTE, CHARLES LUCIEN.
    1832. American ornithology, vols. 1, 3.
WILSON, RALPH R.
    1923. An unusual experience. Bird-Lore, vol. 25, p. 124.
WITHERBY, HARRY FORBES.
    1938. The handbook of British birds, vol. 1. (By H. F. Witherby, F. C. R.
        Jourdain, Norman F. Ticehurst, and Bernard W. Tucker.)
WOLFE, LLOYD RAYMOND.
    1931. Ground nesting of crows in Wyoming. Condor, vol. 33, pp. 124-125.
WOODCOCK, JOHN.
    1913. A friendly chickadee. Bird-Lore, vol. 15, pp. 373-375.

WRIGHT, HORACE WINSLOW.
  1909. Birds of the Boston Public Garden.
  1912. Morning awakening and even-song. Auk, vol. 29, pp. 307-327.
  1914. Acadian chickadees *(Penthestes hudsonicus littoralis)* in Boston and
        vicinity in the fall of 1913. Auk, vol. 31, pp. 236-242.
  1917. Labrador chickadee *(Penthestes hudsonicus nigricans)* in Boston and
        vicinity in the fall of 1916. Auk, vol. 34, pp. 164-170.
  1918. Labrador chickadee *(Penthestes hudsonicus nigricans)* in its return
        flight from the fall migration of 1916. Auk, vol. 35, pp. 37-40.
YARRELL, WILLIAM.
  1882. A history of British birds, ed. 4, vol. 2. Revised and enlarged by
        Alfred Newton.
YEATES, GEORGE KIRKBY.
  1934. The life of the rook.
ZIMMER, JOHN TODD.
  1911. Some notes on the winter birds of Dawes County. Proc. Nebraska Orn.
        Union, vol. 5, pp. 19-30.
ZIMMERMANN, RUDOLF.
  1931. Einiges über das Brutgeschäft deutscher Rabenvögel. Orn. Monatsb.
        Jahrg. 39, pp. 99-102.

INDEX

PLATE 37

Cochise County, Ariz., May 29, 1922.   A. C. Bent.

Cochise County, Ariz., May 31, 1922.   A. C. Bent.

NESTING OF THE WHITE-NECKED RAVEN.

PLATE 38

Cochise County, Ariz.

F. C. Willard.

NESTING OF THE WHITE-NECKED RAVEN.

PLATE 39

Rehoboth, Mass., May 5, 1906.

A. C. Bent.

NESTING OF THE EASTERN CROW.

PLATE 40

Manitoba   C. L. Broley.

Chatham, Mass., May 28, 1904.   A. C. Bent

LOW NESTING OF THE EASTERN CROW.

PLATE 41

W. E. Shore.

YOUNG EASTERN CROWS

Near Toronto, Ontario, May 29, 1941.

PLATE 42

Less than a week old.   Cordelia J. Stanwood.

Erie County, N. Y.

YOUNG EASTERN CROWS.   S. A. Grimes.

PLATE 43

J. G. Suthard.

Orange County, Calif., April 14, 1940.

UNUSUALLY LARGE SET OF EGGS OF THE WESTERN CROW.

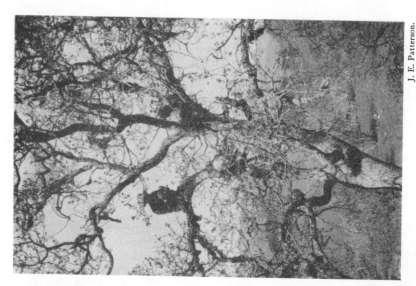

J. E. Patterson.

Near Ashland, Oreg., April 26, 1924.

NESTING OF THE WESTERN CROW.

PLATE 45

Southern California.   W. L. and Irene Finle

Packed in.

Salinas River, Calif., May 1934.   E. C. Aldrich.

Nest-leaving age.

YOUNG WESTERN CROWS.

PLATE 46

Duval County, Fla.   S. A. Grimes.

Nesting site.

Bulls Island, S. C.   A. D. Cruickshank.

Adult

FISH CROW.

PLATE 47

Altitude 12,000 feet, Mount Clark.

HABITAT OF CLARK'S NUTCRACKER.

PLATE 48

Adult on nest.

Mono County, Calif.

E. N. Harrison.

NEST OF CLARK'S NUTCRACKER.

PLATE 49

H. D. and Ruth Wheeler.

CLARK'S NUTCRACKER.

PLATE 50

Ithaca, N. Y., May 26, 1918.

A. A. Allen.

Nest in a bird box.

Harwich, Mass., July 13, 1923.

A. C. Bent.

Adult feeding young.

BLACK-CAPPED CHICKADEE.

PLATE 51

Near Toronto, Ontario.   H. M. Halliday.

BLACK-CAPPED CHICKADEES.

PLATE 52

Ashland, Oreg., June 5, 1913.   J. E. Patterson.

NEST OF OREGON CHICKADEE.

PLATE 53

W. L. Finley and H. T. Bohlman.

ADULT AND YOUNG OF OREGON CHICKADEE.

PLATE 54

Duval County, Fla., April 1933.   S. A. Grimes.

Nesting site.

Greene County, Pa., May 9, 1919.   S. S. Dickey.

CAROLINA CHICKADEE.

PLATE 55

Klamath County, Oreg., May 27, 1925.   J. E. Patterson.

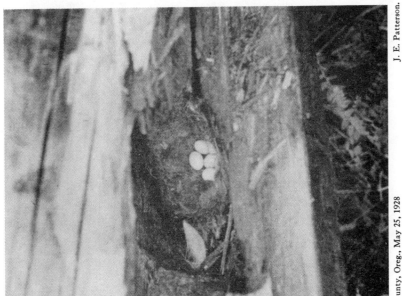

J. E. Patterson.

Jackson County, Oreg., May 25, 1928

NESTING OF SHORT-TAILED CHICKADEES.

PLATE 56

Near Gilroy, Calif., April 25, 1926.　　　　　　　　　J. E. Patterson

NEST OF BARLOW'S CHICKADEE.

PLATE 57

Yellowstone National Park July 1935.   A. A. Allen.

Adult at nest in a bird box.

California   H. D. and Ruth Wheeler.

MOUNTAIN CHICKADEES, TWO SUBSPECIES

PLATE 58

Maine, June 20, 1940.

ACADIAN CHICKADEE.

Eliot Porter.

PLATE 59

June 10, 1928.

Matane County, Quebec, June 1927.   W. J. Brown.

NESTING STUB AND YOUNG OF ACADIAN CHICKADEE.

PLATE 60

S. S. Dickey.

Greene County, Pa.

Waynesburg, Pa.

S. S. Dickey.

NESTING SITES OF TUFTED TITMOUSE.

PLATE 61

Nest in a syrup can in a pear tree.

Duval County, Fla., April 24, 1931.

S. A. Grimes.

NESTING OF THE TUFTED TITMOUSE.

PLATE 62

PLAIN TITMICE, ADULTS.

H. D. and Ruth Wheeler.

PLATE 63

W. L. and Irene Finley.

F C. Willard

NESTS OF THE EASTERN VERDIN.

PLATE 64

S. A. Grimes.

EASTERN VERDIN AND NEST.

Cameron County, Tex., June 1940.

PLATE 65

COAST BUSHTIT AND NEST

PLATE 66

COAST BUSHTIT AND NEST

PLATE 67

L. L. Haskin.

NEST OF COAST BUSHTIT IN BUSH OF HOLIDISCUS DISCOLOR.

PLATE 68

Portland, Oreg.   W. L. and Irene Finley.

Portland, Oreg.   W. L. Finley and H. T. Bohlman.

COAST BUSHTITS.

A CATALOG OF SELECTED
DOVER BOOKS
IN ALL FIELDS OF INTEREST

A CATALOG OF SELECTED DOVER
BOOKS IN ALL FIELDS OF INTEREST

DRAWINGS OF REMBRANDT, edited by Seymour Slive. Updated Lippmann, Hofstede de Groot edition, with definitive scholarly apparatus. All portraits, biblical sketches, landscapes, nudes. Oriental figures, classical studies, together with selection of work by followers. 550 illustrations. Total of 630pp. 9⅛ × 12¼.
21485-0, 21486-9 Pa., Two-vol. set $25.00

GHOST AND HORROR STORIES OF AMBROSE BIERCE, Ambrose Bierce. 24 tales vividly imagined, strangely prophetic, and decades ahead of their time in technical skill: "The Damned Thing," "An Inhabitant of Carcosa," "The Eyes of the Panther," "Moxon's Master," and 20 more. 199pp. 5⅜ × 8½. 20767-6 Pa. $3.95

ETHICAL WRITINGS OF MAIMONIDES, Maimonides. Most significant ethical works of great medieval sage, newly translated for utmost precision, readability. Laws Concerning Character Traits, Eight Chapters, more. 192pp. 5⅜ × 8½.
24522-5 Pa. $4.50

THE EXPLORATION OF THE COLORADO RIVER AND ITS CANYONS, J. W. Powell. Full text of Powell's 1,000-mile expedition down the fabled Colorado in 1869. Superb account of terrain, geology, vegetation, Indians, famine, mutiny, treacherous rapids, mighty canyons, during exploration of last unknown part of continental U.S. 400pp. 5⅜ × 8½. 20094-9 Pa. $6.95

HISTORY OF PHILOSOPHY, Julián Marías. Clearest one-volume history on the market. Every major philosopher and dozens of others, to Existentialism and later. 505pp. 5⅜ × 8½. 21739-6 Pa. $8.50

ALL ABOUT LIGHTNING, Martin A. Uman. Highly readable non-technical survey of nature and causes of lightning, thunderstorms, ball lightning, St. Elmo's Fire, much more. Illustrated. 192pp. 5⅜ × 8½. 25237-X Pa. $5.95

SAILING ALONE AROUND THE WORLD, Captain Joshua Slocum. First man to sail around the world, alone, in small boat. One of great feats of seamanship told in delightful manner. 67 illustrations. 294pp. 5⅜ × 8½. 20326-3 Pa. $4.50

LETTERS AND NOTES ON THE MANNERS, CUSTOMS AND CONDI-TIONS OF THE NORTH AMERICAN INDIANS, George Catlin. Classic account of life among Plains Indians: ceremonies, hunt, warfare, etc. 312 plates. 572pp. of text. 6⅛ × 9¼. 22118-0, 22119-9 Pa. Two-vol. set $15.90

ALASKA: The Harriman Expedition, 1899, John Burroughs, John Muir, et al. Informative, engrossing accounts of two-month, 9,000-mile expedition. Native peoples, wildlife, forests, geography, salmon industry, glaciers, more. Profusely illustrated. 240 black-and-white line drawings. 124 black-and-white photographs. 3 maps. Index. 576pp. 5⅜ × 8½. 25109-8 Pa. $11.95

THE BOOK OF BEASTS: Being a Translation from a Latin Bestiary of the Twelfth Century, T. H. White. Wonderful catalog real and fanciful beasts: manticore, griffin, phoenix, amphivius, jaculus, many more. White's witty erudite commentary on scientific, historical aspects. Fascinating glimpse of medieval mind. Illustrated. 296pp. 5⅜ × 8¼. (Available in U.S. only) 24609-4 Pa. $5.95

FRANK LLOYD WRIGHT: ARCHITECTURE AND NATURE With 160 Illustrations, Donald Hoffmann. Profusely illustrated study of influence of nature—especially prairie—on Wright's designs for Fallingwater, Robie House, Guggenheim Museum, other masterpieces. 96pp. 9¼ × 10¾. 25098-9 Pa. $7.95

FRANK LLOYD WRIGHT'S FALLINGWATER, Donald Hoffmann. Wright's famous waterfall house: planning and construction of organic idea. History of site, owners, Wright's personal involvement. Photographs of various stages of building. Preface by Edgar Kaufmann, Jr. 100 illustrations. 112pp. 9¼ × 10. 23671-4 Pa. $7.95

YEARS WITH FRANK LLOYD WRIGHT: Apprentice to Genius, Edgar Tafel. Insightful memoir by a former apprentice presents a revealing portrait of Wright the man, the inspired teacher, the greatest American architect. 372 black-and-white illustrations. Preface. Index. vi + 228pp. 8¼ × 11. 24801-1 Pa. $9.95

THE STORY OF KING ARTHUR AND HIS KNIGHTS, Howard Pyle. Enchanting version of King Arthur fable has delighted generations with imaginative narratives of exciting adventures and unforgettable illustrations by the author. 41 illustrations. xviii + 313pp. 6⅛ × 9¼. 21445-1 Pa. $5.95

THE GODS OF THE EGYPTIANS, E. A. Wallis Budge. Thorough coverage of numerous gods of ancient Egypt by foremost Egyptologist. Information on evolution of cults, rites and gods; the cult of Osiris; the Book of the Dead and its rites; the sacred animals and birds; Heaven and Hell; and more. 956pp. 6⅛ × 9¼. 22055-9, 22056-7 Pa., Two-vol. set $20.00

A THEOLOGICO-POLITICAL TREATISE, Benedict Spinoza. Also contains unfinished *Political Treatise*. Great classic on religious liberty, theory of government on common consent. R. Elwes translation. Total of 421pp. 5⅜ × 8½. 20249-6 Pa. $6.95

INCIDENTS OF TRAVEL IN CENTRAL AMERICA, CHIAPAS, AND YUCATAN, John L. Stephens. Almost single-handed discovery of Maya culture; exploration of ruined cities, monuments, temples; customs of Indians. 115 drawings. 892pp. 5⅜ × 8½. 22404-X, 22405-8 Pa., Two-vol. set $15.90

LOS CAPRICHOS, Francisco Goya. 80 plates of wild, grotesque monsters and caricatures. Prado manuscript included. 183pp. 6⅝ × 9⅝. 22384-1 Pa. $4.95

AUTOBIOGRAPHY: The Story of My Experiments with Truth, Mohandas K. Gandhi. Not hagiography, but Gandhi in his own words. Boyhood, legal studies, purification, the growth of the Satyagraha (nonviolent protest) movement. Critical, inspiring work of the man who freed India. 480pp. 5⅜ × 8½. (Available in U.S. only) 24593-4 Pa. $6.95

ILLUSTRATED DICTIONARY OF HISTORIC ARCHITECTURE, edited by Cyril M. Harris. Extraordinary compendium of clear, concise definitions for over 5,000 important architectural terms complemented by over 2,000 line drawings. Covers full spectrum of architecture from ancient ruins to 20th-century Modernism. Preface. 592pp. 7½ × 9⅝. 24444-X Pa. $14.95

THE NIGHT BEFORE CHRISTMAS, Clement Moore. Full text, and woodcuts from original 1848 book. Also critical, historical material. 19 illustrations. 40pp. 4⅝ × 6. 22797-9 Pa. $2.25

THE LESSON OF JAPANESE ARCHITECTURE: 165 Photographs, Jiro Harada. Memorable gallery of 165 photographs taken in the 1930's of exquisite Japanese homes of the well-to-do and historic buildings. 13 line diagrams. 192pp. 8⅜ × 11¼. 24778-3 Pa. $8.95

THE AUTOBIOGRAPHY OF CHARLES DARWIN AND SELECTED LETTERS, edited by Francis Darwin. The fascinating life of eccentric genius composed of an intimate memoir by Darwin (intended for his children); commentary by his son, Francis; hundreds of fragments from notebooks, journals, papers; and letters to and from Lyell, Hooker, Huxley, Wallace and Henslow. xi + 365pp. 5⅜ × 8. 20479-0 Pa. $5.95

WONDERS OF THE SKY: Observing Rainbows, Comets, Eclipses, the Stars and Other Phenomena, Fred Schaaf. Charming, easy-to-read poetic guide to all manner of celestial events visible to the naked eye. Mock suns, glories, Belt of Venus, more. Illustrated. 299pp. 5¼ × 8¼. 24402-4 Pa. $7.95

BURNHAM'S CELESTIAL HANDBOOK, Robert Burnham, Jr. Thorough guide to the stars beyond our solar system. Exhaustive treatment. Alphabetical by constellation: Andromeda to Cetus in Vol. 1; Chamaeleon to Orion in Vol. 2; and Pavo to Vulpecula in Vol. 3. Hundreds of illustrations. Index in Vol. 3. 2,000pp. 6⅛ × 9¼. 23567-X, 23568-8, 23673-0 Pa., Three-vol. set $36.85

STAR NAMES: Their Lore and Meaning, Richard Hinckley Allen. Fascinating history of names various cultures have given to constellations and literary and folkloristic uses that have been made of stars. Indexes to subjects. Arabic and Greek names. Biblical references. Bibliography. 563pp. 5⅜ × 8½. 21079-0 Pa. $7.95

THIRTY YEARS THAT SHOOK PHYSICS: The Story of Quantum Theory, George Gamow. Lucid, accessible introduction to influential theory of energy and matter. Careful explanations of Dirac's anti-particles, Bohr's model of the atom, much more. 12 plates. Numerous drawings. 240pp. 5⅜ × 8½. 24895-X Pa. $4.95

CHINESE DOMESTIC FURNITURE IN PHOTOGRAPHS AND MEASURED DRAWINGS, Gustav Ecke. A rare volume, now affordably priced for antique collectors, furniture buffs and art historians. Detailed review of styles ranging from early Shang to late Ming. Unabridged republication. 161 black-and-white drawings, photos. Total of 224pp. 8⅜ × 11¼. (Available in U.S. only) 25171-3 Pa. $12.95

VINCENT VAN GOGH: A Biography, Julius Meier-Graefe. Dynamic, penetrating study of artist's life, relationship with brother, Theo, painting techniques, travels, more. Readable, engrossing. 160pp. 5⅜ × 8½. (Available in U.S. only) 25253-1 Pa. $3.95

HOW TO WRITE, Gertrude Stein. Gertrude Stein claimed anyone could understand her unconventional writing—here are clues to help. Fascinating improvisations, language experiments, explanations illuminate Stein's craft and the art of writing. Total of 414pp. 4⅜ × 6⅜. 23144-5 Pa. $5.95

ADVENTURES AT SEA IN THE GREAT AGE OF SAIL: Five Firsthand Narratives, edited by Elliot Snow. Rare true accounts of exploration, whaling, shipwreck, fierce natives, trade, shipboard life, more. 33 illustrations. Introduction. 353pp. 5⅜ × 8½. 25177-2 Pa. $7.95

THE HERBAL OR GENERAL HISTORY OF PLANTS, John Gerard. Classic descriptions of about 2,850 plants—with over 2,700 illustrations—includes Latin and English names, physical descriptions, varieties, time and place of growth, more. 2,706 illustrations. xlv + 1,678pp. 8½ × 12¼. 23147-X Cloth. $75.00

DOROTHY AND THE WIZARD IN OZ, L. Frank Baum. Dorothy and the Wizard visit the center of the Earth, where people are vegetables, glass houses grow and Oz characters reappear. Classic sequel to *Wizard of Oz*. 256pp. 5⅜ × 8. 24714-7 Pa. $4.95

SONGS OF EXPERIENCE: Facsimile Reproduction with 26 Plates in Full Color, William Blake. This facsimile of Blake's original "Illuminated Book" reproduces 26 full-color plates from a rare 1826 edition. Includes "The Tyger," "London," "Holy Thursday," and other immortal poems. 26 color plates. Printed text of poems. 48pp. 5¼ × 7. 24636-1 Pa. $3.50

SONGS OF INNOCENCE, William Blake. The first and most popular of Blake's famous "Illuminated Books," in a facsimile edition reproducing all 31 brightly colored plates. Additional printed text of each poem. 64pp. 5¼ × 7. 22764-2 Pa. $3.50

PRECIOUS STONES, Max Bauer. Classic, thorough study of diamonds, rubies, emeralds, garnets, etc.: physical character, occurrence, properties, use, similar topics. 20 plates, 8 in color. 94 figures. 659pp. 6⅛ × 9¼. 21910-0, 21911-9 Pa., Two-vol. set $14.90

ENCYCLOPEDIA OF VICTORIAN NEEDLEWORK, S. F. A. Caulfeild and Blanche Saward. Full, precise descriptions of stitches, techniques for dozens of needlecrafts—most exhaustive reference of its kind. Over 800 figures. Total of 679pp. 8⅛ × 11. Two volumes. Vol. 1 22800-2 Pa. $10.95
Vol. 2 22801-0 Pa. $10.95

THE MARVELOUS LAND OF OZ, L. Frank Baum. Second Oz book, the Scarecrow and Tin Woodman are back with hero named Tip, Oz magic. 136 illustrations. 287pp. 5⅜ × 8½. 20692-0 Pa. $5.95

WILD FOWL DECOYS, Joel Barber. Basic book on the subject, by foremost authority and collector. Reveals history of decoy making and rigging, place in American culture, different kinds of decoys, how to make them, and how to use them. 140 plates. 156pp. 7⅞ × 10¾. 20011-6 Pa. $7.95

HISTORY OF LACE, Mrs. Bury Palliser. Definitive, profusely illustrated chronicle of lace from earliest times to late 19th century. Laces of Italy, Greece, England, France, Belgium, etc. Landmark of needlework scholarship. 266 illustrations. 672pp. 6⅛ × 9¼. 24742-2 Pa. $14.95

ILLUSTRATED GUIDE TO SHAKER FURNITURE, Robert Meader. All furniture and appurtenances, with much on unknown local styles. 235 photos. 146pp. 9 × 12. 22819-3 Pa. $7.95

WHALE SHIPS AND WHALING: A Pictorial Survey, George Francis Dow. Over 200 vintage engravings, drawings, photographs of barks, brigs, cutters, other vessels. Also harpoons, lances, whaling guns, many other artifacts. Comprehensive text by foremost authority. 207 black-and-white illustrations. 288pp. 6 × 9. 24808-9 Pa. $8.95

THE BERTRAMS, Anthony Trollope. Powerful portrayal of blind self-will and thwarted ambition includes one of Trollope's most heartrending love stories. 497pp. 5⅜ × 8½. 25119-5 Pa. $8.95

ADVENTURES WITH A HAND LENS, Richard Headstrom. Clearly written guide to observing and studying flowers and grasses, fish scales, moth and insect wings, egg cases, buds, feathers, seeds, leaf scars, moss, molds, ferns, common crystals, etc.—all with an ordinary, inexpensive magnifying glass. 209 exact line drawings aid in your discoveries. 220pp. 5⅜ × 8½. 23330-8 Pa. $3.95

RODIN ON ART AND ARTISTS, Auguste Rodin. Great sculptor's candid, wide-ranging comments on meaning of art; great artists; relation of sculpture to poetry, painting, music; philosophy of life, more. 76 superb black-and-white illustrations of Rodin's sculpture, drawings and prints. 119pp. 8⅜ × 11¼. 24487-3 Pa. $6.95

FIFTY CLASSIC FRENCH FILMS, 1912–1982: A Pictorial Record, Anthony Slide. Memorable stills from Grand Illusion, Beauty and the Beast, Hiroshima, Mon Amour, many more. Credits, plot synopses, reviews, etc. 160pp. 8¼ × 11. 25256-6 Pa. $11.95

THE PRINCIPLES OF PSYCHOLOGY, William James. Famous long course complete, unabridged. Stream of thought, time perception, memory, experimental methods; great work decades ahead of its time. 94 figures. 1,391pp. 5⅜ × 8½. 20381-6, 20382-4 Pa., Two-vol. set $19.90

BODIES IN A BOOKSHOP, R. T. Campbell. Challenging mystery of blackmail and murder with ingenious plot and superbly drawn characters. In the best tradition of British suspense fiction. 192pp. 5⅜ × 8½. 24720-1 Pa. $3.95

CALLAS: PORTRAIT OF A PRIMA DONNA, George Jellinek. Renowned commentator on the musical scene chronicles incredible career and life of the most controversial, fascinating, influential operatic personality of our time. 64 black-and-white photographs. 416pp. 5⅜ × 8¼. 25047-4 Pa. $7.95

GEOMETRY, RELATIVITY AND THE FOURTH DIMENSION, Rudolph Rucker. Exposition of fourth dimension, concepts of relativity as Flatland characters continue adventures. Popular, easily followed yet accurate, profound. 141 illustrations. 133pp. 5⅜ × 8½. 23400-2 Pa. $3.50

HOUSEHOLD STORIES BY THE BROTHERS GRIMM, with pictures by Walter Crane. 53 classic stories—Rumpelstiltskin, Rapunzel, Hansel and Gretel, the Fisherman and his Wife, Snow White, Tom Thumb, Sleeping Beauty, Cinderella, and so much more—lavishly illustrated with original 19th century drawings. 114 illustrations. x + 269pp. 5⅜ × 8½. 21080-4 Pa. $4.50

SUNDIALS, Albert Waugh. Far and away the best, most thorough coverage of ideas, mathematics concerned, types, construction, adjusting anywhere. Over 100 illustrations. 230pp. 5⅜ × 8½. 22947-5 Pa. $4.00

PICTURE HISTORY OF THE NORMANDIE: With 190 Illustrations, Frank O. Braynard. Full story of legendary French ocean liner: Art Deco interiors, design innovations, furnishings, celebrities, maiden voyage, tragic fire, much more. Extensive text. 144pp. 8⅜ × 11¾. 25257-4 Pa. $9.95

THE FIRST AMERICAN COOKBOOK: A Facsimile of "American Cookery," 1796, Amelia Simmons. Facsimile of the first American-written cookbook published in the United States contains authentic recipes for colonial favorites—pumpkin pudding, winter squash pudding, spruce beer, Indian slapjacks, and more. Introductory Essay and Glossary of colonial cooking terms. 80pp. 5⅜ × 8½. 24710-4 Pa. $3.50

101 PUZZLES IN THOUGHT AND LOGIC, C. R. Wylie, Jr. Solve murders and robberies, find out which fishermen are liars, how a blind man could possibly identify a color—purely by your own reasoning! 107pp. 5⅜ × 8½. 20367-0 Pa. $2.00

THE BOOK OF WORLD-FAMOUS MUSIC—CLASSICAL, POPULAR AND FOLK, James J. Fuld. Revised and enlarged republication of landmark work in musico-bibliography. Full information about nearly 1,000 songs and compositions including first lines of music and lyrics. New supplement. Index. 800pp. 5⅜ × 8¼. 24857-7 Pa. $14.95

ANTHROPOLOGY AND MODERN LIFE, Franz Boas. Great anthropologist's classic treatise on race and culture. Introduction by Ruth Bunzel. Only inexpensive paperback edition. 255pp. 5⅜ × 8½. 25245-0 Pa. $5.95

THE TALE OF PETER RABBIT, Beatrix Potter. The inimitable Peter's terrifying adventure in Mr. McGregor's garden, with all 27 wonderful, full-color Potter illustrations. 55pp. 4¼ × 5½. (Available in U.S. only) 22827-4 Pa. $1.75

THREE PROPHETIC SCIENCE FICTION NOVELS, H. G. Wells. *When the Sleeper Wakes, A Story of the Days to Come* and *The Time Machine* (full version). 335pp. 5⅜ × 8½. (Available in U.S. only) 20605-X Pa. $5.95

APICIUS COOKERY AND DINING IN IMPERIAL ROME, edited and translated by Joseph Dommers Vehling. Oldest known cookbook in existence offers readers a clear picture of what foods Romans ate, how they prepared them, etc. 49 illustrations. 301pp. 6⅛ × 9¼. 23563-7 Pa. $6.00

SHAKESPEARE LEXICON AND QUOTATION DICTIONARY, Alexander Schmidt. Full definitions, locations, shades of meaning of every word in plays and poems. More than 50,000 exact quotations. 1,485pp. 6½ × 9¼. 22726-X, 22727-8 Pa., Two-vol. set $27.90

THE WORLD'S GREAT SPEECHES, edited by Lewis Copeland and Lawrence W. Lamm. Vast collection of 278 speeches from Greeks to 1970. Powerful and effective models; unique look at history. 842pp. 5⅜ × 8½. 20468-5 Pa. $10.95

THE BLUE FAIRY BOOK, Andrew Lang. The first, most famous collection, with many familiar tales: Little Red Riding Hood, Aladdin and the Wonderful Lamp, Puss in Boots, Sleeping Beauty, Hansel and Gretel, Rumpelstiltskin; 37 in all. 138 illustrations. 390pp. 5⅜ × 8½. 21437-0 Pa. $5.95

THE STORY OF THE CHAMPIONS OF THE ROUND TABLE, Howard Pyle. Sir Launcelot, Sir Tristram and Sir Percival in spirited adventures of love and triumph retold in Pyle's inimitable style. 50 drawings, 31 full-page. xviii + 329pp. 6½ × 9¼. 21883-X Pa. $6.95

AUDUBON AND HIS JOURNALS, Maria Audubon. Unmatched two-volume portrait of the great artist, naturalist and author contains his journals, an excellent biography by his granddaughter, expert annotations by the noted ornithologist, Dr. Elliott Coues, and 37 superb illustrations. Total of 1,200pp. 5⅜ × 8.
Vol. I 25143-8 Pa. $8.95
Vol. II 25144-6 Pa. $8.95

GREAT DINOSAUR HUNTERS AND THEIR DISCOVERIES, Edwin H. Colbert. Fascinating, lavishly illustrated chronicle of dinosaur research, 1820's to 1960. Achievements of Cope, Marsh, Brown, Buckland, Mantell, Huxley, many others. 384pp. 5¼ × 8¼. 24701-5 Pa. $6.95

THE TASTEMAKERS, Russell Lynes. Informal, illustrated social history of American taste 1850's–1950's. First popularized categories Highbrow, Lowbrow, Middlebrow. 129 illustrations. New (1979) afterword. 384pp. 6 × 9.
23993-4 Pa. $6.95

DOUBLE CROSS PURPOSES, Ronald A. Knox. A treasure hunt in the Scottish Highlands, an old map, unidentified corpse, surprise discoveries keep reader guessing in this cleverly intricate tale of financial skullduggery. 2 black-and-white maps. 320pp. 5⅜ × 8½. (Available in U.S. only) 25032-6 Pa. $5.95

AUTHENTIC VICTORIAN DECORATION AND ORNAMENTATION IN FULL COLOR: 46 Plates from "Studies in Design," Christopher Dresser. Superb full-color lithographs reproduced from rare original portfolio of a major Victorian designer. 48pp. 9¼ × 12¼. 25083-0 Pa. $7.95

PRIMITIVE ART, Franz Boas. Remains the best text ever prepared on subject, thoroughly discussing Indian, African, Asian, Australian, and, especially, Northern American primitive art. Over 950 illustrations show ceramics, masks, totem poles, weapons, textiles, paintings, much more. 376pp. 5⅜ × 8. 20025-6 Pa. $6.95

SIDELIGHTS ON RELATIVITY, Albert Einstein. Unabridged republication of two lectures delivered by the great physicist in 1920–21. *Ether and Relativity* and *Geometry and Experience*. Elegant ideas in non-mathematical form, accessible to intelligent layman. vi + 56pp. 5⅜ × 8½. 24511-X Pa. $2.95

THE WIT AND HUMOR OF OSCAR WILDE, edited by Alvin Redman. More than 1,000 ripostes, paradoxes, wisecracks: Work is the curse of the drinking classes, I can resist everything except temptation, etc. 258pp. 5⅜ × 8½. 20602-5 Pa. $3.95

ADVENTURES WITH A MICROSCOPE, Richard Headstrom. 59 adventures with clothing fibers, protozoa, ferns and lichens, roots and leaves, much more. 142 illustrations. 232pp. 5⅜ × 8½. 23471-1 Pa. $3.95

PLANTS OF THE BIBLE, Harold N. Moldenke and Alma L. Moldenke. Standard reference to all 230 plants mentioned in Scriptures. Latin name, biblical reference, uses, modern identity, much more. Unsurpassed encyclopedic resource for scholars, botanists, nature lovers, students of Bible. Bibliography. Indexes. 123 black-and-white illustrations. 384pp. 6 × 9. 25069-5 Pa. $8.95

FAMOUS AMERICAN WOMEN: A Biographical Dictionary from Colonial Times to the Present, Robert McHenry, ed. From Pocahontas to Rosa Parks, 1,035 distinguished American women documented in separate biographical entries. Accurate, up-to-date data, numerous categories, spans 400 years. Indices. 493pp. 6½ × 9¼. 24523-3 Pa. $9.95

THE FABULOUS INTERIORS OF THE GREAT OCEAN LINERS IN HISTORIC PHOTOGRAPHS, William H. Miller, Jr. Some 200 superb photographs capture exquisite interiors of world's great "floating palaces"—1890's to 1980's: *Titanic, Ile de France, Queen Elizabeth, United States, Europa,* more. Approx. 200 black-and-white photographs. Captions. Text. Introduction. 160pp. 8⅜ × 11¼. 24756-2 Pa. $9.95

THE GREAT LUXURY LINERS, 1927–1954: A Photographic Record, William H. Miller, Jr. Nostalgic tribute to heyday of ocean liners. 186 photos of Ile de France, Normandie, Leviathan, Queen Elizabeth, United States, many others. Interior and exterior views. Introduction. Captions. 160pp. 9 × 12. 24056-8 Pa. $9.95

A NATURAL HISTORY OF THE DUCKS, John Charles Phillips. Great landmark of ornithology offers complete detailed coverage of nearly 200 species and subspecies of ducks: gadwall, sheldrake, merganser, pintail, many more. 74 full-color plates, 102 black-and-white. Bibliography. Total of 1,920pp. 8⅜ × 11¼. 25141-1, 25142-X Cloth. Two-vol. set $100.00

THE SEAWEED HANDBOOK: An Illustrated Guide to Seaweeds from North Carolina to Canada, Thomas F. Lee. Concise reference covers 78 species. Scientific and common names, habitat, distribution, more. Finding keys for easy identification. 224pp. 5⅜ × 8½. 25215-9 Pa. $5.95

THE TEN BOOKS OF ARCHITECTURE: The 1755 Leoni Edition, Leon Battista Alberti. Rare classic helped introduce the glories of ancient architecture to the Renaissance. 68 black-and-white plates. 336pp. 8⅜ × 11¼. 25239-6 Pa. $14.95

MISS MACKENZIE, Anthony Trollope. Minor masterpieces by Victorian master unmasks many truths about life in 19th-century England. First inexpensive edition in years. 392pp. 5⅜ × 8½. 25201-9 Pa. $7.95

THE RIME OF THE ANCIENT MARINER, Gustave Doré, Samuel Taylor Coleridge. Dramatic engravings considered by many to be his greatest work. The terrifying space of the open sea, the storms and whirlpools of an unknown ocean, the ice of Antarctica, more—all rendered in a powerful, chilling manner. Full text. 38 plates. 77pp. 9¼ × 12. 22305-1 Pa. $4.95

THE EXPEDITIONS OF ZEBULON MONTGOMERY PIKE, Zebulon Montgomery Pike. Fascinating first-hand accounts (1805–6) of exploration of Mississippi River, Indian wars, capture by Spanish dragoons, much more. 1,088pp. 5⅜ × 8½. 25254-X, 25255-8 Pa. Two-vol. set $23.90

A CONCISE HISTORY OF PHOTOGRAPHY: Third Revised Edition, Helmut Gernsheim. Best one-volume history—camera obscura, photochemistry, daguerreotypes, evolution of cameras, film, more. Also artistic aspects—landscape, portraits, fine art, etc. 281 black-and-white photographs. 26 in color. 176pp. 8⅜ × 11¼.   25128-4 Pa. $12.95

THE DORÉ BIBLE ILLUSTRATIONS, Gustave Doré. 241 detailed plates from the Bible: the Creation scenes, Adam and Eve, Flood, Babylon, battle sequences, life of Jesus, etc. Each plate is accompanied by the verses from the King James version of the Bible. 241pp. 9 × 12.   23004-X Pa. $8.95

HUGGER-MUGGER IN THE LOUVRE, Elliot Paul. Second Homer Evans mystery-comedy. Theft at the Louvre involves sleuth in hilarious, madcap caper. "A knockout."—Books. 336pp. 5⅜ × 8½.   25185-3 Pa. $5.95

FLATLAND, E. A. Abbott. Intriguing and enormously popular science-fiction classic explores the complexities of trying to survive as a two-dimensional being in a three-dimensional world. Amusingly illustrated by the author. 16 illustrations. 103pp. 5⅜ × 8½.   20001-9 Pa. $2.00

THE HISTORY OF THE LEWIS AND CLARK EXPEDITION, Meriwether Lewis and William Clark, edited by Elliott Coues. Classic edition of Lewis and Clark's day-by-day journals that later became the basis for U.S. claims to Oregon and the West. Accurate and invaluable geographical, botanical, biological, meteorological and anthropological material. Total of 1,508pp. 5⅜ × 8½.
21268-8, 21269-6, 21270-X Pa. Three-vol. set $25.50

LANGUAGE, TRUTH AND LOGIC, Alfred J. Ayer. Famous, clear introduction to Vienna, Cambridge schools of Logical Positivism. Role of philosophy, elimination of metaphysics, nature of analysis, etc. 160pp. 5⅜ × 8½. (Available in U.S. and Canada only)   20010-8 Pa. $2.95

MATHEMATICS FOR THE NONMATHEMATICIAN, Morris Kline. Detailed, college-level treatment of mathematics in cultural and historical context, with numerous exercises. For liberal arts students. Preface. Recommended Reading Lists. Tables. Index. Numerous black-and-white figures. xvi + 641pp. 5⅜ × 8½.
24823-2 Pa. $11.95

28 SCIENCE FICTION STORIES, H. G. Wells. Novels, *Star Begotten* and *Men Like Gods*, plus 26 short stories: "Empire of the Ants," "A Story of the Stone Age," "The Stolen Bacillus," "In the Abyss," etc. 915pp. 5⅜ × 8½. (Available in U.S. only)
20265-8 Cloth. $10.95

HANDBOOK OF PICTORIAL SYMBOLS, Rudolph Modley. 3,250 signs and symbols, many systems in full; official or heavy commercial use. Arranged by subject. Most in Pictorial Archive series. 143pp. 8⅜ × 11.   23357-X Pa. $5.95

INCIDENTS OF TRAVEL IN YUCATAN, John L. Stephens. Classic (1843) exploration of jungles of Yucatan, looking for evidences of Maya civilization. Travel adventures, Mexican and Indian culture, etc. Total of 669pp. 5⅜ × 8½.
20926-1, 20927-X Pa., Two-vol. set $9.90

DEGAS: An Intimate Portrait, Ambroise Vollard. Charming, anecdotal memoir by famous art dealer of one of the greatest 19th-century French painters. 14 black-and-white illustrations. Introduction by Harold L. Van Doren. 96pp. 5⅜ × 8½.
25131-4 Pa. $3.95

PERSONAL NARRATIVE OF A PILGRIMAGE TO ALMANDINAH AND MECCAH, Richard Burton. Great travel classic by remarkably colorful personality. Burton, disguised as a Moroccan, visited sacred shrines of Islam, narrowly escaping death. 47 illustrations. 959pp. 5⅜ × 8½.   21217-3, 21218-1 Pa., Two-vol. set $17.90

PHRASE AND WORD ORIGINS, A. H. Holt. Entertaining, reliable, modern study of more than 1,200 colorful words, phrases, origins and histories. Much unexpected information. 254pp. 5⅜ × 8½.   20758-7 Pa. $4.95

THE RED THUMB MARK, R. Austin Freeman. In this first Dr. Thorndyke case, the great scientific detective draws fascinating conclusions from the nature of a single fingerprint. Exciting story, authentic science. 320pp. 5⅜ × 8½. (Available in U.S. only)   25210-8 Pa. $5.95

AN EGYPTIAN HIEROGLYPHIC DICTIONARY, E. A. Wallis Budge. Monumental work containing about 25,000 words or terms that occur in texts ranging from 3000 B.C. to 600 A.D. Each entry consists of a transliteration of the word, the word in hieroglyphs, and the meaning in English. 1,314pp. 6⅜ × 10.
23615-3, 23616-1 Pa., Two-vol. set $27.90

THE COMPLEAT STRATEGYST: Being a Primer on the Theory of Games of Strategy, J. D. Williams. Highly entertaining classic describes, with many illustrated examples, how to select best strategies in conflict situations. Prefaces. Appendices. xvi + 268pp. 5⅜ × 8½.   25101-2 Pa. $5.95

THE ROAD TO OZ, L. Frank Baum. Dorothy meets the Shaggy Man, little Button-Bright and the Rainbow's beautiful daughter in this delightful trip to the magical Land of Oz. 272pp. 5⅜ × 8.   25208-6 Pa. $4.95

POINT AND LINE TO PLANE, Wassily Kandinsky. Seminal exposition of role of point, line, other elements in non-objective painting. Essential to understanding 20th-century art. 127 illustrations. 192pp. 6½ × 9¼.   23808-3 Pa. $4.50

LADY ANNA, Anthony Trollope. Moving chronicle of Countess Lovel's bitter struggle to win for herself and daughter Anna their rightful rank and fortune—perhaps at cost of sanity itself. 384pp. 5⅜ × 8½.   24669-8 Pa. $6.95

EGYPTIAN MAGIC, E. A. Wallis Budge. Sums up all that is known about magic in Ancient Egypt: the role of magic in controlling the gods, powerful amulets that warded off evil spirits, scarabs of immortality, use of wax images, formulas and spells, the secret name, much more. 253pp. 5⅜ × 8½.   22681-6 Pa. $4.00

THE DANCE OF SIVA, Ananda Coomaraswamy. Preeminent authority unfolds the vast metaphysic of India: the revelation of her art, conception of the universe, social organization, etc. 27 reproductions of art masterpieces. 192pp. 5⅜ × 8½.
24817-8 Pa. $5.95

CHRISTMAS CUSTOMS AND TRADITIONS, Clement A. Miles. Origin, evolution, significance of religious, secular practices. Caroling, gifts, yule logs, much more. Full, scholarly yet fascinating; non-sectarian. 400pp. 5⅜ × 8½.
23354-5 Pa. $6.50

THE HUMAN FIGURE IN MOTION, Eadweard Muybridge. More than 4,500 stopped-action photos, in action series, showing undraped men, women, children jumping, lying down, throwing, sitting, wrestling, carrying, etc. 390pp. 7⅞ × 10⅝.
20204-6 Cloth. $19.95

THE MAN WHO WAS THURSDAY, Gilbert Keith Chesterton. Witty, fast-paced novel about a club of anarchists in turn-of-the-century London. Brilliant social, religious, philosophical speculations. 128pp. 5⅜ × 8½. 25121-7 Pa. $3.95

A CEZANNE SKETCHBOOK: Figures, Portraits, Landscapes and Still Lifes, Paul Cezanne. Great artist experiments with tonal effects, light, mass, other qualities in over 100 drawings. A revealing view of developing master painter, precursor of Cubism. 102 black-and-white illustrations. 144pp. 8¾ × 6⅝. 24790-2 Pa. $5.95

AN ENCYCLOPEDIA OF BATTLES: Accounts of Over 1,560 Battles from 1479 B.C. to the Present, David Eggenberger. Presents essential details of every major battle in recorded history, from the first battle of Megiddo in 1479 B.C. to Grenada in 1984. List of Battle Maps. New Appendix covering the years 1967–1984. Index. 99 illustrations. 544pp. 6½ × 9¼. 24913-1 Pa. $14.95

AN ETYMOLOGICAL DICTIONARY OF MODERN ENGLISH, Ernest Weekley. Richest, fullest work, by foremost British lexicographer. Detailed word histories. Inexhaustible. Total of 856pp. 6½ × 9¼.
21873-2, 21874-0 Pa., Two-vol. set $17.00

WEBSTER'S AMERICAN MILITARY BIOGRAPHIES, edited by Robert McHenry. Over 1,000 figures who shaped 3 centuries of American military history. Detailed biographies of Nathan Hale, Douglas MacArthur, Mary Hallaren, others. Chronologies of engagements, more. Introduction. Addenda. 1,033 entries in alphabetical order. xi + 548pp. 6½ × 9¼. (Available in U.S. only)
24758-9 Pa. $11.95

LIFE IN ANCIENT EGYPT, Adolf Erman. Detailed older account, with much not in more recent books: domestic life, religion, magic, medicine, commerce, and whatever else needed for complete picture. Many illustrations. 597pp. 5⅜ × 8½.
22632-8 Pa. $8.50

HISTORIC COSTUME IN PICTURES, Braun & Schneider. Over 1,450 costumed figures shown, covering a wide variety of peoples: kings, emperors, nobles, priests, servants, soldiers, scholars, townsfolk, peasants, merchants, courtiers, cavaliers, and more. 256pp. 8⅜ × 11¼. 23150-X Pa. $7.95

THE NOTEBOOKS OF LEONARDO DA VINCI, edited by J. P. Richter. Extracts from manuscripts reveal great genius; on painting, sculpture, anatomy, sciences, geography, etc. Both Italian and English. 186 ms. pages reproduced, plus 500 additional drawings, including studies for *Last Supper, Sforza* monument, etc. 860pp. 7⅞ × 10¾. (Available in U.S. only) 22572-0, 22573-9 Pa., Two-vol. set $25.90

THE ART NOUVEAU STYLE BOOK OF ALPHONSE MUCHA: All 72 Plates from "Documents Decoratifs" in Original Color, Alphonse Mucha. Rare copyright-free design portfolio by high priest of Art Nouveau. Jewelry, wallpaper, stained glass, furniture, figure studies, plant and animal motifs, etc. Only complete one-volume edition. 80pp. 9⅜ × 12¼. 24044-4 Pa. $8.95

ANIMALS: 1,419 COPYRIGHT-FREE ILLUSTRATIONS OF MAMMALS, BIRDS, FISH, INSECTS, ETC., edited by Jim Harter. Clear wood engravings present, in extremely lifelike poses, over 1,000 species of animals. One of the most extensive pictorial sourcebooks of its kind. Captions. Index. 284pp. 9 × 12.
23766-4 Pa. $9.95

OBELISTS FLY HIGH, C. Daly King. Masterpiece of American detective fiction, long out of print, involves murder on a 1935 transcontinental flight—"a very thrilling story"—NY Times. Unabridged and unaltered republication of the edition published by William Collins Sons & Co. Ltd., London, 1935. 288pp. 5⅜ × 8½. (Available in U.S. only) 25036-9 Pa. $4.95

VICTORIAN AND EDWARDIAN FASHION: A Photographic Survey, Alison Gernsheim. First fashion history completely illustrated by contemporary photographs. Full text plus 235 photos, 1840–1914, in which many celebrities appear. 240pp. 6½ × 9¼. 24205-6 Pa. $6.00

THE ART OF THE FRENCH ILLUSTRATED BOOK, 1700–1914, Gordon N. Ray. Over 630 superb book illustrations by Fragonard, Delacroix, Daumier, Doré, Grandville, Manet, Mucha, Steinlen, Toulouse-Lautrec and many others. Preface. Introduction. 633 halftones. Indices of artists, authors & titles, binders and provenances. Appendices. Bibliography. 608pp. 8⅜ × 11¼. 25086-5 Pa. $24.95

THE WONDERFUL WIZARD OF OZ, L. Frank Baum. Facsimile in full color of America's finest children's classic. 143 illustrations by W. W. Denslow. 267pp. 5⅜ × 8½. 20691-2 Pa. $5.95

FRONTIERS OF MODERN PHYSICS: New Perspectives on Cosmology, Relativity, Black Holes and Extraterrestrial Intelligence, Tony Rothman, et al. For the intelligent layman. Subjects include: cosmological models of the universe; black holes; the neutrino; the search for extraterrestrial intelligence. Introduction. 46 black-and-white illustrations. 192pp. 5⅜ × 8½. 24587-X Pa. $6.95

THE FRIENDLY STARS, Martha Evans Martin & Donald Howard Menzel. Classic text marshalls the stars together in an engaging, non-technical survey, presenting them as sources of beauty in night sky. 23 illustrations. Foreword. 2 star charts. Index. 147pp. 5⅜ × 8½. 21099-5 Pa. $3.50

FADS AND FALLACIES IN THE NAME OF SCIENCE, Martin Gardner. Fair, witty appraisal of cranks, quacks, and quackeries of science and pseudoscience: hollow earth, Velikovsky, orgone energy, Dianetics, flying saucers, Bridey Murphy, food and medical fads, etc. Revised, expanded In the Name of Science. "A very able and even-tempered presentation."—The New Yorker. 363pp. 5⅜ × 8.
20394-8 Pa. $5.95

ANCIENT EGYPT: ITS CULTURE AND HISTORY, J. E Manchip White. From pre-dynastics through Ptolemies: society, history, political structure, religion, daily life, literature, cultural heritage. 48 plates. 217pp. 5⅜ × 8½. 22548-8 Pa. $4.95

SIR HARRY HOTSPUR OF HUMBLETHWAITE, Anthony Trollope. Incisive, unconventional psychological study of a conflict between a wealthy baronet, his idealistic daughter, and their scapegrace cousin. The 1870 novel in its first inexpensive edition in years. 250pp. 5⅜ × 8½. 24953-0 Pa. $4.95

LASERS AND HOLOGRAPHY, Winston E. Kock. Sound introduction to burgeoning field, expanded (1981) for second edition. Wave patterns, coherence, lasers, diffraction, zone plates, properties of holograms, recent advances. 84 illustrations. 160pp. 5⅜ × 8¼. (Except in United Kingdom) 24041-X Pa. $3.50

INTRODUCTION TO ARTIFICIAL INTELLIGENCE: SECOND, EN-LARGED EDITION, Philip C. Jackson, Jr. Comprehensive survey of artificial intelligence—the study of how machines (computers) can be made to act intelligently. Includes introductory and advanced material. Extensive notes updating the main text. 132 black-and-white illustrations. 512pp. 5⅜ × 8½. 24864-X Pa. $8.95

HISTORY OF INDIAN AND INDONESIAN ART, Ananda K. Coomaraswamy. Over 400 illustrations illuminate classic study of Indian art from earliest Harappa finds to early 20th century. Provides philosophical, religious and social insights. 304pp. 6⅜ × 9⅜. 25005-9 Pa. $8.95

THE GOLEM, Gustav Meyrink. Most famous supernatural novel in modern European literature, set in Ghetto of Old Prague around 1890. Compelling story of mystical experiences, strange transformations, profound terror. 13 black-and-white illustrations. 224pp. 5⅜ × 8½. (Available in U.S. only) 25025-3 Pa. $5.95

ARMADALE, Wilkie Collins. Third great mystery novel by the author of *The Woman in White* and *The Moonstone*. Original magazine version with 40 illustrations. 597pp. 5⅜ × 8½. 23429-0 Pa. $7.95

PICTORIAL ENCYCLOPEDIA OF HISTORIC ARCHITECTURAL PLANS, DETAILS AND ELEMENTS: With 1,880 Line Drawings of Arches, Domes, Doorways, Facades, Gables, Windows, etc., John Theodore Haneman. Sourcebook of inspiration for architects, designers, others. Bibliography. Captions. 141pp. 9 × 12. 24605-1 Pa. $6.95

BENCHLEY LOST AND FOUND, Robert Benchley. Finest humor from early 30's, about pet peeves, child psychologists, post office and others. Mostly unavailable elsewhere. 73 illustrations by Peter Arno and others. 183pp. 5⅜ × 8½. 22410-4 Pa. $3.95

ERTÉ GRAPHICS, Erté. Collection of striking color graphics: *Seasons, Alphabet, Numerals, Aces* and *Precious Stones*. 50 plates, including 4 on covers. 48pp. 9⅜ × 12¼. 23580-7 Pa. $6.95

THE JOURNAL OF HENRY D. THOREAU, edited by Bradford Torrey, F. H. Allen. Complete reprinting of 14 volumes, 1837–61, over two million words; the sourcebooks for *Walden*, etc. Definitive. All original sketches, plus 75 photographs. 1,804pp. 8½ × 12¼. 20312-3, 20313-1 Cloth., Two-vol. set $80.00

CASTLES: THEIR CONSTRUCTION AND HISTORY, Sidney Toy. Traces castle development from ancient roots. Nearly 200 photographs and drawings illustrate moats, keeps, baileys, many other features. Caernarvon, Dover Castles, Hadrian's Wall, Tower of London, dozens more. 256pp. 5⅜ × 8¼. 24898-4 Pa. $5.95

AMERICAN CLIPPER SHIPS: 1833–1858, Octavius T. Howe & Frederick C. Matthews. Fully-illustrated, encyclopedic review of 352 clipper ships from the period of America's greatest maritime supremacy. Introduction. 109 halftones. 5 black-and-white line illustrations. Index. Total of 928pp. 5⅜ × 8½.
25115-2, 25116-0 Pa., Two-vol. set $17.90

TOWARDS A NEW ARCHITECTURE, Le Corbusier. Pioneering manifesto by great architect, near legendary founder of "International School." Technical and aesthetic theories, views on industry, economics, relation of form to function, "mass-production spirit," much more. Profusely illustrated. Unabridged translation of 13th French edition. Introduction by Frederick Etchells. 320pp. 6⅛ × 9¼.
(Available in U.S. only) 25023-7 Pa. $8.95

THE BOOK OF KELLS, edited by Blanche Cirker. Inexpensive collection of 32 full-color, full-page plates from the greatest illuminated manuscript of the Middle Ages, painstakingly reproduced from rare facsimile edition. Publisher's Note. Captions. 32pp. 9⅜ × 12¼. 24345-1 Pa. $4.50

BEST SCIENCE FICTION STORIES OF H. G. WELLS, H. G. Wells. Full novel *The Invisible Man*, plus 17 short stories: "The Crystal Egg," "Aepyornis Island," "The Strange Orchid," etc. 303pp. 5⅜ × 8½. (Available in U.S. only)
21531-8 Pa. $4.95

AMERICAN SAILING SHIPS: Their Plans and History, Charles G. Davis. Photos, construction details of schooners, frigates, clippers, other sailcraft of 18th to early 20th centuries—plus entertaining discourse on design, rigging, nautical lore, much more. 137 black-and-white illustrations. 240pp. 6⅛ × 9¼.
24658-2 Pa. $5.95

ENTERTAINING MATHEMATICAL PUZZLES, Martin Gardner. Selection of author's favorite conundrums involving arithmetic, money, speed, etc., with lively commentary. Complete solutions. 112pp. 5⅜ × 8½. 25211-6 Pa. $2.95
THE WILL TO BELIEVE, HUMAN IMMORTALITY, William James. Two books bound together. Effect of irrational on logical, and arguments for human immortality. 402pp. 5⅜ × 8½. 20291-7 Pa. $7.50

THE HAUNTED MONASTERY and THE CHINESE MAZE MURDERS, Robert Van Gulik. 2 full novels by Van Gulik continue adventures of Judge Dee and his companions. An evil Taoist monastery, seemingly supernatural events; overgrown topiary maze that hides strange crimes. Set in 7th-century China. 27 illustrations. 328pp. 5⅜ × 8½. 23502-5 Pa. $5.00

CELEBRATED CASES OF JUDGE DEE (DEE GOONG AN), translated by Robert Van Gulik. Authentic 18th-century Chinese detective novel; Dee and associates solve three interlocked cases. Led to Van Gulik's own stories with same characters. Extensive introduction. 9 illustrations. 237pp. 5⅜ × 8½.
23337-5 Pa. $4.95

Prices subject to change without notice.
Available at your book dealer or write for free catalog to Dept. GI, Dover Publications, Inc., 31 East 2nd St., Mineola, N.Y. 11501. Dover publishes more than 175 books each year on science, elementary and advanced mathematics, biology, music, art, literary history, social sciences and other areas.